Non-extractable Polyphenols and Carotenoids
Importance in Human Nutrition and Health

Food Chemistry, Function and Analysis

Series editors:
Gary Williamson, *University of Leeds, UK*
Alejandro G. Marangoni, *University of Guelph, Canada*
Juliet A. Gerrard, *University of Auckland, New Zealand*

Titles in the series:

1: Food Biosensors
2: Sensing Techniques for Food Safety and Quality Control
3: Edible Oil Structuring: Concepts, Methods and Applications
4: Food Irradiation Technologies: Concepts, Applications and Outcomes
5: Non-extractable Polyphenols and Carotenoids: Importance in Human Nutrition and Health

How to obtain future titles on publication:
A standing order plan is available for this series. A standing order will bring delivery of each new volume immediately on publication.

For further information please contact:
Book Sales Department, Royal Society of Chemistry, Thomas Graham House, Science Park, Milton Road, Cambridge, CB4 0WF, UK
Telephone: +44 (0)1223 420066, Fax: +44 (0)1223 420247,
Email: booksales@rsc.org
Visit our website at www.rsc.org/books

Non-extractable Polyphenols and Carotenoids

Importance in Human Nutrition and Health

Edited by

Fulgencio Saura-Calixto
Nutriantioxidants and Science FDI S.L., Oviedo, Spain;
Former Research Professor, Spanish National Research Council (CSIC), Madrid, Spain
Email: fsaura@nutriantioxidants.com

and

Jara Pérez-Jiménez
ICTAN-CSIC, Madrid, Spain
Email: jara.perez@ictan.csic.es

ROYAL SOCIETY OF CHEMISTRY

Food Chemistry, Function and Analysis No. 5

Print ISBN: 978-1-78801-106-8
PDF ISBN: 978-1-78801-320-8
EPUB ISBN: 978-1-78801-447-2
ISSN: 2398-0656

A catalogue record for this book is available from the British Library

The Royal Society of Chemistry is a charity, registered in England and Wales, Number 207890, and a company incorporated in England by Royal Charter (Registered No. RC000524), registered office: Burlington House, Piccadilly, London W1J 0BA, UK, Telephone: +44 (0) 207 4378 6556.

Visit our website at www.rsc.org/books

Printed in the United Kingdom by CPI Group (UK) Ltd, Croydon, CR0 4YY, UK

Preface

Polyphenols (PP) are the most abundant antioxidants in the human diet and their bioavailability, metabolism and health effects are currently the focus of wide-ranging and intense research. In particular, numerous clinical trials and observational studies have shown that PP may play a significant role in the reduction of chronic disease risks, including those of cardiovascular diseases and certain kinds of cancer.

Current knowledge of the biological and health-related properties of PP has been established through a huge research effort exclusively focused on extractable PP (EPP) or low-molecular-weight phenolic compounds (flavonoids, stilbenes, phenolic acids and other classes) that can be extracted from plant foods using aqueous–organic solvents.

However, recent findings have shown that plant foods contain abundant amounts of another type of phenolic structures called non-extractable PP (NEPP), which are made up of polymeric PP or plant macromolecules containing bound single PP and carotenoids. These exhibit significant biological activity and potentially promising health effects. Indeed, they are the major phenolics in plant foods and in diets.

NEPP have so far been ignored due to the fact that they are not extracted from plant foods and vegetable materials by the procedures used to obtain EPP and, in addition, they are not detected by the usual methods of PP analysis. However, we ingest significant amounts of both EPP and NEPP daily.

As far as functional ingredients are concerned, there is a huge global market for EPP as plant food extracts or single compounds such as resveratrol and hydroxytyrosol. About 5000 EPP-related patents have been registered in the last 20 years, completely ignoring the potential applications of NEPP.

This book has a double objective: (a) to provide a state-of-the-art and scientific review of NEPP in plant foods and their health-related properties; and (b) to encourage research into this emerging topic.

Food Chemistry, Function and Analysis No. 5
Non-extractable Polyphenols and Carotenoids: Importance in Human Nutrition and Health
Edited by Fulgencio Saura-Calixto and Jara Pérez-Jiménez

Published by the Royal Society of Chemistry, www.rsc.org

We must bear in mind that the scientific evidence of the effects of low-molecular-weight antioxidants (vitamins A, C and E, selenium, EPP and carotenoids) has been established by hundreds of research teams over a century of work. This research was driven forward by the awarding of two Nobel Prizes in 1922 and 1924 for the discovery of vitamins E and C. Research into NEPP and related unavailable carotenoids began just two decades ago and there is still only quite a small number of scientists working in this field.

The story of NEPP, from the origin of this term to our current knowledge, is briefly reported in Chapter 1, along with a short overview of the research highlights and perspectives. The following 17 chapters, written by highly qualified scientists, the pioneers in this field, provide a detailed and thorough study of the key issues concerning NEPP and related carotenoids.

Chapters 2 and 3 address the strong relation between dietary fibre and NEPP, and also the chemical and physiological approaches adopted in order to study NEPP. Chemistry and NEPP databases are the subjects of Chapters 4–6, which report on extraction procedures and chromatographic techniques for the analysis of the NEPP content of plant foods and beverages. They also address specific studies of NEPP in different fruits, including both traditional and non-traditional tropical fruits. The effects of food processing on NEPP is the topic of Chapter 7.

Carotenoids are also important dietary antioxidants. Chapter 8 provides data on carotenes and xanthophylls in plant foods and in different diets, together with structure–activity relationships for antioxidant capacity and associated health-related benefits. Chapter 9, after summarising the factors that influence the fate of absorbed and non-absorbed carotenoids, highlights an important gap in carotenoid research: a major part of ingested carotenoids are not bioavailable and they are fermented by colonic microbiota to become unknown metabolites. That is, they have a similar physiological fate to that of PP, but we have less knowledge of their colonic metabolites.

Physiological fate, metabolism and health-related properties are the focuses of Chapters 10–14. Together, they comprise an evaluation of *in vivo* and *in vitro* models for studying bioavailability and colonic fermentation. They also consider specific studies of the metabolic fates of EPP and NEPP, NEPP–microbiome interactions and the nutritional significance of antioxidant capacity as key dietary parameter.

Chapters 15 and 16 then present two case studies performed on grape products and sorghum as representative sources of the two main types of NEPP: polymeric proanthocyanidins and phenolics bound to cell wall polymers. Finally, a study of NEPP as food ingredients is presented in Chapter 17.

It is expected that chemists, nutritionists, doctors, pharmacists, food scientists and technologists, as well as health writers, will be interested in this book.

My thanks go to the Royal Society of Chemistry for the invitation to edit this, the first book on NEPP, and also to the different chapter authors and my fellow editor, Dr Pérez-Jiménez, for all their efforts towards achieving this task.

I hope this book will serve to promote research on this topic and its applications in the fields of nutrition and health.

Fulgencio Saura-Calixto

Contents

Food Chemistry, Function and Analysis No. 5
Non-extractable Polyphenols and Carotenoids: Importance in Human Nutrition and Health
Edited by Fulgencio Saura-Calixto and Jara Pérez-Jiménez

Published by the Royal Society of Chemistry, www.rsc.org

CHAPTER 1

The Story of the Introduction of Non-extractable Polyphenols into Polyphenol Research: Origin, Development and Perspectives

F. SAURA-CALIXTO

[a]Nutriantioxidants and Science FDI S.L., Avda Muñoz Degrain 17, 33007 Oviedo, Asturias, Spain; [b]Former Research Professor, Spanish National Research Council (CSIC), Madrid, Spain
*E-mail: fsaura@nutriantioxidants.com

1.1 Introduction

Research into polyphenols (PP) has undergone massive and increasing development in recent decades, as indicated by the large number of articles—over 51 000—published in the period 1997–2017 *versus* around 12 000 in the period 1977–1996. Most of those papers report chemical, biological, nutritional, epidemiological and intervention studies that focus exclusively on phenolic compounds extracted from plant material using aqueous–organic solvents, ignoring an important fraction of PP in food and diets, the so-called non-extractable PP (NEPP).

Food Chemistry, Function and Analysis No. 5
Non-extractable Polyphenols and Carotenoids: Importance in Human Nutrition and Health
Edited by Fulgencio Saura-Calixto and Jara Pérez-Jiménez

Published by the Royal Society of Chemistry, www.rsc.org

This chapter presents an overview of what we know about NEPP, including the main milestones in the research and perspectives in this field, based mainly on the studies developed by my research team. It addresses the gap in the field of PP research, from the origin of the concept of NEPP to the current understanding of their physiological behaviour, metabolism and health-related properties. Other aspects such as differences between the chemical and physiological approaches to NEPP, the importance of their interactions with dietary fibre (DF) and their potential relevance in nutrition and health are discussed.

The aim of the chapter is to provide scientists currently working on PP, and especially scientists from other fields, with an overall idea of the topic. It may be seen as an introduction to the following chapters of the book, written by pioneering scientists working in this field, which describe in detail the main aspects of our current knowledge of NEPP.

1.2 Why is There a Gap in Dietary Polyphenols?

A quick look at the methods typically used to analyse PP in food (Figure 1.1) can provide an answer to the question of why there is a significant gap in the current predominant approach in the field of dietary PP. The contents and composition of PP are determined by analysing aqueous–organic extracts from plant foods, and such analysis is the only source of current databases

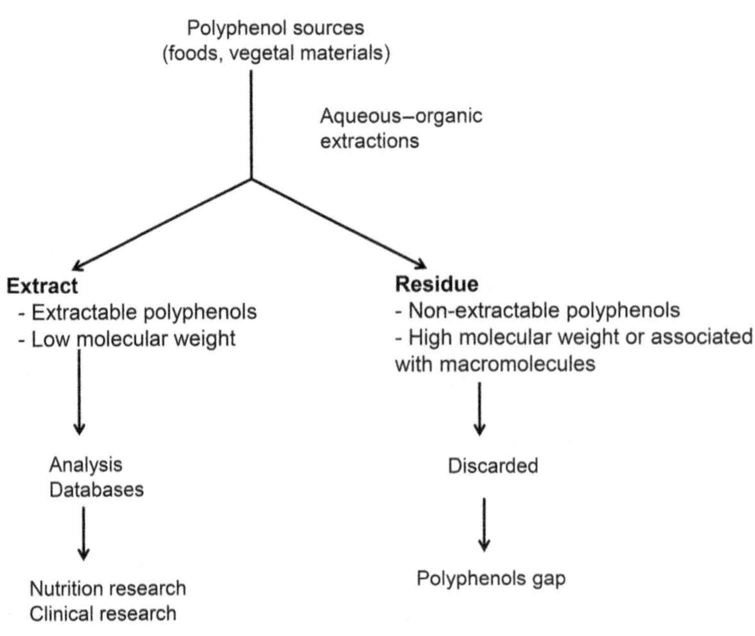

Figure 1.1 Common procedure for polyphenol analysis and the derived gap in polyphenols research.

of PP in foods. The general assumption that the total PP content of a food is in these extracts, neglecting the possible presence of PP in the extraction residues, is what leads to the current gap. However, simple analysis of the residues shows this assumption to be incorrect.

Indeed, for the analysis of other food constituents, the assumption that the entire constituent is present in an aqueous–organic extract is not held, in contrast to the PP field. Thus, the most common methods used for the analysis of plant macromolecules such as polysaccharides and protein require, in addition to initial extraction of sugars or free amino acids, treatment of the samples with a strong acid or enzymes to obtain hydrolysates containing the monomers (monosaccharides and amino acids) that form the polymeric structures. In these cases, if the same partial approach had been followed as in the field of PP, we would now have almost complete knowledge of soluble sugars and free amino acids, but we would not know of the existence of insoluble proteins and polysaccharides in plant foods.

In this context, after 1988, the analysis of hydrolysates obtained *via* the acid treatment of the extraction residues of numerous materials and plant foods showed the presence of appreciable amounts of polymeric proanthocyanidins and simple PP linked to cell walls, as described below. These PP were named NEPP. Unfortunately, this complete approach to PP analysis has not been generally adopted in the field of PP, a methodological error that has led to the current situation: virtually complete knowledge of extractable PP (EPP) but a lack of knowledge of NEPP.

1.3 Polyphenols Associated with Dietary Fibre: Origin and Development of Research into Non-extractable Polyphenols

After the first physiological definition of DF by Trowell,[1] intense research carried out in the 1980s and 1990s demonstrated the key role of DF in gastrointestinal health and in reducing chronic disease risks. As a consequence, increased DF intake is currently recommended by the World Health Organization (WHO) and numerous national and international health agencies, especially in Western countries where DF intake is below dietary recommendations. At the same time, massive development of foods enriched with DF has occurred.

The first studies on DF described it as consisting of two fractions, depending on their solubility in intestinal fluids: insoluble DF (IDF), constituted by the polysaccharides cellulose and hemicellulose; and soluble DF (SDF), consisting of polysaccharides such as beta-glucans, arabinoxylans, galactomannans and pectins. Nevertheless, later research showed that DF is a complex matrix containing numerous indigestible substances, in addition to the polysaccharides indicated in the original definition of Trowell. Presently, DF is considered the edible part of plant food that is resistant to digestion and absorption in the human small intestine, which undergoes complete or partial fermentation in the large intestine. It contains polysaccharides as major constituents, along with other substances such as lignin, resistant starch, oligosaccharides, Maillard compounds, cutin, PP and carotenoids.

Interestingly, a certain similarity was observed in the physiological behaviour of DF and PP.[2,3] Thus, EPP and SDF can be solubilized in intestinal fluids during digestion, while IDF and NEPP remain insoluble. Once in the colon, SDF and the EPP that are not absorbed in the small intestine are mostly fermented by the action of the microbiome, while IDF and NEPP are only partially fermented. This suggested that dietary PP may have a significant contribution to some health effects attributed exclusively to DF.

The close relationship between DF and NEPP is also seen in the fact that the concept of NEPP arose from the analysis of DF in some plant materials. Following the Association of Official Analytical Chemists (AOAC) enzymatic–gravimetric method (Figure 1.2a), which was the most common procedure available in the 1980s, a paradoxical result was found in the analysis of DF from carob pods (*Ceratonia siliqua*), a Mediterranean legume.[4] Their DF content was high (36.9% of dry matter) and we initially assumed that most of this corresponded to indigestible polysaccharides, as in other legumes and in most vegetable foods. However, analysis of the DF showed that the actual amount of polysaccharides and lignin was only 18.9% of dry matter and therefore there was a gap of 18.0%! This missing matter was identified, after treatment of the DF with HCl/butanol, as non-extractable proanthocyanidins (NEPA), with a basic flavan-3-ol structure and a molecular weight of over 30 000 Da. In a later analysis, we also found NEPA in other plant materials (Table 1.1), and this can be considered the origin of NEPP.

Since the year 2000, we have used a new procedure for DF analysis to avoid the use of the non-physiological temperatures and other sources of error of the AOAC method.[5] Our procedure of DF analysis (Figure 1.2b) mimics physiological conditions.[6,7] In general terms, it consists of eliminating nutrients (proteins, starch, sugars, minerals and fat) from foods using enzymatic treatments, which allows the DF to be obtained as a solid residue, and then analysing the composition of this solid residue, including NEPP. This procedure also allows the determination of NEPP associated with DF.

1.4 Development of Research

After finding NEPP in some plant material, research developed in two different parallel ways: chemical studies of the occurrence of NEPP in foods and diets; and *in vitro* and *in vivo* studies to elucidate the significance of NEPP for nutrition and health.

Following a specific chemical method for the analysis of NEPP (Figure 1.3), systematic analyses of NEPP were carried out on plant foods, including the determination of NEPA and hydrolysable PP (HPP) and low-molecular-weight phenolic compounds strongly associated with cell wall. This method is faster

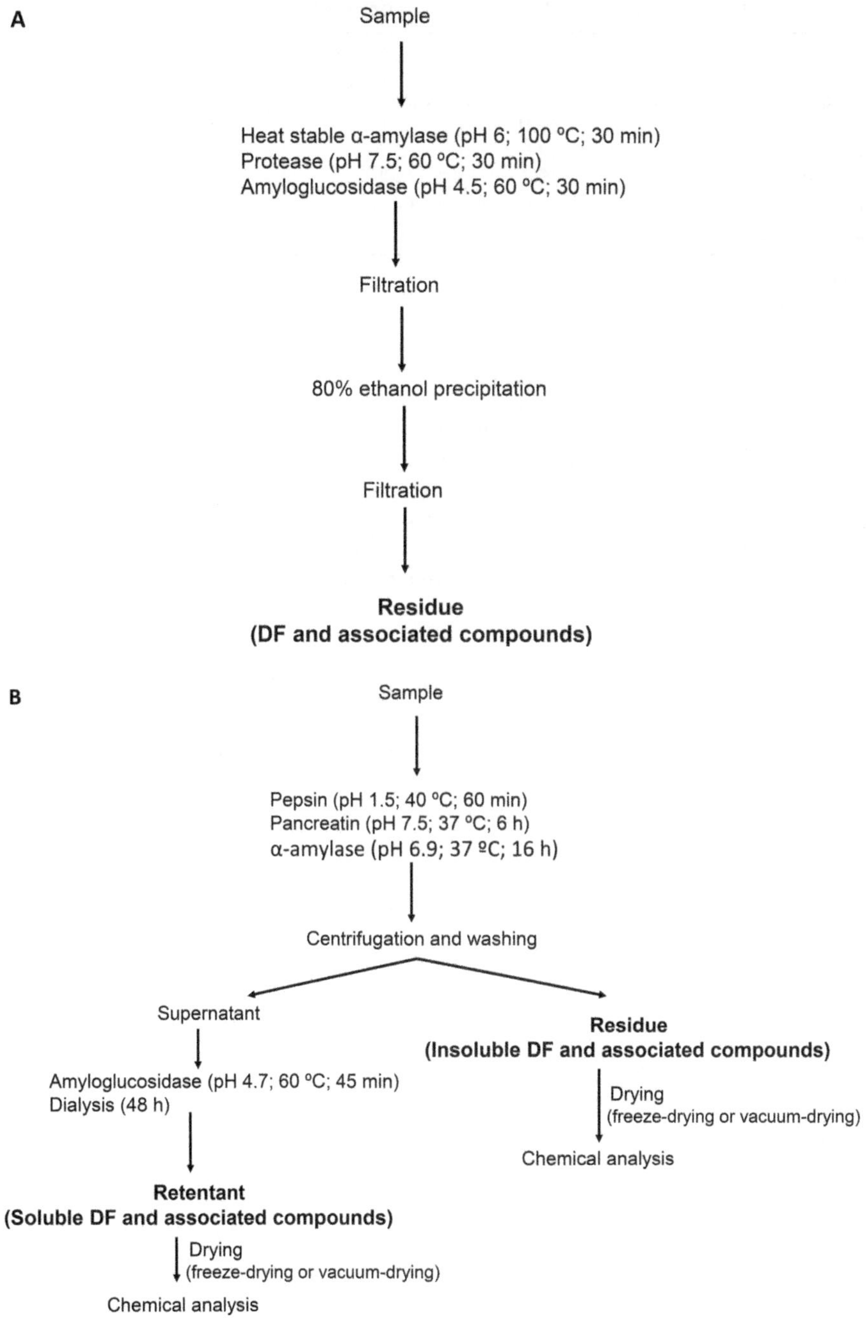

Figure 1.2 Methodologies for dietary fibre analysis. (A) AOAC method; (B) indigestible fraction method. DF, dietary fibre.

Table 1.1 Dietary fibre (DF) and associated non-extractable proanthocyanidins (NEPA) in plant materials (g/100 g dry matter).

	DF[a]	NEPA in DF residue	Corrected DF[b]
Carob pod	36.9	18.0	18.9
Spanish sainfoin	68.4	5.9	62.5
Cider wastes	65.5	3.1	62.4
White grape pomace	59.0	14.5	44.5
Red grape pomace	60.8	36.4	24.4

[a]Determined by the AOAC method.
[b]Corrected DF = DF − NEPA.

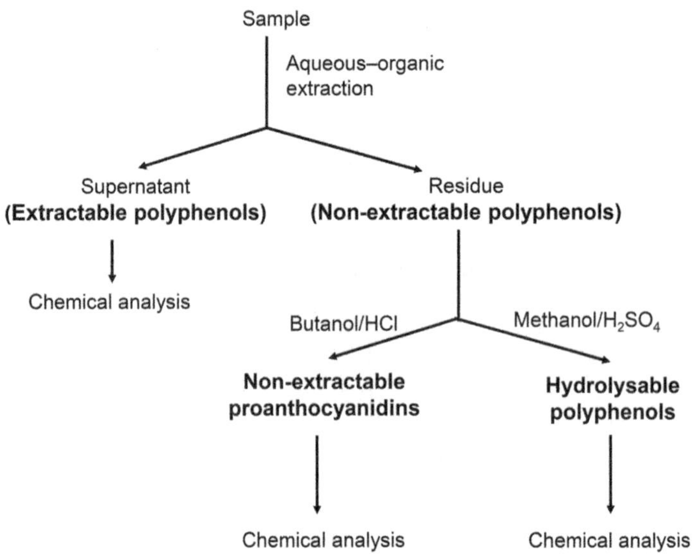

Figure 1.3 Procedure for chemical analysis of extractable and non-extractable polyphenols.

than the enzymatic procedure (Figure 1.2b) and enables the determination of both the content and composition of the NEPP.

The application of this methodology to a large number of samples showed that, in the DF of some plant materials, NEPP were found to be abundant, which meant that their antioxidant properties and potential health effects were exceptional. This type of DF was named antioxidant DF to differentiate it from common DF, and its health-related properties and their potential applications as a functional ingredient were reported.[8–11]

These studies also enabled the determination of the occurrence of NEPP in different food groups. It should be highlighted that NEPP were found in all types of plant foods (fruit, vegetables, cereals, legumes and nuts), leading to the conclusion that NEPP are a natural constituent of plant foods. In

Table 1.2 Soluble dietary fibre (SDF) and polyphenols (PP) in coffee espresso and red wine (mg/100 mL).[a]

	Coffee expresso	Red wine
SDF	650.0	137.0
Total PP	198.8	167.9
PP associated with DF	103.7	100.1

[a]Adapted from ref. 16 and 17.

a second step, NEPP intake *via* the Spanish Mediterranean diet was determined and it was found that NEPP make up the major part of dietary PP.[12–14]

The study of dietary sources of NEPP included beverages. Traditionally, it was considered that drinks do not contain DF. However, it has been shown that appreciable amounts of indigestible polysaccharides (pectins, galactomannans, beta-glucans, xylans and others; *i.e.*, DF) can be solubilised in the processing of vegetable foods. The content of SDF in fruit juices, some alcoholic beverages, coffee and infusions has been reported.[15] At the same time, some common drinks contain DF and are also rich in PP. In such cases, the SDF contains associated PP, so the total PP content can be divided into free PP and NEPP linked to DF by weak bounds. As an example, the PP distribution in coffee beverages and in red wine[16,17] is shown in Table 1.2.

Meanwhile, *in vitro* and *in vivo* studies, including animal experiments and clinical trials, determined the physiological fate and health-related properties of NEPP. It was found that NEPP are partially fermented in the colon, where they increase the antioxidant environment and produce absorbable metabolites with potential systemic effects.[18–22]

The results of these preclinical and clinical studies were summarised in a recent review, which reports that NEPP have potential health-related properties in relation to gastrointestinal health (increases in antioxidant and antiproliferative capacities, reduction of intestinal tumorigenesis and modification of gene expression), as well as systemic effects derived from their colonic metabolites.[23] However, these results should be considered preliminary as the number of studies performed so far is still limited, and it should be noted that we are in the initial stages of a promising research field.

1.5 Non-extractable Polyphenols: Chemical and Physiological Approaches

The analysis of NEPP can be performed through two different approaches (*i.e.*, chemical and physiological). They are based on the fact that the solubility of PP is different in aqueous–organic solvents and in intestinal fluids. Schemes of the two procedures used for NEPP analysis illustrate the differences between the two approaches (Figures 1.2b and 1.3).

1.5.1 Chemical Approach

NEPP are phenolic compounds that remain in the residues of the aqueous-organic extraction performed on plant foods. They are constituted by the two classes previously mentioned: NEPA and HPP. These structures arise due to the capacity of the aromatic rings and hydroxyl groups of single phenolics to form polymers and also to link as single molecules to polysaccharides and proteins *via* esters and ether covalent bonds. Additionally, NEPP may be included in insoluble complexes formed by food processing (thermal treatment, the ageing of beverages and the tanning of leather). Maillard compounds in cereal products and coffee are examples of these insoluble complexes.[24,25]

The chemical procedure for complete analysis of PP (Figure 1.3) allows EPP and NEPP to be separated and enables quantification of the total content of food PP as EPP plus NEPP. It should be noted that the NEPP values may vary slightly depending on the type of solvent used and the experimental conditions of extraction (temperature, particle size, agitation, liquid/solid ratio, *etc.*), but not the total content (NEPP + EPP).

A concept linked to NEPP is that of 'bound PP', being commonly determined in cereals. It should be emphasised that NEPP is a broader concept than 'bound PP', which corresponds to phenolic compounds bound to cell walls. In particular, the contents of bound PP reported in foods, especially in cereals and pulses, correspond to the PP that the polymer matrix releases due to alkaline hydrolysis, and therefore they may only be part of the actual content of bound PP. In this way, comparative HPLC-MS analysis of the alkali and sulphuric acid hydrolysates of cereal products determined the amounts of several PP (hydroxybenzoic, caffeic, cinnamic and protocatechuic acids) in the acid NEPP hydrolysates to be higher than in the alkali hydrolysates.[26]

1.5.2 Physiological Approach

The physiological approach considers that NEPP are phenolic compounds that are insoluble in intestinal fluids after the action of digestive enzymes (instead of aqueous organic solvents, as in the chemical approach). These structures are simple PP covalently bound to each other or to polysaccharides and proteins, along with small amounts associated with the DF matrix by weak links (H-bonds and hydrophobic and Van der Waals bonds). NEPP—and the part of EPP that are not absorbed in the small intestine—together with DF reach the colon, where the fermentation process provides the organic matter and energy necessary for the maintenance and development of the bacterial mass, in an antioxidant environment. The action of the microbiome produces metabolites from DF (short-chain fatty acids and gases) and from PP (hydroxyphenylacetic and hydroxyphenylvaleric acids or urolithin, among others) (Figure 1.4). A specific procedure for the determination of

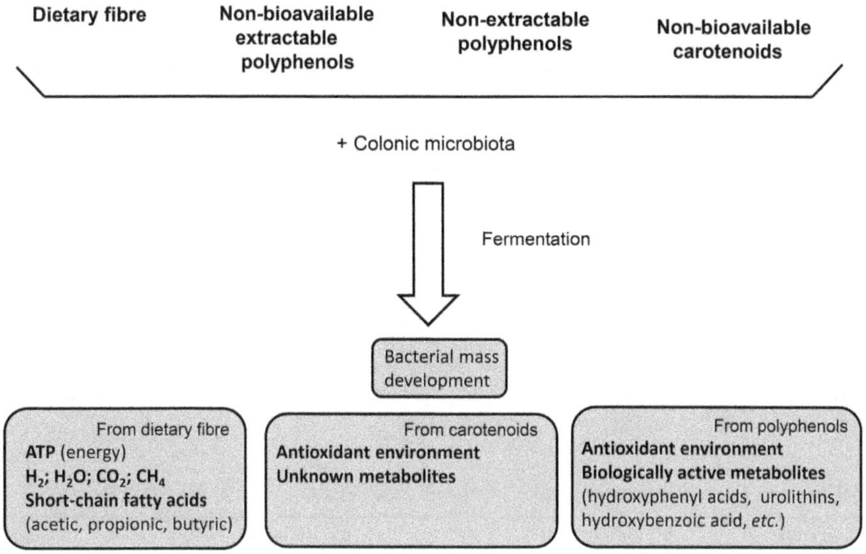

Figure 1.4 Colonic fermentation of dietary fibre and polyphenols.

NEPP in foods by this physiological approach has been developed by our team (Figure 1.2b).

Antioxidants acting in the colonic lumen can be called colonic antioxidants to differentiate them from the antioxidants absorbed in the small intestine. In this context, some comments on research on the metabolism and bioavailability of PP in the colon may be called for. Generally, *in vitro* and *in vivo* assays, including cell cultures, use single PP, and there is no doubt that such studies are useful for discovering mechanisms of action. However, they are probably physiologically irrelevant, as it seems logical to think that the results would be different were we to introduce these single PP in a real environment; that is, the target PP together with a balanced amount of the remaining dietary PP at physiological doses and also with DF and other indigestible substrates. As examples of this, Tables 1.3 and 1.4 show the daily amounts of substrates entering the human colon and the estimated intake of antioxidants in European countries.[27–32]

Moreover, phenolic compounds are not the only bioactive compounds that produce metabolites in the colon, as carotenoids also do so. The terms 'extractable' and 'non-extractable' can also be applied to carotenoids, constituents of the lipid fraction of plant foods. From a chemical point of view, there are no non-extractable carotenoids because the analytical method involves their complete extraction. However, from a physiological point of view, there are non-extractable carotenoids because their bioavailability in the small intestine is only partial, and it is estimated that a major part of ingested carotenoids reaches the colon. In fact, appreciable amounts of

Table 1.3 Estimated daily amount of substrates available for colonic fermentation in Western diets.[a]

Substrate	g day^{-1}
Dietary fibre (as non-starch polysaccharides and lignin)	20.0
Other carbohydrates (resistant starch, oligosaccharides)	7.0
Undigested protein and fat	8.0
Polyphenols	
Extractable polyphenols	0.8
Non-extractable polyphenols	1.0
Polyols, phytosterols, additives, organic acids	6.0
Endogenous substances (mucus, bacteria, pancreatic enzymes, bile secretions, urea, epithelial cells)	4.0
TOTAL	~47.0

[a]Adapted from ref. 27 and 28.

Table 1.4 Ranges of daily intake of dietary antioxidants in European countries.[b]

Constituent	Daily intake
Vitamin A	0.6–1.4 μg
Vitamin C	65–161 mg
Vitamin E	7.8–8.3 mg
Selenium	3.8–8.3 μg
Carotenoids	9.5–16.0 mg
β-carotene	2.96–5.55 mg
Lutein	1.56–3.25 mg
Polyphenols[a]	1822 mg
Extractable polyphenols	790 mg
Non-extractable polyphenols	1032 mg

[a]Only one available study for the Spanish diet.[13]
[b]Adapted from ref. 13 and 29.

beta-carotene, lutein and lycopene were found in DF using the methodology described above, and partial bacterial fermentation was also observed.[33,34] This is a topic that requires further research, as we currently know little of the process of colonic fermentation of carotenoids and their metabolites.

Finally, apart from DF, there is little information on the interactions between NEPP and other colonic substrates, such as protein, fat, bile acids, phytosterols and others, and the effects derived from them. This is another topic for future research, which is beyond the scope of this book.

1.6 From Non-extractable Polyphenols to Macromolecular Antioxidants

Research into NEPP is currently restricted to the field of PP, and we believe that placing this subject within a broader context, such as that of antioxidants, may favour the development of knowledge about NEPP and their applications in nutrition and health. Therefore, since 2015, we have been using the term 'macromolecular antioxidants' (MACAN) instead of NEPP in scientific papers and presentations.[35–39] This is justified by several factors.

First, NEPP is an appropriate term for chemists and biologists, but it is difficult to understand within, and therefore difficult to extend to, the sector of health professionals and the general public: the main target audience. Certainly, NEPP are currently ignored in medical research, despite being the most abundant antioxidants in food and diets. And obviously, the term NEPP would be unintelligible and unattractive to consumers. Also, the term MACAN may favour research within a broader framework, including it with other non-phenolic antioxidants such as carotenoids and other non-digestible substances that contain antioxidants.

In this context, daily intakes of dietary antioxidants (Table 1.4) may be an important aspect to consider in a more holistic approach by nutritionists and physicians in both research and applications. It should be noted that PP represent over 93% of the antioxidants ingested daily, with NEPP being the major dietary antioxidant fraction.

A suitable amount balanced between the different classes of dietary antioxidants may be a contributing factor to the lower all-cause mortality associated with diets rich in antioxidants.[40] In fact, numerous clinical trials and observational studies emphasise that antioxidants are key substances that can explain the better overall health status and reduced risk of major chronic diseases associated with the Mediterranean diet:[41–43] a dietary pattern that is characterised by high consumption of plant foods and beverages rich in antioxidants.

However, there are substantial quantitative and qualitative differences between dietary antioxidants and the antioxidants used in medical research. These mainly are related to three aspects: intakes, site of absorption and synergism.

The daily intake of dietary antioxidants in European countries is estimated to be around 1.8 g, and amounts of the same order can be expected in other Western diets. However, clinical and epidemiological studies generally focus on a few milligrams of single dietary antioxidants (vitamins C and E, selenium or beta-carotene, among others), which represents a minor part of the total amount of antioxidants present in human biological fluids, tissues and cells. Thus, PP and particularly NEPP should be systematically included in that kind of study.

The site of absorption is also a differential characteristic of dietary antioxidants. The compounds with a small molecular size (vitamins, selenium, carotenoids and a part of EPP) are mostly absorbed in the small intestine, whilst microbial-derived metabolites from EPP and NEPP are absorbed in the colon, and they are not addressed in clinical studies. One example may be illustrative: a systematic review and meta-analysis[44] concluded that clinical trials have failed to show any benefit from antioxidant supplements (vitamins A, C and E and selenium) in the prevention of gastrointestinal cancer. However, this study ignored the large amount of antioxidants acting in the colon—the primary site of the development of intestinal cancer—so only a small part of the antioxidants was considered.

Regarding synergism, it should be noted that no antioxidant reaches tissues and cells on its own, but it does so together with other dietary antioxidants.[45] Synergic effects are derived from interactions between hydrophilic and lipophilic compounds at water–lipid interfaces (vitamin C–vitamin E, carotenoids–vitamin E and vitamins C and E with EPP and NEPP) and occur: when one antioxidant is regenerated by others; when one antioxidant protects another through sacrificial oxidation; and when several antioxidants execute different antioxidant mechanisms. Synergism may also affect the results of clinical trials. For instance, the detrimental effects observed when using a high dose of individual antioxidants (*e.g.*, beta-carotene)[46] may be related to a disruption of the synergism corresponding to a balanced composition of dietary antioxidants.

On this basis, the classification of dietary antioxidants into two groups is shown in Figure 1.5, which includes the different classes of compounds that make up the two groups of antioxidants. Structures of specific antioxidants are also shown.

1.7 Potential Applications of Macromolecular Antioxidants

An updated review of MACAN properties suggests potential applications for them aimed at gastrointestinal health and the prevention of risk factors associated with oxidative stress. The development of research in this field could enhance applications in food, dietetics and medicine.

Current technology for obtaining antioxidants is based primarily on extracting them from various plant materials, but obviously this is not applicable to MACAN, being non-extractable substances. After some attempts to obtain MACAN using steam explosion and pressurised water methods with poor results, two patents licensed to food companies address this issue.

In the first patent,[47] fruit pulps are treated with ethanol : water under specific experimental conditions to remove sugars, PP and soluble substances. After dehydration and micronisation of the corresponding residue, an ingredient with a moderately high content of MACAN (up to 30%) is obtained, which has been used for the production of low-calorie antioxidant drinks and cosmetic creams currently on the market.

The second patent[48] is based on a biotechnological method that avoids the use of organic solvents. This provides ingredients with a high concentration (50–99%) of NEPA, the main class of MACAN with recognised health-related properties to be obtained from previously selected plant materials. Such concentrates can be used as ingredient in functional foods and dietetic products, or possibly in the near future in pharmaceuticals.

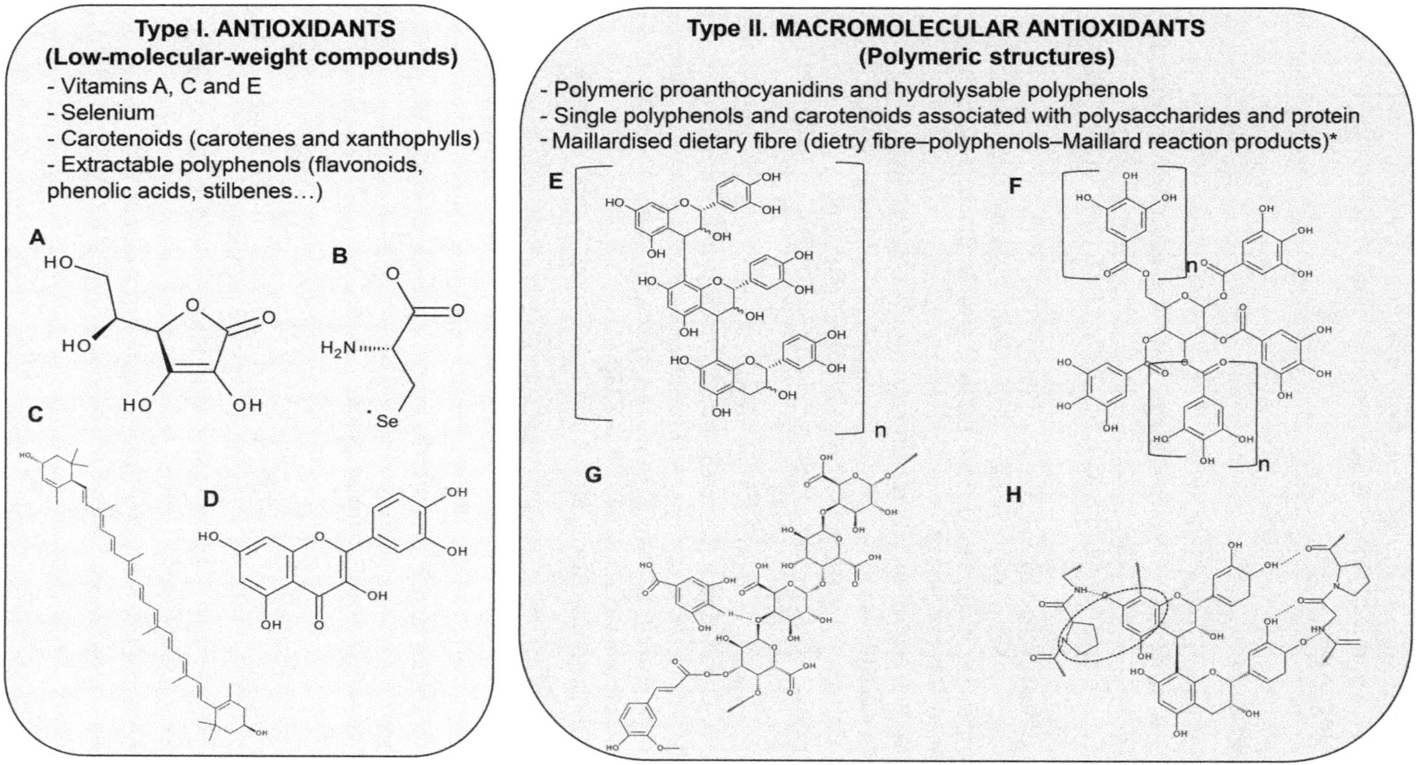

Figure 1.5 Classification of dietary antioxidants and examples of each category. (A) ascorbic acid; (B) selenocysteine; (C) lutein; (D) quercetin; (E) polymeric proanthocyanidin; (F), polymeric gallotannin; (G) pectin chain with ferulic acid associated by ester bonds and gallic acid associated by hydrogen bonds; and (H) protein chain with proanthocyanidin associated by hydrogen bonds and hydrophobic bond (dashed circle). * Unknown structure.

These are preliminary steps in the development of products from MACAN, but more research into obtaining them from new plant materials and using new technologies may lead to successful new applications of MACAN in food, dietetics and medicine.

References

1. H. Trowell, *Lancet*, 1974, **1**, 503.
2. L. Bravo, R. Abia and F. Saura-Calixto, *J. Agric. Food Chem.*, 1994, **42**, 1481–1487.
3. L. Bravo, R. Abia, I. Goni and F. Saura-Calixto, *Eur. J. Clin. Nutr.*, 1995, **49**, S211–S214.
4. F. Saura Calixto, *J. Food Sci.*, 1988, **53**, 1769–1771.
5. E. Manas and F. Saura-Calixto, *Eur. J. Clin. Nutr.*, 1995, **49**, S158–S162.
6. F. Saura-Calixto, A. Garcia-Alonso, I. Goni and L. Bravo, *J. Agric. Food Chem.*, 2000, **48**, 3342–3347.
7. I. Goñi, M. E. Díaz-Rubio, J. Pérez-Jiménez and F. Saura-Calixto, *Food Res. Int.*, 2009, **42**, 840–846.
8. F. Saura-Calixto, *J. Agric. Food Chem.*, 1998, **46**, 4303–4306.
9. I. Sánchez-Alonso, A. Jiménez-Escrig, F. Saura-Calixto and A. J. Borderías, *Food Chem.*, 2006, **101**, 372–378.
10. J. Pérez-Jiménez, J. Serrano, M. Tabernero, S. Arranz, M. E. Díaz-Rubio, L. García-Diz, I. Goñi and F. Saura-Calixto, *Nutrition*, 2008, **24**, 646–653.
11. S. Sánchez-Tena, D. Lizárraga, A. Miranda, M. P. Vinardell, F. García-García, J. Dopazo, J. L. Torres, F. Saura-Calixto, G. Capellà and M. Cascante, *Carcinogenesis*, 2013, **34**, 1881–1888.
12. S. Arranz, F. Saura-Calixto, S. Shaha and P. A. Kroon, *J. Agric. Food Chem.*, 2009, **57**, 7298–7303.
13. S. Arranz, J. M. Silván and F. Saura-Calixto, *Mol. Nutr. Food Res.*, 2010, **54**, 1646–1658.
14. P. Pinto and C. N. Santos, *Eur. J. Nutr.*, 2017, **56**, 1393–1408.
15. M. E. Diaz-Rubio and F. Saura-Calixto, *Int. J. Food Sci. Nutr.*, 2011, **62**, 715–718.
16. M. E. Díaz-Rubio and F. Saura-Calixto, *J. Agric. Food Chem.*, 2007, **55**, 1999–2003.
17. F. Saura-Calixto and M. E. Díaz-Rubio, *Food Res. Int.*, 2007, **40**, 613–619.
18. F. Saura-Calixto, J. Pérez-Jiménez, S. Touriño, J. Serrano, E. Fuguet, J. L. Torres and I. Goñi, *Mol. Nutr. Food Res.*, 2010, **54**, 939–946.
19. M. L. Mateos-Martín, J. Pérez-Jiménez, E. Fuguet and J. L. Torres, *Br. J. Nutr.*, 2012, **108**, 290–297.
20. J. Pérez-Jiménez, J. Serrano, M. Tabernero, S. Arranz, M. E. Díaz-Rubio, L. García-Diz, I. Goñi and F. Saura-Calixto, *Plant Foods Hum. Nutr.*, 2009, **64**, 102–107.
21. I. Goñi and J. Serrano, *J. Sci. Food Agric.*, 2005, **85**, 1877–1881.
22. M. E. López-Oliva, A. Agis-Torres, I. Goñi and E. Muñoz-Martínez, *Br. J. Nutr.*, 2010, **103**, 1110–1117.

23. J. Pérez-Jiménez, M. E. Díaz-Rubio and F. Saura-Calixto, *Nutr. Res. Rev.*, 2013, **26**, 118–129.
24. J. M. Silván, F. J. Morales and F. Saura-Calixto, *J. Agric. Food Chem.*, 2010, **58**, 12244–12249.
25. J. Pérez-Jiménez, M. E. Díaz-Rubio, M. Mesías, F. J. Morales and F. Saura-Calixto, *Food Res. Int.*, 2014, **55**, 391–396.
26. S. Arranz and F. Saura Calixto, *J. Cereal Sci.*, 2010, **51**, 313–318.
27. J. H. Cummings and G. T. Macfarlane, *J. Appl. Bacteriol.*, 1991, **70**, 443–459.
28. F. Saura-Calixto, *J. Agric. Food Chem.*, 2011, **59**, 43–49.
29. *European Nutrition and Health Report 2004*, ed. I. Elmadfa and E. Weichselbaum, European Commission, Brussels, Belgium, 2005.
30. G. Maiani, M. J. P. Castón, G. Catasta, E. Toti, I. G. Cambrodón, A. Bysted, F. Granado-Lorencio, B. Olmedilla-Alonso, P. Knuthsen, M. Valoti, V. Böhm, E. Mayer-Miebach, D. Behsnilian and U. Schlemmer, *Mol. Nutr. Food Res.*, 2009, **53**, 194–218.
31. M. E. Díaz-Rubio, Dietary fibre in beverages from diet. Determination, composition and contribution to dietary fiber intake, PhD Thesis, Autonomous University of Madrid, Spain, 2008.
32. F. Saura-Calixto and I. Goñi, *Food Chem.*, 2006, **94**, 442–447.
33. J. Serrano, I. Goñi and F. Saura-Calixto, *J. Agric. Food Chem.*, 2005, **53**, 2936–2940.
34. I. Goñi, J. Serrano and F. Saura-Calixto, *J. Agric. Food Chem.*, 2006, **54**, 5382–5387.
35. J. Pérez-Jiménez, M. Elena Díaz-Rubio and F. Saura-Calixto, *Plant Foods Hum. Nutr.*, 2015, **70**, 365–370.
36. J. Pérez-Jiménez and F. Saura Calixto, *Food Res. Int.*, 2015, **74**, 315–323.
37. N. Sanz-Pintos, J. Pérez-Jiménez, A. H. Buschmann, J. R. Vergara-Salinas, J. R. Pérez-Correa and F. Saura-Calixto, *J. Food Sci.*, 2017, **82**, 289–295.
38. F. Saura-Calixto and J. Pérez-Jiménez, Fruit sources of macromolecular antioxidants, a novel ingredient, and applications in food and beverages, in *18th IUFoST World Congress of Food Science and Technology*, Dublin, Ireland, 2016.
39. F. Saura-Calixto and J. Pérez-Jiménez, Properties and applications of macromolecular antioxidants: the hidden face of dietary antioxidants, *ACS Spring 2017 National Meeting*, San Francisco, USA, 2017.
40. N. Bastide, L. Dartois, V. Dyevre, L. Dossus, G. Fagherazzi, M. Serafini and M. C. Boutron-Ruault, *Eur. J. Nutr.*, 2017, **56**, 1233–1243.
41. M. J. Amiot, C. Riva and A. Vinet, *Obes. Rev.*, 2016, **17**, 573–586.
42. J. Godos, D. Sinatra, I. Blanco, S. Mulè, M. La Verde and M. Marranzano, *Nutrients*, 2017, **9**, 1069.
43. A. Medina-Remón, R. Casas, A. Tressserra-Rimbau, E. Ros, M. A. Martínez-González, M. Fitó, D. Corella, J. Salas-Salvadó, R. M. Lamuela-Raventos and R. Estruch, *Br. J. Clin. Pharmacol.*, 2017, **83**, 114–128.
44. G. Bjelakovic, D. Nikolova, R. G. Simonetti and C. Gluud, *Lancet*, 2004, **364**, 1219–1228.
45. S. Wang and F. Zhu, *Crit. Rev. Food Sci. Nutr.*, 2017, **57**, 2343–2357.

46. O. P. Heinonen and D. Albanes, *N. Engl. J. Med.*, 1994, **330**, 1029–1035.
47. F. Saura-Calixto, Antioxidant ingredient with low calorie content, method for obtaining same and use thereof, WO 2013.093161, 2012.
48. F. Saura-Calixto and J. Pérez-Jiménez, Method for producing a concentrate of polymer proanthocyanidins with a concentration of more than 40% by means of enzymatic treatments and without using organic solvents, WO 2017013299 A1, 2015.

CHAPTER 2

Dietary Fiber and Non-extractable Polyphenols

F. T. MACAGNAN*, A. B. B. BENDER AND C. S. SPERONI

Federal University of Santa Maria, Department of Food Technology and Science, Av. Roraima, no. 1000, Santa Maria, 97105-900, RS, Brazil
*E-mail: fe.tmacagnan@gmail.com

2.1 Introduction

Dietary fiber (DF) is naturally present in vegetable foods such as fruits, vegetables and cereals, and also is one of the main active ingredients of functional foods. The incentive for proper consumption of DF is related to several benefits such as regulating intestinal transit and the prevention or treatment of diabetes, cardiovascular diseases, colon cancer and even hepatic steatosis.[1–3] However, such benefits and the intensity of its effects are directly related to DF composition, its physicochemical properties and the antioxidant compounds associated with it.[4]

DF and antioxidant compounds are two recognized dietary factors in the prevention of chronic diseases. Probably due to structural, physicochemical and biological differences between these components, there is a certain tendency for the scientific community to treat them separately (*i.e.* as non-related compounds). However, DF and a considerable number of antioxidant compounds follow a common physiological process with synergic effects inside the gastrointestinal tract.[5]

Food Chemistry, Function and Analysis No. 5
Non-extractable Polyphenols and Carotenoids: Importance in Human Nutrition and Health
Edited by Fulgencio Saura-Calixto and Jara Pérez-Jiménez

Published by the Royal Society of Chemistry, www.rsc.org

Dietary antioxidants are a diverse group of chemical compounds with different action mechanisms and solubility in biological fluids or biomembranes. Among them are the carotenoids, vitamins C and E and polyphenols (PP).[6] The PP are the most abundant antioxidants of our diet, corresponding to about 90% of dietary antioxidant intake. They encompass a complex group of substances with different molecular weights that share one or more phenol groups and may be free or connected to the cell wall in plants and foods.[7,8]

Foods PP can be classified into two groups: extractable PP (EPP) or compounds solubilized by aqueous–organic solvents (methanol, acetone, ethanol, *etc.*), and non-extractable PP (NEPP) or compounds remaining in their corresponding extraction residues. According to Saura-Calixto,[9] this classification is also physiological and not just a concept derived from chemical analyses.

EPP is a complex mixture of low- and medium-molecular-weight compounds that include extractable proanthocyanidins (EPA) and hydrolysable tannins, along with other flavonoids and phenolic acids. A large proportion of EPP are bioaccessible in the small intestine, where they can be absorbed by the intestinal mucosa, be metabolized and exert systemic effects. Regarding NEPP, they are polymeric PP—including the non-EPA (NEPA) and hydrolysable PP—or low-molecular-weight PP that are linked to cell wall constituents, such as proteins and polysaccharides present in the DF matrix or trapped inside the food matrix.[5,10,11] Therefore, NEPP are not released significantly from the food matrix by chewing, by acid pH in the stomach or by the action of digestive enzymes,[11] and neither are they extracted from foods by the organic aqueous solvents that are traditionally used. NEPP are generally not included in the analysis of PP of foodstuffs, although their content is greater than the EPP fraction. Studies show that NEPP can play an important role in maintaining the antioxidant status of the organism. However, NEPP are ignored in most chemical and biological studies that focus only on EPP. As a result, there are plenty of articles in the literature that approach the physiological properties and potential effects on the health of EPP, making scientific data available on PP in foods and their dietary, partial and incomplete intake.[12]

With regards to their structure, NEPP have hydrophobic aromatic rings and hydrophilic hydroxyl groups with an ability to bind to polysaccharides and proteins at several sites on the cell wall surface. This interaction can occur during the maturation of fruit, food processing or gastrointestinal transit and can be attributed to the ability of polysaccharides to unite and capture phenolic compounds at several sites.[13] Interaction with the DF matrix occurs through different mechanisms: hydrogen bonds (between the hydroxyl group of PP and the oxygen atoms of the glycosidic links of polysaccharides), hydrophobic interactions (including possible encapsulation in hydrophobic pockets with NEPA) and covalent bonds (such as ester bonds between phenolic acids and polysaccharides). Moreover, a fraction of EPP may be associated with NEPP. The interaction can also depend on specific porosity and surface properties, which can restrict the size of molecules that penetrate.[5,11]

Another important aspect is that, because they are associated or trapped in the insoluble food matrix and not easily extracted from it, NEPP may be relevant constituents of DF. However, they are treated as non-related compounds to DF in traditional methods of fiber analysis in foods. The presence of bioactive compounds associated with the cell wall exerts significant effects on the physicochemical properties of DF, which influence its physiological properties.[4]

It is understood that despite the importance of NEPP and its strong relationship with DF, scientific circles still lack research that explores their presence in food, as well as the mechanisms of their biological function in the human organism, thus becoming a promising area for research. Our approach aims to show that DF and NEPP are interrelated compounds that present a common physiological fate and complementary effects on human health, with DF being considered an important carrier of antioxidant compounds into the human organism.

2.2 Non-extractable Polyphenols as Constituents of Dietary Fiber Matrix

DF consists of a complex and heterogeneous group of substances that present different physicochemical and physiological properties, which hinders its definition and the development of methodologies capable of efficiently reflecting its biological effects. According to the Official Methods of Analysis of the Association of Official Analytical Chemists (AOAC) recommended for the analysis of dietary fibers in foods (AOAC 985.29 and AOAC 991.43, enzymatic gravimetric and dietary fiber quantitated as resistant insoluble and soluble polysaccharides, respectively), there are two types of DF that coexist in varied proportions in food. They are classified based on the solubility of their constituents in water, named soluble DF (SDF) and insoluble DF (IDF). Studies show that the ratio of these fractions of DF has nutritional and technological significance.[4,14] The majority of scientific research on the benefits of DF for human health and on methodologies for their quantification in foods focuses on its polysaccharide constituents (soluble and insoluble) such as cellulose, hemicelluloses, pectin and β-glucan, and also on oligosaccharides and resistant starch.[4] However, many SDF and IDF present specific and relevant characteristics due to the presence of significant quantities of compounds with high biological activity, such as PP and carotenoids. Such bioactive compounds (called NEPP) are associated with the DF matrix and have attracted attention from researchers by giving important properties to DF, such as the derivatives of their antioxidant capacity (AC).[5]

According to the current definition of DF from the *Codex Alimentarius*, the compounds associated with polysaccharides of the cell wall can be included in the definition of DF when it is derived from plants. However, when extracted or even reintroduced into a food containing non-digestible carbohydrates, the associated compounds cannot be defined as DF.[15] Although they are included in the definition of DF from the *Codex Alimentarius*, the associated antioxidant compounds are omitted in the usual methods of determining.[4] Thus, the scientific community requires research that explores their presence in food.

According to previous studies, it is expected that in the residue of the gravimetric method of determination of DF contains not only indigestible polysaccharides, but also a fraction of NEPP. The interaction between these two constituents should be considered both in the realization of DF analyses (since it will include NEPP) and PP (because if NEPP are not released from the food matrix and analyzed, a relevant fraction of PP will be ignored and the result and its interpretations will be incomplete).[16] In this context, since NEPP are constituents of DF, updated methods for analysis of DF should include the analysis of these compounds. This approach has already been used in analysis of DF of various foods.[8,17,18]

Traditionally, DF is considered an independent food component of PP. However, scientific evidence proves that NEPP are relevant components of DF[8,19,20] and constitute approximately 50% of total dietary antioxidants.[5] A study by Goñi *et al.*[8] shows that PP are significant constituents of DF, representing from 1.4% to 50.7% of the IDF dry weight in plant foods and from 2.9% to 62.8% of the SDF in common beverages. The results of this study also demonstrate that fruits present the highest proportion of associated PP (followed by vegetables and cereals). Furthermore, PP appear in greater proportions in the IDF fraction.

According to Saura-Calixto,[5] most of the PP associated with DF of fruits consist of tannins, which are also present in high concentrations in vegetables and nuts. Cereals are important dietary sources of hydrolysable PP,[5] which are mainly associated (95%) with cell wall polysaccharides.[21] Saura-Calixto and Díaz-Rubio[20] reported that PP linked to DF represented 30–60% of total PP in red wine and approximately 9% in white wine. The results obtained in the study by Arranz *et al.*[22] show that the content of NEPP (112–126 mg/100 g) was greater than the content of EPP (18.8–28.0 mg/100 g) in fresh fruits like apples, peaches and nectarines. Similarly, Arranz *et al.*[12] evaluated EPP and NEPP in the Spanish diet and the results demonstrated that foods derived from plants contain substantially more NEPP than EPP, both in dry and fresh matter. Moreover, this study showed that fruits and vegetables contribute to the highest proportion of NEPP in the Spanish diet (more than 60%), followed by cereals. Table 2.1 presents the quantity of NEPP and their contribution to DF, showing their relevance as antioxidant constituents of DF.

The official analytical methods for DF have been specially developed for solid foods, making food composition tables and most of the data contained in the literature consider as zero the content of DF in beverages. However, certain SDF amounts can pass into beverages during maceration and/or extraction of solid plant material, allowing them to contain appreciable associated contents of SDF and PP that need to be considered.[5,8]

Table 2.2 presents SDF values in some beverages (from 0.3 to 8.67 g L^{-1}) and shows that PP contributes a substantial portion to traditional beverages

Table 2.1 Contribution of polyphenols (PP) to the constitution of insoluble dietary fiber (IDF) in plant foods.

Samples	IDF[a] (g L^{-1})	% of PP in IDF
Onions[8]	18.83	2.18
Green beans[8]	27.51	2.91
Lettuce[8]	19.76	3.39
Plums[8]	14.60	2.33
Strawberries[8]	21.51	2.60
Apples with peel[8]	14.89	2.42
Pears with peel[8]	16.63	4.69
Cashew apples[8]	20.70	50.72
Almonds[8]	17.34	7.38
Dark chocolate[8]	11.47	39.49
Red grape pomace[8]	71.50	25.89

[a]IDF content including associated polyphenols.

Table 2.2 Contribution of polyphenols (PP) to the constitution of soluble dietary fiber (SDF) in beverages.

Beverages	SDF[a] (g L^{-1})	% of PP in SF
Red wine[8]	2.29	38.86
Lager beer[8]	2.74	24.09
Brewed mate[8]	2.19	36.07
Cocoa milk drink[8]	8.67	17.30
Red tea[8]	0.35	62.86
Orange juice[8]	0.95	16.84
Apple juice[8]	1.72	2.91
Cider[8]	0.30	43.33
Brewed coffee[5]	5.57	15.62
Peach juice[5]	2.19	8.22
Red wine[5]	1.37	38.69
Lager beer[5]	2.04	4.90
Cocoa milk drink[5]	8.21	4.14

[a]SDF content including associated polyphenols.

such as wine, beer and coffee (38%, 24% and 15%, respectively). Compounds associated with DF in beverages and other foods interfere with their functionality and confer specific physiological properties. This supports the inclusion and analysis of NEPP as constituents of DF and joint evaluation of them in clinical and biological trials.

2.3 Dietary Fiber as a Carrier of Phenolic and Carotenoid Antioxidants

The AC of DF is reinforced by the presence of associated compounds (such as PP and carotenoids) and its determination can be helpful in complementing DF characterization and estimating its potential health effects and applications as a functional ingredient.[5] The study by Larrauri *et al.*[23] proposed for the first time the measuring AC in order to evaluate the potential effects of DF on health in association with other nutritional parameters, such as fermentability and rate of delay of glucose absorption. In recent years, the search for new sources of DF with antioxidant properties in vegetable food products has been boosted.[24]

Unfortunately, there is no solvent that is entirely satisfactory for the extraction of all antioxidants present in a food, especially those associated with the cell wall. In this way, the AC of food ends up being underestimated, mainly in DF-rich samples. The most widely used procedure in the research of antioxidant compounds is solid–liquid extraction with different combinations of organic solvents (ethanol, methanol or acetone) with water. However, compounds associated with DF are not extracted from food by these solvents, remaining in solid residues of extraction. This residue is generally ignored in the antioxidant analysis and considered devoid of PP, although it is a rich source of another PP fraction with specific biological activities (NEPP).[11] If the residue is treated with acids (HCl and H_2SO_4) or enzymes (cellulases, proteases, digestive enzymes and bacterial enzymes), significant amounts of PP can be released from the food matrix. This shows that the complete determination of PP in foods requires analysis of both extracts and extraction residues.[9] In this sense, the AC associated with DF can be measured in DF and hydrolyzed extracts by the most common AC methods (ABTS - 2,2-azinobis 3-ethyl-benzothiazoline-6-sulfonic acid, DPPH - 2,2-diphenyl-1-picryhydrazyl, FRAP - ferric reducing antioxidant power, ORAC - oxygen radical absorbance capacity).[5,11]

The importance of determining the AC of the non-extractive fraction is evidenced in the study by Matsumura *et al.*,[25] where 81.7% and 96.1% of the values of ORAC and DPPH, respectively, were derived from the NEPP fraction of dried persimmon. The difference between chemical and physiological quantification indicates that human digestive enzymes exhibit a capacity to solubilize PP that is higher than that of organic solvents.[9] Rufino *et al.*[18] point out that, in the case of proanthocyanidins, only the ABTS method was able to evaluate the AC of açaí. Other methods tested (DPPH, FRAP and ORAC) were not suitable for running on media containing butanol, requiring the release of anthocyanins from proanthocyanidins. Considerable amounts of NEPP associated with DF after acid treatment were first reported in some specific vegetable materials such as grape pomace, apple pomace and carob pods.[5]

From the physiological standpoint, the measurement of the DF-related AC provides an integrated measure that can represent the amount of antioxidant units (Trolox equivalents) carried by DF in the human gut. In this context, it may be considered that this is an essential physiological function of DF, since it ends up acting as an important carrier of bioactive compounds through the gastrointestinal tract, providing additional and continuous protection to the organism. DF with an exceptional AC, called antioxidant DF (ADF), was found in mango peel, pineapple shell, guava pulp, grape pomace, acerola fruits and some seaweed.[5] DF concentrates obtained from mango peel (*Mangifera indica* L.) were the first materials obtained with a high AC, this property being similar to butylhydroxyanisole, synthetic antioxidant widely used in the food industry, and much higher than vitamin E, in an approximate concentration.[26]

2.4 Influence of Dietary Fiber on the Bioaccessibility and Bioavailability of Dietary Antioxidants

PP (flavonoids, phenolic acids, stilbenes and tannins) and carotenoids (carotenes and xanthophylls) are the main groups of dietary antioxidants associated with DF. The biological properties of these compounds depend on their bioaccessibility and bioavailability,[5] since it is not the amount of phytochemicals in the food that determines their functional properties, but the fraction that reaches its biological target.[27]

The bioaccessibility of these compounds has been greatly studied. Bioaccessibility is related to the capacity of PP to be released from the food during gastrointestinal digestion (by the action of digestive enzymes) and is available for intestinal absorption.[28] PP are not much absorbed in the gastrointestinal tract, although the variation in the absorption rate is related to the different chemical structures presented by these compounds, and also to the conditions of the environment in the gastrointestinal tract.[29,30] During digestion, changes may occur in the chemical structure of PP (due to metabolic disorders or physiological conditions) that may influence their bioaccessibility, such as modification in molecular weight, glycosylation and molecular esterification.[31,32] In a study by Gouw *et al.*,[33] the bioaccessibility of phenolic compounds of residues of apple, blueberry, cranberry and red raspberry was evaluated, and it was observed that this property is closely related to chemical and morphological properties of DF. Thus, the raspberry PP presented themselves as more bioaccessible under *in vitro* digestion conditions because they were in a free form. PP of apple, blueberry and cranberry residues were more abundant in the linked form and hence less bioavailable. Calvache *et al.*,[34] when assessing the bioaccessibility of antioxidant compounds of papaya DF concentrates, concluded that 66% and 64% of pulp and peel, respectively, were bioaccessible in the small intestine. According to these authors, considerable amounts of PP (for each 100 g, 0.112 g and 0.209 g—soluble fraction; 0.049 g and 0.155 g—insoluble fraction) remain associated with indigestible fractions (soluble and insoluble) and reach the large intestine. Subsequently, it could be released by fermentation processes carried out by bacterial flora by exercising some antioxidant activity.

Bioavailability refers to the potential absorption of EPP and NEPP to be available for its metabolic pathways and to follow specific routes, and it depends on the release of these compounds from the food matrix during digestion.[35] In other words, it refers to the proportion of an antioxidant that is digested, absorbed and used in normal metabolism.[36] The biological properties of PP are closely linked to bioavailability, since the chemical structure determines the intestinal absorption rate, as well as the nature of the circulating metabolites in the bloodstream.[37] The bioavailability of PP is closely related to digestive processes, such as the release of the food matrix, changes occurring during gastric and large intestine digestion, cell absorption of aglycone and PP being conjugated by enterocytes. Furthermore, fermentation of NEPP by intestinal microbiota, enzymatic modifications after absorption in the small intestine/colon, transportation through blood circulation, tissue redistribution, renal and intestinal excretion through the urine and association of bile with pancreatic juices are also related to bioavailability.[38] In addition, the bioavailability of EPP and NEPP (in relation to the food matrix) is also influenced by other factors such as the environment (storage conditions, degree of maturation), processing conducted, interactions with other food constituents, structure and concentration of PP in the food matrix and the intestinal and systemic conditions of the organism that will ingest these compounds.[39] One of these factors can be observed in a study by Zhao *et al.*,[40] in which less bioavailability of corn bran in relation to wheat bran was observed. This fact is due to maize having the most complex cell wall structure and being composed of large amounts of NEPP. The decoction process can also increase the bioaccessibility of PP, making them more available for absorption after the gastric and duodenal phase.[41]

Throughout the absorption process, PP undergo several modifications and may be first conjugated by the intestinal cells and then metabolized in the liver, producing different metabolites, which may be present in plasma and even be excreted in the urine.[42] Monomeric and dimeric epicatechins were identified in the feces of rats fed with NEPA, and ten phase II metabolites (epicatechin) were identified in the urine of the animals, indicating that partial depolymerization of NEPA occurs throughout intestinal transit. A portion of large polymers of proanthocyanidins appears to be depolymerized along the intestinal tract, resulting in the release of monomers of epicatechins and, possibly, oligomers too.[43] NEPP can be catabolized by intestinal microbiota and, because they are associated with the DF matrix, particularly with insoluble polymers, the conversion should be slower and the produced metabolites can stay for 3–4 days in the bloodstream, where they can exert anti-inflammatory and antioxidant effects.[44] The detection of epicatechin metabolites in feces indicates that active species remain in contact with intestinal epithelium for at least 24 hours after ingestion, indicating a potential role in the prevention of colorectal cancer.[43]

The presence of DF in fruits and vegetables rich in phenolic or carotenoid compounds may interfere with the bioavailability of these antioxidants during the digestion process.[36] Free phenolic compounds (EPP) are rapidly absorbed during gastrointestinal digestion, while linked compounds (NEPP) are less absorbed, because they are associated with a complex food matrix in their intact forms.[29] Thus, it is estimated that most of these compounds (95%) are not absorbed, accumulating in the colon,[45] which may be fermented or excreted in feces.[41]

In the same way as NEPP (main dietary antioxidants), carotenoids exhibit incomplete digestion in the small intestine and a portion of them reaches the large intestine in association with DF.[5] Cell rupture appears to be a prerequisite for releasing carotenoids, since carotenoids that are within intact cells are inaccessible during digestion in the upper tract.[27] Non-absorbable carotenoids can be active in the gastrointestinal tract and can also be metabolized during intestinal transit and exert biological actions. For example, β-cryptoxanthin can be absorbed in the colon, where it and its metabolites may be biologically active.[46] The liver stores and distributes the carotenoids to other tissues, but the mechanisms are not well elucidated yet. However, intervention studies showed that the increase in carotenoid ingestion influences the serum concentrations more than colon concentrations, suggesting that absorption occurs primarily in the small intestine.[47]

A study by Serrano *et al.*[48] indicated that between 20% and 70% of β-carotene and lutein present in green leafy vegetables was associated with DF. The presence of these carotenoids was correlated inversely with bioavailability in the small intestine. The reduction of the bioavailability of carotenoids associated with DF is due to the fact that the DF interferes with micelle formation between lipids and bile salts in the small intestine, blocking the passive absorption of these compounds.[36] Due to the lipophilic nature of the carotenoids, the incorporation of these micelles during digestion is essential for bioavailability.[27] Moreover, with the increase in the viscosity of intestinal content, the enzymatic activity is reduced and contact with intestinal enterocytes becomes difficult. Thus, carotenoids, lipids and bile salts are not absorbed into the small intestine and reach the large intestine, where polysaccharides are hydrolyzed and carotenoids contribute to the formation of a healthy antioxidant environment.[36]

By assessing the influence of structural barriers in the bioaccessibility of anthocyanins present in black carrots, Carrillo *et al.*[27] reported a reduction of bioaccessibility when both organelles or cell membranes and cell walls were present. The authors attribute this result to the fact that when cell rupture occurs, bioactive compounds are released, but at the same time they become potentially available to associate themselves with DF. This leads to modulation in their relative bioaccessibility. As mentioned, DF is the main carrier of phenolic compounds and influences its bioaccessibility.

2.5 Colonic Metabolism of Non-extractable Polyphenols and Dietary Fiber: Relationship and Interaction

One of the main roles of DF is the supply of substrates for fermentation by bacteria in the large intestine. The fermentation of DF (mainly of the soluble fraction) produces energy, biomass and important final compounds such as short-chain fatty acids (SCFA). The main SCFA produced are acetic, propionic and butyric acids, which are fundamental to intestinal metabolism and capable of exerting influence on the metabolism of lipids and carbohydrates.[49,50] This fermentative process also has a function that is essential to the intestinal ecosystem: maintaining the microbiota of the colon and improving the immune system. In addition to DF, other indigestible compounds are available for fermentation due to antioxidant compounds not being absorbed in the small intestine. The fermentation of these compounds seems not to produce SCFA, a fact evidenced by a fermentation test that uses the monomer epicatechin (basic structure of NEPA) as a control. In this test, SCFA were not detected in the fermentation supernatant; rather, significant amounts of the typical final products of PA were detected (such as hydroxyphenylacetic acid).[10] However, the production of fermentation metabolites from PP present in the colon can be quantitative and biologically important. Human studies showed that phenolic compounds available for colonic fermentation in Western diets varied from 0.9 to 2.7 g day^{-1}.[5] This fact also contributes to an updated definition of DF, which considers it a combination of chemically heterogeneous substances not absorbed in the upper gastrointestinal tract, but that are interrelated and present in a common physiological fate in the human organism.

As already mentioned, a portion of dietary antioxidants (vitamins, low-molecular-weight PP and carotenoids) are at least partially bioaccessible in the small intestine for exerting their systemic effects in organism. These compounds (EPP) and their potential beneficial health effects are already well reported in the literature. However, the antioxidant compounds associated with the insoluble food matrix (NEPP) are not absorbed into the small intestine. Thus, they arrive intact to the colon, where they become fermentable substrates by the action of bacterial enzymes, along with carbohydrates and indigestible proteins.[5,8,11,20,29] These non-bioaccessible antioxidants in the upper gastrointestinal tract are mainly phenolic compounds (polymers of PP and low-molecular-weight PP linked or trapped in cores of the fiber matrix), with lesser amounts of other antioxidant compounds such as carotenoids.[5] Because they are not readily available in the small intestine and are not extracted from the food matrix by solvents traditionally used in the extraction techniques of the antioxidants of foods, the compounds associated with DF are ignored in most studies. This is probably due to the belief that they are irrelevant constituents from the quantitative and physiological standpoints.[4] However, as already reported, NEPP are important dietary antioxidants and constituents of DF that follow a common physiological process with synergic effects inside the gastrointestinal tract.[5,13] Moreover, the hypothesis that food DF serves as a carrier of significant amount of NEPP along the human intestine should not be disregarded.[35]

Little information is available on the effect of added DF in the absorption of PP in humans. It is speculated that it decreases their bioavailability, mainly due to some factors such as physical imprisonment, increased viscosity and volume.[38] While studies advocate that fiber hinders the absorption of PP and consequently its functionality, evidence increasingly suggests that this interaction is biologically important. In recent years, several authors have studied the health-related properties of phenolic compounds present in DF, using isolated PP or PP-rich matrices combining *in vitro* models, animal models or human trials.[4] These studies have shown that NEPP play an important role in health, particularly in relation to gastrointestinal health.

Most of the ingested NEPP arrive almost intact to the colon, mainly associated with DF, and can become accessible through microbial fermentation of the molecules to which they are associated (carbohydrates or proteins) or by the action of some intestinal enzymes (such as esterases). These enzymes are capable of breaking existing covalent bonds between the DF–PP complex. Intestinal bacteria develop important metabolic and immune functions and are crucial in disrupting the matrix of DF and releasing the dietary antioxidants, a fact stemming from the wide variety and genomic complexity of the human microbiota. Intestinal microbiota is able to hydrolysate, reduce, decarboxylate, demethylate and dehydroxylate PP and thus produce several metabolites with potential health effects.[5] While a small portion of ingested flavonoids is absorbed in intact form, the largest percentage is absorbed as phenolic metabolites, primarily phenolic acids. Glycosides of quercetin can be metabolized by intestinal microorganisms to simple phenolic acids, such as the 3,4-dihydroxyphenylacetic acid.[51] Thus, daily, significant amounts of different classes of NEPP become bioaccessible in the large intestine, along with a fraction of solubilized EPP, but not previously absorbed in the small intestine.[11] It is estimated that about 48% of PP from the Spanish diet is bioaccessible in the small intestine, whereas 42% becomes bioaccessible in the large intestine.[20] PP activity released later in the intestine may be different from that of PP that are absorbed faster, and additional metabolites with altered properties can be formed.[38]

The results of the study by Arranz *et al.*[12] suggest that about 1 g of NEPP can reach the human colon daily and it is hoped that these compounds have a significant physiological impact on the large intestine. These compounds can influence the development of microbiota in intestinal antioxidant status and produce metabolites that can be absorbed by the mucosa. According to Vitaglione *et al.*,[21] PP released by the action of intestinal and microbial esterases are absorbed to varying degrees, passing into the bloodstream, where they can exercise their health benefits by reducing, for example, the oxidation of low-density lipoprotein. Because the SDF is less reticulate than IDF, the release of PP becomes easier. Therefore, it is interesting to convert the IDF into SDF to maximize the benefits that the associated PP (NEPP) present in cereals can entail in health, which can be accomplished through chemical, physical or enzymatic treatments.

Phenylacetic and phenylpropionic acids are examples of absorbable metabolites generated by the colonic fermentation of PP and that can exert systemic effects.[5,52] Saura-Calixto *et al.*[10] investigated the colonic fermentation of NEPA associated with DF of two products prepared from materials rich in these compounds (carob pods and red grapes) using a model of *in vitro* small intestine digestion and colonic fermentation. The main metabolite identified after 24 hours of fermentation colonic was hydroxyphenylacetic acid for both samples, which can be a result of dehydroxylation of dihydroxyphenylacetic acid, typically formed during the *in vitro* colonic fermentation of flavonols. Hydroxyphenylvaleric acid was the second metabolite detected in the specimens, followed by hydroxyphenylpropionic acid. The main metabolite of NEPA in human plasma was also dihydroxyphenylacetic acid. The presence in plasma of the same metabolites that were detected after *in vitro* colonic fermentation of proanthocyanidins provides evidence that such dietary PP undergo fermentation in the colon and release absorbable metabolites with potential health effects.

Similarly, Pérez-Jiménez *et al.*[53] measured the antioxidant capability of plasma, in which the phenolic compounds associated with DF are partially bioavailable in humans, although DF seems to delay its absorption. Mateos-Martín *et al.*[43] evaluated the colonic fermentation of NEPA derived from grapes (preparation free from EPP) in rats and reported evidence that these proanthocyanidins behave *in vivo* as a source of phenolic acids, which are progressively released and deliver phenolic species that come into contact with the intestinal walls and are bioavailable for at least 24 hours after ingestion. In fact, the kinetics of NEPP differ from EPP, presenting delayed absorption by DF, and may vary up to 50 postprandial hours.[38] In a rat experiment, ferulic acid of wheat bran (bound form) presented a concentration in plasma in 30 minutes that was more than two times smaller than that of the diet containing its free form (0.8% *vs.* 1.9% of the ingested dose). However, its presence in plasma is constant for 24 hours, while it decreases to zero after 4.5 hours for its free form.[54] In this way, NEPP continually provide to the colonic microbiota with a significant amount of fermentable substrates that need more time to be released and fermented when compared to EPP due to their specific nature and, thus, they result in a more prolonged circulation of beneficial metabolites.[11] According to Vitaglione *et al.*,[21] the hypothesis of the slow and continuous release of PP associated with DF of cereals in the intestinal lumen can exert important *in vivo* antioxidant activity. This enables constant consumer protection against pathologies whose etiology and progression are governed by a state of redox imbalance. Moreover, this slow and continuous release of PP may be more advantageous than peaks of very high antioxidant concentrations in blood, which, in some cases, also determine a pro-oxidant environment.

Saura-Calixto *et al.*[20] estimated the intestinal bioaccessibility of dietary PP through an *in vitro* gastrointestinal model, and they showed that 10% of PP present in the Spanish diet is inaccessible and remained in the food matrix after the whole digestion process. In addition to exercising its biological effects as a fermentable substrate for bacterial microbiota, NEPP can produce local effects in the gastrointestinal tract in the form of a complex structure with binding properties that is not absorbable. Metabolites of colonic fermentation of PP that have not been absorbed and the unfermented PP remain in the colon for a long period of time, where they can contribute to the formation of a healthy antioxidant environment by the elimination of free radicals and the neutralization of the pro-oxidant effects in the diet.[5,8,12,21,55] Goñi and Serrano[55] studied the ADF of grapes (GADF, concentrated from grape seeds) and reported that PP not absorbed in the small intestine of rats (EPP and NEPP) provided considerable antioxidant status to the large intestine, which were 89% and 80% larger in animals fed with GADF, respectively. Similarly, our recent study (not published) assessed the impact of apple DF consumption and its modifications by acid hydrolysis and extrusion on the antioxidant status of the cecum of rats. The study showed that the animals treated with apple DF presented significantly higher amounts (>100%) of PP in the cecum when compared to controls treated with cellulose. The formation of the antioxidant environment in the intestine has been confirmed by the greatest AC of the cecal contents of the animals of these groups. When assessing the AC of non-extractable fractions of dried persimmon in an *in vitro* model of gastrointestinal digestion, Matsumura *et al.*[25] observed that ORAC values were low after the oral phase and increased gradually after the stomach stage and intestinal digestion. However, subsequent digestion in the large intestine led to significant increases in the ORAC values of non-extractable fractions. In this way, it is suggested that the fermentation decomposition of non-extractable fractions of dried persimmon by the intestinal microbiota produced the present antioxidant properties. These antioxidants present in the digest can protect the exposed surface of the intestinal lumen or even enter the intestine cells and provide intracellular protection.[55]

Non-absorbed PP from the diet can also exert an effect on the modulation of intestinal microbiota, modifying the numbers and types of bacteria.[5] Some authors suggest that PP present in the colon can act as a metabolic prebiotic. This prebiotic potential is verified in studies that show the action of PP and its metabolites in the growth of beneficial bacteria (*Lactobacillus* and *Bifidobacterium*) and the inhibition of pathogenic bacteria (*Staphylococcus aureus*, *Escherichia coli* and *Candida albicans*).[56–59] For example, GADF has demonstrated the potential to improve the gastrointestinal health of the host by stimulating the proliferation of *Lactobacillus* species in *in vitro* tests and in the cecum of rats (increase of approximately one log compared to the control group).[60] However, the effects of PP on the modulation of intestinal ecology and the growth of intestinal microbial species are still little understood.

When arriving intact at the colon, NEPP can play a preventive role in colorectal cancer by different biological mechanisms.[11] Several studies report that tannins, the main structure of NEPP, can produce chemopreventive effects through various different mechanisms such as induction of apoptosis, inhibition of the inflammation signaling pathway or generation of an antioxidant environment in the colon.[12] GADF (obtained from red grapes) was able to reduce intestinal tumorigenesis in rats with colon cancer, reducing significantly the total number (76%) and the size (65–87%) of the polyps.[61] Lizarraga *et al.*[62] also reported that the consumption of a lyophilized red grape pomace containing proanthocyanidin-rich DF (GADF) by female mice induced changes in the expression of tumor suppressor genes and proto-oncogenes, as well as the modulation of genes from pathways, including lipid biosynthesis, energetic metabolism, the cell cycle and apoptosis. This research provides evidence that GADF protects the colon tissue against tumor development and reduces the risk of cancer. Similarly, non-absorbable compounds (triterpenes and some diterpenes) of rosemary (*Rosmarinus officinalis*) can display their biological activity in the large intestine through interactions with intestinal microbiota and by a direct effect on the colonocytes at the onset of cancer or in its progression. These compounds are not absorbed or metabolized, reaching the colon intact, exercising antitumor effect and acting as supplementary chemoprotective agents against colorectal cancer.[63]

Furthermore, NEPP (through its absorbable bioactive metabolites in the colon) can also play a role in preventing other chronic diseases, especially in reducing risk factors for the development of cardiovascular diseases, such as hyperlipidemia and high blood pressure. Studies show that a matrix rich in NEPP can mitigate certain risk factors for the development of cardiovascular diseases such as hyperlipidemia and high blood pressure.[11] The behavior of a food rich in DF, EPP and NEPP in the gastrointestinal tract is demonstrated in Figure 2.1.

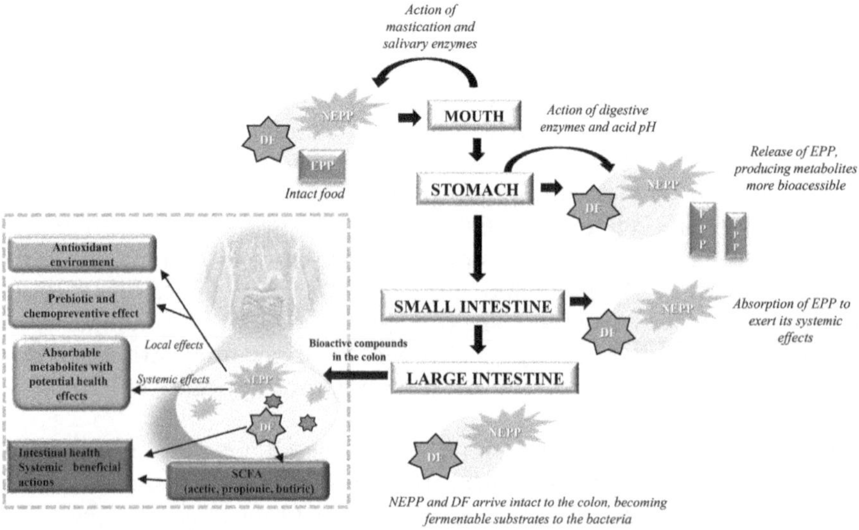

Figure 2.1 Overview of the metabolic fate of a food rich in dietary fiber (DF), non-extractable polyphenols (NEPP) and extractable polyphenols (EPP). SCFA, short-chain fatty acid.

2.6 Dietary Fiber Rich in Non-extractable Polyphenols: Potential Application in Food Science and Nutrition

Studies report that the use of DF-rich ingredients and phenolic compounds in the production of functional foods can result in loss of absorption of the antioxidant, because the DF may trap these substances, reducing the functionality of the proposed food. However, these interactions with DF can give a very different role to PP.[4] From the nutritional point of view, it is important to consider the association between DF and PP as the compounds associated with the cell wall can be responsible for some of the benefits traditionally attributed only to the DF.[13] In addition to playing an important role in maintaining the antioxidant status of the organism, these compounds may interfere with the physicochemical properties of DF due to increased reactivity with components in complex food matrices,[64,65] influencing their physiological properties and their effects on health.[5]

The fact that a food matrix presents a significant content of DF and PP can potentiate the biological effects of food, which makes them especially relevant as sources of bioactive compounds.[16] Thus, it should be considered that the presence of NEPP in the DF matrix can give rise to specific synergies. This was observed in a study by Pérez-Jiménez *et al.*[66] in which the daily supplementation of hypercholesterolemic individuals with 7.5 g of a product rich in NEPP from grapes for 16 weeks resulted in significant reductions of plasma cholesterol and blood pressure. These reductions were larger than those described separately for DF and EPP in various meta-analyses. A study by Saura-Calixto *et al.*[10] also demonstrated that DF can influence the fermentation rate of proanthocyanidins, probably by stimulating bacterial enzymatic activity. Therefore, the fermentation of NEPP in the colon would be accompanied by a synergism with the other components of the food matrix to which they are associated, an aspect that is not present in the fermentation of EPP, which are free in the food matrix. This indicates that a food rich in DF can favor the fermentation of this type of PP in the colon, releasing absorbable metabolites with potential health effects. Similarly, other researchers have already reported the largest fermentability of apple PP when the fermentation was performed in the presence of soluble pectins, since the combined fermentation of this SDF and PP still provided greater improvements in the lipid profiles of rat plasma than supplementation with these compounds separately.[67]

According to surveys, dietary PP that reach the colon can act as metabolic prebiotics.[56–58] This effect could be potentialized when a substantial amount of PP is associated with DF. Wallace *et al.*[68] reported significant changes in the bacterial population when consuming fruit juice beverages containing fruit DF, fruit PP and their combination. The hypothesis for the study was that the combination of PP and DF would have a greater benefit to intestinal health than the placebo product or the DF and PP separately. In this sense, regular consumption of a diet rich in fibrous vegetable foods with high levels of PP could balance intestinal microbial ecology, helping to prevent gastrointestinal disorders and thereby improve the health of the host.[69] Researchers also showed an inverse association between the consumption of food rich in DF with associated antioxidants and colon cancer.[13,62] When this complex reaches the colon, the fraction of polysaccharides ferments, releasing PP and forming beneficial components (SCFA). PP, DF and its beneficial metabolites (such as butyric acid and free antioxidants) remain in contact with the intestinal surface for several hours, forming a favorable and protective environment that could modulate the incidence of certain types of degenerative diseases such as colon cancer. The absorption of bioactive metabolites generated in the colon (both from the fermentation of NEPP itself and of its associated polysaccharides) may also give rise to systemic effects, manifesting in other organs and tissues.[11] In a study in humans, the acute ingestion of GADF (a natural product obtained from the skins and seeds of red grapes), rich in associated phenolic antioxidants, increased the AC of plasma in relation to a control group. This increase was significant after 8 hours of the ingestion (area below the curve in ABTS test 3.6% lower in a control group).[53]

The transportation of dietary antioxidants through the gastrointestinal tract and the production of antioxidant metabolites of fermentation in the colon appear to be essential physiological functions of DF rich in NEPP. The association between these components of the foods suggests positive effects on gastrointestinal health and also on reducing the risk of developing cardiovascular diseases and colon cancer. As food ingredients, these DF prevent lipid oxidation in meat products and restructured fish products, maintaining nutritional quality and extending their shelf life.[5]

2.7 Conclusions

PP and DF are two dietary factors recognized in the prevention of chronic diseases and are fundamental to maintaining human health. Although they present structural, physicochemical and biological differences, these compounds are strongly related. Parts of the dietary PP (NEPP) are associated with DF and therefore are not readily absorbed in the small intestine. Moreover, they are not extracted from the insoluble food matrix by solvents traditionally used in the techniques of extraction of food antioxidants. In this sense, some studies advocate that the use of DF-rich ingredients and PP in the production of functional foods can result in loss of absorption of the antioxidants and lower functionality of the proposed products. However, there is evidence suggesting that this interaction is biologically important and provides a very different role to PP. Thus, the fact that a food matrix presents significant amounts of DF and PP can give rise to specific synergies that enhance the biological effects of food.

Studies also show that NEPP are the main dietary antioxidants and relevant constituents of DF. Such bioactive compounds need to be considered in chemical analyses and biological studies involving DF because they exert significant effects on the physicochemical properties of the DF, which directly influence their physiological and functional properties.

Most of the ingested NEPP arrive almost intact to the colon, mainly associated with DF, and can become accessible through microbial fermentation and thus exert systemic beneficial actions (reduction of low-density lipoprotein oxidation and constant antioxidant protection), acting in the prevention of the development of cardiovascular diseases. PP and non-absorbable metabolites remain in the colon, where they can exert important local action (antioxidant protection and prebiotic and chemopreventive effects) and contribute to health and intestinal integrity. In this context, the DF can be considered an important carrier of antioxidant compounds into the gastrointestinal tract, providing additional and continuous protection to the body.

Evidence on the health effects of naturally occurring antioxidant compounds in the DF matrix increasingly encourages the search for new sources of DF rich in NEPP with potential for application in food science and nutrition. The development of foodstuffs with these antioxidant DF would enable greater consumer protection. It is also crucial that the studies related to DF take into consideration the PP associated with it, because, as reported, the presence of these compounds interferes with DF's functional and biological properties, providing DF with special features and additional beneficial effects that need to be considered.

References

1. C. W. C. Kendall, A. Esfahani and D. J. A. Jenkins, *Food Hydrocolloids*, 2010, **24**(1), 42.
2. M. M. Kaczmarczyk, M. J. Miller and G. G. Freund, *Metab., Clin. Exp.*, 2012, **61**(8), 1058.
3. D. M. Amaya-Cruz, S. Rodríguez-González, I. F. Pérez-Ramírez, G. Loarca-Piña, S. Amaya-Llano and M. A. Gallegos-Corona, *et al.*, *J. Funct. Foods*, 2015, **17**, 93.
4. F. T. Macagnan, L. P. Silva and L. H. Hecktheuer, *Food Res. Int.*, 2016, **85**, 144.
5. F. Saura-Calixto, *J. Agric. Food Chem.*, 2011, **59**(1), 43.
6. B. Halliwell, *Annu. Rev. Nutr.*, 1996, **16**, 33.
7. F. Saura-Calixto and I. Goñi, *Food Chem.*, 2006, **94**, 442.
8. I. Goñi, M. E. Díaz-Rubio, J. Pérez-Jiménez and F. Saura-Calixto, *Food Res. Int.*, 2009, **42**(7), 840.
9. F. Saura-Calixto, *J. Agric. Food Chem.*, 2012, **60**, 11195.
10. F. Saura-Calixto, J. Pérez-Jiménez, S. Touriño, J. Serrano, E. Fuguet and J. L. Torres, *et al.*, *Mol. Nutr. Food Res.*, 2010, **54**, 939.
11. J. Pérez-Jiménez, M. E. Díaz-Rubio and F. Saura-Calixto, *Nutr. Res. Rev.*, 2013, **26**(2), 118.
12. S. Arranz, J. M. Silván and F. Saura-Calixto, *Mol. Nutr. Food Res.*, 2010, **54**(11), 1646.

13. A. E. Quirós-Sauceda, H. Palafox-Carlos, S. G. Sáyago-Ayerdi, J. F. Ayala-Zavala, L. A. Bello-Perez and E. Alvarez-Parrilla, *et al.*, *Food Funct.*, 2014, **5**(6), 1063.
14. S. Westenbrink, K. Brunt and J. W. Van der Kamp, *Food Chem.*, 2013, **140**(3), 562.
15. Codex Alimentarius, *Report of the 30th Session of the Codex Committee on Nutrition and Foods for Special Dietary Uses*, Cape Town, South Africa, 3–7 November, 2008, ALINORM 09/32/26.
16. N. Sanz-Pintos, J. Pérez-Jiménez, A. H. Buschmann, J. R. Vergara-Salinas, J. R. Pérez-Correa and F. Saura-Calixto, *J. Food Sci.*, 2017, **82**(2), 289.
17. M. D. S. M. Rufino, J. Pérez-Jiménez, M. Tabernero, R. E. Alves, E. S. Brito and F. Saura-Calixto, *Int. J. Food Sci. Technol.*, 2010, **45**, 2227.
18. M. D. S. M. Rufino, J. Pérez-Jiménez, S. Arranz, R. E. Alves, E. S. Brito and M. S. P. Oliveira, *et al.*, *Food Res. Int.*, 2011, **44**, 2100.
19. F. Saura-Calixto, *J. Agric. Food Chem.*, 1998, **46**(10), 4303.
20. F. Saura-Calixto and M. E. Díaz-Rubio, *Food Res. Int.*, 2007, **40**(5), 613.
21. P. Vitaglione, A. Napolitano and V. Fogliano, *Trends Food Sci. Technol.*, 2008, **19**(9), 451.
22. S. Arranz, F. Saura-Calixto, S. Shaha and P. A. Kroon, *J. Agric. Food Chem.*, 2009, **57**, 7298.
23. J. A. Larrauri, I. Goñi, N. Martín-Carrón, P. Rupérez and F. Saura-Calixto, *J. Sci. Food Agric.*, 1996, **71**, 515.
24. B. A. Acosta-Estrada, J. A. Gutiérrez-Uribe and S. O. Serna-Saldívar, *Food Chem.*, 2014, **152**, 46.
25. Y. Matsumura, T. Ito, H. Yano, E. Kita, K. Mikasa and M. Okada, *et al.*, *Food Chem.*, 2016, **202**, 99.
26. J. A. Larrauri, P. Rupérez and F. Saura-Calixto, *Z. Lebensm.-Unters.–Forsch. A*, 1997, **205**, 39.
27. C. Carrillo, C. Buvé, A. Panozzo, T. Grauwet and M. Hendrickx, *Food Chem.*, 2017, **227**, 271.
28. C. Cardoso, C. Afonso, H. Lourenço, S. Costa and M. L. Nunes, *Trends Food Sci. Technol.*, 2015, **41**(1), 5.
29. C. Manach, G. Williamson, C. Morand and A. Scalbert, *Am. J. Clin. Nutr.*, 2005, **81**, 230.
30. D. Tagliazucchi, E. Verzelloni, D. Bertolini and A. Conte, *Food Chem.*, 2010, **120**(5), 599.
31. A. Scalbert, C. Morand, C. Manach and C. Rémésy, *Biomed. Pharmacother.*, 2002, **56**(6), 276.
32. M. H. Ahmad-Qasem, J. Cánovas, E. Barrajón-Catalán, J. E. Carreres, V. Micol and J. V. García-Pérez, *J. Agric. Food Chem.*, 2014, **62**(26), 6190.
33. V. P. Gouw, J. Jung and Y. Zhao, *LWT–Food Sci. Technol.*, 2017, **80**, 136.
34. J. N. Calvache, M. Cueto, A. Farroni, M. Escalada Pla and L. N. Gerschenson, *J. Funct. Foods*, 2016, **27**, 319.
35. G. A. González-Aguilar, F. J. Blancas-Benítez and S. G. Sáyago-Ayerdi, *Curr. Opin. Food Sci.*, 2017, **13**, 84.

36. H. Palafox-Carlos, J. F. Ayala-Zavala and G. A. González-Aguilar, *J. Food Sci.*, 2011, **76**(1), 6.
37. A. Scalbert and G. Williamson, *J. Nutr.*, 2000, **130**, 2073.
38. T. Bohn, *Nutr. Rev.*, 2014, **72**(7), 429.
39. M. D'Archivio, C. Filesi, R. Vari, B. Scazzocchio and R. Masella, *Int. J. Mol. Sci.*, 2010, **11**(4), 1321.
40. Z. Zhao, Y. Egashira and H. Sanada, *J. Agric. Food Chem.*, 2005, **53**, 5030.
41. G. Mercado-Mercado, F. J. Blancas-Benitez, G. R. Velderrain-Rodríguez, E. Montalvo-González, G. A. González-Aguilar and E. Alvarez-Parrilla, *et al.*, *J. Funct. Foods*, 2015, **18**, 171.
42. M. D'Archivio, C. Filesi, R. Di Benedetto, R. Gargiulo, C. Giovannini and R. Masella, *Ann. Ist. Super. Sanita*, 2007, **43**(4), 348.
43. M. L. Mateos-Martín, J. Pérez-Jiménez, E. Fuguet and J. L. Torres, *Br. J. Nutr.*, 2012, **108**, 290.
44. A. González-Sarrías, J. C. Espín and F. A. Tomás-Barberán, *Trends Food Sci. Technol.*, 2017, **69**, 281.
45. M. N. Clifford, *Planta Med.*, 2004, **70**, 1103.
46. F. Granado-Lorencio, I. Blanco-Navarro, B. Pérez-Sacristán and E. Hernández-Álvarez, *Food Res. Int.*, 2017, **99**, 902–916.
47. T. Bohn, G. J. McDougall, A. Alegría, M. Alminger, E. Arrigoni and A.-M. Aura, *et al.*, *Mol. Nutr. Food Res.*, 2015, **59**, 1307.
48. J. Serrano, I. Goñi and F. Saura-Calixto, *J. Agric. Food Chem.*, 2005, **53**, 2936.
49. F. Guillon and M. Champ, *Food Res. Int.*, 2000, **33**(3–4), 233.
50. J. Slavin, *Nutrients*, 2013, **5**, 1417.
51. K. D. Croft, *Arch. Biochem. Biophys.*, 2015, **595**, 120.
52. A. R. Rechner, M. A. Smith, G. Kuhnle, G. R. Gibson, E. S. Debnam and S. K. S. Srai, *et al.*, *Free Radical Biol. Med.*, 2004, **36**(2), 212.
53. J. Pérez-Jiménez, J. Serrano, M. Tabernero, S. Arranz, M. E. Díaz-Rubio and L. García-Diz, *et al.*, *Plant Foods Hum. Nutr.*, 2009, **64**, 102.
54. L. Rondini, M. N. Peyrat-Maillard, A. Marsset-Baglier, G. Fromentin, P. Durand and D. Tomé, *et al.*, *J. Agric. Food Chem.*, 2004, **52**, 4338.
55. I. Goñi and J. Serrano, *J. Sci. Food Agric.*, 2005, **85**(11), 1877.
56. D. Hervert-Hernández, C. Pintado, R. Rotger and I. Goñi, *Int. J. Food Microbiol.*, 2009, **136**(1), 119.
57. C. Cueva, M. V. Moreno-Arribas, P. J. Martín-Alvarez, G. Bills, M. F. Vicente and A. Basilio, *et al.*, *Res. Microbiol.*, 2010, **161**(5), 372.
58. M. Dueñas, I. Muñoz-González, C. Cueva, A. Jiménez-Girón, F. Sánchez-Patán and C. Santos-Buelga, *et al.*, *BioMed Res. Int.*, 2015, 850902.
59. X. Tzounis, A. Rodriguez-Mateos, J. Vulevic, G. R. Gibson, C. Kwik-Uribe and J. P. E. Spencer, *Am. J. Clin. Nutr.*, 2011, **93**, 62.
60. M. J. Pozuelo, A. Agis-Torres, D. Hervert-Hernández, M. E. López-Oliva, E. Muñoz-Martínez and R. Rotger, *et al.*, *J. Food Sci.*, 2012, **77**(2), 59.
61. S. Sánchez-Tena, D. Lizárraga, A. Miranda, M. P. Vinardell, F. García-García and J. Dopazo, *et al.*, *Carcinogenesis*, 2013, **34**(8), 1881.

62. D. Lizarraga, M. P. Vinardell, V. Noé, J. H. Van Delft, G. Alcarraz-Vizán and S. G. Van Breda, *et al.*, *J. Nutr.*, 2011, 1597.
63. A. Fernández-Ochoa, I. Borrás-Linares, A. Pérez-Sánchez, E. Barrajón-Catalán, I. González-Álvarez and D. Arráez-Román, *et al.*, *J. Funct. Foods*, 2017, **33**, 202.
64. D. Sun-Waterhouse, K. Bekkour, S. S. Wadhwa and G. I. N. Waterhouse, *Food Bioprocess Technol.*, 2014, **7**, 409.
65. L. Jakobek, *Food Chem.*, 2015, **175**, 556.
66. J. Pérez-Jiménez, J. Serrano, M. Tabernero, S. Arranz, M. E. Díaz-Rubio and L. García-Diz, *et al.*, *Nutrition*, 2008, **24**, 646.
67. O. Aprikian, V. Duclos, S. Guyot, C. Besson, C. Manach and A. Bernalier, *et al.*, *J. Nutr.*, 2003, **133**, 1860.
68. A. J. Wallace, S. L. Eady, D. C. Hunter, M. A. Skinner, L. Huffman and J. Ansell, *et al.*, *Nutr. Res.*, 2015, **35**(1), 23.
69. D. Hervert-Hernández and I. Goñi, *Food Rev. Int.*, 2011, **27**(2), 154.

CHAPTER 3

Extractable and Non-extractable Polyphenols: an Overview

A. DURAZZO

Consiglio per la ricerca in agricoltura e l'analisi dell'economia agraria – Centro di ricerca CREA – Alimenti e Nutrizione, Via Ardeatina 546, 00178, Rome, Italy
*E-mail: alessandra.durazzo@crea.gov.it

3.1 Definition and Types of Extractable and Non-extractable Polyphenols

Extractable polyphenols (EPP) are those that are readily solubilised by aqueous–organic solvents. They are mostly low-molecular-weight compounds grouped into several classes and subclasses, including proanthocyanidins (usually up to decamers) and hydrolysable tannins of low molecular weight. In contrast, non-EPP (NEPP) remain in the residues of the previous extractions and can only be released by hydrolysis treatments, either chemical or enzymatic. NEPP consist of: (a) polymeric polyphenols (non-extractable proanthocyanidins [NEPA])—these are high-molecular-weight proanthocyanidins that are free in the food matrix and proanthocyanidins that are complexed with protein or cell wall polysaccharides;[1,2] and (b) small phenolic compounds linked to carbohydrates, mainly polysaccharide constituents of dietary fibre, and to proteins (hydrolysable polyphenols [HPP])—these include several classes of bioactive components (*e.g.*, hydrolysable tannins

Food Chemistry, Function and Analysis No. 5
Non-extractable Polyphenols and Carotenoids: Importance in Human Nutrition and Health
Edited by Fulgencio Saura-Calixto and Jara Pérez-Jiménez

Published by the Royal Society of Chemistry, www.rsc.org

or hydroxycinnamic acids) linked to carbohydrates and proteins *via* covalent bonds, hydrogen bonds and/or hydrophobic interactions.[3,4] Examples of the most common constituents of HPP are phenolic acids[5] such as ferulic acid, caffeic acid and sinapic acid, ellagic acids and other derivatives such as hydrolysable tannins.[6]

3.2 Chemical Analysis of Extractable and Non-extractable Polyphenols

It is known that EPP extraction is influenced by different factors such as the type of solvent used, extraction time and temperature, as well as by the chemical composition and physical characteristics of the sample analysed.[7] At the same time, to release NEPP, different hydrolytic treatments must be carried out. These can be adjusted depending on the specific fractions of NEPP (HPP and NEPA), the food source, the type of bonds with the food matrix or the exact structure of the target compounds.[4,8–10] For instance, Cheng *et al.*,[11] by studying acid and alkaline hydrolysis extraction of NEPP from blueberries, optimised the extraction conditions including the solvent type, liquid/solid ratio, extraction time and temperature, by applying the response surface methodology and the Box–Behnken response surface design.

Overall, the main steps for isolating EPP and NEPP from plant foods are as described in what follows.[8,9]

3.2.1 Extractable Polyphenols

An amount of each sample is placed in a test tube and 20 mL acid methanol/water (50 : 50 v/v, pH 2) is added. The tubes are vortexed at room temperature for 3 minutes then mildly shaken for 1 hour in a water bath at room temperature. The tubes are then centrifuged at 2500 *g* for 10 minutes and the supernatants are recovered. A total of 20 mL of acetone/water (70 : 30 v/v) is added to the residues, then all the operations (vortexing, shaking, centrifugation) are repeated. Methanolic and acetonic extracts are combined and centrifuged at 2800 *g* for 15 minutes.

The residue left after this extraction is then moved into a ventilated and heated apparatus (maximum temperature 25 °C) or under nitrogen until dry. The residue is then subjected to two different acidic treatments in order to isolate the HPP and NEPA fractions.

3.2.2 Hydrolysable Polyphenols

The residue is mixed with 20 mL methanol and 2 mL concentrated sulphuric acid (18 M). The samples are gently stirred for 1 minute and shaken in a water bath at 85 °C for 20 hours; the samples are then centrifuged (2500 *g* for

10 minutes) and the supernatant is recovered. After two washings with minimum volumes of distilled water and re-centrifuging as required, the final volume is taken up to 50 mL.[12]

3.2.3 Non-extractable Proanthocyanidins

This process is based on depolymerisation by butanolysis, which leads to anthocyanidin monomers and xanthylium compounds. The samples are treated with butanol/HCl containing $FeCl_3$ (100 °C, 60 minutes), and absorbances at 555 nm (anthocyanidins) and 450 nm (xanthylium compounds) are measured in the hydrolysates.[13]

Currently, alternative extraction methods (ultrasound, microwave, ohmic heating)[14–17] are emerging that could reduce both extraction times and the use of solvents and avoid oxidation.[18] Although these techniques may be capable of releasing a fraction of NEPP into the extraction supernatant, it should be noted that relevant proportions of NEPP remain in the corresponding residues.[19] Finally, a recent review[20] summarises the conventional and advanced extraction methods, as well as analytical techniques for qualitative and quantitative analysis for EPP and NEPP.

3.3 Dietary Sources of Extractable and Non-extractable Polyphenols

As has been pointed out,[21] due to all the factors affecting the polyphenol profile and concentration, it is difficult to categorise the main trends of the distribution of EPP and NEPP in the major food groups. Notwithstanding this, several studies have shown that NEPP represent a significant portion in some groups of foods (*i.e.*, cereals, fruit, vegetables, nuts, and legumes).[8,22–24]

A significant number of studies have been performed on different grains[25–27] and derivatives such as bread or pasta,[28–31] as well as pulses.[8] For instance, when studying the effect of cooking experimental pastas made with different wholegrain cereals, it was found that in cooked pasta, the total polyphenol content varied from 128.84 to 226.38 mg/100 g dry weight in aqueous–organic extracts (corresponding to EPP) and from 701.36 to 1041.10 mg/100 g dry weight in the residues (corresponding to NEPP).[28] Also, the effect of cooking on the distribution of EPP and NEPP depends on the nature of the food matrix.[28] For maize products, it was found that the NEPP content was higher than the EPP content,[32] and it has been reported that NEPA were more abundant in a tortilla made with a blend of maize protein and black beans than in the control.[33]

Regarding other food groups, several studies have found that in fruit, the NEPP content is higher than that of EPP.[24,34] Similar results were recently reported for several chicory varieties.[35] Furthermore, the presence of NEPP has been reported in several nut crops.[36]

Particular attention should be paid to the study of polyphenols in fatty food matrices, where previous defatting is required before evaluating NEPP content.[37,38]

Regarding less common foods, it has been reported that in soy-based beverages, the contribution of NEPP to total polyphenols was over 80%.[39] A recent study also investigated macromolecular antioxidants or NEPP in several edible Chilean seaweed species (*Gracilaria chilensis*, *Callophyllis concepcionensis*, *Macrocystis pyrifera*, *Scytosyphon lomentaria*, *Ulva* sp. and *Enteromorpha compressa*), finding that NEPP constituted a major polyphenol fraction (averaging 42% of the total polyphenol content), with hydroxycinnamic acids, hydroxybenzoic acids and flavonols being the main constituents. This fraction also showed a remarkable antioxidant capacity.[19]

All the preceding data refer to individual foods. However, since many foods are consumed after being cooked or processed and as part of dishes containing many ingredients, it is important to study the effects of processing and interactions between ingredients on EPP and NEPP content.[40–42] Thus, EPP in some traditional Italian dishes ranged between 28.08 mg/100 g d.m. (dry matter) and 2841.44 mg/100 g d.m., whereas NEPP ranged between 946.86 mg/100 g d.m. (*spaghetti alle vongole*) and 3801.95 mg/100 g d.m. (*carciofi alla romana*).[42] Therefore, NEPP (in the form of HPP) represented a significant fraction of total polyphenols, accounting for between 57% and 97%.[42]

3.4 Food By-products as Sources of Extractable and Non-extractable Polyphenols

During the last few years, there has been increasing interest in the valorisation of food by-products as sources of bioactive compounds in the area of the so-called circular economy, and in this context our attention is given to antioxidants. Socaci *et al.*[43] provide an update based on the scientific evidence regarding the possibilities of extraction and purification, the health benefits and the envisaged applications of antioxidants recovered from these wastes. Thus, several studies have explored these by-products as sources of EPP and particularly of NEPP. For instance, mandarin waste dried at different temperatures[44] was shown to be rich in both extractable and non-extractable hesperidin and eriocitrin, as well as non-extractable gallic acid. Other types of fruit and vegetable wastes were also shown to have important contents of NEPP.[45–48] In particular, by-products from tropical fruits, such as flour prepared from pequi fruit (*Caryocar brasiliense* Camb.) have high NEPP contents.[49] Also, different discarded fractions of peanuts were found to have high NEPP contents, and these were always higher than those of EPP.[50]

However, it should be kept in mind that, depending on the treatment used on these residues, most of the EPP and/or NEPP may be removed or degraded, as observed in spent coffee grounds after alkaline hydrogen peroxide treatment.[51]

New procedures are also being applied to isolate NEPP from food by-products. Thus, a recent study[52] applied an enzyme-assisted eco-friendly process for the extraction of NEPP from black tea leftovers—a little-used tea waste. The pre-treatment effectively released NEPP by hydrolysing the cellulosic and hemicellulosic framework of black tea leftovers.

3.5 Physiological Concept of Extractable and Non-extractable Polyphenols

The classification into EPP and NEPP is relevant not only from an analytical point of view (when aiming to completely characterise the polyphenol profile of a sample), but also from a physiological point of view. Thus, there are differences in the metabolic fates of EPP and NEPP, which may give rise to different physiological effects. Although many studies have examined the bioaccessibility and bioavailability of EPP,[53] only a few recent studies have focused on NEPP. These latter studies have shown that dietary NEPP are not significantly released from the food matrix by mastication, the acid pH in the stomach or the action of digestive enzymes, so they reach the colon nearly intact, where they are subjected to extensive transformation by the colonic microflora.[4,54]

Indeed, a general scheme of the metabolic fate of NEPP has been proposed:[4] most NEPP pass through the small intestine without being subjected to transformations. After reaching the colon, they are subjected to the action of the microbiota, with new metabolites being produced and released. Some of these metabolites can be absorbed throughout the portal vein and reach the liver, where they are subjected to several processes that lead to the production of phase II metabolites. Then, they may return to the digestive tube *via* bile, pass into the bloodstream and finally reach target tissues or be excreted in urine.[4]

It is worth mentioning here the study by Saura-Calixto *et al.*[54] that evaluated the bioaccessibility (considered to be the amount of a food constituent that is solubilised in intestinal fluids as a consequence of the physicochemical conditions and the action of either digestive enzymes or bacterial microbiota) of total polyphenols in a complete diet using an *in vitro* gastrointestinal model followed by *in vitro* colonic fermentation. They highlighted the difference in the behaviour of EPP, HPP and NEPA: most EPP were bioaccessible in the small intestine, 50% of HPP became bioaccessible in the small intestine and most NEPA arrived nearly intact in the colon, as reported in other studies.[55,56] Once in the colon, these NEPP may become accessible through fermentation by colonic microbiota of the carbohydrates or proteins, the molecules to which they are linked, as well as through the action of some intestinal enzymes such as esterases, which break covalent bonds.[57] A previous study[58] estimated that 95% of NEPA arriving in the colon are released from the food matrix by these two mechanisms.

3.6 Evaluation of the Intake and Health Effects of Non-extractable Polyphenols

Besides an increasing number of studies aimed at determining the NEPP content in food and their antioxidant properties over the last decade, most *in vitro* and *in vivo* studies of the physiological effects of polyphenols still focus on EPP.

However, preliminary studies have produced promising results for this class of dietary polyphenols. Thus, recent work[11] evaluated the effects of EPP and NEPP from blueberries on an *in vitro* model of inflammation and concluded that both polyphenol fractions exerted anti-inflammatory actions *via* the inhibition of iNOS and COX-2 mRNA expression through the suppression of NF-κB. Concerning the interaction between gut microbiota and polyphenols,[59,60] it is worth mentioning the recent work by González-Sarrías *et al.*[61] that describes how NEPP produce gut microbiota metabolites that persist in circulation and show anti-inflammatory effects and free radical scavenging.

Also, a proper evaluation of NEPP intake is needed in order to consider these compounds in clinical and epidemiological studies. However, as mentioned, most studies of the topic still focus only on EPP, with very few studies dealing with NEPP intake, despite their significant contribution to total polyphenol intake.[62] Among these few studies, NEPP contents and profiles were evaluated for the 24 most consumed fruit and vegetables in four European countries (France, Germany, The Netherlands and Spain) in order to determine NEPP intake. This intake was estimated at about 200 mg; Spain had the highest daily per capita NEPP intake from fruit, while The Netherlands had the highest derived from vegetables.[63] A recent review[62] offers an overview of polyphenol intake in the world by gathering information on polyphenol intake in different countries around the whole world in order to compare methodologies and the value of polyphenol intake. The authors underline the importance of taking into account both EPP and NEPP for the correct estimation of polyphenol intake.

One reason that explains this gap in studies evaluating polyphenol intake is related to the databases currently available for this determination. There are three major core public databases that gather extensive data on the polyphenol content of foods and beverages: the United States Department of Agriculture (USDA) databases (https://ndb.nal.usda.gov/ndb/) for flavonoids, proanthocyanidins and isoflavones, the Phenol-Explorer database (http://phenol-explorer.eu/) for polyphenols[64] and, in recent times, eBASIS (Bioactive Substances in Food Information Systems; http://ebasis.eurofir.org/Default.asp) for generally bioactive compounds, including polyphenols, published through the EuroFIR project.[65,66] These databases mostly focus on EPP, with only scarce data on NEPP. In this context, the need to include NEPP in bioactive compound databases is emerging in order to allow for better dietary intake assessments.

References

1. W. Huemmer and P. Schereier, *Mol. Nutr. Food Res.*, 2008, **52**, 1381.
2. J. Pérez-Jiménez, S. Arranz and F. Saura-Calixto, *Food Res. Int.*, 2009, **42**, 1381.
3. J. Pérez-Jiménez and J. L. Torres, *J. Agric. Food Chem.*, 2011, **59**, 12713.
4. J. Pérez-Jiménez, M. E. Diaz-Rubio and F. Saura-Calixto, *Nutr. Res. Rev.*, 2013, **26**, 118.
5. K. K. Adom and R. H. Liu, *J. Agric. Food Chem.*, 2002, **50**, 6182.
6. J. A. John and F. Shahidi, *J. Funct. Foods*, 2010, **2**, 196.
7. D. L. Luthria, *J. Sci. Food Agric.*, 2006, **86**, 2266.
8. A. Durazzo, V. Turfani, E. Azzini, G. Maiani and M. Carcea, *Food Chem.*, 2013, **140**, 666.
9. J. Perez-Jimenez, S. Arranz, M. Tabernero, M. E. Diaz-Rubio, J. Serrano and I. Goni, *et al.*, *Food Res. Int.*, 2008, **41**, 274.
10. J. Pérez-Jiménez, M. E. Díaz-Rubio and F. Saura-Calixto, Non-extractable Polyphenols in Plant Foods: Nature, Isolation, and Analysis, in *Polyphenols in Plants Isolation, Purification and Extract Preparation*, ed. R. R. Watson, 2014, ch. 10, pp. 203–218, ISBN: 978-0-12-397934-6.
11. A. Cheng, C. Han, X. Fang, J. Sun, X. Chen and F. Wan, *J Food Sci. Agric.*, 2016, **96**, 3393.
12. P. W. Hartzfeld, R. Forkner, D. M. Hunter and A. E. Hagerman, *J. Agric. Food Chem.*, 2002, **50**, 1785.
13. L. Porter, L. Hrstich and B. Chan, *Phytochemistry*, 1985, **25**, 223.
14. A. Carrera, A. Ruiz-Rodriguez, M. Palma and C. G. Barroso, *Anal. Chim. Acta*, 2012, **732**, 100.
15. M. L. Soto, A. Moure, H. Domínguez and J. C. Parajó, *J. Food Eng.*, 2011, **105**, 1.
16. S. Martins, S. I. Mussatto, G. Martinez-Avila, J. Montanez-Saenz, C. N. Aguilar and J. A. Teixeira, *Rev. Biotechnol. Adv.*, 2011, **29**, 365.
17. B. A. Acosta-Estrada, J. A. Gutiérrez-Uribe and S. O. Serna-Saldívar, *Food Chem.*, 2014, **152**, 46.
18. A. E. L. Fernández, E. N. Obledo-Vázquez, M. A. Vivar-Vera, S. G. Sáyago Ayerdi and E. Montalvo-González, *Rev. Bras. Frutic.*, 2017, **39**, e-358.
19. N. Sanz-Pintos, J. Pérez-Jiménez, A. H. Buschmann, J. R. Vergara-Salinas, J. R. Pérez-Correa and F. Saura-Calixto, *J. Food Sci.*, 2017, **82**, 289–295.
20. G. Domínguez-Rodríguez, M. L. Marina and M. Plaza, *J. Chromatogr. A*, 2017, **1514**, 1.
21. A. Durazzo, *Foods*, 2017, **6**, 17.
22. S. Arranz, F. Saura-Calixto, S. Shaha and P. A. Kroon, *J. Agric. Food Chem.*, 2009, **57**, 7298.
23. S. Arranz, J. M. Silván and F. Saura-Calixto, *Mol. Nutr. Food Res.*, 2010, **54**, 1646.
24. I. Tarascou, J. M. Souquet, J. P. Mazauric, S. Carrillo, S. Coq, F. Canon, H. Fulcrand and V. Cheynier, *Archi. Biochem. Biophys.*, 2010, **501**, 16.

25. J. Perez-Jimenez and F. Saura-Calixto, *J. Agric. Food Chem.*, 2005, **53**, 5036.
26. A. Chandrasekara and F. Shahidi, *J. Agric. Food Chem.*, 2010, **58**, 6706.
27. G. A. Camelo-Méndez and L. A. Bello, *J. Chem., Biol. Phys. Sci.*, 2014, **4**, 6.
28. A. Durazzo, V. Turfani, E. Azzini, G. Maiani and M. Carcea, *J. Food Res.*, 2014, **3**, 4.
29. A. Durazzo, G. Casale, V. Melini, G. Maiani and R. Acquistucci, *Foods*, 2015, **4**, 391.
30. A. Durazzo, G. Casale, V. Melini, G. Maiani and R. Acquistucci, *J. Food Sci.*, 2016, **28**, 221.
31. V. Turfani, V. Narducci, A. Durazzo, V. Galli and M. Carcea, *LWT–Food Sci. Technol.*, 2017, **78**, 361.
32. E. M. Grajales-García, P. Osorio-Díaz, I. Goñi, D. Hervert-Hernández, S. H. Guzmán-Maldonado and L. A. Bello-Pérez, *Int. J. Mol. Sci.*, 2012, **13**, 286.
33. E. M. Grajales-García, P. Osorio-Díaz, I. Goñi, D. Hervert-Hernández, S. H. Guzmán-Maldonado and L. A. Bello-Pérez, *Int. J. Mol. Sci.*, 2012, **13**, 286.
34. J. K. Hellstrom and P. H. Mattila, *J. Agric. Food Chem.*, 2008, **56**, 7617.
35. Y. Sahan, O. Gurbuz, M. Guldas, N. Degirmencioglu and A. Begenirbas, *Food Chem.*, 2017, **217**, 483.
36. O. Ogah, C. S. Watkins, B. E. Ubi and N. C. Oraguzie, *J. Agric. Food Chem.*, 2014, **62**, 9369.
37. S. Arranz, J. Perez-Jimenez and F. Saura-Calixto, *Eur. Food Res. Technol.*, 2008, **227**, 425.
38. M. d. S. M. Rufino, J. Pérez-Jiménez, S. Arranz, R. Elesbão Alves, E. S. de Brito, M. S. P. Oliveira and F. Saura-Calixto, *Food Res. Int.*, 2011, **44**, 2100.
39. A. Durazzo, P. Gabrielli and P. Manzi, *Antioxidants*, 2015, **4**, 523.
40. L. Greco, R. Riccio, S. Bergero, A. A. M. Del Re and M. Trevisan, *Food Chem.*, 2007, **103**, 1127.
41. I. Ioannou, *J. Food Res.*, 2012, **1**, 143.
42. A. Durazzo, S. Lisciani, E. Camilli, P. Gabrielli, S. Marconi, L. Gambelli, A. Aguzzi, M. Lucarini, G. Maiani and G. Casale, *et al.*, *Food Chem.*, 2017, **218**, 70.
43. S. A. Socaci, D. O. Rugină, Z. M. Diaconeasa, O. L. Pop, A. C. Fărcaş, A. Păucean, M. Tofană and A. Pintea, Antioxidant Compounds Recovered from Food Wastes, in *Functional Food – Improve Health through Adequate Food*, ed. M. Chavarri Hueda, InTech, ISBN 978-953-51-3440-4, Print ISBN 978-953-51-3439-8, 316 Pages, Chapters published August 02, 2017 under CC BY 3.0 license, DOI: 10.5772/66263.
44. F. J. Esparza-Martínez, R. Miranda-López, S. M. Mata-Sánchez and S. H. Guzmán-Maldonado, *Plant Foods Hum. Nutr.*, 2016, **71**, 294–300.
45. F. J. Esparza-Martínez, R. Miranda-López and S. H. Guzman-Maldonado, *Ind. Crops Prod.*, 2016a, **84**, 1.
46. W. W. Tow, R. Premier, H. Jing and S. Ajlouni, *J. Food Sci.*, 2011, **76**, T163.
47. P. Y. Toh, F. S. Leong, S. K. Chang, H. E. Khoo and H. S. Yim, *Acta Sci. Pol., Technol. Aliment.*, 2016, **15**, 65.

48. G. B. Gonzales, K. Raes, H. Vanhoutte, S. Coelus, G. Smagghe and J. Van Camp, *J. Chromatogr. A*, 2015, **1402**, 60.
49. D. P. Leão, A. S. Franca, L. S. Oliveira, R. astos and M. A. Coimbra, *Food Chem.*, 2017, **225**, 146.
50. F. S. Taha, S. M. Wagdy and F. A. Singer, *Life Sci. J.*, 2012, **9**, 207.
51. W. F. Vilela, D. P. Leão, A. S. Franca and L. S. Oliveira, *Int. J. Food Eng.*, 2016, **2**, 42.
52. M. Mushtaq, B. Sultana, S. Akram, F. Anwar, A. Adnan and S. S. H. Rizvi, *Anal. Bioanal. Chem.*, 2017, **409**, 3645.
53. A. Manach, G. Williamson, C. Morand, A. Scalbert and C. Rémésy, *Am. J. Clin. Nutr.*, 2005, **81**, 230S.
54. F. Saura-Calixto, J. Pérez-Jiménez, S. Touriño, J. Serrano, E. Fuguet, J. L. Torres and I. Goñi, *Mol. Nutr. Food Res.*, 2010, **54**, 939.
55. S. Touriño, J. Pérez-Jiménez, M. L. Mateos-Martín, E. Fuguet, M. P. Vinardell, M. Cascante and J. L. Torres, *J. Agric. Food Chem.*, 2011, **59**, 5955.
56. M. L. Mateos-Martín, J. Pérez-Jiménez, E. Fuguet and J. L. Torres, *Br. J. Nutr.*, 2012, **108**, 290.
57. P. A. Kroon, C. B. Faulds and P. Ryden, *et al.*, *J. Agric. Food Chem.*, 1997, **45**, 661.
58. F. Saura-Calixto, J. Serrano and I. Goni, *Food Chem.*, 2007, **101**, 492.
59. F. A. Tomás-Barberán, M. V. Selma and J. C. Espín, *Curr. Opin. Clin. Nutr. Metab. Care*, 2016, **69**, 281.
60. J. C. Espín, A. González-Sarrías and F. A. Tomás-Barberán, *Biochem Pharmacol.*, 2017, **139**, 82.
61. A. González-Sarrías, J. C. Espín and F. A. Tomás-Barberán, *Trends Food Sci. Technol.*, 2017, **69**, 281.
62. P. Pinto and C. N. Santo, *Eur. J. Nutr.*, 2017, **56**, 1393.
63. J. Pérez-Jiménez and F. Saura-Calixto, *Food Res. Int.*, 2015, **74**, 315.
64. V. Neveu, J. Perez-Jiménez, F. Vos, V. Crespy, L. du Chaffaut, L. Mennen, C. Knox, R. Eisner, J. Cruz, D. Wishart and A. Scalbert, Phenol-Explorer: an online comprehensive database on polyphenol contents in foods, *Database (Oxford)*, 2010, **2010**, bap.024.
65. M. Kiely, L. J. Black, J. Plumb, P. A. Kroon, P. C. Hollman, J. C. Larsen, G. J. Speijers, M. Kapsokefalou, D. Sheehan, J. Gry and P. Finglas, *Eur. J. Clin. Nutr.*, 2010, **3**, S101.
66. J. Plumb, S. Pigat, F. Bompola, M. Cushen, H. Pinchen, E. Nørby, S. Astley, J. Lyons, M. Kiely and P. Finglas, *Nutrients*, 2017, **9**, 320.

CHAPTER 4

Analysis of Non-extractable Polyphenols (NEPP)

T. WU[†a,b,c], N. N. M. PHUONG[†a,b,d,e], J. VAN CAMP[b], G. SMAGGHE[c], T. T. LE[d] AND K. RAES*[a]

[a]Ghent University, Department of Industrial Biological Sciences, Graaf Karel de Goedelaan 5, 8500 Kortrijk, Belgium; [b]Ghent University, Department of Food Technology, Safety and Health, Coupure Links 653, 9000 Ghent, Belgium; [c]Ghent University, Department of Plants and Crops, Coupure Links 653, 9000 Ghent, Belgium; [d]Nong Lam University, Department of Food Engineering, Faculty of Food Science and Technology, Ho Chi Minh City, Vietnam; [e]Can Tho University, Department of Food Technology, College of Agriculture and Applied Biology, Can Tho City, Vietnam
*E-mail: Katleen.Raes@ugent.be

4.1 Introduction

Polyphenols are normally divided into two groups based on their solubility properties: extractable polyphenols (EPP) and non-extractable polyphenols (NEPP). EPP are compounds that can be extracted from foods with aqueous–organic solvents, while compounds remaining in the residues after extraction are NEPP,[1] including: (i) macromolecular polyphenols such as high-molecular-weight tannins and condensed tannins or high-molecular-weight

[†]Shared first authors.

Food Chemistry, Function and Analysis No. 5
Non-extractable Polyphenols and Carotenoids: Importance in Human Nutrition and Health
Edited by Fulgencio Saura-Calixto and Jara Pérez-Jiménez

Published by the Royal Society of Chemistry, www.rsc.org

proanthocyanidins (PA); and (ii) low-molecular-weight polyphenols such as phenolic acids (PhA) and some flavonoids linked to cell wall constituents (*e.g.* polysaccharides and proteins) or trapped within the food matrix.[1,2]

Saura-Calixto noticed that most studies of dietary polyphenols have focused on EPP, which may be only the tip of the iceberg.[3] Despite the limitation of references of NEPP contributions in dietary polyphenols, the available information indicates that they may be regarded as the primary fraction of plant food polyphenols. The classes of NEPP have been found in different families of plant foods such as cereals, fruits, vegetables, nuts and legumes.[4] For instance, the contribution of PA in apple pomace was 48%.[2,5] Similarly, non-extractable PhA contributions in millet and wheat flour were 91% and 70%, respectively,[6,7] but in great contrast, macromolecular antioxidants and non-extractable PA were not detected in some fruits and vegetables like (water)melon, broccoli and tomato.[8] According to Pérez-Jiménez and co-authors, NEPP as high-molecular-weight molecules or compounds that are weakly linked with the food matrix are called class I NEPP, while those that are strongly linked to the different constituents of the cell wall such as neutral polysaccharides, acidic pectic substances and structural proteins are called class II NEPP.[2,9] Due to the complex nature of NEPP, the isolation of NEPP requires different steps, extracting EPP or soluble compounds and disrupting macromolecules by enzymatic treatment and chemical hydrolysis.[2] As regards the existing analysis of NEPP, spectrophotometry and chromatography with different detection methods can be proposed in procedures. To date, application of polyphenols mainly focuses on EPP; however, increasing evidence supports the notion that NEPP can be applied in several fields such as the food industry (natural antioxidants), animal nutrition (supplementation to improve antioxidant status) and human health (gastrointestinal health).[2] Therefore, to better understand their potential health effects, procedures for extraction and isolation as well as methods for analysis of NEPP are reviewed in this chapter.

4.2 Current Extraction Methods of NEPP

4.2.1 Alkaline and Acid Hydrolysis

NEPP can be released by different methods, of which alkaline (saponification) and acid hydrolysis are frequently used for extracting NEPP at room or accelerated temperature (Table 4.1). Concentration of base or acid used, reaction temperature and hydrolysis time are the main variables.[10] Normally, after an alkaline or acid pre-treatment, phenolic compounds (PC) are extracted by conventional solvents such as methanol, ethanol, acetone, ethyl acetate or diethyl ether.

In the acid hydrolysis method, inorganic acids are used such as HCl and H_2SO_4.[7,16] Acid treatment can hydrolyse glycosidic bonds of (hemi)cellulose and solubilise sugars.[17] However, some PC can be lost due to acid hydrolysis (*e.g.* 92% losses for *trans*-sinapic acid, 87% for caffeic acid and 15% for

Table 4.1 Extraction of bound phenolic compounds using acidic, alkaline and enzymatic hydrolysis.[a,b]

Substrate	Acid/alkaline/enzymatic hydrolysis: Conditions	Acid/alkaline/enzymatic hydrolysis: Remarks	Solvent extraction: Conditions	Solvent extraction: Remarks	Reference
Acid hydrolysis					
Wheat flour, wheat bran	$MeOH/H_2SO_4$ 90 : 10 (v/v) at 85 °C for 20 h	Polyphenol content in: —Wheat flour: 262 ± 13.2 mg (100 g)$^{-1}$ (f.w.) —Wheat bran: 1588.8 ± 30.7 mg (100 g)$^{-1}$ (f.w.)	Acidic MeOH (HCl)/water (50 : 50, v/v; pH 2) for 1 h + acetone/water (70 : 30, v/v) at room temperature for 1 h	Polyphenol content in: —Wheat flour: 112.2 ± 7 mg (100 g)$^{-1}$ (f.w.) —Wheat bran: 161.7 ± 5.3 mg (100 g)$^{-1}$ (f.w.)	Arranz and Saura (2010)[7]
Cactus pear peels (*Opuntia ficus-indica*)	1 M HCl in MeOH:water (50 : 50) at 90 °C for 1 h	—Only isorhamnetin found with 90 ± 4 µg (100 mg)$^{-1}$ sample (d.m.)	MeOH:water (80 : 20) + sonication for 5 min	—Mainly isorhamnetin glycosides (data not shown)	Moussa-Ayoub *et al.* (2011)[11]
Alkaline hydrolysis					
Cranberry pomace	2 N NaOH at 60 °C for 15 min	Concentration of dimeric procyanidin: ≈700 mg/100 g d.m. (whole pomace treated with NaOH) ≈450 mg/100 g d.m. (residue treated with NaOH)	Homogenisation with acetone/water/acetic acid (70 : 29.5 : 0.5 v/v/v) for 1 min	Concentration of dimeric procyanidin: ≈100 mg/100 g d.m.	White *et al.* (2010)[12]
Cranberry beans (*Phaseolus vulgaris* L.) Cultivar: Red Rider	2 N NaOH + sonication for 15 min, rotational mixing at room temperature for 16 h	Ferulic acid: 29.20 ± 0.86 µg g^{-1} whole raw bean	70% MeOH + 1% HCl (v/v) + sonication for 15 min, rotational mixing at room temperature for 2 h	Ferulic acid: 3.57 ± 0.46 µg g^{-1} whole raw bean	Chen *et al.* (2015)[13]

Enzymatic hydrolysis					
Winemaking by-products	Viscozyme: 2% in 0.1 M phosphate buffer (pH 4) at 37 °C for 12 h	TPC of insoluble-bound phenolics: 184.6 ± 7.88 mg GAE g^{-1} d.m.	70% (v/v) acetone at 30 °C for 20 min	TPC of insoluble-bound phenolics: 256 ± 5.69 mg GAE g^{-1} d.m.	de Camargo *et al.* (2016)[14]
	Extraction solvent: diethyl ether and ethyl acetate (1:1, v/v)	Reducing power: 471.5 ± 4.61 μmol TE g^{-1} d.m.		Reducing power: 292.2 ± 35.4 μmol TE g^{-1} d.m.	
Grape wastes	Novoferm® at 40 °C for 12 h	—TPC: ≈40 mg/100 g dried sample —Scavenging capacity (DPPH): ≈90% —Release of *o*-coumaric acid from the cell wall fibre (compared to the control)	Control (not clearly shown)	—TPC: ≈30 mg/100 g dried sample —Scavenging capacity (DPPH): ≈65%	Gómez-García *et al.* (2012)[15]

[a]Within the last decade for agricultural by-products.
[b]f.w., fresh weight; d.m., dry matter; MeOH, methanol; TPC, total phenolic content; GAE, gallic acid equivalent; TE, Trolox equivalent; DPPH, 2,2-diphenyl-1-picrylhydrazyl.

o-coumaric acid were observed in oilseeds).[18] Moussa-Ayoub and co-authors reported that only isorhamnetin and no flavonols were found after acidic hydrolysis at 90 °C for 1 hour with 1 M HCl in cactus pear peel.[11] This may indicate that factors such as type of substrate and flavonol structure, concentration of acid, treatment temperature and time may cause degradation of flavonols.

For alkali extraction, the ester bond linkage of PhA to the cell wall can be broken and thus this is an effective way to release PhA from the polysaccharide backbone.[19] NaOH, NH_4OH, $Ca(OH)_2$ and H_2O_2 are mostly used as catalysts for the hydrolysis. Alkaline treatment is a common technique to extract PhA (*e.g.* from corn cobs or wheat bran).[16,19] Such a treatment also results in some degradation of certain PhA. For example, caffeic acid, which was in the extract after acid hydrolysis of wheat bran, was not detected in the alkali fraction.[16] This was in agreement with previous results of Krygier and co-authors, who reported a 66.7% loss of caffeic acid of oilseeds (*Brassica napus* and *Brassica campestris*) when hydrolysed with 2 N NaOH at room temperature for 4 hours.[18] There are several studies using an antioxidant (*e.g.* ascorbic acid [AA]) and a metal chelator (*e.g.* EDTA) to prevent the degradation of PhA during alkaline hydrolysis. Ross and co-authors used these protectors during aggressive base hydrolysis of dry beans with 10 N NaOH for 16 hours.[20] They also implied that the protection of PhA by AA and EDTA under aggressive base hydrolysis condition was more effective than under mild conditions. In another study, AA and EDTA were used in the prevention of losses of PhA during alkaline treatment (4 N NaOH, room temperature for 4 hours, under nitrogen), showing that the addition of AA and EDTA increased the recovery of NEPP.[21]

4.2.2 Enzymatic Hydrolysis

Some phenolics are retained in the polysaccharide–lignin network by hydrogen or hydrophobic linkage, thus they are not accessible for solvents in normal extraction processes.[22] Besides alkaline and acid treatments, enzymatic treatment is another specific and effective way to release NEPP from solid materials. Cellulases, hemicellulases, pectinases, amylases and glucanases are usually applied to extract NEPP. Enzymes catalyse specific biochemical reactions that are efficient at low concentrations, active under mild pH and temperature conditions, have high substrate specificity and low toxicity and it is easily to stop their reaction.[23]

Several studies on the application of enzymes for release of NEPP are available in Table 4.1. It is clear that the extraction efficiency largely differs among studies as the type of PC released depends on the type of enzyme used, treatment time, sample particle size and type of extraction solvent used afterwards.[24] Although enzymatic treatment has been considered as a novel and an effective way to extract bound PC, some drawbacks related to the use of enzymes are, for example, their high cost and slow reaction rates.

4.2.3 Microwave-assisted Hydrolysis

Microwave-assisted extraction (MAE) is also one of the green solvent extraction methods as it utilises water or alcohols at elevated temperatures and controlled pressure conditions.[25] The efficiency of the extraction process by MAE may be the result of a synergistic combination of two transport phenomena: heat and mass gradients working in the same direction.[26] Microwaves operate in the frequency range from 300 MHz to 300 GHz, while 915 MHz is considered most useful for industrial applications due to its great penetration depth.[22,27] In the microwave heating process, energy transfer occurs by two mechanisms: dipole rotation and ionic conduction through reversals of dipoles and displacement of charged ions present in the solute and the solvent.[26] The extraction mechanism of this technique undergoes three steps: (i) the desorption of solutes from active sites in the sample matrix under the pressurised and elevated temperature conditions; (ii) the diffusion of solvents into the matrix; and (iii) the liberation of solutes from the sample matrix to solvents.[25]

Variables such as extraction time, temperature, solvent type, solvent/solid matrix ratio, microwave power and duration are mostly considered as factors to be optimised. Papoutsis and co-authors[28] studied the effect of MAE on the levels of total PC, total flavonoids and PA of dried lemon pomace, showing that MAE in a power from 120 to 480 W for 2–5 minutes increased the extraction yield by 2.5- to 5.5-fold. However, treatment beyond 480 W for 5 minutes resulted in the decrease of these compounds. These results are in agreement with the results obtained by Hayat and co-authors,[29] explaining that the liberation and cleavage of PhA and flavonoids due to electromagnetic fields may result in an increased phenolic content. However, it is important to know that higher power and longer duration caused degradation of some flavonol compounds. In some studies, the effective release of bound PhA may result from a combination between alkali treatment and MAE.[30] It may be concluded that MAE is a faster and easier method in comparison to conventional extraction processes. MAE not only increases yields, but also results in a decreased extraction cost due to the reduction of time and solvent consumption needed compared to traditional methods. For instance, Chen *et al.* compared the extraction yield of flavonoids in *Platycladus orientalis* (L.) Franco using MAE and heat reflux extraction. The results showed that the total flavonoid content obtained in MAE extracts (1.72%) was higher than in heat reflux extracts (1.46%) with shorter extraction times and less solvent consumption (5 minutes, 5 mL of 80% aqueous methanol for MAE and 2 hour, 40 mL of 50% aqueous methanol for heat reflux extraction).[31]

4.2.4 Ultrasound-assisted Hydrolysis

Ultrasound is well known as a green extraction technique that uses high-frequency sound waves (from 20 kHz to 100 MHz). It passes through a medium, such as water, by creating compression and expansion. When the ultrasound waves are strong enough, bubbles are produced, grow and collapse, also known as cavitation.[22] A change in temperature and pressure within the bubbles due to collapse leads to a large amount of energy being produced. According to Wijngaard and co-authors,[32] these bubbles may have an extremely high temperature (about 5000 °C) and pressure (1000 bar). The efficiency of ultrasound-assisted extraction (UAE) results in ultrasound intensification of mass transfer and accelerated access of solvent to plant cell walls. Herein, the mechanism of extraction consists of two main phenomena: (i) diffusion of the medium through the cell wall; and (ii) releasing the contents of cells into the medium after disrupting the walls. Important factors that have an impact on efficiency are the moisture content of the sample, particle size, type of solvents used and the solvent : material ratio, as well as variables related to the ultrasound technique itself such as temperature, pressure, time and frequency of sonication.[22] Normally, UAE operates along with other techniques like solid–liquid extraction, maceration, chemical hydrolysis, pressurized liquid and enzymatic treatment. The advantages of using UAE include its short treatment time, low energy and small amount of solvent.

Various studies used UAE for the extraction of NEPP. For instance, a combination of ultrasound with alkaline treatment was an effective way to release and extract NEPP from cauliflower waste.[33] The results showed that alkaline treatment with 2 M NaOH at 60 °C for 30 minutes under sonication (frequency of 37 kHz, power of 180 W) resulted in a higher amount of NEPP being extracted (7.3 mg gallic acid equivalent (GAE) g^{-1} dry waste) in comparison with conventional extraction (6 mg GAE g^{-1} dry waste). In addition, kaempferol and quercetin glucosides along with several PhA such as caffeic, coumaric, ferulic and sinapic acids were found in this fraction from cauliflower waste. Another recent study compared the extraction efficiency of anthocyanins from blackberry, blueberry and grumixama residues using UAE and combined with pressurised liquid extraction (PLE).[34] The results showed that extraction in combination with UAE (37 kHz, 580 W, 90 minutes, 80 °C, ambient pressure) and PLE (10 MPa, 80 °C, 30 minutes) using hydroethanolic mixtures as the solvent had the highest amount of total PC and antioxidant capacity in comparison with other techniques. The results also demonstrated the inefficiency of ultrasound if it was performed alone.

Other remarkable observations are that UAE may degrade compounds during the extraction process when using high intensities or frequencies, long time periods and high temperatures. Furthermore, depending on the process, the chemical and physical effects of cavitation can be positive or negative. The negative effect is often due to reactions of free radicals formed during sonication with molecules in the medium or the rupture of intermolecular bonds that are essential to the molecule's stability.[34,35]

4.2.5 Other Assisted Hydrolysis Techniques

4.2.5.1 Pulsed Electric Field

Pre-treatment of solid materials using a pulsed electric field (PEF) prior to conventional extraction helps the selective recovery of valuable intracellular compounds without damaging the structure of the matrix itself and without any significant increase of temperature.[36] The principles of PEF is the intermittent application (<300 Hz) of electric fields at moderate to high intensity (0.1–20.0 kV cm^{-1}) and short duration (from a few microseconds to several milliseconds).[37] These electric fields cause a phenomenon called electroporation, meaning the formation of pores located in cell membranes and causing an intense increased permeability. The efficiency of the PEF-assisted extraction depends on the process parameters, extraction variables and physicochemical characteristics of the (cell wall) matrix.[38] There are several studies using PEF-assisted extraction to release bioactive compounds from various materials (*e.g.* anthocyanins from grape by-products,[39] polyphenols from orange peel[40] and nutritionally valuable compounds from mango peels[41]). Recently, mild-intensity PEF (2 kV cm^{-1} intensity, 875 μs treatment time for sorghum flour and 2 kV cm^{-1} intensity, 500 μs treatment time for apple pomace) was studied regarding the release of NEPP in sorghum flour and apple pomace.[42] This resulted in respective increases of total phenolic content and antioxidant activity of 24.8% and 33.9% compared to 80% aqueous methanol extraction at 25 °C for 1 hour without PEF application for sorghum flour and of 37.4% and 86% for apple pomace.

4.2.5.2 Far-infrared Radiation

Far-infrared radiation (FIR) is one of the techniques using electromagnetic waves with wavelengths ranging from 4 μm to less than 0.1 cm to assist for extraction.[43,44] NEPP can be effectively released by FIR rays because they can transfer heat to the centre of materials and break the covalent bonds of these compounds.[44,45] The mechanisms of FIR-assisted solvent extraction involve provoking vibrations in molecules (such as rutin, gentisic acid, quercetin and solvents) in stretching, bending, rocking and twisting ways. Friction between molecules results in heat generation so that the temperature of the sample as well as of the solvent increases significantly until the extracting solution boils. As a result, heated compounds will be released and dissolved in the extraction solvents.[45] According to the results from existing studies, FIR may be considered as a suitable treatment to preserve bioactive compounds and antioxidant activities. For example, PhA such as gallic acid (GA) and protocatechuic, vanillic, chorogenic, caffeic, *p*-coumaric, ferulic and sinapic acid extraction from rice bran and rice husks were significantly increased using FIR-assisted extraction in comparison with hot air and enzyme treatments.[46] However, extraction with FIR may lead to thermal degradation of some active compounds because this method resulted in drastic temperature increases of materials and extraction solvents.[47]

4.2.5.3 Supercritical Fluids

Supercritical fluid extraction (SFE) has provided an ideal alternative to conventional solvent extraction. Supercritical fluid can be defined as a fluid possessing attributes of both a gas and a liquid. Carbon dioxide (CO_2), methanol,

ethanol and water are solvents used in SFE.[36] CO_2 is the most widely used solvent due to its particular features and economic inexpensiveness. In addition, the European Food Safety Authority and the United States Food and Drug Administration have assigned CO_2 a generally recognized as safe (GRAS) status.[48] Supercritical CO_2 has the unique ability to penetrate through solids like a gas and dissolve materials like a liquid. The major advantages of SFE are providing ideal conditions for extracting compounds with a high degree of recovery in a short time and allowing the maximum degree of separation of solvent from the target extract. Factors influencing SFE efficiency are supercritical fluid selection, particle size, moisture content, flow rate of CO_2, type and quantity of co-solvents or modifiers (methanol, ethanol and water) and solvent-to-feed ratio.[36] The successful extraction of PC from solid materials using SFE is often studied. For example, extraction of PC from grape bagasse using supercritical CO_2 containing 10% (w/w) co-solvent (96% ethanol) at 40 °C and 20–35 MPa produced extracts with higher concentrations of phenolics (20–23 g kg^{-1} extract) than extracts produced using conventional techniques (1.8 g kg^{-1} extract for soxhlet and 1.4 g kg^{-1} extract for agitated-bed extraction). Although these finding results generally mentioned on PC extraction, SFE are possible for NEPP extraction.[49]

4.2.5.4 Pressurised Liquid

PLE, high-pressure solvent extraction or accelerated solvent extraction is performed at high pressures (4–20 MPa) and high temperatures (up to 200 °C) to ensure the solvent is liquid beyond its boiling point at atmospheric pressure.[50,51] Under high-pressure conditions, the difference of pressure between the inner part and the exterior of the plant cells is very large, leading to breakage of the membranes. Herein, solvent can permeate very fast into plant cells and the mass transfer rate increases, thus improving the extraction rate and shortening the extraction time. Factors influencing the extraction efficiency are type of fluid media (solvents), pressurization cycles, pressure, time and temperature.[36] In recent studies, PLE has been successfully applied to release bioactive compounds from different plant materials (*e.g.* PC [anthocyanins, flavonoids] from Brazilian pepper[52] or PC and anthocyanins from jucara residues[53]). Therefore, application of PLE in a single step and/or in combination with other methods (*i.e.* alkali extraction) could be considered as a promising method for NEPP extraction of plant matrices; however, this needs to be demonstrated by future studies.

4.2.5.5 Ionic Liquids

Ionic liquids (IL), also called 'green' solvents, are a group of new organic salts that exist as liquids at a low temperature (<100 °C). The main types of room temperature IL are cations such as alkylammonium, tetraalkylammonium and bis-(*N*,*N*-dialkyl)-dimethylguanidinium and anions such as bis-(trifluoromethylsulfonyl)-imide, hexafluorophosphate and tetrafluoroborat.[54]

The important features of IL are their relatively high viscosities, they exhibit very low vapour pressures under ambient conditions and they are immiscible with numerous organic solvents.[55] In addition, their thermal stability, high conductivity, high heat capacity and low flammability are also special characteristics of IL. The most favourable feature is their ability to dissolve a wide range of inorganic and organic compounds.[56] Therefore, IL are suitable for conventional liquid–liquid extraction or combining with other techniques like ultrasound and microwave. For example, flavonoids, stilbenes and coumarins in pigeon pea leaves were effectively extracted using 1-butyl-3-methylimidazolium bromide IL as a solvent in MAE.[57] However, the main disadvantage of using these IL for extraction is probably that they may be difficult to remove completely and/or may contaminate the final solutions. Because not all IL are safe and non-toxic, future use of these IL should be considered. Furthermore, although the number of studies using IL in NEPP extraction is limited, the use of IL in combination with other techniques could be responsible for bound phenolic extraction.

4.2.5.6 Irradiation

Irradiation has been widely used for food preservation. In recent years, irradiation has been mentioned as a pre-treatment for the extraction of PC. In accordance with the EU (1999), irradiation is only allowed with gamma-rays, X-rays and electron beams. Gamma irradiation induced enhancement in isoflavones from soybeans with a dose ranging from 0.5 to 5.0 kGy.[58] In another study, gamma irradiation with doses up to 10 kGy also improved isoflavone contents and increased the antioxidant activities of soybean.[59] Morata and co-authors proved that anthocyanin extraction using electron beams with a dose of 10 kGy increased by more than 70% compared to non-irradiated samples.[60] The proposed mechanism for the increase was the breakdown of high-molecular-weight into low-molecular-weight components and a change from insoluble to soluble components.[61] In addition, the degradation of hydrolysable tannins (HT) in the cellular matrix can release PhA. The negative effect of irradiation is that PC may be degraded or destroyed at high irradiation doses.[62]

4.3 Current Analysis Methods of NEPP

The methods in this section are not exclusively for the analysis of NEPP, but for all polyphenols. The difference between the analysis of NEPP and other polyphenols is that when measuring EPP, these methods are directly carried out on the extractions, while the analysis of NEPP is always conducted after treatments (hydrolysis mentioned in the previous section) of the residues of EPP extracts, performed as described above. As NEPP are mainly PA and HT associated with dietary fibre and protein,[55] the analysis methods of these two classes will be stressed.

4.3.1 Spectrophotometry

Spectrophotometric methods are the main techniques for analysing NEPP due to their simplicity and low cost. The most widely used method is the Folin–Ciocalteu assay. In alkaline solution, NEPP reduce the phosphomolybdic–phosphotungstic reagent to produce a blue-coloured complex.[63] The main defect of this assay is the lack of selectivity, as the reagent reacts with all reducing substances, which may lead to an overestimation of the (N)EPP contents. To improve selectivity, the use of solid-phase extraction to purify the polyphenols[64] and the pre-treatment of samples with H_2O_2 prior to the Folin–Ciocalteu assay were recommended.[65] Since the assay cannot distinguish among different phenolic classes, more specific methods for PA and HT will be discussed.

4.3.1.1 Assays of Proanthocyanidins

A number of colorimetric methods for PA quantification have been developed, among which the most commonly used are the vanillin assay, dimethylaminocinnamaldehyde (DMACA) assay and butanol–HCl assay.

The vanillin assay relies on the condensation of the nucleophilic site of the phloroglucinol structure (A ring) with the aldehyde reagent in strong acid conditions, producing a coloured compound detectable at 500 nm. This assay is specific to flavan-3-ols, dihydrochalcones and PA, having a single bond at the 2,3-position and free meta-hydroxy groups on the B ring.[66] As the absorption of anthocyanins coincides with that of the coloured compounds in this assay, correction by using a suitable blank is necessary.[67] The use of catechin as standard may lead to inaccurate estimation of PA content, since the reactivity of vanillin with flavan-3-ol monomers (*e.g.* catechin) is different from that of vanillin with PA.[68] Sun *et al.*[67] reported that if the H_2SO_4 concentration is more than 10%, the utilisation of catechin as standard underestimates the PA content, while catechin overestimates the PA content if the H_2SO_4 concentration is less than 10%.

The DMACA reacts with PC to form a green chromophore at 640 nm, which is specific to PC that show a phloroglucinol or resorcinol type.[69,70] The DMACA prefers the C8 position of the A ring and reacts only with the terminal units of PA,[71] and as a consequence, the molar absorption of PA decreases when increasing the degree of polymerisation (DP).[72] As a result, the PA content may be underestimated if the PA monomers and dimers are used as standards to measure the content of PA polymers.[73] Compared to the vanillin assay, the DMACA assay excludes interferences with anthocyanins and is more accurate and sensitive.[73] The presence of methanol, acetone, ethyl acetate and dimethylformamide does not affect the rate and intensity of colour development.[74] However, the colour formed is not stable,[75] and therefore it was suggested that the reaction time should be between 20 and 30 minutes.[76]

The butanol–HCl assay is based on the cleavage of PA into the extension unit and the terminal unit through the breakdown of interflavanic bonds in an acidic heat solution.[77] The extension units are converted into anthocyanins with a specific absorption at 550 nm by *autoxidation*. This assay is fast and efficiently releases PA due to the drastic conditions employed.[78] However, partial degradation of the anthocyanins may occur.[78] Another shortcoming is that the formation of coloured anthocyanins depends on the structure of PA.[79] The 4β–6 linkage between the monomer units is more resistant than the 4β–8 linkage.[80] The substitution pattern of the A ring influences the stability of the interflavanoid bond, which thus affects the colour formation.[81] In addition, structural differences in PA units are likely to affect both the colour intensity and the absorbance maximum of the product.[82] To improve the reproducibility and the yield of anthocyanidins, ferrous and ferric ions were recommended as the most effective catalysts in the formation of anthocyanidins by accelerating autoxidation.[83]

4.3.1.2 Assays of Hydrolysable Tannins

The most widely used method for determining HT content is the KIO_3 assay. In strong acidic conditions, methylgallates are formed from HT through a methanolysis step, followed by the reaction with KIO_3, producing a red compound having a maximum absorption at 525 nm.[84] This assay can measure both gallo- and ellagitannins, as the reaction relies on GA, the common structure of both gallo- and ellagitannins.[85] However, it was reported that the assay can accurately measure gallotannins, but it underestimates ellagitannins as cross-linked and condensed GA structures are missed.[78] Moreover, Hagerman *et al.*[86] noted that this method was not suitable for complex mixtures of HT as these tended to give brown rather than red chromophores. The assay is not robust as the development of the chromophore is highly dependent on the temperature and duration of the reaction.[87] Methanol and acetone can be used to stabilise the chromophore.[85]

The rhodanine assay can specifically quantify gallotannins as the reagent reacts only with GA and not with galloylesters, ellagic acid and ellagitannins.[88] Gallotannins are measured by determining the GA content before and after acid hydrolysis of the sample under anaerobic conditions. The drawback is that the assay cannot provide an absolute quantification of gallotannins as the number of GA units varies between different gallotannins.[89]

The $NaNO_2$ assay is selective for the determination of ellagitannins. After hydrolysis, the ellagic acid is measured, while GA, gallotannins and PA do not interfere.[90] This assay also needs to be carried out under anaerobic conditions and it is solvent consuming.[85]

Spectrophotometric methods are cheap and fast, but they cannot provide separation nor quantification of individual compounds. Therefore, for analysing samples with complex matrices, it is advisable to use chromatographic methods.

4.3.2 Chromatography

4.3.2.1 *Thin-layer Chromatography*

Among all chromatographic techniques, thin-layer chromatography (TLC) is an especially low-cost and powerful method for screening purposes. The quality of compound separation depends on the positioning of the sample, size, homogeneity and shape of the application zones.[91] The separated compounds can be identified by the retention parameter (R_f) and colour under UV light before and after spraying with a reagent that reveals polyphenols. Mueller-Harvey *et al.*[92] separated PA and HT by using two-dimensional TLC, and Sharma *et al.*[93] was able to separate 13 polyphenols including GA and flavan-3-ols.

TLC is the only chromatographic technique that visualises results as picture images. Samples can be compared and analysed at the same time on the same plate side by side, which is time and solvent saving. As the plate is only used once, complicated clean-up of the samples is not needed. Mobile and stationary phases can be flexibly changed during the process, which influences the resolution of the samples. TLC is always combined with spectrophotometric methods for preliminary screening of antioxidant capacity of PC.[94] The main limitation of TLC is its low separation capacity and poor efficiency compared to other chromatographic techniques.[95]

4.3.2.2 *High-performance Liquid Chromatography*

High-performance liquid chromatography (HPLC) is the dominating method for the separation and quantification of (N)EPP. Reversed phase HPLC (RP-HPLC) is the most widely used method for analysing low-molecular-weight PC, but it is limited in its ability to analyse high-molecular-weight polyphenols. On a C18 column, the separation of large-polymer PA is limited up to tetramers, while higher-polymer PA and HT are eluted as unresolved humps.[96] Thus, it is important to carry out hydrolysis treatments of the extraction residue before HPLC injection. The hydrolysis procedure transforms polymer PA into extension and terminal units of flavan-3-ols, while HT are converted into ellagic acid or GA and their derivatives, making it possible to quantify the NEPP based on the concentration of the monomeric units.[97,98] For PA, the mean DP of the sample can also be calculated by estimating the proportion of total units to terminal units. It is worth noting that during the hydrolysis, as degradation of NEPP may take place, the content of NEPP may be underestimated. Also, as hydrolysis causes depolymerisation, no information of individual NEPP can be derived.[99]

Compared to RP-HPLC, normal phase HPLC (NP-HPLC) is more suitable for polymer quantification, as PA can be separated according to the DP from dimers up to decamers[100] and HT can be separated by their weight.[101] Although this method gives rather detailed information of individual compounds, the DP distribution of PA is not representative of the original sample as the hydrolysis procedure causes depolymerisation.[12] Moreover, unmodified high-polymer NEPP are often not completely eluted due to their strong adsorption to the stationary phase and the relatively low solubility in the mobile phase,[73] underestimating the total content of PA.

4.3.3 Detection After Chromatographic Separation

The most common methods for (N)EPP detection after HPLC separation are UV-DAD, fluorescence and mass spectrometry (MS).

4.3.3.1 UV-DAD

(N)EPP are strong chromophores due to their multiple conjugated double bonds in the structure.[102] As specific (N)EPP exhibit a characteristic UV spectrum, (N)EPP from different subclasses can be identified using an ultraviolet diode array detector (UV-DAD).

The UV spectra of PA exhibit two major absorption peaks (at 200–220 nm and at 278 nm).[103] In addition, the second band is always characteristic of homogeneous procyanidin polymers.[104] In most studies, 280 nm is chosen as the absorption peak for PA, while Hümmer and Schreier[73] suggested that a higher sensitivity in the quantification of PA can be reached at 210 nm because of the higher molar absorption coefficients of PA in comparison to 280 nm. The molar absorption of PA is dependent on their structure. Procyanidin polymers have a higher molar absorbance than prodelphinidin polymers.[105]

For HT, gallotannins present two peaks at 210–220 nm and at 275–285 nm,[106] while ellagitannins have their maxima at 254 nm and 360 nm.[106,107] Different numbers of C–C linkages between moieties can cause differences in UV spectra among individual HT. Salminen *et al.*[108] concluded that the intensity of absorbance was dependent on the number of galloyl groups in relation to hexahydroxydiphenoyl groups present in the molecule. Gallotannins always exhibited greater values of absorbance maxima in comparison to ellagitannins.

After HPLC separation, compounds of NEPP are mostly hydroxybenzoic and hydroxycinnamic acids. Hydroxybenzoic acids present absorbance maxima at around 280 nm due to the A ring benzoyl system, while the absorption peak of hydroxycinnamic acids is around 320 nm, involving the B ring cinnamoyl system.[102]

As NEPP from different classes have characteristic spectra, specific NEPP can be identified by checking the proper wavelength and comparing the retention time of the component to the standard compound, after which the quantification can be achieved with a calibration curve. The main drawback of UV-DAD in the analysis of NEPP is that it cannot distinguish the monomers, oligomers, polymers nor derivatives of the same class, as they all exhibit the same absorption bands. As commercial standards for NEPP are rare and expensive, in most cases it is only possible to carry out semi-quantitative analyses, expressing contents of the NEPP as equivalents to a component from the same class, normally the monomeric unit.

4.3.3.2 Fluorescence

As most of NEPP have their UV absorbance maxima at around 280 nm, it is ineffective to identify the individual compounds in that band due to the complexity of the profile. Compared to UV-DAD, fluorescence detection is more selective due to the specificity of the excitation and emission spectra of NEPP from different classes, which makes it possible to quantify peaks that appear as overlapping peaks in the chromatograms when UV-DAD is used.[109]

Because of the higher fluorescence of flavan-3-ols than other polyphenols, fluorescence detection has been shown to be more suitable for analysing PA.[110] PA are normally detected by fluorescence at the excitation wavelength of 280 nm and an emission wavelength of 315–325 nm.[110,111] It also enabled the correct separation and quantification of procyanidin B1, B2 and B3.[112]

4.3.3.3 Mass Spectrometry

MS is a very important tool for the identification and quantification of (N) EPP, as it provides structural information on individual compounds in their native forms, even in the absence of commercial standards. A MS is composed of three parts: an ion source, a mass analyser and a detector. Selecting the first two parts is crucial to characterising the profile of a sample efficiently.[102]

Electrospray ionisation (ESI) is the most used ion source in NEPP analysis. It permits the detection of the molecular ion with little or no fragmentation, which is a more gentle and sensitive way compared to other methods (*e.g.* atmospheric pressure chemical ionisation).[113] However, it is not suitable for the detection of complex NEPP due to the generation of multiple charged ions, which makes the interpretation of the spectra difficult.[114] Thus, for the analysis of polymer PA and high-molecular-weight HT, matrix-assisted laser desorption/ionisation (MALDI) is preferred, which offers advantages over ESI including higher resolution, greater tolerance for impurities and single charged ions.[114,115]

The MS analyser is a device that separates ions based on their *m*/*z* values by electrical or magnetic fields. Among all analysers, the time of fight (TOF) is always combined with MALDI. MALDI-TOF-MS is a very powerful method for determining complex NEPP, especially PA, as it enables the interpretation of mass signals corresponding to differences in the degree of hydroxylation of the flavan-3-ol unit, presence of galloylation, nature of the interflavanic linkage and DP.[116] The combination of MS analysers, called tandem MS (MS/MS or MS^n), allows us to obtain more specific structural information by further fragmentation of the original molecule fragments, providing the exact structure of the component. For example, both quercetin and ellagitannin have the same fragment of *m*/*z* 301, but this can be distinguished after MS^2, as quercetin *m*/*z* 301 further produces characteristic ions at *m*/*z* 179 and 151, while ellagitannin *m*/*z* 301 forms ions at *m*/*z* 257 and 229.[117]

4.3.4 Other Techniques

4.3.4.1 Nuclear Magnetic Resonance Spectroscopy

Nuclear magnetic resonance (NMR) is demonstrated as an excellent tool for qualitative and quantitative analysis of complex NEPP.[115,118,119] Briefly, in a magnetic field, the resonances and absorptions of protons (^{1}H-NMR) and carbons (^{13}C-NMR) can provide detailed information about the chemical environment of each individual atom.[73,87] For example, in the analysis of PA, the ratio of the 2,3-*cis* to 2,3-*trans* stereoisomers can be achieved through ^{13}C-NMR due to the distinct differences in their respective C2 chemical shifts.[120] The average molecular weight can be obtained by integrating the areas of the C3 resonances for the extender and terminal flavan-3-ol.[121] The main limitation of NMR is its relatively low sensitivity compared to HPLC.[122] The cross-polarization of ^{1}H magnetisation to ^{13}C is applied in order to enhance the sensitivity of NMR.[123]

4.3.4.2 Infrared Spectroscopy

Infrared (IR) spectroscopy is a powerful, fast, non-destructive technique that can be used for the identification and quantification of NEPP. With different sample presentation methods, IR spectroscopy can be used for any type of liquid, slurry, powdered or solid food sample.[124] Characterisation is based on determining the functional groups of the molecule through the vibrations of atoms, while the quantification relies on the correlation between sample composition and the absorption of light at different wavelengths in the IR region measured by either reflectance or transmission.[125] The Fourier transform mid-IR spectroscopy technique was applied to differentiate HT and PA. The HT show intense absorptions in the region at 1720–1700 cm^{-1}, whereas PA show intense absorptions between 1400 and 1100 cm^{-1}.[126] Near-IR and mid-IR spectroscopies combined with multivariate analyses were suitable for evaluating the total phenolic content of by-products in coffee and wine processing.[127,128]

4.4 Conclusions and Future Trends

The techniques for NEPP mentioned above could be applied either as pre-treatments or simultaneously with the extraction methodology applied. Determining how to combine different extraction methods would be interesting in order to optimise the extraction of NEPP on a large scale. However, environmental awareness, economic issues and quality of bioactive compounds need to be taken into account while choosing an extraction method.

There is a need for standardising the analytical methods of NEPP, making it possible to compare results from different laboratories and to build a worldwide dietary NEPP database. The development of a technique suitable for all the groups of NEPP (PA, HT and other NEPP) will always be a major task. As the current analysis of NEPP is mostly combined with pre-treatments and is always accompanied by some degradation and as there is no standard reference method to be used to compare the efficiency of the extraction methods, it will always be difficult to find out what the original content of the PC was in the starting product.

References

1. S. Arranz, J. M. Silván and F. Saura-Calixto, *Mol. Nutr. Food Res.*, 2010, **54**, 1646–1658.
2. J. Pérez-Jiménez, M. E. Díaz-Rubio and F. Saura-Calixto, *Nutr. Res. Rev.*, 2013, **26**, 118–129.
3. F. Saura-Calixto, *J. Agric. Food Chem.*, 2012, **60**, 11195–11200.
4. J. Pérez-Jiménez, M. E. Díaz-Rubio and F. Saura-Calixto, *Nutr. Res. Rev.*, 2013, **26**, 118–129.
5. W. W. Tow, R. Premier, H. Jing and S. Ajlouni, *J. Food Sci.*, 2011, **76**, T163–T172.
6. A. Chandrasekara and F. Shahidi, *J. Agric. Food Chem.*, 2010, **58**, 6706–6714.
7. S. Arranz and F. Saura Calixto, *J. Cereal Sci.*, 2010, **51**, 313–318.
8. J. Pérez-Jiménez and F. Saura-Calixto, *Food Res. Int.*, 2015, **74**, 315–323.
9. M. Pinelo, A. Arnous and A. S. Meyer, *Trends Food Sci. Technol.*, 2006, **17**, 579–590.
10. B. A. Acosta-Estrada, J. A. Gutiérrez-Uribe and S. O. Serna-Saldívar, *Food Chem.*, 2014, **152**, 46–55.
11. T. E. Moussa-Ayoub, S. K. El-Samahy, L. W. Kroh and S. Rohn, *Food Chem.*, 2011, **124**, 1177–1184.
12. B. L. White, L. R. Howard and R. L. Prior, *J. Agric. Food Chem.*, 2010, **58**, 7572.
13. P. X. Chen, Y. Tang, M. F. Marcone, P. K. Pauls, B. Zhang, R. Liu and R. Tsao, *Food Chem.*, 2015, **185**, 298–308.
14. A. C. de Camargo, M. A. B. Regitano-d'Arce, A. C. T. Biasoto and F. Shahidi, *Food Chem.*, 2016, **212**, 395–402.
15. R. Gómez-García, G. C. Martínez-Ávila and C. N. Aguilar, *3 Biotech*, 2012, **2**, 297–300.
16. B. Verma, P. Hucl and R. N. Chibbar, *Food Chem.*, 2009, **116**, 947–954.
17. H. Chen, in *Biotechnology of Lignocellulose*, ed. H. Chen, Springer, Netherlands, 2014, pp. 25–71.
18. K. Krygier, F. Sosulski and L. Hogge, *J. Agric. Food Chem.*, 1982, **30**, 330–334.
19. P. Torre, B. Aliakbarian, B. Rivas, J. M. Domínguez and A. Converti, *Biochem. Eng. J.*, 2008, **40**, 500–506.
20. K. Ross, T. Beta and S. Arntfield, *Food Chem.*, 2009, **113**, 336–344.
21. D. Peričin, V. Krimer, S. Trivić and L. Radulović, *Food Chem.*, 2009, **113**, 450–456.

22. J. Azmir, I. S. M. Zaidul, M. M. Rahman, K. M. Sharif, A. Mohamed, F. Sahena, M. H. A. Jahurul, K. Ghafoor, N. A. N. Norulaini and A. K. M. Omar, *J. Food Eng.*, 2013, **117**, 426–436.
23. G. B. Sarita and R. K. Sukumaran, in *Valorization of Food Processing By-products*, ed. M. Chandrasekaran, CRC Press, Taylor & Francis Group, 2013, pp. 233–259.
24. A. S. Meyer, S. M. Jepsen and N. S. Sørensen, *J. Agric. Food Chem.*, 1998, **46**, 2439–2446.
25. A. Alupului, I. Calinescu and V. Lavric, *UPB Sci. Bull., Series B*, 2012, **74**, 1454–2331.
26. P. C. Veggi, J. Martinez and M. A. A. Meireles, in *Fundamentals of Microwave Extraction*, ed. F. Chemat and G. Cravotto, Springer, New York, 2012, pp. 15–52.
27. W. Routray and V. Orsat, *Food Bioprocess Technol.*, 2012, **5**, 409–424.
28. K. Papoutsis, P. Pristijono, J. B. Golding, C. E. Stathopoulos, M. C. Bowyer, C. J. Scarlett and Q. V. Vuong, *J. Food Process. Preserv.*, 2016, 1–8.
29. K. Hayat, X. Zhang, H. Chen, S. Xia, C. Jia and F. Zhong, *Sep. Purif. Technol.*, 2010, **73**, 371–376.
30. C. Chiremba, L. W. Rooney and T. Beta, *J. Agric. Food Chem.*, 2012, **60**, 4735–4742.
31. L. Chen, L. Ding, A. Yu, R. Yang, X. Wang, J. Li, H. Jin and H. Zhang, *Anal. Chim. Acta*, 2007, **596**, 164–170.
32. H. Wijngaard, M. B. Hossain, D. K. Rai and N. Brunton, *Food Res. Int.*, 2012, **46**, 505–513.
33. G. B. Gonzales, G. Smagghe, K. Raes and J. Van Camp, *J. Agric. Food Chem.*, 2014, **62**, 3371–3376.
34. A. P. D. F. Machado, A. L. D. Pereira, G. F. Barbero and J. Martínez, *Food Chem.*, 2017, **231**, 1–10.
35. R. Kidak and N. H. Ince, *Ultrason. Sonochem.*, 2006, **13**, 195–199.
36. F. Lebovka, N. Vorobiev and E. Chemat, *Enhancing Extraction Processes in the Food Industry*, CRC Press, Taylor & Francis Group, 2011.
37. E. Puértolas and F. J. Barba, *Food Bioprod. Process.*, 2016, **100**, 172–184.
38. F. J. Barba, E. Puértolas, M. Brncic, I. N. Panchev, D. A. Dimitrov, V. Athes-Dutour, M. Mousaa and I. Souchon, in *Food Waste Recovery: Processing Technologies and Industrial Techniques*, ed. C. M. Galanakis, Elsevier, 2015, pp. 249–272.
39. M. Corrales, S. Toepfl, P. Butz, D. Knorr and B. Tauscher, *Innovative Food Sci. Emerging Technol.*, 2008, **9**, 85–91.
40. E. Luengo, I. Álvarez and J. Raso, *Innovative Food Sci. Emerging Technol.*, 2013, **17**, 79–84.
41. O. Parniakov, F. J. Barba, N. Grimi, N. Lebovka and E. Vorobiev, *Food Chem.*, 2016, **192**, 842–848.
42. U. C. Lohani and K. Muthukumarappan, *Innovative Food Sci. Emerging Technol.*, 2016, **35**, 29–35.
43. S. Inoué and M. Kabaya, *Int. J. Biometeorol.*, 1989, **33**, 145–150.

44. S.-C. Lee, J.-H. Kim, S.-M. Jeong, D.-R. Kim, J.-U. Ha, K. Nam and D. Ahn, *J. Agric. Food Chem.*, 2003, **51**, 4400–4403.
45. Z. Gan, Q. Chen, Y. Fu and G. Chen, *Food Chem.*, 2012, **130**, 1122–1126.
46. P. Wanyo, N. Meeso and S. Siriamornpun, *Food Chem.*, 2014, **157**, 457–463.
47. H. Duan, Y. Chen and G. Chen, *J. Chromatogr.*, 2010, **1217**, 4511–4516.
48. K. Ameer, H. M. Shahbaz and J. H. Kwon, *Compr. Rev. Food Sci. Food Saf.*, 2017, 1–21.
49. A. M. Farías-Campomanes, M. A. Rostagno and M. A. A. Meireles, *J. Supercrit. Fluids*, 2013, **77**, 70–78.
50. A. Nieto, F. Borrull, E. Pocurull and R. M. Marcé, *TrAC, Trends Anal. Chem.*, 2010, **29**, 752–764.
51. A. Hubert, K.-D. Wenzel, M. Manz, L. Weissflog, W. Engewald and G. Schüürmann, *Anal. Chem.*, 2000, **72**, 1294–1300.
52. M. M. Feuereisen, M. Gamero Barraza, B. F. Zimmermann, A. Schieber and N. Schulze-Kaysers, *Food Chem.*, 2017, **214**, 564–571.
53. M. d. P. Garcia-Mendoza, F. A. Espinosa-Pardo, A. M. Baseggio, G. F. Barbero, M. R. Maróstica Junior, M. A. Rostagno and J. Martínez, *J. Supercrit. Fluids*, 2017, **119**, 9–16.
54. P. Wasserscheid and T. Welton, *Ionic Liquids in Synthesis*, John Wiley & Sons, 2008.
55. D. Han and K. H. Row, *Molecules*, 2010, **15**, 2405–2426.
56. C. F. Poole and S. K. Poole, *J. Chromatogr.*, 2010, **1217**, 2268–2286.
57. Z. Wei, Y. Zu, Y. Fu, W. Wang, M. Luo, C. Zhao and Y. Pan, *Sep. Purif. Technol.*, 2013, **102**, 75–81.
58. D. A. Kumar, D. Bhatnagar, V. Kumar, A. Rani, J. G. Manjaya and D. Bhatnagar, *J. Agric. Food Chem.*, 2010, **58**, 4298–4302.
59. B. M. Popović, D. Štajner, A. Mandić, J. Čanadanović-Brunet and S. Kevrešan, *Sci. World J.*, 2013, **2013**, 1–5.
60. A. Morata, M. A. Bañuelos, W. Tesfaye, I. Loira, F. Palomero, S. Benito, M. J. Callejo, A. Villa, M. C. González and J. A. Suárez-Lepe, *Food Bioprocess Technol.*, 2015, **8**, 1845–1853.
61. K. F. Khattak, T. J. Simpson and Ihasnullah, *Food Chem.*, 2008, **110**, 967–972.
62. A. S. Teets, C. S. Minardi, M. Sundararaman, C. Hughey and L. M. Were, *J. Food Sci.*, 2009, **74**, 298–305.
63. O. Folin and W. Denis, *J. Biol. Chem.*, 1912, **12**, 239–243.
64. S. Georgé, P. Brat, P. Alter and M. J. Amiot, *J. Agric. Food Chem.*, 2005, **53**, 1370–1373.
65. B. L. Ford, J. Bai, J. Manthey and E. A. Baldwin, *Proceedings of Florida State Horticultural Society*, 2010.
66. S. K. Sarkar and R. E. Howarth, *J. Agric. Food Chem.*, 1976, **24**, 317–320.
67. B. Sun, J. M. Ricardo-da-Silva and I. Spranger, *J. Agric. Food Chem.*, 1998, **46**, 4267–4274.
68. S. S. Deshpande, M. Cheryan, D. Salunkhe and B. S. Luh, *Crit. Rev. Food Sci. Nutr.*, 1986, **24**, 401–449.

69. M. J. Payne, W. J. Hurst, D. A. Stuart, B. Ou, E. Fan, H. Ji and Y. Kou, *J. AOAC Int.*, 2010, **93**, 89–96.
70. M. S. Marles, H. Ray and M. Y. Gruber, *Phytochemistry*, 2003, **64**, 367–383.
71. J. Oszmianski and M. Bourzeix, *Pol. J. Food Nutr. Sci.*, 1996, **5**, 43–50.
72. R. P. Feliciano, M. P. Shea, D. Shanmuganayagam, C. G. Krueger, A. B. Howell and J. D. Reed, *J. Agric. Food Chem.*, 2012, **60**, 4578–4585.
73. W. Hümmer and P. Schreier, *Mol. Nutr. Food Res.*, 2008, **52**, 1381–1398.
74. I. McMurrough and J. McDowell, *Anal. Biochem.*, 1978, **91**, 92–100.
75. F. Lang and E. Wilhelm, *Pharmazie*, 1996, **51**, 734–737.
76. Y.-G. Li, G. Tanner and P. Larkin, *J. Sci. Food Agric.*, 1996, **70**, 89–101.
77. Y. G. Li, G. Tanner and P. Larkin, *J. Sci. Food Agric.*, 1996, **70**, 89–101.
78. J. Pérez-Jiménez and J. L. Torres, *J. Agric. Food Chem.*, 2011, **59**, 12713–12724.
79. A. Scalbert, *Plant Polyphenols*, Springer, 1992, pp. 259–280.
80. P. Schofield, D. Mbugua and A. Pell, *Anim. Feed Sci. Technol.*, 2001, **91**, 21–40.
81. B. I. Giner-Chavez, P. J. Van Soest, J. B. Robertson, C. Lascano, J. D. Reed and A. N. Pell, *J. Sci. Food Agric.*, 1997, **74**, 359–368.
82. T. Kraus, Z. Yu, C. Preston, R. Dahlgren and R. Zasoski, *J. Chem. Ecol.*, 2003, **29**, 703–730.
83. L. J. Porter, L. N. Hrstich and B. G. Chan, *Phytochemistry*, 1985, **25**, 223–230.
84. P. W. Hartzfeld, R. Forkner, M. D. Hunter and A. E. Hagerman, *J. Agric. Food Chem.*, 2002, **50**, 1785–1790.
85. M. Naczk and F. Shahidi, *J. Chromatogr.*, 2004, **1054**, 95–111.
86. A. E. Hagerman, Y. Zhao and S. Johnson, *ACS Symp. Ser.*, ACS Publications, 1997.
87. I. Mueller-Harvey, *Anim. Feed Sci. Technol.*, 2001, **91**, 3–20.
88. L. Porter, *Methods Plant Biochem.*, 1989, **1**, 389–419.
89. K. H. Inoue and A. E. Hagerman, *Anal. Biochem.*, 1988, **169**, 363–369.
90. T. C. Wilson and A. E. Hagerman, *J. Agric. Food Chem.*, 1990, **38**, 1678–1683.
91. C. Cimpoiu, *J. Liq. Chromatogr. Relat. Technol.*, 2006, **29**, 1125–1142.
92. I. Mueller-Harvey, R. D. Hartley and J. D. Reed, *J. Sci. Food Agric.*, 1987, **39**, 1–14.
93. O. P. Sharma, T. K. Bhat and B. Singh, *J. Chromatogr. A*, 1998, **822**, 167–171.
94. S. Agatonovic-Kustrin, C. G. Hettiarachchi, D. W. Morton and S. Razic, *J. Pharm. Biomed. Anal.*, 2015, **102**, 93–99.
95. Ł. M. Cieśla, M. Waksmundzka-Hajnos, K. A. Wojtunik and M. Hajnos, *Phytochem. Lett.*, 2015, **11**, 445–454.
96. E. E. D. U. E. Proantocianidinas and D. Vinho, *Cienc. Tec. Vitivinic.*, 2005, **20**, 59–89.
97. S. Guyot, N. Marnet and J.-F. Drilleau, *J. Agric. Food Chem.*, 2001, **49**, 14–20.
98. L. Li, R. Tsao, R. Yang, C. Liu, H. Zhu and J. C. Young, *J. Agric. Food Chem.*, 2006, **54**, 8033–8040.

99. J. K. Hellström and P. H. Mattila, *J. Agric. Food Chem.*, 2008, **56**, 7617–7624.
100. L. Gu, M. Kelm, J. F. Hammerstone, G. Beecher, D. Cunningham, S. Vannozzi and R. L. Prior, *J. Agric. Food Chem.*, 2002, **50**, 4852–4860.
101. A. E. Hagerman, C. T. Robbins, Y. Weerasuriya, T. C. Wilson and C. McArthur, *J. Range Manage.*, 1992, 57–62.
102. S. G. Sáyago-Ayerdi, G. Mercado-Mercado, S. Ramos-Romero, J. L. Torres and J. Pérez-Jiménez, *Adv. Technol. Prod. Food-Relevant Polyphenols*, 2016, 132.
103. S. Scharbert, N. Holzmann and T. Hofmann, *J. Agric. Food Chem.*, 2004, **52**, 3498–3508.
104. B. Vennat, A. Pourrat, O. Texier, H. Pourrat and J. Gaillard, *Phytochemistry*, 1986, **26**, 261–263.
105. G. Rohr, B. Meier and O. Sticher, *Stud. Nat. Prod. Chem.*, 2000, **21**, 497–570.
106. B. Yang, M. Kortesniemi, P. Liu, M. Karonen and J.-P. Salminen, *J. Agric. Food Chem.*, 2012, **60**, 8672–8683.
107. M. K. Seikel and W. Hillis, *Phytochemistry*, 1970, **9**, 1115–1128.
108. J.-P. Salminen, V. Ossipov, J. Loponen, E. Haukioja and K. Pihlaja, *J. Chromatogr. A*, 1999, **864**, 283–291.
109. M. Rodríguez-Delgado, S. Malovana, J. Perez, T. Borges and F. G. Montelongo, *J. Chromatogr. A*, 2001, **912**, 249–257.
110. P. Viñas, C. López-Erroz, J. J. Marín-Hernández and M. Hernández-Córdoba, *J. Chromatogr. A*, 2000, **871**, 85–93.
111. J. Obón, M. Díaz-García and M. Castellar, *J. Food Compos. Anal.*, 2011, **24**, 760–771.
112. S. Gómez-Alonso, E. García-Romero and I. Hermosín-Gutiérrez, *J. Food Compos. Anal.*, 2007, **20**, 618–626.
113. S. de Pascual-Teresa and J. Rivas-Gonzalo, *Application of LC-MS for the Identification of Polyphenols*, The Royal Society of Chemistry: Cambridge, UK, 2003.
114. M. Monagas, J. E. Quintanilla-López, C. Gómez-Cordovés, B. Bartolomé and R. Lebrón-Aguilar, *J. Pharm. Biomed. Anal.*, 2010, **51**, 358–372.
115. P. Arapitsas, *Food Chem.*, 2012, **135**, 1708–1717.
116. J. D. Reed, C. G. Krueger and M. M. Vestling, *Phytochemistry*, 2005, **66**, 2248–2263.
117. W. Mullen, T. Yokota, M. E. Lean and A. Crozier, *Phytochemistry*, 2003, **64**, 617–624.
118. F. Melone, R. Saladino, H. Lange and C. Crestini, *J. Agric. Food Chem.*, 2013, **61**, 9316–9324.
119. V. Cheynier, F. A. Tomas-Barberan and K. Yoshida, ACS Publications, 2015.
120. N.-E. Es-Safi, S. Guyot and P.-H. Ducrot, *J. Agric. Food Chem.*, 2006, **54**, 6969–6977.
121. Z. Czochanska, L. Y. Foo, R. H. Newman and L. J. Porter, *J. Chem. Soc., Perkin Trans. 1*, 1980, 2278–2286.

122. I. Ignat, I. Volf and V. I. Popa, *Food Chem.*, 2011, **126**, 1821–1835.
123. D. G. Reid, S. L. Bonnet, G. Kemp and J. H. van der Westhuizen, *Phytochemistry*, 2013, **94**, 243–248.
124. B. G. Osborne, *Encyclopedia of Analytical Chemistry*, John Wiley & Sons, Ltd, 2006.
125. D. Cozzolino, M. Kwiatkowski, M. Parker, W. Cynkar, R. Dambergs, M. Gishen and M. Herderich, *Anal. Chim. Acta*, 2004, **513**, 73–80.
126. G. Tondi and A. Petutschnigg, *Ind. Crops Prod.*, 2015, **65**, 422–428.
127. R. N. Páscoa, L. M. Magalhães and J. A. Lopes, *Food Res. Int.*, 2013, **51**, 579–586.
128. B. Lorrain, I. Ky, L. Pechamat and P.-L. Teissedre, *Molecules*, 2013, **18**, 1076–1100.

CHAPTER 5

Dietary Sources and Intakes of Non-extractable Polyphenols

S. ARRANZ*[a] AND M. E. DIAZ[b]

[a]AZTI, Food Research Division, Astondo Bidea, Building 609, 48160, Derio, Spain; [b]Metabolomics Facility, Biotechnology Resource Center, Cornell University, Itaca, NY, USA
*E-mail: sarranz@azti.es

5.1 Introduction

Polyphenols (PP) are common constituents of foods of plant origin. The main dietary sources of PP are fruit and beverages. Fruit contains up to 200–300 mg PP per 100 g fresh weight, whereas a glass of red wine or a cup of tea or coffee contains about 100 mg PP.[1] However, and despite their wide distribution, only in recent decades have nutritionists started to pay attention to the health benefits of these compounds.

The main factor responsible for the delay in research into PP is their variety and complexity. Recent findings suggest that there are many potential mechanisms of action of PP in preventing diseases that may be independent of their conventional antioxidant activity. To collect evidence of the effects of PP consumption, it is essential to determine the nature and distribution of these compounds in our diet, as well as to gain a better understanding of their bioavailability and metabolism.[2]

Research into PP mainly focuses on the fraction of dietary PP corresponding to those that can be extracted from food with aqueous–organic solvents,

Food Chemistry, Function and Analysis No. 5
Non-extractable Polyphenols and Carotenoids: Importance in Human Nutrition and Health
Edited by Fulgencio Saura-Calixto and Jara Pérez-Jiménez

Published by the Royal Society of Chemistry, www.rsc.org

identified as extractable polyphenols (EPP). These EPP have, to date, been considered in most studies to be the total PP content, and they are therefore used as the basis for calculations of dietary intake, bioavailability studies and the design of intervention or observation studies. However, it is known that a significant fraction of food PP remains in the corresponding residues after extraction: the so-called non-extractable polyphenols (NEPP).[3]

Only a few studies have reported a significant fraction of NEPP, mainly from fruit and vegetables, as contributing to total PP intake.[4,5] In 2009, Arranz *et al.*[6] reported that the NEPP content of apple, nectarine and peach (112–126 mg/100 g fresh fruit) was far higher than the EPP content (18.8–28.0 mg/100 g fresh fruit), suggesting that determination of total PP content in fruit requires specific analysis of hydrolysates of the corresponding extraction residues, to avoid underestimation of the actual PP content of plant foods. NEPP are not bioavailable in the small intestine; however, they reach the colon, where they may be fermented by colonic microflora, releasing some absorbable compounds and some antioxidant metabolites, as well as increasing the production of beneficial short-chain fatty acids.[7]

Perez-Jimenez *et al.*[8] report the macromolecular antioxidant contents and profiles of the 24 most consumed fruit and vegetables in four European countries (France, Germany, The Netherlands and Spain). Those results show that hydrolysable PP (HPP) and polymeric proanthocyanins are major contributors (mean value 57%) to the total PP content of fruit and vegetables.[8] Per capita macromolecular antioxidant intake from fruit and vegetables in the four countries studied was about 200 mg day^{-1}; Spain had the highest intake from fruit, while The Netherlands had the highest intake derived from vegetables.

A high degree of heterogeneity was observed between countries in terms of average intake of PP. This may reflect not only different diet patterns, but also different methods of collecting food consumption data and estimating PP content. Thus, criteria of harmonisation are necessary to assess food intake, determine PP content in foods and validate the results with biomarkers.

Analysis of NEPP in individual foodstuffs should serve as the basis for calculating NEPP content in complete diets in different populations and therefore for estimating NEPP intake. A few papers have dealt with the analysis and dietary intake of NEPP in the Spanish diet as an example of the Mediterranean diet,[9,10] and data have also been published for a rural Mexican population.[11]

Although only a minor part of the existing literature on PP in plant foods deals with NEPP, the existing data show that they are even more abundant than EPP in many plant foods. The present chapter aims to provide an overview of NEPP in plant foods and beverages, including their nature, procedures for their isolation, their main food sources and typical dietary intake in the Spanish diet.

As we have said, to date, studies of the main dietary sources of NEPP, as well as their intake in different populations, are scarce. The most comprehensive data have been provided for the Spanish diet. Thus, in this chapter, the studies carried out of this particular diet (an example of the Mediterranean dietary pattern) will be presented as a case study, including all the experimental details, in order to encourage other researchers to carry out similar studies of other diets.

5.2 Experimental

5.2.1 Sample Selection and Preparation for NEPP Analysis

Fruit, vegetables, cereal foods, nuts and legumes of the kinds most commonly consumed in the typical Spanish diet were purchased at local supermarkets at the optimal stage of maturity. Table 5.1 shows the amount of the edible part of each food consumed daily per capita, as reported in the National Dietary Survey,[11] which was weighed out and mixed into five samples: one per group. Some vegetables, cereals and legumes were previously boiled as this is how they are usually consumed. The samples were freeze dried and milled to a particle size of 0.5 mm in a Rest ZM 200 centrifugal mill (Haan, Germany). Nuts were then defatted at room temperature with

Table 5.1 Intake of plant foods in the Spanish diet.

Food group	Sample type	g edible portion day^{-1a}
Sample 1: fruits	Yellow apple with skin	12.8
	Red apple with skin	12.8
	Apricot with skin	2.4
	Avocado	0.4
	Banana	16.3
	Cherry	3.2
	Grape	5.8
	Kiwi fruit	7.1
	Lemon	1.9
	Mandarin	12.8
	Melon	13.4
	Green table olive	2.7
	Black table olive	2.7
	Orange	38.4
	Peach with skin	12.3
	Pear	16.8
	Pineapple	2.0
	Green plum	1.8
	Red plum	1.8
	Strawberry	5.4
	Watermelon	9.0
	Other	
	Blackberry	2.4
	Chirimoya	1.5
	Fig	2.0
	Mango	1.6
	Pomegranate	1.5
Total		**192.8**

Food group	Sample type	g edible portion day^{-1a}
Sample 2: vegetables	Green bean[b]	6.8
	Chard[b]	1.7
	Lettuce	12.4
	Onion	22.1
	Potato[b]	84.7
	Red pepper	12.4
	Tomato	42.8
	Other	
	Garlic	4.5
	Cucumber	4.6
	Mushroom	4.7
	Asparagus[b]	3.5
	Carrot	4.9
	Aubergine[b]	5.0
	Artichoke[b]	2.1
	Celery[b]	4.1
	Leek[b]	5.0
	Pumpkin[b]	4.1
	Cabbage[b]	4.7
Total		**230.1**
Sample 3: cereals	Rice[b]	6.7
	White bread	96.9
	Spaghetti[b]	9.1
	Biscuits	12.5
	Croissant	10.7
	Breakfast cereals	3.8
Total		**139.8**
Sample 4: nuts	Walnut	0.8
	Almond	1.2
	Peanut	0.3
	Cashew nut	1.3
	Pistachio	0.6
Total		**4.2**
Sample 5: legumes	Chickpeas[b]	4.4
	Red beans[b]	1.9
	White beans[b]	1.9
	Lentils[b]	3.6
Total		**12.1**

[a]Data from National Dietary Survey (Ministry of Agricultural, Fisheries and Food 2006).[12]
[b]Boiled.

petroleum ether. These five groups of foods represent 97% of the total daily per capita intake of plant-derived foods in the Spanish diet and were the five test samples used for high-performance liquid chromatography (HPLC) analysis of NEPP.

Beverages were selected considering the amount consumed by the Spanish population, based on data published by the Spanish Ministry of Agriculture Fisheries and Food 2006.[12] To obtain this consumption data, the

Ministry uses a method based on three types of research data. One is *Household Research* and covers 6000 households; another is *Research in Restaurants/Catering*, covering some 898 establishments; and the third type is *Research in Institutions*, which covers 375 different establishments.

The drinks that were most consumed in Spain in 2006 were: alcoholic beverages (90 L person^{-1} year^{-1}), soft drinks (61 L person^{-1} year^{-1}), coffee (45 L person^{-1} year^{-1}) and juices (18 L person^{-1} year^{-1}, with orange juice being the most consumed juices: 3.9 L person^{-1} year^{-1}). Different types of tea and herbal tea infusions were also consumed in large quantities.

Consequently, the drinks selected for this study were divided into four groups: Group 1, juices and nectars; Group 2, alcoholic beverages; Group 3, coffee and teas; Group 4, soft drinks and other beverages (Table 5.2).

For analysis, alcoholic beverages are distilled and concentrated by vacuum distillation; this avoids interference due to the presence of alcohol. Juices and carbonated drinks were concentrated to increase the concentration of the soluble dietary fibre (SDF) complex (above the limit of

Table 5.2 Beverages intake in the Spanish diet.

Group sample	Type of beverage	Intake in Spanish diet (mL person^{-1} day^{-1})[a]
Group 1: juices	Apple	0.54
	Commercial orange juice	10.01
	Natural orange juice	1.51
	Orange with soya	2.19
	Pineapple	11.42
	Peach	10.85
	Pineapple and grape	3.07
	Peach and grape	3.08
	Grape juice	3.09
	Tomato	2.20
Group 2: alcoholic drinks	Beer with alcohol	127.00
	Beer without alcohol	32.01
	Black beer	1.64
	White wine	5.61
	Red wine	14.81
	Cider	4.30
Group 3: coffee and infusions	Red tea	1.67
	Coffee[b]	123.00
	Mint tea	7.33
	Chamomile tea	10.00
Group 4: soft drinks and others	Cocoa drink	82.10
	Cola drink	77.81
	Isotonic drink	10.12
	Orange carbonated drink	31.80
	Carbonated water	6.22
	Tiger nut beverage	1.12
Total		**444.08**

[a]Data from National Dietary Survey (Ministry of Agricultural, Fisheries and Food 2006).[12]
[b]Coffee includes expresso, instantaneous and filter coffee.

detection). In the case of teas, it was sufficient to prepare more concentrated beverages than are usually consumed.

5.2.2 Determination of NEPP in Plant Foods

For decades, the literature has reported a classification of PP types that focuses exclusively on EPP, based on the different chemical structures that they present. However, classification that takes account of the extractability of PP has been associated with their bioavailability in our organism. Several procedures have been reported for extracting PP from plant foods; the most common is solid–liquid extraction with different combinations of organic solvents (methanol, ethanol or acetone) with water. When performing such chemical extractions on foodstuffs, a solid residue remains. In PP research, the residue is usually ignored and considered to be devoid of PP. However, as we have indicated, the residue is actually a rich source of another fraction of PP with specific biological activities: NEPP. Independent residues from EPP extraction can be subjected to different treatments in order to obtain HPP and non-extractable proanthocyanidin (NEPA) solutions (see Figure 5.1).

NEPP were determined as HPP plus NEPA after subjecting the residue of EPP extractions to two different hydrolysis treatments: methanol/H_2SO_4, 90 : 10 v/v at 85 °C for 20 hours to obtain HPP, including hydrolysable tannins; and butanol/HCl 97.5 : 2.5 v/v at 100 °C for 60 minutes to obtain cyanidin/delphinidin solutions derived from NEPA.

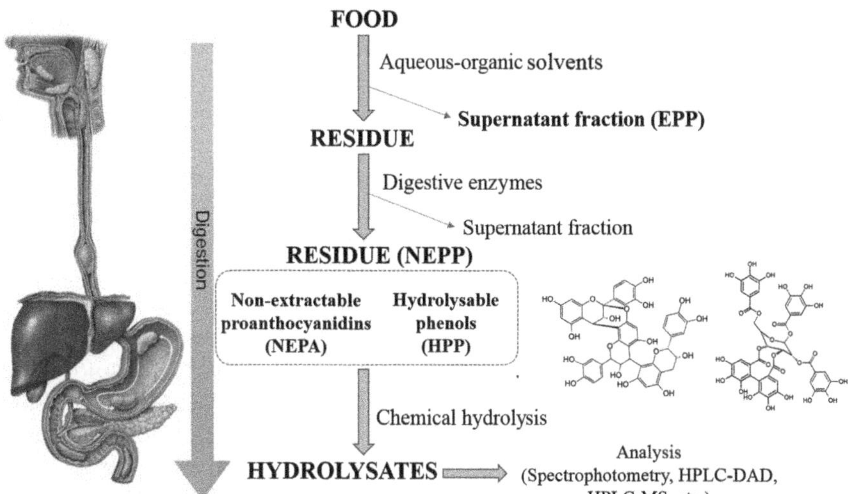

Figure 5.1 Scheme of analysis of non-extractable polyphenols (NEPP) and extractable polyphenols (EPP) from plant foods. HPLC, high-performance liquid chromatography; MS, mass spectrometry.

For HPP, a reverse-phase HPLC C18 column (Gemini, 250; 4.6 mm, 5 mm) appropriate for acid samples was used, applying the method previously reported.[13] Quantification was achieved by comparison with known standards when possible.

NEPA were quantified by measuring the anthocyanidin concentration in butanol/HCl/FeCl hydrolysates at 555 nm and referenced to a calibration curve of a carob pod proanthocyanidin concentrate.[14,15] The determinations were performed in triplicate in extracts and hydrolysates. All results were reported as mean ± SD on a dry matter basis.

Most studies of PP in cereals referred to the compounds determined in aqueous–organic extracts and alkali hydrolysates. Alkali hydrolysis is carried out to obtain free PP, avoiding degradation of hydroxycinnamic acids.[16] However, an appreciable amount of PP trapped in cores or bound to cell wall constituents remains insoluble after aqueous–organic extraction and/or alkali treatment. Acidic hydrolysis, usually omitted in analyses of PP in cereals, is proposed here for enhanced release of NEPP associated with cell walls in cereals.[17]

5.2.3 Determination of NEPP in Beverages

The different beverages were concentrated under a vacuum to obtain suitable indigestible polysaccharide concentrations for further chemical analysis, and also to remove the alcohol from alcoholic drinks. A concentration of around 2.5-fold was deemed to be enough for most dietary beverages. Beverages containing gas (beer and soft drinks) were sonicated before concentration. The samples were then treated enzymatically with pepsin, α-amylase and amyloglucosidase to hydrolyse the digestible constituents (protein, carbohydrates, *etc.*). The products of this hydrolysis and small-molecule constituents were removed by dialysis (intestinal absorption), while the soluble dietary fibre (SDF) complex remained inside the dialysis bag (was not absorbed). This complex contained non-digestible polysaccharides and associated compounds (protein and phenolic compounds) (see Figure 5.2).[18]

Although different types of hydrolysis were tested to release PP from the SDF complex, finally enzymatic hydrolysis with cellulase was deemed the most convenient for separating associated PP from the complex and was the one that was used in a systematic way with all the beverages. Cellulase was added to the SDF complex after adjusting the pH to 4 (a pH suitable for enzyme performance) and, after 16 hours of incubation at 37 °C, the samples were centrifuged and the hydrolysed phenolic compounds were determined both spectrophotometrically and by HPLC[19] following the method described by Lamuela-Raventós and Waterhouse,[20] with some modifications. The different beverages and their corresponding SDF complexes were also analysed.

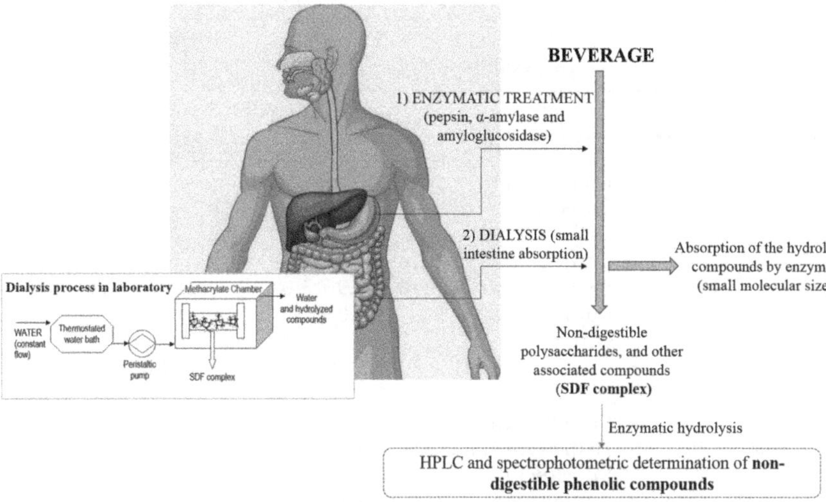

Figure 5.2 Scheme of beverage digestion and non-digestible polyphenol determination. HPLC, high-performance liquid chromatography; SDF, soluble dietary fibre.

5.3 Results and Discussion

5.3.1 Dietary Sources of NEPP

NEPP comprise PP belonging to different classes, such as macromolecular PP or single PP associated with cell wall macromolecules (proteins or polysaccharides).

The literature offers scattered and incomplete data on the content of bioactive compounds in certain foods, while estimations of dietary PP intake from complete diets is practically non-existent.[1,21,22] Nonetheless, focusing on the Spanish diet, there have been some estimations of the intake of PP compounds (1209 mg person^{-1} day^{-1}).[21]

Table 5.3 shows the NEPP contents, estimated as the sum of HPP and NEPA, of solid plant foods classified by food group. Fruits show the highest concentrations of NEPP (880 mg/100 g dry weight), followed by legumes (568 mg/100 g), vegetables (326 mg/100 g), nuts (333 mg/100 g) and cereals (210 mg/100 g). The absence of NEPA in vegetables and cereals and the appreciable amount of these compounds in fruit (514 mg/100 g), legumes (312 mg/100 g) and nuts (199 mg/100 g) determine the differences in total NEPP contents.

Our results indicate that fruit is also an important source of HPP, much of which remains in the residue from EPP extractions. Ferulic acid, *p*-hydroxybenzoic acid, catechin, gallocatechin, gallic acid and caffeic acid were identified as the main non-extractable HPP in fruit. Protocatechuic

Table 5.3 Non-extractable polyphenol content of plant foods in the Spanish diet analysed by high-performance liquid chromatography–diode array detector–mass spectrometry and spectrophotometric methods.[c,d]

	Hydrolysable polyphenols[a]			Non-extractable proanthocyanins[b]
Sample	$[M-H]^+$	Structure	mg/100 g dry weight	mg/100 g dry weight
Sample 1: fruits	195	Ferulic acid	224.0 ± 17.0	
	290	Catechin	43.0 ± 6.0	
	307	Gallocatechin	11.0 ± 2.0	
	171	Gallic acid derivative	10.8 ± 0.9	
	139	Hydroxybenzoic acid	6.9 ± 5.0	
	181	Caffeic acid	4.0 ± 1.0	
		Others	3.1 ± 1.0	
Total			**370.0 ± 20.0**	**514.0 ± 12.0**
Sample 2: vegetables	155	Protocatechuic acid	150.5 ± 4.1	
	139	Hydroxybenzoic acid	116.3 ± 16.5	
	291	Epicatechin	58.9 ± 3.7	
	181	Caffeic acid	Trace	
	165	Coumaric acid	Trace	
	195	Ferulic acid	Trace	
Total			**325.7 ± 17.4**	**n.d.**
Sample 3: cereals	181	Caffeic acid	77.9 ± 4.1	
	291	Catechin	70.1 ± 7.0	
	171	Benzoic acid	48.4 ± 2.7	
	139	*p*-Hydroxybenzoic acid	13.5 ± 0.7	
Total			**209.9 ± 5.4**	**n.d.**
Sample 4: nuts	307	Gallocatechin	27.0 ± 1.0	
	443	Epicatechin gallate	25.0 ± 1.0	
	181	Caffeic acid	17.0 ± 1.0	
	259, 634	Hexahydroxydiphenic acid	16.0 ± 1.0	
	141, 171	Gallate	12.0 ± 1.0	
	303	Ellagic acid	12.0 ± 1.0	
	290	(−)-Epicatechin	8.7 ± 1.9	
	171	Gallic acid	8.0 ± 0.5	
	367	Malvidin	8.0 ± 0.1	
Total			**124.0 ± 3.0**	**199.0 ± 8.0**
Sample 5: legumes	290	Catechin	139.0 ± 9.0	
	181	Caffeic acid	60.0 ± 4.0	
	290	(−)-Epicatechin	36.0 ± 2.0	
	141, 171	Gallate	12.0 ± 1.0	
	171	Gallic acid	7.0 ± 0.5	
Total			**250.0 ± 10.0**	**310.0 ± 50.0**

[a]Hydrolysable polyphenols determined in methanol/H_2SO_4 hydrolysates by high-performance liquid chromatography–mass spectrometry.
[b]Non-extractable proanthocyanins determined in butanol/HCl hydrolysates by spectrophotometric assay.
[c]n.d., not detected.
[d]Triplicate high-performance liquid chromatography–mass spectrometry and spectrophotometric analyses were developed in triplicate acidic hydrolysates. Results are expressed as mean values ± SD.

acid, *p*-hydroxybenzoic acid and epicatechin were the major compounds identified in vegetables. Cereals are a rich source of NEPP associated with dietary fibre.[17] Caffeic acid, catechin, gallic acid derivatives and *p*-hydroxybenzoic acid were the main HPP identified in dietary cereals. Legumes are similarly a rich source of NEPP containing high concentrations of both HPP (256 mg/100 g) and NEPA (312 mg/100 g). Catechin, caffeic acid and epicatechin were the main compounds identified. Finally, the main HPP identified in nuts were gallocatechin, epicatechin gallate, caffeic acid, hexahydroxy diphenic acid, gallate and ellagic acid. Caffeic acid (37%) and catechin (33%) were the main HPP identified in the cereal mixture, while hydroxybenzoic acid and gallic acid derivatives were also identified, together contributing 29%.

Table 5.4 shows the extractable phenolic content analysed in beverages consumed in the Spanish diet and the NEPP associated with the SDF complex, both quantified by the Folin–Ciocalteau method. Table 5.4 also shows the values of the associated PP that are hydrolysed after enzymatic reaction from the SDF complex that have been identified by HPLC. It must be noted that the Folin–Ciocalteau method quantifies all phenolic compounds—those that are free and those that can be found associated with different macromolecules—whereas chromatographic methods (HPLC) only determine free PP. Most of the beverages show values of hydrolysed PP higher than values of associated PP, both determined by Folin, due the fact that enzymatic treatment release those PP that are associated with dietary fibre.

Coffee (3613 mg L^{-1}), red wine (2312 mg L^{-1}), cocoa (1114 mg L^{-1}), apple juice (903 mg L^{-1}) and pineapple juice (926 mg L^{-1}) are the beverages with the highest amounts of EPP. Table 5.4 also shows that beverages contain an appreciable amount of associated phenolic compounds (from 5% in apple or pineapple juice to 30% in red wine or coffee, 60% in cocoa or 70% in blonde beer, considering the total of extractable and associated PP). In addition, results for HPP revealed that flavonoids (mainly catechins, proanthocyanins, anthocyanins and flavonols) are the most common class of PP; in most of these beverages, they represent between 80% and 90% of these PP.

5.3.2 Dietary Intake of NEPP in the Spanish Diet

This chapter summarises some results previously published by the authors in 2010,[10] reporting for the first time a complete characterisation by HPLC of EPP and NEPP intake in the Spanish diet (Table 5.5). It was found that in all of the plant-based food groups (cereals, vegetables, legumes, fruit and nuts) the NEPP content was higher than that of EPP. When the data were transformed into estimates of intake, total PP intake in the Spanish diet was estimated to be about 1218 mg $person^{-1}$ day^{-1}, with NEPP contributing approximately 80% of the total PP intake. Combined fruits are the major contributor to NEPP intake (some 37%), followed by cereals (24%), vegetables (10%), legumes (7%) and nuts (3%). Fruit contributes a considerable intake of NEPA and ferulic acid as the main HPP. Legumes also represent an important source of NEPA intake in the Spanish diet. Vegetables contribute a considerable

Table 5.4 Extractable and dietary fibre-associated polyphenol contents in the main consumed Spanish beverages analysed by high-performance liquid chromatography.[e]

	Beverage	Extractable polyphenols (mg L^{-1})[a]	Associated polyphenols (mg L^{-1})[b]	Hydrolysed phenolics (mg L^{-1})[c]	Main hydrolysed polyphenolic groups (HPLC), %[d] Benzoic acids	Hydroxycinnamic acids	Flavonoids	Anthocyanins
Juices	Apple juice	903 ± 8	48.5 ± 9	197 ± 4	30 ± 3.1	28 ± 2	42 ± 3	n.d.
	Commercial orange juice	529 ± 22	155 ± 2	206 ± 6	11 ± 1.0	57 ± 5	32 ± 2	n.d.
	Natural orange juice	622 ± 12	630 ± 2	275 ± 5	3.7 ± 0.2	23 ± 2	73 ± 6	n.d.
	Orange and soy juice	535 ± 7	733 ± 0.2	239 ± 1	21 ± 2.0	32 ± 2	48 ± 4	n.d.
	Pineapple juice	926 ± 32	45.1 ± 1	141 ± 1	6.8 ± 0.4	49 ± 4	44 ± 3	n.d.
	Peach juice	760 ± 32	180 ± 4	201 ± 3	12 ± 1.0	14 ± 1	74 ± 6	n.d.
	Pineapple and grape juice	400 ± 13	29 ± 6	182 ± 2	8.7 ± 0.6	58 ± 5	33 ± 3	n.d.
	Peach and grape juice	816 ± 8	159 ± 1	181 ± 2	5.1 ± 0.3	45 ± 3	50 ± 4	n.d.
	Grape juice	493 ± 20	95 ± 1	214 ± 10	17 ± 2.0	26 ± 2	57 ± 4	n.d.
	Tomato juice	449 ± 20	139 ± 3	425 ± 12	7.8 ± 0.4	19 ± 1	74 ± 5	n.d.
Alcoholic beverages	Beer without alcohol	299 ± 1	521 ± 1	289 ± 3	12.55 ± 0.9	36.46 ± 2.1	50.99 ± 3.2	n.d.
	Blond beer	312 ± 3	660 ± 2	154 ± 2	29.61 ± 1.3	21.43 ± 0.9	48.95 ± 2.1	n.d.
	Dark beer	585 ± 19	139 ± 1	422 ± 17	4.2 ± 0.2	11.43 ± 0.7	84.37 ± 4.2	n.d.
	White wine	255 ± 11	226 ± 0.2	150 ± 7.6	14.04 ± 0.7	25.35 ± 2.1	60.6 ± 4.1	n.d.
	Red wine	2312 ± 21	890 ± 10	983 ± 11	31.7 ± 1.7	14.8 ± 1.1	46.4 ± 2.7	7.05 ± 0.5
	Cider	376 ± 2	128 ± 1	217 ± 3	21 ± 2	45 ± 4	34 ± 3	n.d.

Coffee and infusions	Red tea	532 ± 26	223 ± 2	491 ± 18	9.4 ± 0.7	n.d.	91 ± 7	n.d.
	Coffee	3613 ± 42	1050 ± 20	1336 ± 31	6.93 ± 0.4	8.46 ± 1.1	15.39 ± 0.8	n.d.
	Mint tea	880 ± 12	208 ± 2	446 ± 8	24 ± 2	24 ± 1	44 ± 4	8.5 ± 0.6
	Chamomile	140 ± 9	31.8 ± 0.9	100 ± 7	16 ± 1	22 ± 2	90 ± 7	n.d.
Soft drinks and others	Cocoa drink	1114 ± 21	1459 ± 30	717 ± 10	11 ± 1	3.1 ± 0.1	86 ± 7	n.d.
	Cola drink	61 ± 1	38 ± 2	57 ± 2	49 ± 3	n.d.	51 ± 4	n.d.
	Isotonic drink	23 ± 0.7	2.4 ± 0.1	18 ± 1	25 ± 2	n.d.	75 ± 6	n.d.
	Orange soft drink	150 ± 5	6.9 ± 0.3	122 ± 3	16 ± 1	10 ± 1	74 ± 6	n.d.
	Tonic water	6.7 ± 0.3	5.1 ± 0.2	4.0 ± 0.2	19 ± 2	n.d.	81 ± 7	n.d.
	Tigger nut beverage	532 ± 11	334 ± 1	253 ± 2	29 ± 2	12 ± 1	59 ± 5	n.d.

[a]Extractable polyphenols analysed in beverages by the Folin–Ciocalteau method.
[b]Associated polyphenols to the soluble dietary fibre complex analysed by the Folin–Ciocalteau method in a dialysed sample.
[c]Associated polyphenols hydrolysed with cellulase and quantified by the Folin–Ciocalteau method.
[d]Main groups of hydrolysed polyphenols expressed as percentages from total hydrolysed polyphenols quantified by HPLC.
[e]HPLC, high-performance liquid chromatography; n.d., not detected.

Table 5.5 Total polyphenol intake of plant foods in the Spanish diet (mg person^{-1} day^{-1}).[b]

Sample (intake)[a]	EPP Structure	EPP Intake (mg person^{-1} day^{-1})	NEPP (HPP + NEPA) Structure	NEPP (HPP + NEPA) Intake (mg person^{-1} day^{-1})	Total intake (mg person^{-1} day^{-1})
Fruits (192.8)	Hesperidin	54.1 ± 2.5	Ferulic acid	112 ± 8.4	
	Naringin	19.3 ± 3.9	Cinnamic acid	34.7 ± 2.3	
	Epicatechin	12.1 ± 4.6	Catechin	21.7 ± 2.9	
	Gallocatechingallate	5.6 ± 3	Gallic acid	5.4 ± 0.4	
	Chlorogenic acid	4.4 ± 1.3	Gallocatechin	5.4 ± 0.9	
	Pelargonidin-3-glucoside	4.2 ± 0.9	Caffeic acid	1.9 ± 0.7	
	Sinapic acid	3 ± 0.6	Protocatechuic acid	1.3 ± 0.6	
	Other	7.7 ± 1.1	Vanillic acid	0.2 ± 0.07	
			NEPA	256.4 ± 5.8	
Total		**110.4 ± 7.6**		**439 ± 8.1**	**549.5 ± 5.9**
Vegetables (230.1)	Catechin	36.9 ± 3.8	Protocatechuic acid	56.8 ± 1.5	
	Sinapic acid	15.3 ± 0.5	Hydroxybenzoic acid	43.8 ± 6.2	
	Quercein-3-glucoside	9.1 ± 1	Epicatechin	22.2 ± 1.4	
	Caffeoylquinic acid	5.8 ± 0.06	NEPA	n.d.	
	Caffeic acid derivative	2.7 ± 0.1			
	Caffeoylsinapylquinic acid	1.7 ± 0.02			
	Caffeic acid	2 ± 0.06			
Total		**73.6 ± 4**		**122.9 ± 6.5**	**196.5 ± 7.6**
Cereals (139.8)	(−)-Epicatechin	26.6–3.3	Caffeic acid	108.2 ± 5.7	
	Ellagic acid	17.2 ± 1.2	Catechin	97.9 ± 9.1	
	Hydroxybenzoic acid	4.2 ± 1	Gallate	67.2 ± 3.4	
	Caffeoylquinic acid	1.4 ± 0.1	Hydroxybenzoic acid	18.2 ± 0.9	
	Others	0.5 ± 0.07	NEPA	n.d.	
Total		**49.9 ± 3.7**		**291.7 ± 11.3**	**341.6 ± 11.8**

Nuts (4.3)	Epicatechin	1.9 ± 0.03	Gallocatechin	1.8 ± 0.1	
	Protocatechuic	1.2 ± 0.1	Epicatechin gallate	1.7 ± 0.1	
	Gentisic acid	1.6 ± 0.2	Caffeic acid	1.2 ± 0.06	
	Vanillic acid	0.6 ± 0.1	Hexahydroxy diphenic acid	1.1 ± 0.1	
	Procyanidin B2	0.6 ± 0.1	(−)-Epicatechin	0.6 ± 0.1	
	Epicatechin gallate	0.5 ± 0.01	Ellagic acid	0.85 ± 0.08	
	Other	0.41 ± 0.06	Gallic acid	0.59 ± 0.03	
			Malvidin	0.5 ± 0.008	
			NEPA	20.2 ± 0.5	
Total		**6.8 ± 1.1**		**28.8 ± 1.2**	**35.3 ± 2.3**
Legumes (12.1)	Hydroxybenzoic acid	3.8 ± 0.1	Catechin	14.7 ± 0.9	
	Protocatechuic	1.5 ± 0.1	Caffeic acid	6.4 ± 0.4	
	Ferulic acid	0.6 ± 0.1	(−)-Epicatechin	3.8 ± 0.2	
	Epicatechin	0.3 ± 0.005	Gallic acid derivative	1.3 ± 0.1	
			Protocatechuic acid	0.8 ± 0.06	
			NEPA	62.1 ± 1.5	
Total		**6.2 ± 0.8**		**89.1 ± 2.3**	**95.3 ± 2.8**
Total intake		**246.9 ± 9.4**		**971.5 ± 15.5**	**1218.2 ± 15.6**

[a]Daily intake of plant food in the Spanish diet (mg day^{-1}).
[b]EPP, extractable polyphenols; HPP, hydrolysed polyphenols; NEPA, non-extractable proanthocyanins; NEPP, non-extractable polyphenols.

Table 5.6 Total polyphenol intake of beverages in the Spanish diet (mg person^{-1} day^{-1}).

	Beverage	Intake (mL day^{-1})[a]	Intake of extractable polyphenols (mg person^{-1} day^{-1})[b]	Intake of associated polyphenols (mg person^{-1} day^{-1})[c]
Juices	Apple juice	0.54	$0.49 \pm 2 \times 10^{-5}$	0.026 ± 0.005
	Commercial orange juice	10	5.29 ± 0.05	1.55 ± 0.02
	Natural orange juice	1.5	0.93 ± 0.0003	0.09 ± 0.003
	Orange and soy juice	2.19	1.17 ± 0.0002	0.16 ± 0.0004
	Pineapple juice	11.42	10.57 ± 0.13	0.51 ± 0.01
	Peach juice	10.85	8.25 ± 0.12	1.95 ± 0.04
	Pineapple and grape juice	3.07	1.23 ± 0.0016	0.089 ± 0.018
	Peach and grape juice	3.08	2.51 ± 0.0006	0.490 ± 0.003
	Grape juice	3.09	1.52 ± 0.004	0.294 ± 0.003
	Tomato juice	2.2	0.99 ± 0.002	0.305 ± 0.007
Total juices		**47.94**	**32.96 ± 0.55**	**5.48 ± 0.05**
Alcoholic beverages	Beer without alcohol	32	9.57 ± 0.001	16.67 ± 0.032
	Blond beer	127	39.62 ± 0.14	83.82 ± 0.25
	Dark beer	1.64	0.96 ± 0.001	0.23 ± 0.002
	White wine	5.6	1.43 ± 0.004	1.26 ± 0.001
	Red wine	14.8	34.22 ± 0.10	13.17 ± 0.15
	Cider	4.3	$1.62 \pm 7 \times 10^{-5}$	0.55 ± 0.004
Total alcoholic beverages		**185.34**	**87.41 ± 0.49**	**115.71 ± 0.30**
Coffee and infusions	Red tea	6.67	3.55 ± 0.17	1.48 ± 0.01
	Coffee	123	444.4 ± 5.17	129.64 ± 2.70
	Mint tea	7.33	6.45 ± 0.1	1.52 ± 0.01
	Chamomile	10	1.40 ± 0.1	0.32 ± 0.01
Total coffee and infusions		**147**	**455.8 ± 5.17**	**132.97 ± 2.7**
Soft drinks and others	Cocoa drink	82	93.48 ± 2.96	119.64 ± 2.46
	Cola drink	77.8	4.75 ± 0.006	2.95 ± 0.15
	Isotonic drink	10.12	$0.23 \pm 5 \times 10^{-5}$	0.024 ± 0.001
	Orange soft drink	31.8	4.77 ± 0.02	0.22 ± 0.01
	Tonic water	6.22	$0.04 \pm 3 \times 10^{-6}$	0.032 ± 0.001
	Tigger nut beverage	1.12	$0.60 \pm 1 \times 10^{-5}$	0.37 ± 0.001
Total soft drinks and others		**127.06**	**10.39 ± 0.17**	**3.61 ± 0.38**
Total		**589.34**	**680.03 ± 6.01**	**377.40 ± 3.67**

[a]Daily intake of beverage in the Spanish diet (mL person^{-1} day^{-1}).
[b]Intake of extractable polyphenols from beverages (mg person^{-1} day^{-1}).
[c]Intake of soluble dietary fibre-associated polyphenols from beverages (mg person^{-1} day^{-1}).

amount of protocatechuic and hydroxybenzoic acids as HPP. Cereals are the main source of caffeic acid, catechin and gallic acid derivatives.

Table 5.6 shows the contributions of beverages to the EPP and NEPP intake in the Spanish diet. We have included values of associated PP, instead of hydrolysed PP, because these values represent both those PP that are free and those that can be found associated with different macromolecules in the SDF complex.

If we add the intake of non-digestible polyphenolic compounds, provided by the consumption of beverages, to the intake of PP from plant foods, total intake of PP in the Spanish diet is approximately 2200 mg day^{-1}, of which 900 mg would be EPP and 1300 mg NEPP. These values are higher than the values cited in the literature for common diets (around 1 g), because of the important contribution of NEPP.[22–25]

Figure 5.3 represents the contribution of plant foods and beverages to total (extractable and non-extractable) PP intake in the Spanish diet. Beverages, mainly coffee, cocoa and fermented alcoholic beverages (beer and wine), represent the major contributors to total EPP intake (73%). Although fruit and cereals represent 41% of total NEPP intake, beverages, mainly coffee and cocoa, contribute a significant amount (28%) to the total NEPP in the diet; although we have not included the PP intake from oils, their contribution has previously been reported as 8 mg $person^{-1}$ day^{-1} of EPP.[21]

The important contribution of NEPP (72%) to total PP intake in the Spanish diet could change the current approach to the nutrition and health properties of dietary PP. In fact, most studies in this field almost exclusively address the biological properties, bioavailability, metabolism and systemic effects of EPP- or PP-rich foods and beverages. Our results suggest that more than 1 g of NEPP may reach the human colon daily, and NEPP can be expected to have a significant physiological impact within the large intestine, affecting microflora development and intestinal antioxidant status by producing metabolites that can be absorbed through the mucosa.[26,27]

Over the last few years, different studies have reported dietary PP intakes in whole diets. In 2014, Grosso *et al.*[28] reported a detailed description of the classes of and individual PP intakes in a diverse nutritional cohort in Poland. The mean PP intake of that study (1756 mg day^{-1}) was higher than in previous reports for the Polish population (1172 ± 354 mg day^{-1} for men and 1031 ± 320 mg day^{-1} for women),[29] as well as those reported in other studies of French (1193 ± 510 mg day^{-1}),[30] Spanish (820 ± 323 mg day^{-1})[25] and Finnish (863 ± 415 mg day^{-1})[22] populations. The main dietary PP groups reported in those studies were flavonoids and phenolic acids. Dietary data were collected using food frequency questionnaires (FFQ), and the PP contents of foods included in the FFQ would corresponded to the Phenol-Explorer database.[22]

Most of those studies only reported the intake of EPP. Saura-Calixto *et al.*[21] reported for the first time the mean daily intake of total PP in the Spanish diet, which was estimated by spectrophotometric methods to be between 2590 and 3016 mg $person^{-1}$ day^{-1}. The amount of NEPP was almost double that of EPP. It was estimated that about 48% of dietary PP are bioaccessible in the small intestine, while 42% become bioaccessible in the large intestine.

A recent study evaluating the content of macromolecular antioxidants in the 24 most consumed fruit and vegetables in four European countries (France, Germany, The Netherlands and Spain) showed that HPP and polymeric proanthocyanins are major contributors (mean value 57%) to the total PP content of fruit and vegetables.[8]

A)

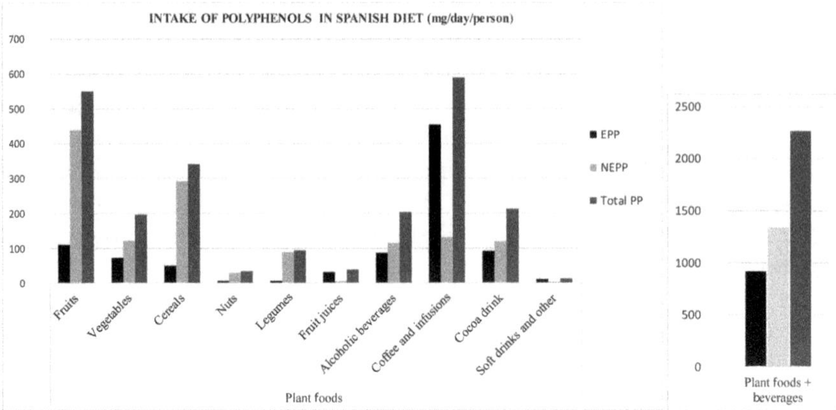

B)

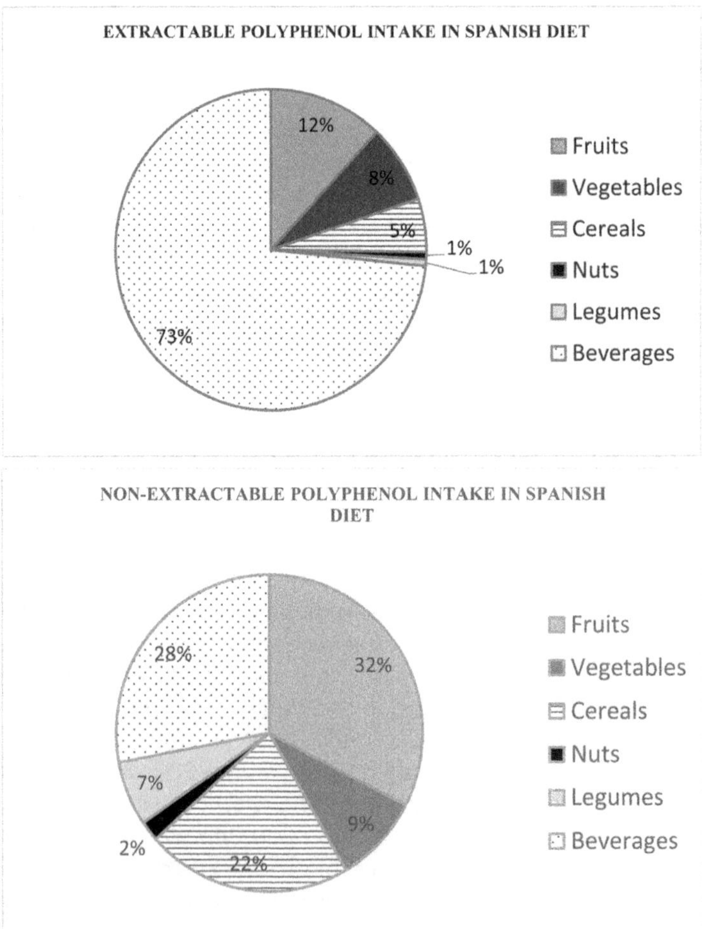

Figure 5.3 (A) Intake of polyphenols in Spanish diet (mg person $^{-1}$ day^{-1}). (B) Contribution of plant food and beverages consumption to EPP and NEPP intake in the Spanish diet. EPP, extractable polyphenols; NEPP, non-extractable polyphenols, including non-extractable proanthocyanins (NEPA) and hydrolysed polyphenols (HPP); Total, sum of EPP and NEPP.

The values shown in this work indicate that about 1 g day^{-1} of NEPP may reach the colon where, in addition to increasing antioxidant status, they serve as a substrate for the colonic microflora that is capable of metabolising them into compounds that can be absorbed through the mucosa.[31,32] Studies of the metabolism of hydrolysable tannins and NEPA have demonstrated their potential inhibitory effect through different mechanisms such as induction of apoptosis, inhibition of inflammation signals or an increase in the antioxidant environment in the colon.[33,34]

However, this study has some limitations. First, the analysis of NEPP requires strong acidic treatment that may degrade some phenolic compounds, especially flavonoids. Therefore, it is possible that our results may even underestimate the actual NEPP content. Second, different analytical procedures used to determine PP in extracts and acidic hydrolysates (HPLC with different columns and detectors and spectrophotometry) make direct comparison of quantitative results difficult.

It is reasonable to consider the possibility that in some cases, especially in NEPA analysis, we may have only an estimation and not an accurate determination. Be that as it may, there is evidence of the importance of the presence of NEPP in plant-derived foods.

5.4 Conclusion

The importance of NEPP compounds must be emphasised, as is justified by the results presented in this chapter, since they represent the most important fraction of total PP intake in typical diets, with fruit, legumes and nuts being the main sources of NEPP, including NEPA and HPP. These results provide greater knowledge of the composition of the food consumed in the diet. This complements the extensive information that already exists about PP, and it could contribute to our understanding of their bioavailability and physiological properties and therefore of the health effects that they have. The total intake of PP in the typical Spanish diet is about 2300 mg day^{-1}, of which 900 mg would be EPP and 1350 mg NEPP.

Acknowledgement

This work was supported by the Spanish Ministry of Science (project AGL2008-02541-ALI).

References

1. A. Scalbert, C. Manach, C. Morand, C. Remesy and L. Jimenez, *Crit. Rev. Food Sci. Nutr.*, 2005, **45**, 287–306.
2. M. D'Archivio, *et al.*, *Ann. Ist. Super. Sanita*, 2007, **43**, 348–361.
3. L. Bravo, R. Abia and F. Saura-Calixto, *J. Agric. Food Chem.*, 1994, **42**, 1481–1487.

4. G. Domínguez-Rodríguez, M. L. Marina and M. S. Plaza, *J. Chromatogr. A*, 2017, **1514**, 1–15.
5. J. Pérez-Jiménez, S. Arranz and F. Saura-Calixto, *Food Res. Int.*, 2009, **42**, 1381–1388.
6. S. Arranz, F. Saura-Calixto, S. Shaha and P. A. Kroon, *J. Agric. Food Chem.*, 2009, **57**, 7298–7303.
7. M.-P. Gonthier, *et al.*, *Free Radical Biol. Med.*, 2003, **35**, 837–844.
8. J. Perez-Jimenez and F. Saura-Calixto, *Food Res. Int.*, 2015, **74**, 315–323.
9. F. D. Saura-Calixto and I. Goñi, *Eur. J. Clin. Nutr.*, 2004, **58**, 1078–1082.
10. S. Arranz, J. M. Silván and F. Saura-Calixto, *Mol. Nutr. Food Res.*, 2010, **54**, 1646–1658.
11. D. Hervert-Hernández, O. P. García, J. L. Rosado and I. Goñi, *Food Res. Int.*, 2011, **44**, 1182–1189.
12. http//www.mapama.gob.es_es_alimentacion_temas_consumo-y-comercializacion-y-distribucion-alimentaria_panel-de-consumo-alimentario_resumen-anual-de-la-alimen.pdf.
13. R. N. Bennett, F. A. Mellon, E. A. S. Rosa, L. Perkins and P. A. Kroon, *J. Agric. Food Chem.*, 2004, **52**, 5856–5862.
14. F. Saura-Calixto, I. Goñi, E. Mañas and R. Abia, *Food Chem.*, 1991, **39**, 299–309.
15. L. J. Porter, L. N. Hrstich and B. G. Chan, *Phytochemistry*, 1985, **25**, 223–230.
16. K. Krygier, F. Sosulski and L. Hogge, *J. Agric. Food Chem.*, 1982, **30**, 334–336.
17. S. Arranz and F. Saura Calixto, *J. Cereal Sci.*, 2010, **51**, 313–318.
18. F. Saura-Calixto, A. Garcia-Alonso, I. Goni and L. Bravo, *J. Agric. Food Chem.*, 2000, **48**, 3342–3347.
19. A. L. Waterhouse, *Ann. N. Y. Acad. Sci.*, 2002, **957**, 21–36.
20. R. M. Lamuela-Raventós and A. L. Waterhouse, *Am. J. Enol. Vitic.*, 1994, **45**, 1 LP-5.
21. F. Saura-Calixto, J. Serrano and I. Goñi, *Food Chem.*, 2007, **101**, 492–501.
22. M.-L. Ovaskainen, *et al.*, *J. Nutr.*, 2008, **138**, 562–566.
23. C. Taguchi, *et al.*, *Nutrients*, 2015, **7**, 10269–10281.
24. A. M. Witkowska, M. E. Zujko, A. Waskiewicz, K. M. Terlikowska and W. Piotrowski, *Nutrients*, 2015, **7**, 9299–9308.
25. A. Tresserra-Rimbau, *et al.*, *Nutr., Metab. Cardiovasc. Dis.*, 2013, **23**, 953–959.
26. G. Williamson and M. N. Clifford, *Biochem. Pharmacol.*, 2017, **139**, 24–39.
27. F. Saura-Calixto, *J. Agric. Food Chem.*, 2012, **60**, 11195–11200.
28. G. Grosso, U. Stepaniak, R. Topor-Madry, K. Szafraniec and A. Pajak, *Nutrition*, 2014, **30**, 1398–1403.
29. M. Zujko, A. Witkowska, A. Waśkiewicz and E. Sygnowska, *Adv. Med. Sci.*, 2012, **57**, 375–384.
30. J. Perez-Jimenez, *et al.*, *Am. J. Clin. Nutr.*, 2011, **93**, 1220–1228.

31. M. V. Selma, A. González-Sarrías, J. Salas-Salvadó, C. Andrés-Lacueva, C. Alasalvar, A. Örem, F. A. Tomás-Barberán and J. C. Espín, *Clin. Nutr.*, 2017, DOI: 10.1016/j.clnu.2017.03.012.
32. A. Crozier, D. Del Rio and M. N. Clifford, *Mol. Aspects Med.*, 2010, **31**, 446–467.
33. A. González-Sarrías, J. C. Espín and F. A. Tomás-Barberán, *Trends Food Sci. Technol.*, 2017, **69**, 281–288.
34. P. Mena, L. Calani and R. Bruni, Bioactivation of High-Molecular-Weight Polyphenols by the Gut Microbiome, in *Diet-Microbe Interactions in the Gut*, 2015, ch. 6, pp. 73–101, DOI: 10.1016/B978-0-12-407825-3.00006-X.

CHAPTER 6

Non-extractable Polyphenols in Tropical Fruits: Occurrence and Health-related Properties

R. REYNOSO-CAMACHO[a], M. S. M. RUFINO[b], D. M. AMAYA-CRUZ[a] AND A. M. PÉREZ*[c]

[a]Universidad Autónoma de Querétaro (UAQ), Departamento de Investigación y Posgrado en Alimentos, Facultad de Química, C.U., Cerro de las Campanas S/N, 76010, Santiago de Querétaro, Mexico; [b]Universidade da Integração Internacional da Lusofonia Afro-Brasileira (UNILAB), Instituto de Desenvolvimento Rural, Campus da Liberdade, Avenida da Abolição 3, 62.790-000, Redenção, Ceará, Brazil; [c]Universidad de Costa Rica (UCR), Centro Nacional de Ciencia y Tecnología de Alimentos (CITA), Ciudad Universitaria Rodrigo Facio, 11501-2060, San José, Costa Rica
*E-mail: ana.perez@ucr.ac.cr

6.1 Introduction

In 2009, the world production of tropical fruits reached 82.2 million tons, of which 10% was traded internationally. Asia was the main producing region for tropical fruits, followed by Latin America and the Caribbean, Africa and Oceania. The USA was the largest importer of fresh tropical fruits, followed by the European Union (EU), Japan and China.[1] Banana (*Musa* AAA) and plantain (*Musa* AAB) jointly represent 46.9% of the world fruit importation.[2]

Food Chemistry, Function and Analysis No. 5
Non-extractable Polyphenols and Carotenoids: Importance in Human Nutrition and Health
Edited by Fulgencio Saura-Calixto and Jara Pérez-Jiménez

Published by the Royal Society of Chemistry, www.rsc.org

Tropical fruits represent approximately 8.1% of the world fruit market; the most commercial are the traditional tropical fruits such as mango (39%), pineapple (23%), papaya (13%) and avocado (4%). Other less traded tropical fruits represent 22% of the total production and include litchi, rambutan, guava, mangosteen and passion fruit.[1] In 2009, the value of international trade of fresh tropical fruits was US$13.3 billion (US$7.9 billion for bananas and US$5.4 billion for other tropical fruits).[1] Other tropical and subtropical fruits include tamarind and watermelon, respectively. Watermelon production is similar to that of banana,[3–5] while the harvest and production of tamarind are low.[6] Cashew apple is a tropical fruit native to the Amazon, and Brazil is one of the major producing countries. This fruit has been naturalised in many other tropical countries such as Vietnam, India, Nigeria, Tanzania, Ivory Coast, Mozambique and Benin.[7]

Among the main producing countries of the aforementioned tropical fruits are India, Thailand, Ecuador, Brazil, Costa Rica, China, Spain and Mexico.[3–5] Table 6.1 shows the worldwide area and production of some tropical and subtropical fruits.

Non-traditional tropical fruits are those underexploited and exotic fruit species with a reduced market and for which production data are scarce or unavailable. These fruits have been studied for their high content of bioactive compounds with health-related properties. Efforts have been made in Brazil and other countries to rescue these local fruits and promote their cultivation and commercialisation.[11–13] Among these underutilised fruits are acerola (*Malpighia emarginata*), assai (*Euterpe oleracea*), bacuri (*Platonia*

Table 6.1 Worldwide tropical and subtropical fruit production and export data.[a]

	World area harvested (Ha)	World production (ton)	World gross production value (US$ million)
Traditional tropical fruits			
Avocado (*Persea americana*)	547 849 [4]	5 028 756 [4]	4 543.87 [8]
Banana (*Musa* AAA)	5 393 811 [4]	114 130 151 [4]	40 452.64 [8]
Cantaloupe (*Cucumis melo*)	1 178 808 [4]	29 626 335 [4]	14 620.87 [8]
Cashew apple (*Anacardium occidentale*)	675 605 [4]	1 889 934 [4]	760.70 [8]
Mango (*Mangifera indica*), guava (*Psidium guajava*) and mangosteen (*Garcinia mangostana*)	5 642 359 [4]	45 225 211 [4]	29 110.59 [8]
Papaya (*Carica papaya*)	411 355 [4]	12 671 038 [4]	4 657.07 [8]
Pineapple (*Ananas comosus*)	1 022 319 [4]	25 439 366 [4]	12 761.06 [8]
Tamarind (*Tamarindus indica*)	208 599 [6,9]	338 790 [6,9]	NA
Watermelon (*Citrullus lanatus*)	3 477 439 [4]	111 009 149 [4]	38 559.37 [8]
Non-traditional tropical fruits			
Acerola (*Malpighia emarginata*)	NA	150 000 [10]	NA
Assai (*Euterpe oleracea*)	NA	1 080 639 [7]	NA

[a]NA: not available.

insignis), camu–camu (*Myrciaria dubia*), cherimoya (*Annona cherimola*), chicozapote (*Manilkara zapote*), Costa Rican guava (*Psidium friedrichsthalianum*), dragon fruit or pitahaya (*Hylocereus* spp.), jocote or siriguela (*Spondias purpurea*), mangaba (*Hancornia speciosa*), murta (*Blepharocalyx salicifolius*), nanche (*Byrsonima crassifolia*), naranjilla (*Solanum quitoense*), peach palm (*Bactris gasipaes*), puçá-preto (*Mouriri pusa*), soursop (*Annona muricata*), tree tomato or tamarillo (*Solanum betaceum*) and yellow mombin (*Spondias mombin*). Assai is an Amazonian native fruit that is in rapid expansion. Its exportation, for use mainly in fruit juices, has increased over the years. Currently, 90% of assai production is destined for Brazilian consumption and the remaining 10% is for the foreign market.[14] Acerola is a fruit from the Antilles. Brazil is now its largest producer, consumer and exporter in the world.[15] Production is mainly focused on the domestic market and is divided between the processing industry and fresh consumption, with the latter having a higher percentage.[10] Camu–camu is a fruit tree typically found along the Amazon basin. It is commercialised on a small scale and marketed in the form of pulp, which is widely appreciated in Japan, the USA and Europe.[16]

6.2 Tropical Fruits as Sources of Polyphenols: Contribution of the Extractable and Non-extractable Fractions

Fruits and vegetables are rich in bioactive compounds, including dietary fibre and resistant starch, carotenoids and especially polyphenolic compounds.[17–20] Phenolics are secondary metabolites in plants that protect against predators, microbial pathogens, parasites and UV light and help in plant reproduction by attracting pollinators and seed-dispersing animals.[21] Agro-climatic conditions in the tropics may favour higher accumulations of phenolic compounds in plants than conditions in temperate climates because of higher UV-B radiation and temperature levels, among other factors.[22]

Polyphenolic compounds can be classified as extractable polyphenols (EPP), or those that are solubilised in aqueous–organic solvents, and non-extractable polyphenols (NEPP), which are retained in the resulting residue after aqueous–organic extraction.[23] Different studies have reported the content of polyphenols in plant foods, usually flavonoids, and several databases on polyphenol content in foods can be found in the literature; however, most of these databases refer to the analysis of EPP, ignoring NEPP.[23] Fruits in general represent one of the most consumed vegetable foods in the diet, and they possess the widest diversity of polyphenols, including EPP and NEPP.

6.2.1 Extractable Polyphenols

The EPP contents of different tropical and subtropical fruits are shown in Table 6.2. Fruits with the highest reported EPP contents are camu–camu and acerola, followed by assai, tropical blackberry, murta and puçá-preto.

Lower EPP contents are seen in avocado, cherimoya, chicozapote and watermelon. Camu–camu contains mostly flavan-3-ol compounds and ellagic acid derivatives, as well as kaempferol, quercetin, myricetin, catechin, rutin, delphinidin-3-glucoside and cyanidin-3-glucoside.[84] The presence of pelargonidin-3-rhamnoside has been reported in acerola, with cyanidin-3-rhamnoside and quercetin-3-rhamnoside being the most abundant anthocyanins.[83] Glycosidic ellagitannins (ET) such as vescalagin and castalagin are also found in acerola.[85] The main phenolic compounds reported in assai pulp are cyanidin-3-*O*-glucoside, cyanidin-3-*O*-rutinoside, homoorientin, orientin and isovitexin.[81] The anthocyanin composition of tropical blackberries (cyanidin-3-glucoside, cyanidin-3-rutinoside, cyanidin-3-malonyl-glucoside) and the presence or absence of kaempferol glycosides can be used to differentiate some *Rubus* species.[82] Banana contains dopamine, a phenolic compound found in the peel (80–560 mg 100 g^{-1}) and pulp (2.5–10 mg 100 g^{-1}).[87]

6.2.2 Non-extractable Polyphenols

Because NEPP were recognised relatively recently, there are few reports of their content in tropical fruits. Several standards and methods are used for the quantification of these compounds. The highest NEPP concentrations have been reported for banana, tropical blackberry and cashew apple, followed by acerola, cantaloupe, mango, papaya and plantain (Table 6.2). The lowest NEPP concentration is seen in assai.

Phenolic acids detected in avocado pulp after alkaline and acid hydrolysis include protocatechuic, *p*-hydroxybenzoic, vanillic, caffeic, ferulic,

Table 6.2 Extractable and non-extractable polyphenol content in some tropical and subtropical fruits.[a]

Fruit	Extractable polyphenols	Non-extractable polyphenols
Traditional tropical fruits		
Avocado (*Persea americana*)	344.09[b,12]	62.1–73.4[b] (PC)[28]
	40–57[c,24]	110[b] (PC)[29]
	1.1–6.1[c,25]	
	61–504[b,26]	
	2.94[b,27]	
	76–175[b,28]	
Banana (*Musa* AAA)	72.2 ± 2.03[c,30]	1415.1 ± 206.5[b] (HPP)[23]
	66 ± 2[c,31]	1868.3 ± 63.4[b] (NEPA)[23]
	54.8 ± 1.3[c,32]	980[b] (NEPA)[33]
Tropical blackberry (*Rubus* spp.)	520 ± 20[c,34]	1480.2[d,34]
	3500 ± 100[b,34]	
	4250[b,35]	
Cashew apple (*Anacardium occidentale*)	118 ± 3.7[c,11]	1210[b] (HT)[37]
	830 ± 26.5[b,11]	1690[b] (HPP-IDF)[37]
	71.43–236.97[c,36]	5200[e] (NEPA)[37]
		8800[e] (NEPA-IDF)[37]

(*continued*)

Table 6.2 (*continued*)

Fruit	Extractable polyphenols	Non-extractable polyphenols
Guava (*Psidium guajava*)	750[b] (BP)[38] 1163.5[b,12] 158.9–231.6[c,39] ≈240[b] (BP)[40] 5–10[b,41] 55–516[b,42]	69[f] (HT, BP)[38] 132[g] (NEPA, BP)[38] 38[h] (NEPA)[43] 14–17[c] (HT)[43] 62.6[i] (T)[44]
Mango (*Mangifera indica*)	10 200[b] (BP)[38] 297.60[b,39] 19.5–130.8[c,45] 217.6[b,46] 48.4–208.7[c,47] 1457.0[b,48] 546[b] (BP)[40] 144.34 ± 1.37[c,49]	338[f] (HT, BP)[38] 940[g] (NEPA, BP)[38] 27–708[j] (GT, pulp)[45] 20[b] (GT, pulp)[50] 129.95 ± 8.41[c] (HT)[49]
Melon (*Cucumis melo*)	379.0[c,51] 1278.0[b,52] 100–420[c,53]	211.9 ± 16.2[b] (HPP)[23]
Papaya (*Carica papaya*)	53.2 ± 3.6[c,13] 93.18 ± 8.66[c,49]	118.98 ± 3.92[c] (HT)[49]
Pineapple (*Ananas comosus*)	54.1 ± 0.8[c,30] 78.1 ± 0.8[c,32] 56.83 ± 0.60[c,49]	50.79 ± 0.65[c] (HT)[49]
Plantain (*Musa paradisiaca* AAB)	39.4 ± 1.0[c,54] 112 ± 2.8 (unripe boiled)[d,54]	400[b] (HPP, unripe boiled)[55] 369 ± 8[b] (HPP, ripe boiled)[55]
Tamarind (*Tamarindus indica*)	83.8[c,13] 77–585[c,56] 328–424[b,57]	NA
Watermelon (*Citrullus lanatus*)	15.1 ± 0.2[c,31] 14.1 ± 0.3[c,32]	32.7 ± 3.8[b] (HPP)[23]
Non-traditional tropical fruits		
Acerola (*Malpighia emarginata*)	1063 ± 53.1[c,11] 10 280 ± 77.7[b,11] 451.0–1104.7[c,58] 1060 ± 53.0[c,59]	390[b] (HT)[37] 380[b] (HPP-IDF)[37]
Assai (*Euterpe oleracea*)	454 ± 44.6[c,11] 3268 ± 527[b,11] 518 ± 115.5[c,60] 528.0–691.2[c,61]	1.59[b] (HT)[62] 1.24[e] (NEPA)[62]
Bacuri (*Platonia insignis*)	23.8 ± 0.7[c,11]	NA
Black sapote (*Diospyros digyna*)	864.92[b,12] 652.72[b,63] 247.81[c,64] 3.45–28.94[c,65]	NA
Camu-camu (*Myrciaria dubia*)	1 176 ± 14.8[c,11] 11 615 ± 384[b,11]	NA
Dragon fruit (*Hylocereus* spp.)	42.4 ± 0.04[c,66]	7.21 ± 0.02[h] (NEPA, flesh)[66]
Cherimoya (*Annona cherimola*)	3.06 ± 0.03–12.0 ± 0.04[c,66]	NA

Table 6.2 (*continued*)

Fruit	Extractable polyphenols	Non-extractable polyphenols
Chicozapote (*Manilkara zapota*)	18.32[b,67] 196.78[b,39] 86.09[b,63] 13.5[c,13]	0.30% tannic acid[68]
Jocote (*Spondias purpurea*)	55.0 ± 2.1[c,13] 1350 ± 130[b,69]	NA
Mangaba (*Hancornia speciosa*)	169 ± 21.5[c,11] 935 ± 37[b,11] 171.8 ± 31.1[c,59]	NA
Mamey zapote (*Pouteria sapota*)	284.64[b,39] 56.81[b,63] 23.4[c,70] 247.81[c,64]	NA
Murta (*Blepharocalyx salicifolius*)	610 ± 17.7[c,11] 2055 ± 75.7[b,11]	NA
Nance (*Byrsonima crassifolia*)	1174.4[b,63] 937[b,60] 2847.17[b,71] 254.7[b,72] 2219.3–2919.3[b,73] 4860[b,74] 159.9[c,13] 80[c,75] 290[b,76]	NA
Naranjilla (*Solanum quitoense*)	48 ± 3 6[c,77] 58.3 ± 2.4[c,78] 650[b,35]	NA
Peach palm (*Bactris gasipaes*)	70 ± 4[b] (cooked and peeled)[79] 65.7 ± 2.1[c,78]	NA
Puçá-preto (*Mouriri pusa*)	868 ± 51.0[c,11] 2638 ± 48.9[b,11]	NA
Soursop (*Annona muricata*)	54.8 ± 2.7[c,13] 86.5 ± 3.5[c,80]	NA
Tamarillo (*Solanum betaceum*)	308–570[b,35]	NA
Yellow mombim (*Spondias mombin*)	72.0[c,11] 579[b,11]	NA

[a]NA: not available; FW: fresh weight; DW: dry weight; BP: by-product; PC: procyanidins; T: tannins; GT: gallotannins; HT: hydrolysable tannins; HPP: hydrolysable polyphenols; NEPA: non-extractable proanthocyanidins; CE: catechin equivalents; GAE: gallic acid equivalents; EA: ellagic acid equivalents; MGE: methyl gallate equivalents; PA: proanthocyanidins; HPP-IDF: hydrolysable tannins associated with insoluble dietary fibre; NEPA-IDF: condensed tannins associated with insoluble dietary fibre.

[b]mg GAE 100 g^{-1} DW.

[c]mg GAE 100 g^{-1} FW.

[d]mg EA 100 g^{-1} DW.

[e]mg CE 100 g^{-1} DW.

[f]MGE mg 100 g^{-1} DW.

[g]PA mg 100 g^{-1} DW.

[h]mg CE 100 g^{-1} FW.

[i]mg tannin 100 g^{-1} DW.

[j]mg 100 g^{-1} FW.

anisic and sinapic acids.[26] Gallocatechin was identified in banana, with a higher content in the peel (158 mg 100 g^{-1} dry weight (DW)) than in pulp (29.6 mg 100 g^{-1} DW).[86]

NEPP content has been more widely studied in mango than in other tropical fruits. The NEPP quantified in mango by-products formed by peel and pulp remnants contained three times more non-extractable proanthocyanidins (NEPA) than hydrolysable polyphenols (HPP), the two classes of NEPP.[38] HPP may be constituted by phenolic acids linked to the cell wall or by high-molecular-weight hydrolysable tannins (HT), as is the case in mango. The non-extractable HT contents in the peel of two Mexican mango varieties, Ataulfo and Tommy Atkins, were quantified and the NEPP contents were found to be similar for both varieties.[88] Another study[50] characterised 17 extractable HT compounds, specifically gallotanins, in pulp and peel of mango cv. Tommy Atkins and identified tetra-*O*-galloyl-glucoses, penta-*O*-galloyl-glucose, hexa-*O*-galloyl-glucoses and hepta-, octa- and nona-*O*-galloyl-glucoses; it would be expected that, in the non-extractable fraction, HT exhibited higher molecular weight, although this has not been evaluated.

Information related to NEPP in papaya is scarce. In a study of papaya, mango and pineapple, the NEPP content of papaya was 0.1-times less than that of mango and 2.5-times more than that of pineapple. The NEPP content of papaya was 118.98 ± 3.92 mg 100 g^{-1} fresh weight and consisted mainly of HT.[49]

Regarding non-traditional tropical fruits, Brito and Narain studied the tannin content of chicozapote at three stages of maturity. Tannic acid decreased during ripening and ranged from 4.5% at the mature-green stage to 0.2% in ripe fruit.[68] ET, another type of HT, were the main constituents of NEPP identified and quantified in two tropical highland blackberries (*Rubus glaucus* and *Rubus adenotrichos*) after acid or alkaline hydrolysis. Sanguiin H-6 and lambertianin C were the major ET found in *R. adenochitros.*[82]

6.2.3 Antioxidant Capacities

The majority of EPP are bioaccessible in the small intestine and a fraction is bioavailable and crosses the intestinal barrier (mucosa), where it is metabolised and exerts systemic effects at different levels. In contrast, NEPP are transported through the gastrointestinal tract associated with the insoluble food matrix (indigestible polysaccharides from the cell wall) and reach the colon unaltered. NEPP can be metabolised by the colonic microbiota to become absorbable metabolites or non-absorbable metabolites that remain in the colonic lumen where they provide antioxidant and protective effects against pro-oxidant substances.[89]

Table 6.3 shows the antioxidant capacities (AC) of NEPP for some tropical and subtropical fruits. It is important to highlight that the available information is still scarce. According to the 2,2-diphenyl-1-picrylhydrazyl (DPPH) method, AC values for both total HPP and NEPA associated with insoluble dietary fibre

Table 6.3 Antioxidant capacities of non-extractable polyphenols of some tropical and subtropical fruits according to DPPH, ABTS, FRAP and ORAC assays.[a]

Fruit	DPPH (EC50 g DW g^{-1})	ABTS (μmol TE g^{-1} DW)	FRAP (μmol TE g^{-1} DW)	ORAC (μmol TE g^{-1} DW)
Traditional tropical fruits				
Banana (*Musa* AAA)	NR	NR	NR	38.2 (HPP)[23]
Cashew apple (*Anacardium occidentale*)	5.95 (HT)[37] 28.28 (HT-IDF)[37]	60.53 (NEPA)[37] 111.23 (CT-IDF)[37]	56.37 (HT)[37] 82.35 (HT-IDF)[37]	344.65 (HT)[37] 469.27 (HT-IDF)[37]
Melon (*Cucumis melo*)	NR	NR	NR	36.0 (HPP)[23]
Watermelon (*Citrullus lanatus*)	NR	NR	NR	11.1 (HPP)[23]
Non-traditional tropical fruits				
Acerola (*Malpighia emarginata*)	28.65 (HT)[37] 262.28 (HT-IDF)[37]	NR	14.87 (HT)[37] 8.37 (HT-IDF)[37]	173.92 (HT)[37] 146.21 ± 15.14 (HT-IDF)[37]
Assai (*Euterpe oleracea*)	4.92 (HT)[62] 47.31 (HT-IDF)[62]	20.73 (HT)[62] 24.62 (NEPA)[62] 20.51 (NEPA-IDF)[62]	109.87 (HT)[62] 11.64 (HT-IDF)[62]	1 514.46 (HT)[62] 155.61 (HT-IDF)[62]

[a]NR: not reported; DW: dry weight; HT: hydrolysable tannins; IDF: insoluble dietary fibre; HPP: hydrolysable polyphenols; TE: trolox equivalents; NEPA: non-extractable proanthocyanidins; HT-IDF: hydrolysable tannins associated with insoluble dietary fibre; NEPA-IDF: non-extractable proanthocyanidins associated with insoluble dietary fibre.

(IDF) were lower in acerola than in assai and cashew apple. In general, the AC of HPP associated with IDF was lower than that of total HPP. Using the 2,2-azinobis(3-ethyl-benzothiazoline-6-sulfonic acid) (ABTS) method, NEPA and NEPA associated with IDF from cashew apple showed higher AC than assai.

According to the ferric ion reducing antioxidant power (FRAP) assay, the AC of the HPP extracts showed the following order: assai > cashew apple > acerola for dry weight. As for the AC of HPP associated with IDF, the value assessed by the FRAP assay was higher for cashew apple than for assai and acerola. For assai, the AC of HPP was higher than that of HPP bound to IDF, in contrast to cashew apple, where the HPP associated with IDF had a higher AC. The AC of NEPP was also evaluated by the oxygen radical absorbance capacity (ORAC) assay. The HPP of assai presented the highest AC, followed by cashew apple and acerola. Applying the ORAC method for HPP associated with IDF, cashew apple had a higher AC than assai and acerola. Also using the ORAC method, the AC of HPP of banana and melon were similar, while the AC of watermelon was lower. The AC values for HPP from banana, melon and watermelon were lower than those of HPP of tropical fruits from Brazil as assessed by the ORAC assay (Table 6.3). For each method, AC values were highest for HPP of assai and NEPA or HPP associated with IDF of cashew apple.

6.3 Biological Activities of Some Tropical Fruits and Traditional Uses

Polyphenolic compounds may contribute to the prevention of chronic diseases.[90] However, most of the health benefits reported in the literature have focused on extractable polyphenols, and information about NEPP is scarce. Studies using animal models indicate the positive effects of NEPP on gastrointestinal health.[91] These studies examined the effects of ingestion of NEPP-rich products, mainly grape pomace or grape antioxidant dietary fibre; however, few reports are found on tropical fruits. The different biological activities of some tropical and subtropical fruits are shown in Table 6.4, focusing on the health effects of the polyphenolic constituents.

6.3.1 Traditional Tropical Fruits

An ethanolic extract of avocado pulp showed a hypoglycaemic effect as well as protective activity on the liver and kidneys in diabetic animals.[92] These effects were related to antioxidant properties and insulin secretion enhancement associated with phenolic compounds, organic acids and alkaloids.[92] In addition, EPP from avocado pulp inhibited *in vitro* enzymes involved in carbohydrate digestion.[93] Furthermore, a phenolic extract of avocado pulp showed antimicrobial activity against Gram-positive bacteria.[28]

Banana fruit has been broadly used in traditional medicine and different components from banana have been associated with health benefits. An aqueous extract of banana (*Musa* spp.) inhibited lipid peroxidation and

Table 6.4 Biological activities of some tropical and subtropical fruits.

Fruit	Biological activity	Model/dose
Traditional tropical fruits		
Avocado (*Persea americana*)	Hypoglycaemic, antioxidant, hepatoprotective, decrease glycosylated haemoglobin, blood urea and serum creatinine[92]	*In vivo*, pulp ethanolic extract, diabetic male rats[92]
	Inhibition of α-glucosidase, α-amylase and lipid peroxidation[93]	*In vitro*, phenolic extract of fruit, chemical assay[93]
	Antioxidant and antimicrobial activities[28]	*In vitro*, ethyl acetate, acetone and methanol extract of pulp, chemical assays[28]
	Promote wound healing[94]	*In vivo*, pulp paste, excision wound model in rats[94]
	Satiety effect[95]	Clinical study, half of a Hass avocado fruit, healthy overweight adults[95]
	Antioxidant capacity[26]	*In vitro*, extractable polyphenol extract, chemical assay[26]
Banana (*Musa* spp.)	Decreased lipid peroxidation and antioxidant effect[96]	*Ex vivo* and *in vitro*, aqueous extract of pulp, different organs of rats and chemical assay[96]
Tropical blackberry (*Rubus* spp.)	Antioxidant activity[32,34]	*In vitro*, fruit extractable polyphenol extract, cells[32] and chemical[34] assays
	Inhibition against lipid peroxidation[97,98] and antioxidant activity[97]	*In vitro*, fruit juice,[97] extractable polyphenol extract,[98] liver homogenates of rats[97] and chemical assays[97,98]
	Photochemopreventive effect[99]	*In vitro*, fruit juice, cell assays[99]
	Antioxidant activity[100]	Clinical study, fruit juice, healthy females[100]
	Anti-inflammatory activity[98,101,102]	*In vitro*,[98,101] polyphenolic extracts,[98,101] anthocyanin-rich ethanolic extract,[102] cell assays,[101] chemical assays[98]
		In vivo, anthocyanin-rich ethanolic extract, model of lung inflammation in rats[102]
	Anti-inflammatory effect[103]	*In vitro*, polyphenol-rich extracts of fruit, chemical assays and liver homogenates[103]
	Hypoglycaemic, hypolipidaemic and antioxidant effects[104]	*In vivo*, fruit beverage, diabetic male rats[104]
	Oxidative markers enhance[105]	*In vivo*, anthocyanin and ellagitannin-enriched fractions of fruit, healthy male rats[105]
	Inhibition of lipid peroxidation and scavenging intracellular radicals[106]	*In vitro*, fruit juice, chemical assays, rat liver homogenates and human erythrocytes[106]

(*continued*)

Table 6.4 (*continued*)

Fruit	Biological activity	Model/dose
Guava (*Psidium guajava*)	Antioxidant capacity and decrease α-amylase activity[38]	*In vitro*, powders and polyphenolic extract, chemical assay[38]
	Decrease body weight gain[38]	*In vivo*, fruit powder, high-fat, high-fructose diet male rats[38]
	Anti-cancer activity[107]	*In vitro*, *ex vivo*, acetone extract, cell assays[107]
Mango (*Mangifera indica*)	Inhibition of α-amylase[38]	*In vitro*, powders and extractable polyphenol extract, chemical assays[38]
	Prevention of the development of hepatic steatosis and hyperglycaemia[38]	*In vivo*, fruit powder, high-fat, high-fructose diet male rats[38]
	Anticarcinogenic effect[46]	*In vivo*, pulp aqueous extract, cell assays and mice cancer model[46]
	Prevention of the development of diabetic cataract[108]	*In vivo*, mango peel powder, diabetic male rats[108]
	Hypoglycaemic effect[109]	*In vivo*, mango flour, diabetic male rats[109]
	Cytoprotective effects[110]	*In vitro*, flesh and peel ethanolic extracts, cell assays[110]
	Antiradical and antioxidant properties[111]	*In vitro*, mangiferin, chemical assays[111]
	Analgesic, antidiabetic, antisclerotic, antimicrobial and antiviral; cardio-, hepato- and neuroprotective, anti-inflammatory, antiallergic, as well as radioprotective against X-ray, gamma and UV radiation[111]	*In vitro* and *in vivo*, mangiferin, chemical assays and animal models[111]
	Inhibit cancerogenesis and cancer cell growth[111]	*In vitro* and *in vivo*, mangiferin, chemical assays and animal models[111]
Melon (*Cucumis melo*)	Antioxidant and anti-inflammatory effect[112]	*In vitro* and *in vivo*, extract rich in superoxide dismutase activity, cell assays and inflammatory model in mice[112]
Papaya (*Carica papaya*)	Decreased oxidative injury and inflammation, improved immune function[113]	*In vitro*, fermented papaya preparation, cell assays[113]
	Antioxidant and inmunostimulant effects[114]	*In vivo*, aqueous extract, male rats supplemented with acrylamide[114]
	Anti-ulcer effects[115]	*In vivo*, aqueous and methanol extracts, indomethacin-induced gastric ulcer models in rats[115]

Pineapple (*Ananas comosus*)	Inhibitory lipid peroxidation[116]	*Ex vivo*, ethanolic and aqueous pulp extract, rat liver[116]
	Antioxidant activity[116]	*In vitro*, pulp ethanolic extract, chemical assay[116]
Non-traditional tropical fruits		
Acerola (*Malpighia emarginata*)	Antimutagenic activity[117]	*In vivo*, ripe and unripe acerola juice, rats exposed to iron[117]
Assai (*Euterpe oleracea*)	Anti-inflammatory activity[118]	*In vitro*, polyphenol-enriched pulp extract, chemical assays[118]
	Antioxidant activity[118]	*In vitro*, extractable polyphenol extract, chemical/cell-based assays[118]
Black sapote (*Diospyros digyna*)	Antioxidant activity[63]	*In vitro*, cetonic-aqueous-acetic pulp extract, chemical assay[63]
Camu–camu (*Myrciaria dubia*)	Antioxidant and anti-inflammatory activities[119]	Clinical study, juice fruit, smoking volunteers[119]
Cashew apple (*Anacardium occidentale*)	Antioxidant activity and reduction of lipoperoxidation in rat brain tissue[120]	*In vitro* and *in vivo*, hydroalcoholic from bagasse chemical, assay and healthy rats[120]
Cherimoya (*Annona cherimola*)	Antioxidant activity and cytoprotective effects[121]	*In vitro*, ethanolic extract, chemical assay and cell assay[121]
Chicozapote (*Manilkara zapota*)	Antidiabetic and anti-lipidaemic effects[122]	*In vivo*, fruit juice, healthy rats[122]
	Anti-ageing properties: Inhibition against elastase and collagenase[123]	*In vitro*, ethanolic pulp and peel extract[123]
	Antioxidant and cytotoxicity activites[124]	*In vitro*, phenolic compounds from fruit, chemical assay and cell assay[124]
	Inhibition of α-amylase and α- glucosidase[125]	*In vitro*, aqueous extract[125]
	Cytotoxicity and inhibition of tumour growth[126]	*In vitro* and *in vivo*, methanolic fruit extract, cell assay and breast cancers in mice[126]
	Antioxidant capacity[63]	*In vitro*, extractable polyphenols, chemical assay[63]
Costa Rican guava (*Psidium friedrichsthalianum*)	Antioxidant activity, protection against haemolysis of human erythrocytes[54]	*In vitro*, aqueous extract[54]
	Antioxidant and anti-inflammatory activities[127]	*In vitro*, ethyl acetate fraction of phenol-rich extract, chemical assay and cell assay[127]
Dragon fruit (*Hylocereus* spp.)	Antioxidant activity, protection against haemolysis of human erythrocytes[54]	*In vitro*, aqueous extract[54]
	Antioxidant and antiproliferative activities[66]	*In vitro*, cetonic pulp extract, chemical assay and cell assay[66]

(*continued*)

Table 6.4 (*continued*)

Fruit	Biological activity	Model/dose
Mamey zapote (*Pouteria sapota*)	Antioxidant capacity[63]	*In vitro*, extractable polyphenols, chemical assay[63]
Nance (*Byrsonima crassifolia*)	Antioxidant activity[128]	*In vitro*, methanolic-aqueous extract from pulp and peel, chemical assay and cell assay[128]
	Promotion of diabetic wound healing[129]	*In vivo*, methanolic extract, diabetic rats[129]
Soursop (*Annona muricata*)	Analgesic and anti-inflammatory properties[130]	*In vivo*, fruit juice administered to mice with induced pain and carrageenan-induced rat paw edema[130]
	Antibacterial and antifungal activities[131]	*In vitro*, methanolic pulp extract[131]
Tamarind (*Tamarindus indica*)	Improved serum lipids[132]	*In vivo*, ethanolic fruit extract, healthy rats[132]
	Reduced fluoride toxicity[133]	*In vivo*, pulp extract, rats received sodium fluoride, hydro-methanolic[133]
	Inhibition of atherogenesis, improved lipid profile serum and defence system[134]	*In vivo*, hamsters fed with high-cholesterol diets, ethanolic pulp extract[134]
	Decrease body weight and improved serum lipids[135]	*In vivo*, ethanolic, pulp extract, rats fed with cafeteria diet-induced obesity[135]
	Antiobesity, decreased hepatosteatosis, enhanced antioxidant defence system[136]	*In vivo*, aqueous pulp extract, diet-induced obese rats[136]
	Reduced serum lipids[137]	*In vivo*, hamsters fed with high-cholesterol diet, ethanolic pulp extract[137]
	Hepatoprotective activity against anti-tuberculosis drugs[138]	*In vivo*, aqueous pulp extract, Wistar rats[138]

generated superoxide anion and hydroxyl radicals. These antioxidant activities were related to phenolic compounds[96] such as gallocatechin, dopamine and flavonoids (mainly catechins), among others.[86,87] Dopamine has anti-inflammatory activity and an endogenous protective effect on the stomach.[139] Also, dopamine is a neurotransmitter and a precursor of norepinephrine and epinephrine.[87] Leucocyanidin is a flavonoid and forms part of the flavan-3,4-diol class. Its content is higher in plantain than in Cavendish bananas. This flavonoid, obtained from unripe plantain, has anti-ulcerogenic properties.[140] The effects of banana consumption in the treatment of constipation may be related to the higher content of bound phenolics that reach the colon.[140]

Tropical blackberry is another well-studied traditional tropical fruit with health benefits associated with phenolic compounds. High amounts of HT, mostly ET of different molecular weights, have been found in this fruit.[141] Metabolites of ET, urolithins A and B, were identified in human urine at the beginning of the third day after ingestion of a tropical blackberry beverage containing lambertianin C and sanguiin H-6 (552 ± 10 mg of ellagic acid equivalents L^{-1}).[142] Reported health benefits of dietary ET and their derived metabolites include antioxidant, anti-inflammatory, antimicrobial, prebiotic and estrogenic and/or anti-estrogenic activities[143] and chemopreventive properties.[144] In addition, the antioxidant activity of blackberry pulp has been shown *in vitro* assessed by the ORAC method (64.00 ± 4.00 μmol Trolox equivalents (TE) g^{-1} fresh weight (FW)),[34] and in fruit extracts using a cellular assay with HepG2 cells (62.21 ± 0.43 μmol TE g^{-1} FW).[32]

A cashew apple hydroalcoholic extract obtained from the peduncle by-product contains natural antioxidants that contribute to the reduction of lipoperoxidation and preservation of long-chain polyunsaturated fatty acids in rat brain tissue.[120] An EPP extract or powder of guava by-product inhibited *in vitro* α-amylase activity; also, the addition of guava by-product to a high-fat, high-fructose diet decreased body weight gain in rats. These effects could be partially attributed to the presence of NEPP in the powder.[38] Likewise, a cetonic extract of guava pulp showed anti-cancer activity through cell cycle arrest and apoptosis due to the presence of the flavonoids guaijavarin and quercetin.[107]

Polyphenolic extracts from a mango by-product inhibited α-amylase enzyme *in vitro*. Furthermore, the consumption of a mango by-product protected against hepatic steatosis and hypoglycaemic effects in rats fed a high-fat, high-fructose diet.[38] Mango by-product contains NEPP (12.78 mg g^{-1}), and these compounds may have contributed to the observed beneficial effects. Likewise, the intake of mango pulp by diabetic animals produced a hypoglycaemic effect related to an increase in hepatic glycogen and serum insulin,[109] and the consumption of mango by-product ameliorated diabetic nephropathy and decreased intestinal digestion and absorption of triglycerides. These anti-diabetic properties could also be related to NEPP.[145] Consumption of mango pulp reduced the growth of tumour cells and prevented the appearance of precancerous lesions in the colon during the onset of carcinogenesis. This effect was attributed to antioxidant compounds.[46]

The consumption of EPP extracts of tamarind fruit flesh decreased lipid peroxidation[132] and serum lipid levels[132,135] and regulated body weight gain in rats.[135] Likewise, a crude extract of tamarind pulp containing sugars, polyphenols and flavonoids exerted a hypolipidaemic effect, inhibited atherogenesis and enhanced antioxidant enzyme activity in hypercholesterolaemic hamsters.[134] Similar effects on serum lipids were observed in hamsters fed with tamarind pulp containing a significant amount of phenolics and flavonoids. Low serum low-density lipoprotein (LDL) cholesterol levels and high serum high-density lipoprotein cholesterol levels were attributed to an increase in the expression of ApoA1, Abcg5 and LDL receptor genes. Low serum cholesterol levels were associated with a decrease in the expression of the HMG-CoA reductase gene and provided a protective effect against oxidative damage.[137]

6.3.2 Non-traditional Tropical Fruits

Acerola juice has been found to exert antimutagenic activity *in vivo*. Studies of the influence of ripe and unripe acerola juices on iron genotoxicity demonstrated that the consumption of acerola juice can be beneficial to health by protecting DNA against damage.[117] These benefits are attributed to the presence of cyanidin-3-rhamnoside, quercetin-3-rhamnoside and yellow flavonoids in acerola fruit.[146] Assai, another non-traditional tropical fruit, has anti-inflammatory and antioxidant capacity, as determined by chemical and cell-based assays, respectively.[118] These benefits are attributed to the presence of cyanidin-3-*O*-glucoside, cyanidin-3-*O*-rutinoside, homoorientin, orientin and isovitexin, the main phenolic compounds present in the assai pulp.[81]

The antioxidant and anti-inflammatory activities of camu–camu juice were evaluated in volunteer male smokers; the authors recommended the use of the juice as a dietary supplement to prevent atherosclerosis.[119]

EPP extracts from chicozapote showed cytotoxic effects in human colon cancer cell lines,[124] inhibited tumour growth and activated the mitochondrial pathway of apoptosis in a mouse model.[126] The aqueous extract from chicozapote containing phenols and flavonoids inhibited enzymes related to digestion of carbohydrates, such as α-amylase and α-glucosidase.[125] An ethanolic extract of chicozapote fruit showed AC due to the presence of EPP compounds and inhibited elastase and collagenase in *in vitro* assays,[123] effects that may protect against skin ageing. Organic extracts from cherimoya pulp exhibited antioxidant properties, decreased lipid peroxidation and showed cytoprotective effects on lymphocytes treated with *tert*-butyl hydroperoxide.[121] These antioxidant effects were linked to polyphenolic compounds, mainly flavanols and procyanidins.[121]

Antioxidant and antiproliferative activities of red dragon peel and pulp were reported on B16F10 melanoma cells. Flavonoids such as myricetin, baicalein and gallic acid significantly inhibited the growth of tumour cells.[66] Phenol-rich extracts from Costa Rican guava pulp, specifically from the methyl acetate fraction containing ellagic acid, quercetin and other identified compounds, showed antioxidant and anti-inflammatory activities on human small airway epithelial cells previously treated with cigarette smoke, thus exhibiting potential for drug development for the therapy of chronic obstructive pulmonary disease.[127]

6.4 Conclusions

Banana and mango are among the traditional tropical fruits with the highest levels of production worldwide with health-related properties. On the other hand, cashew apple and tropical blackberry are underutilised fruits with high potential as functional foods containing NEPP compounds. Broad biological properties have been reported for tropical blackberry. In addition, there are many non-traditional fruits, such as assai, camu–camu and acerola, mainly native to Brazil, whose international market is expanding due to their high content of bioactive compounds. NEPP, which are generally omitted from food analyses and databases, make up an important fraction of phenolic compounds in all fruits. Future studies of NEPP, including dietary intake, are needed, particularly in tropical fruits, to elucidate their contribution to the health effects associated with fruit consumption. There is a great potential for non-traditional tropical fruits, and more research is necessary to select genotypes for cultivation and commercialisation and especially for developing applications in food, cosmetics, nutrition and health.

Acknowledgements

This work was supported by UNILAB projects, CAPES, CNPq and FUNCAP Research Fund from Brazil, and the Vice-Rectory of Research of the University of Costa Rica (Project 735-A6-912).

References

1. FAO, *Intergovernmental Group on Bananas and Tropical Fruits*, Cameron, 2011.
2. Secretaría de Integración Económica (SIECA), *Análisis de la competitividad regional del mercado de frutas*, 2016.
3. Food and Agriculture Organization of the United Nations, *Banana Market Review*, Rome, 2017.
4. FAOSTAT, *Rome, Italy: Food and Agriculture Organization of the United Nations*, http://www.fao.org/faostat/en/edata/QV/visualize, accessed 2 August 2017.
5. European Fruit Juice Association, *AIJN*, Brussels, 2012.
6. M. Orozco-Santos, *El cultivo de tamarindo (Tamarindus indica L.) en el trópico seco de Mexico*, 2001.

7. C. E. dos Santos, B. B. Kist, C. de Carvalho, M. Treichel and C. F. Filter, *Anuário Brasileiro da Fruticultura*, Santa Cruz do Sul, 2017.
8. FAOSTAT, *Rome, Italy: Food and Agriculture Organization of the United Nations*, http://www.fao.org/faostat/es/ecompare, accessed 2 August 2017.
9. J. C. Viveros García, K. A. Figueroa Rodríguez, F. Gallardo López, E. García Pérez, O. Ruiz Rosado and F. Hernández Rosas, *Rev. Mex. Cienc. Agric.*, 2010, **3**, 1217–1230.
10. EMBRAPA, *A cultura da Acerola*, Brasília, 3rd edn, 2012.
11. M. S. M. Rufino, R. E. Alves, E. S. de Brito, J. Pérez-Jiménez, F. Saura-Calixto and J. Mancini-Filho, *Food Chem.*, 2010, **121**, 996–1002.
12. R. D. Corral-Aguayo, E. M. Yahia, A. Carrillo-Lopez and G. Gonzalez-Aguilar, *J. Agric. Food Chem.*, 2008, **56**, 10498–10504.
13. M. M. B. Almeida, P. H. M. de Sousa, Â. M. C. Arriaga, G. M. do Prado, C. E. de C. Magalhães, G. A. Maia and T. L. G. de Lemos, *Food Res. Int.*, 2011, **44**, 2155–2159.
14. L. L. Aguiar, *Açaí. Companhia Nacional de Abastecimento*, 2016.
15. S. A. de Assis, P. F. Fernandes, A. B. Geraldo Martins and O. M. M. De Faria Oliveira, *Fruits*, 2008, **63**, 93–101.
16. K. Yuyama, *Rev. Bras. Frutic.*, 2011, **33**, 335–690.
17. J. L. Ble-Castillo, M. A. Aparicio-Trápala, M. U. Francisco-Luria, R. Córdova-Uscanga, A. Rodríguez-Hernández, J. D. Méndez and J. C. Díaz-Zagoya, *Int. J. Environ. Res. Public Health*, 2010, **7**, 1953–1962.
18. S. Wichienchot, M. Jatupornpipat and R. A. Rastall, *Food Chem.*, 2010, **120**, 850–857.
19. F. A. Hoffmann Sardá, E. B. Giuntini, M. L. P. A. Gomez, M. C. Y. Lui, J. A. E. Negrini, C. C. Tadini, F. M. Lajolo and E. W. Menezes, *J. Funct. Foods*, 2016, **24**, 63–74.
20. D. G. Rodríguez-Sánchez, M. Flores-García, C. Silva-Platas, S. Rizzo, G. Torre-Amione, A. De la Peña-Diaz, C. Hernández-Brenes and G. García-Rivas, *Food Funct.*, 2015, **6**, 192–202.
21. A. Crozier, T. Yokota, I. Jaganath, S. Marks, M. Saltmarsh and M. Clifford, in *Plant Secondary Metabolites: Occurrence, Structure and Role in the Human Diet*, 2006, pp. 208–302.
22. C. Atkinson, R. Nestby, Y. Ford and P. Dodds, *Biofactors*, 2005, **23**, 229–234.
23. J. Pérez-Jiménez and F. Saura-Calixto, *Food Res. Int.*, 2015, **74**, 315–323.
24. M. A. Tremocoldi, E. R. Daiuto, R. L. Vieites and S. M. de Alencar, *Nat. Prod. J.*, 2012, **2**, 164–170.
25. P. Brat, S. Georgé, A. Bellamy, L. Du Chaffaut, A. Scalbert, L. Mennen, N. Arnault and M. J. Amiot, *J. Nutr.*, 2006, **136**, 2368–2373.
26. S. Poovarodom, R. Haruenkit, S. Vearasilp, J. Namiesnik, M. Cvikrová, O. Martincová, A. Ezra, M. Suhaj, P. Ruamsuke and S. Gorinstein, *Int. J. Food Sci. Technol.*, 2010, **45**, 921–929.
27. U. Arukwe, B. A. Amadi, M. K. C. Duru, E. N. Agomuo, E. A. Adindu, P. C. Odika, K. C. Lele, L. Egejuru and J. Anudike, *Int. J. Recent Res. Appl. Stud.*, 2012, **11**, 346–349.

28. J. G. Rodríguez-Carpena, D. Morcuende, M. J. Andrade, P. Kylli and M. Est Evez, *J. Agric. Food Chem.*, 2010, **59**, 5625–5635.
29. W. Wang, T. R. Bostic and L. Gu, *Food Chem.*, 2010, **122**, 1193–1198.
30. M. Alothman, R. Bhat and A. A. Karim, *Food Chem.*, 2009, **115**, 785–788.
31. C. Velázquez, A. M. Pérez and L. A. Prado, in *Biotransformation of Agricultural Waste and By-products: The Food, Feed, Fibre, Fuel (4F) Economy*, ed. P. Poltronieri and O. F. D'Urso, 2016, pp. 21–48.
32. K. L. Wolfe, X. Kang, X. He, M. Dong, Q. Zhang and R. H. Liu, *J. Agric. Food Chem.*, 2008, **56**, 8418–8426.
33. J. Pérez-Jiménez, S. Arranz and F. Saura-Calixto, *Food Res. Int.*, 2009, **42**, 1381–1388.
34. Ó. Acosta-Montoya, F. Vaillant, S. Cozzano, C. Mertz, A. M. Pérez and M. V. Castro, *Food Chem.*, 2010, **119**, 1497–1501.
35. C. Mertz, A. L. Gancel, Z. Gunata, P. Alter, C. Dhuique Mayer, F. Vaillant, A. M. Pérez, J. Ruales and P. Brat, *J. Food Compos. Anal.*, 2009, **22**, 381–387.
36. C. R. A. Abreu, G. A. Maia, R. W. Figueiredo, P. H. M. Sousa, R. E. Alves, E. S. de Brito, C. F. H. Moura and M. do S. M. Rufino, *III Int. Symp. Hum. Heal. Eff. Fruits Veg. - FAVHEALTH 2009*, 2009, vol. 841, pp. 451–454.
37. M. S. M. Rufino, J. Pérez-Jiménez, M. Tabernero, R. E. Alves, E. S. Brito and F. Saura-Calixto, *Int. J. Food Sci. Technol.*, 2010, **45**, 2227–2233.
38. D. M. Amaya-Cruz, S. Rodriguez-Gonzalez, I. F. Perez-Ramirez, G. Loarca-Piña, S. Amaya-Llano, M. A. Gallegos-Corona and R. Reynoso-Camacho, *J. Funct. Foods*, 2015, **17**, 93–102.
39. K. Mahattanatawee, J. A. Manthey, G. Luzio, S. T. Talcott, K. Goodner and E. A. Baldwin, *J. Agric. Food Chem.*, 2006, **54**, 7355–7363.
40. R. Martínez, P. Torres, M. A. Meneses, J. G. Figueroa, J. A. Pérez-Álvarez and M. Viuda-Martos, *Food Chem.*, 2012, **135**, 1520–1526.
41. V. Marquina, A. L. Ruíz, J. Rodríguez-Malaver and P. Vit, *Arch. Latinoam. Nutr.*, 2008, **58**, 98–102.
42. W. N. L. Santos, M. C. Silva Sauthier, A. M. P. Santos, D. Andrade Santana, R. S. Almeida Azevedo and J. Cruz Caldas, *Microchem. J.*, 2017, **133**, 583–592.
43. R. Polinati, L. Oliveira and E. Fialho, *Nat. Prod. J.*, 2012, **2**, 196–204.
44. L. Sukeksi and S. Maya, *ARPN J. Eng. Appl. Sci.*, 2016, **11**, 5209–5216.
45. J. A. Manthey and P. Perkins-Veazie, *J. Agric. Food Chem.*, 2009, **57**, 10825–10830.
46. A. Corrales-Bernal, L. A. Urango, B. Rojano and M. E. Maldonado, *Arch. Latinoam. Nutr.*, 2014, **64**, 16–23.
47. S. M. Rocha Ribeiro, J. H. De Queiroz, M. E. Lopes, R. De Queiroz, F. M. Campos and H. M. Pinheiro Santana, *Plant Foods Hum. Nutr.*, 2007, **62**, 13–17.
48. N. A. Abdul Aziz, L. M. Wong, R. Bhat and L. H. Cheng, *J. Sci. Food Agric.*, 2011, **92**, 557–563.
49. G. Velderrain-Rodríguez, A. Quirós-Sauceda, G. Mercado-Mercado, J. F. Ayala-Zavala, H. Astiazarán-García, R. M. Robles-Sánchez, A. Wall-Medrano, S. Sayago-Ayerdi and G. A. González-Aguilar, *Food Sci. Technol.*, 2016, **36**, 188–194.

50. N. Berardini, R. Carle and A. Schieber, *Rapid Commun. Mass Spectrom.*, 2004, **18**, 2208–2216.
51. M. Martuscelli, C. Di Mattia, F. Stagnari, S. Speca, M. Pisante and D. Mastrocola, *J. Sci. Food Agric.*, 2015, **96**, 2715–2722.
52. X. Wu, G. R. Beecher, J. M. Holden, D. B. Haytowitz, S. E. Gebhardt and R. L. Prior, *J. Agric. Food Chem.*, 2004, **52**, 4026–4037.
53. S. V. Menon and T. V. R. Rao, *J. Food Biochem.*, 2014, **38**, 415–423.
54. E. González, F. Vaillant and A. M. Pérez, *J. Nutr. Food Sci.*, 2012, **2**, 1–8.
55. O. Ebun and K. Santosh, *World J. Dairy Food Sci.*, 2011, **6**, 189–194.
56. D. I. M. Amararathna, W. A. P. Weerakkody, T. Madhujith and J. P. Eeswara, *Acta Hortic.*, 2014, **1106**, 67–72.
57. U. Tril, J. Fernández-López, J. Á. P. Álvarez and M. Viuda-Martos, *Ind. Crops Prod.*, 2014, **55**, 155–162.
58. C. G. Sampaio, S. M. Morais, M. S. M. Rufino, R. E. Alves and E. S. Brito, *III Int. Symp. Hum. Heal. Eff. Fruits Veg. - FAVHEALTH 2009*, 2009, vol. 841, pp. 463–466.
59. M. S. M. Rufino, R. E. Alves, E. S. Brito, J. Pérez-Jiménez and F. D. Saura-Calixto, *III Int. Symp. Hum. Heal. Eff. Fruits Veg. - FAVHEALTH 2009*, 2009, vol. 841, pp. 459–462.
60. M. S. M. Rufino, F. Fernandes, R. Alves and E. Brito, *Food Chem.*, 2009, **114**, 693–695.
61. M. C. Souza, R. W. Figueiredo, G. A. Maia, R. E. Alves, E. S. Brito, C. F. H. Moura and M. S. M. Rufino, *III Int. Symp. Hum. Heal. Eff. Fruits Veg. - FAVHEALTH 2009*, 2009.
62. M. S. M. Rufino, J. Pérez-Jiménez, S. Arranz, R. E. Alves, E. S. Brito, M. S. P. Oliveira and F. Saura-Calixto, *Food Res. Int.*, 2011, **44**, 2100–2106.
63. V. M. Moo-Huchin, I. Estrada-Mota, R. Estrada-León, L. Cuevas-Glory, E. Ortiz-Vázquez, M. de L. Vargas, D. Betancur-Ancona and E. Sauri-Duch, *Food Chem.*, 2014, **152**, 508–515.
64. E. M. Yahia, F. Gutiérrez-Orozco and C. Arvizu-de León, *Food Res. Int.*, 2011, **44**, 2175–2181.
65. L. A. Arellano-Gómez, C. Saucedo-Veloz and L. Arévalo-Galarza, *Agrociencia*, 2005, **39**, 173–181.
66. L. C. Wu, H. W. Hsu, Y. C. Chen, C. C. Chiu, Y. I. Lin and J. A. Ho, *Food Chem.*, 2006, **95**, 319–327.
67. M. de L. Vargas y Vargas, E. Sauri Duch, E. Tamayo Canul, S. A. González Novelo and J. Tamayo Cortez, *Rev. Mex. Agronegocios*, 2015, **36**, 1315–1324.
68. E. S. Brito and N. Narain, *Pesqui. Agropecu. Bras.*, 2002, **37**, 567–572.
69. C. M. B. Omena, I. B. Valentim, G. da S. Guedes, L. A. Rabelo, C. M. Mano, E. J. H. Bechara, A. C. H. F. Sawaya, M. T. S. Trevisan, J. G. da Costa, R. C. S. Ferreira, A. E. G. Sant'Ana and M. O. F. Goulart, *Food Res. Int.*, 2012, **49**, 334–344.
70. A. Torres-Rodríguez, Y. Salinas-Moreno, S. Valle-Guadarrama and I. Alia-Tejacal, *Food Res. Int.*, 2011, **44**, 1956–1961.
71. P. M. C. Silva, L. C. Neves, V. J. Bastos, C. G. B. Lima, K. G. M. de Araújo and S. R. Roberto, *Sci. Hortic. (Amsterdam, Neth.)*, 2016, **200**, 66–72.

72. A. Gordon, E. Jungfer, B. Alexandre Silva, J. Guilherme Maia and F. Marx, *J. Agric. Food Chem.*, 2011, **59**, 7688–7699.
73. L. C. Neves, J. M. Tosin, R. M. Benedette and L. Cisneros-Zevallos, *Food Chem.*, 2015, **174**, 188–196.
74. V. R. Souza, P. A. P. Pereira, F. Queiroz, S. V. Borges and J. Deus Souza Carneiro, *Food Chem.*, 2012, **134**, 381–386.
75. E. M. Silva, J. N. S. Souza, H. Rogez, J. F. Rees and Y. Larondelle, *Food Chem.*, 2007, **101**, 1012–1018.
76. J. N. S. Souza, E. M. Silva, A. Loir, J.-F. Rees, H. Rogez and Y. Larondelle, *Food Chem.*, 2008, **106**, 331–339.
77. Ó. Acosta, A. M. Pérez and F. Vaillant, *Arch. Latinoam. Nutr.*, 2009, **59**, 88–94.
78. J. Contreras-Calderón, L. Calderón-Jaimes, E. Guerra-Hernández and B. García-Villanova, *Food Res. Int.*, 2011, **44**, 2047–2053.
79. C. Rojas-Garbanzo, A. M. Pérez, M. L. Pineda Castro and F. Vaillant, *Fruits*, 2012, **67**, 415–427.
80. K. D. R. R. Silva and M. S. F. Sirasa, *Food Chem.*, 2016, 1–6.
81. S. Gallori, A. R. Bilia, M. C. Bergonzi, W. L. R. Barbosa and F. F. Vincieri, *Chromatographia*, 2004, **59**, 739–743.
82. C. Mertz, V. Cheynier, Z. Günata and P. Brat, *J. Agric. Food Chem.*, 2007, **55**, 8616–8624.
83. V. V. de Rosso, S. Hillebrand, E. Cuevas Montilla, F. O. Bobbio, P. Winterhalter and A. Z. Mercadante, *J. Food Compos. Anal.*, 2008, **21**, 291–299.
84. R. Chirinos, J. Galarza, I. Betalleluz-Pallardel, P. Romina and D. Campos, *Food Chem.*, 2010, **120**, 1019–1024.
85. T. Kaneshima, T. Myoda, K. Toeda, T. Fujimori and M. Nishizawa, *Food Sci. Technol. Res.*, 2013, **19**, 223–228.
86. S. Someya, Y. Yoshiki and K. Okubo, *Food Chem.*, 2002, **79**, 351–354.
87. K. Kanazawa and H. Sakakibara, *J. Agric. Food Chem.*, 2000, **48**, 844–848.
88. M. de L. García-Magaña, H. S. García, L. A. Bello-Pérez, S. G. Sáyago-Ayerdi and M. M.-M. de Oca, *Plant Foods Hum. Nutr.*, 2013, **68**, 254–258.
89. P. Begoña Cerdá, J. Espín and F. Tomás-Barberán, *J. Agric. Food Chem.*, 2005, **53**, 5571–5576.
90. D. Del Rio, J. P. E. Rodriguez-Mateos, A. Spencer, M. Tognolini, G. Borges and A. Crozier, *Antioxid. Redox Signaling*, 2013, **18**, 1818–1892.
91. J. Pérez-Jiménez, M. E. Díaz-Rubio and F. Saura-Calixto, *Nutr. Res. Rev.*, 2013, **26**, 118–129.
92. U. S. Mahadeva Rao and B. Adinew, *Afr. J. Tradit., Complementary Altern. Med.*, 2011, **8**, 210–217.
93. G. Oboh, A. T. Isaac, A. J. Akinyemi and R. A. Ajani, *Int. J. Biomed. Sci.*, 2014, **10**, 208–216.
94. B. S. Nayak, S. S. Raju and A. V. Chalapathi Rao, *J. Wound Care*, 2008, **17**, 123–125.
95. M. Wien, E. Haddad, K. Oda and J. Sabaté, *Nutr. J.*, 2013, **12**, 1–9.
96. E. M. Pérez-Pérez, A. J. Rodríguez-Malaver, N. Padilla, G. Medina-Ramírez and J. Dávila, *J. Med. Food*, 2006, **9**, 517–523.

97. G. Azofeifa, S. Quesada and A. M. Pérez, *Rev. Bras. Farmacogn.*, 2011, **21**, 829–834.
98. C. S. Bowen-Forbes, Y. Zhang and M. G. Nair, *J. Food Compos. Anal.*, 2010, **23**, 554–560.
99. L. Calvo-Castro, D. N. Syed, J. C. Chamcheu, F. M. P. Vilela, A. M. Pérez, F. Vaillant, M. Rojas and H. Mukhtar, *Photochem. Photobiol.*, 2013, **89**, 1199–1207.
100. N. M. A. Hassimotto, M. D. S. Pinto and F. M. Lajolo, *J. Agric. Food Chem.*, 2008, **56**, 11727–11733.
101. E. O. Cuevas-Rodríguez, V. P. Dia, G. G. Yousef, P. A. García-Saucedo, J. López-Medina, O. Paredes-López, E. González de Mejía and M. A. Lila, *J. Agric. Food Chem.*, 2010, **58**, 9542–9548.
102. A. Rossi, I. Serraino, P. Dugo, R. Di Paola, L. Mondello, T. Genovese, D. Morabito, G. Dugo, L. Sautebin, A. P. Caputi and S. Cuzzocrea, *Free Radical Res.*, 2003, **37**, 891–900.
103. G. Azofeifa, S. Quesada, F. Boudard, M. Morena, J.-P. Cristol, A. M. Pérez, F. Vaillant and A. Michel, *J. Agric. Food Chem.*, 2013, **61**, 5798–5804.
104. G. Azofeifa, S. Quesada, L. Navarro, O. Hidalgo, K. Portet, A. M. Pérez, F. Vaillant, P. Poucheret and A. Michel, *J. Funct. Foods*, 2016, **26**, 330–337.
105. N. M. Hassimotto and F. M. Lajolo, *J. Sci. Food Agric.*, 2011, **91**, 523–531.
106. G. Azofeifa, S. Quesada, A. M. Pérez, F. Vaillant and A. Michel, *J. Food Compos. Anal.*, 2015, **42**, 56–62.
107. P. Bontempo, A. Doto, M. Miceli, L. Mita, R. Benedetti, A. Nebbioso, M. Veglione, D. Rigano, M. Cioffi, V. Sica, A. M. Molinari and L. Altucci, *Cell Proliferation*, 2012, **45**, 22–31.
108. M. Gondi, S. A. Basha, P. V. Salimath and U. J. S. P. Rao, *J. Food Biochem.*, 2017, **41**, 1–13.
109. G. F. Perpetuo and J. M. Salgado, *Plant Foods Hum. Nutr.*, 2004, **58**, 1–12.
110. H. Kim, J. Y. Moon, H. Kim, D.-S. Lee, M. Cho, H.-K. Choi, Y. S. Kim, A. Mosaddik and S. K. Cho, *Food Chem.*, 2010, **121**, 429–436.
111. A. Matkowski, P. Kus, E. Goralska and D. Wozniak, *Mini-Rev. Med. Chem.*, 2013, **13**, 439–455.
112. I. Vouldoukis, D. Lacan, C. Kamate, P. Coste, A. Calenda, D. Mazier, M. Conti and B. Dugas, *J. Ethnopharmacol.*, 2004, **94**, 67–75.
113. O. I. Aruoma, Y. Hayashi, F. Marotta, P. Mantello, E. Rachmilewitz and L. Montagnier, *Toxicology*, 2010, **278**, 6–16.
114. K. M. Sadek, *Acta Inform. Med.*, 2012, **20**, 180–185.
115. A. C. Ezike, P. A. Akah, C. O. Okoli, N. A. Ezeuchenne and S. Ezeugwu, *J. Med. Food*, 2009, **12**, 1268–1273.
116. M. Mhatre, J. Tilak-Jain, S. De and T. P. A. Devasagayam, *Food Chem. Toxicol.*, 2009, **47**, 2696–2702.
117. R. Horta, V. Francilia, S. Kahl, M. Da, S. Sarmento, M. Fernanda, S. Nunes, C. Rejane, M. Porto, V. Moraes De Andrade, A. De Barros, F. Ferraz and J. Da Silva, *Genet. Mol. Biol.*, 2016, **39**, 122–128.
118. J. Kang, K. M. Thakali, C. Xie, M. Kondo, Y. Tong, B. Ou, G. Jensen, M. B. Medina, A. G. Schauss and X. Wu, *Food Chem.*, 2012, **133**, 671–677.

119. T. Inoue, H. Komoda, T. Uchida and K. Node, *J. Cardiol.*, 2008, **52**, 127–132.
120. P. R. B. Broinizi, E. R. S. de Andrade-Wartha, A. M. de O. e Silva, R. P. Torres, H. M. C. Azeredo, R. E. Alves and J. Mancini-Filho, *Rev. Bras. Cienc. Farm.*, 2008, **44**, 773–781.
121. D. Barreca, G. Laganá, S. Ficarra, E. Tellone, U. Leuzzi, A. Galtieri and E. Bellocco, *Food Res. Int.*, 2011, **44**, 2302–2310.
122. S. M. Barbalho, P. C. S. Bueno, D. S. Delazari, E. L. Guiguer, D. P. Coqueiro, A. C. Araújo, M. S. S. Souza, F. M. Farinazzi-Machado, C. G. Mendes and M. Groppo, *J. Med. Food*, 2015, **18**, 385–391.
123. R. Sakdawattanakul, P. Panapisal and A. Tansirikongkol, *Thai J. Pharm. Sci.*, 2016, **40**, 108–111.
124. J. Ma, X. D. Luo, P. Protiva, H. Yang, C. Ma, M. J. Basile, I. B. Weinstein and E. J. Kennelly, *J. Nat. Prod.*, 2003, **66**, 983–986.
125. S. Das, S. Das and B. De, *Curr. Nutr. Food Sci.*, 2012, **8**, 19–24.
126. M. Srivastava, M. Hegde, K. K. Chiruvella, J. Koroth, S. Bhattacharya, B. Choudhary and S. C. Raghavan, *Sci. Rep.*, 2014, **4**, 1–9.
127. G. Flores, K. Dastmalchi, S. B. Wu, K. Whalen, A. J. Dabo, K. A. Reynertson, R. F. Foronjy, J. M. D'Armiento and E. J. Kennelly, *Food Chem.*, 2013, **141**, 889–895.
128. L. R. B. Mariutti, E. Rodrigues, R. C. Chisté, E. Fernandes and A. Z. Mercadante, *Food Res. Int.*, 2014, **64**, 618–625.
129. A. Muñiz and R. Pérez, *WIT Trans. Ecol. Environ.*, 2013, **170**, 241–251.
130. I. O. Ishola, O. Awodele, A. M. Olusayero and C. O. Ochieng, *J. Med. Food*, 2014, **17**, 1375–1382.
131. R. Vinothini and L. Growther, *Int. J. Curr. Microbiol. Appl. Sci.*, 2016, **5**, 617–625.
132. E. Sutrisna, D. Usdiana, R. Maidina Taqwin and A. Roni Rosyidi, *Natl. J. Physiol., Pharm. Pharmacol.*, 2015, **5**, 285–290.
133. S. Dey, D. Swarup, A. Saxena and A. Dan, *Res. Vet. Sci.*, 2011, **91**, 422–425.
134. F. Martinello, S. M. Soares, J. J. Franco, A. C. Santos, A. Sugohara, S. B. Garcia, C. Curti and S. A. Uyemura, *Food Chem. Toxicol.*, 2006, **44**, 810–818.
135. V. Jindal, D. Dhingra, S. Sharma, M. Parle and R. K. Harna, *J. Pharmacol. Pharmacother.*, 2011, **2**, 80–84.
136. K. Fairuz Azman, Z. Amom, A. Azlan, N. Mohd Esa, R. Mat Ali, Z. Md Shah and K. K. Abdul Kadir, *J. Nat. Med.*, 2012, **66**, 333–342.
137. C. Yin Lim, S. Mat Junit, M. Ameen Abdulla, A. Abdul Aziz and B. Smith, *PLoS One*, 2013, **8**, 1–12.
138. M. Amir, M. A. Khan, S. Ahmad, M. Akhtar, M. Mujeeb, A. Ahmad, S. A. Khan and F. A. Al-Abbasi, *Nat. Prod. Res.*, 2015, **30**, 715–719.
139. W. K. MacNaughton and J. L. Wallace, *Gastroenterology*, 1989, **96**, 972–980.
140. L. Fahrasmane, B. Parfait and G. Aurore, *Acta Hortic.*, 2014, 75–82.
141. L. R. Howard and T. J. Hager, in *Berry Fruit: Value-added Products for Health Promotion*, CRC Press, 2007, pp. 73–104.

142. C. García-Muñoz, L. Hernández, A. Pérez and F. Vaillant, *Food Res. Int.*, 2014, **55**, 161–169.
143. J. M. Landete, *Food Res. Int.*, 2011, **44**, 1150–1160.
144. C. García-Muñoz and F. Vaillant, *Crit. Rev. Food Sci. Nutr.*, 2014, **54**, 1584–1598.
145. S. Rodríguez-González, I. M. Gutiérrez-Ruíz, I. F. Pérez-Ramírez, O. Mora, M. Ramos-Gomez and R. Reynoso-Camacho, *J. Funct. Foods*, 2017, **37**, 190–199.
146. E. S. de Brito, M. C. P. de Araújo, R. E. Alves, C. Carkeet, B. A. Clevidence and J. A. Novotny, *J. Agric. Food Chem.*, 2007, **55**, 9389–9394.

CHAPTER 7

Fruit and Vegetable Processing: Effects on Extractable and Non-extractable Polyphenols

MARÍA GUADALUPE HERRERA HERNÁNDEZ[a], RITA MIRANDA LÓPEZ[b], FRANCISCO JAVIER ESPARZA MARTÍNEZ[c] AND SALVADOR HORACIO GUZMÁN MALDONADO*[a]

[a]Functional Food Laboratory, Campo Experimental Bajío (INIFAP), Km, 6.5 Carr, Celaya-San Miguel Allende s/n, C.P. 39110, Celaya, Guanajuato, Mexico; [b]Biochemistry Engineering Department, Celaya Institute of Technology, Antonio García Cubas Pte, 600, Fovissste, Celaya, Guanajuato, Mexico; [c]Facultad de Ciencias de la Salud e Ingenierías, Universidad de Guanajuato, Mutualismo 303, Zona Centro, C.P. 38060, Celaya, Guanajuato, Mexico

*E-mail: herrera.guadalupe@inifap.gob.mx, guzman.horacio@inifap.gob.mx, rita.miranda@itcelaya.edu.mx, fj.esparza@ugto.mx

7.1 Introduction

In recent years, there has been an increase in the consumption of fruit and vegetables thanks to reports that associate them with different benefits for human health. Several of these effects have been attributed to the presence of polyphenolic compounds, which are the most abundant bioactive antioxidant compounds in the diet.

Food Chemistry, Function and Analysis No. 5
Non-extractable Polyphenols and Carotenoids: Importance in Human Nutrition and Health
Edited by Fulgencio Saura-Calixto and Jara Pérez-Jiménez

Published by the Royal Society of Chemistry, www.rsc.org

Dietary polyphenols represent a group of secondary metabolites with over 8000 structural variants, which occur widely in different fruits, vegetables, wine, tea, olive oil and chocolate and other cocoa products. For many years, it has been recognised that polyphenols exhibit many biologically relevant functions, especially their powerful free radical scavenging, which have been associated with specific structural features, such as the number and position of hydroxyl and catechol units.[1] Due to the protection afforded by polyphenols against oxidative stress, they can potentially have beneficial effects when used to treat pathological conditions, including age-related and degenerative diseases such as cardiovascular diseases and cancer. Polyphenols may also offer indirect protection by activating endogenous defence systems or by modulating cellular signalling processes. The effective extraction and proper assessment of polyphenols from food and medicinal plants are crucial to exploring their potential use as antioxidant sources and promoting their application in functional foods, pharmaceuticals and food additives.[2]

Phenolic compounds in fruit and vegetables interact with different macromolecules such as carbohydrates, lipids and proteins and can be found in free or conjugated forms. Several studies indicate that dietary fibre (DF) interacts directly with the phenolic compounds present in foods and this may interfere with their proper assimilation. Hence, it is important to know how the phenolic content is affected by the different processes to which fruit and vegetables are subjected. In addition, in the processing of fruit and some vegetables, the waste that is generated is potentially an important source of polyphenols and DF, which could be used in the elaboration of other products or nutritional supplements.

After this introduction, we continue this chapter with a brief historical description of the consumption of fruit and vegetables as a source of nutrients and bioactive compounds, as well as their processing to obtain food products that can be easily prepared and served by the consumer. We also describe the potential of industrial waste generated during the processing of fruit and vegetables as a source of novel products. We then mention some important aspects of the chemical composition of fruit and vegetables, focusing mainly on the DF–phenolic compounds relationship and describing the properties of extractable and non-extractable phenolic compounds. To conclude, different examples of the effects of various processes on these compounds are provided. Finally, several uses of antioxidant fibre from food processing are mentioned.

7.2 Food Processing to Produce Marketable and Healthy Food Products

Food processing is the transformation of agricultural products, such as fruit and vegetables, by physical or chemical means into other forms. It combines agricultural products to produce marketable food products that can be easily prepared and served by the consumer. Food processing typically involves activities such as mincing and macerating, liquefaction, emulsification and cooking by boiling, broiling, frying or grilling. Pickling, pasteurisation and many other kinds of preservation are also used.[3]

In primary processing, common practices are the use of dicing or slicing, drying and freezing, before subjecting fruit and vegetables to secondary processing to generate end products.[4] All of these processes have a variety of purposes, including preservation of food, sanitation and enabling the handling of large volumes of agricultural products. They are also used to enhance the texture, flavour and appearance of foods, among many other purposes. In spite of these benefits, certain processing techniques also have drawbacks that manifest themselves in processed foods. A classic example is the loss of vitamins from fruit and vegetables when they are cooked or dried with heat, and even when they are sundried.[5,6]

7.3 The Use of Industrial Fruit and Vegetable Waste for Novel Products

In recent years, the consumption of fruit and vegetables has increased as knowledge of their nutritive value and potential use as nutraceuticals has expanded. The estimated current consumption of raw fruit and vegetables varies widely from 13 g day^{-1} in Nicaragua to 120 g day^{-1} in Colombia, and from 16 g day^{-1} in Nicaragua to 285 g day^{-1} in Belgium for fruits and vegetables respectively.[7] Industry is probably the largest "consumer" of fruit and vegetables, using them for the production of processed foods such as juices, soups and jellies, or ingredients such as pectin. All this consumption generates large amounts of waste that contribute to environmental pollution, especially in developing countries. Roughly a third of the edible parts of agricultural products used to produce processed food for human consumption is lost or wasted globally, representing some 1300 million tons per year, of which 782 million tons are produced by seven regions in the world (Table 7.1). As a specific example

Table 7.1 Production of fruit and vegetables, percentages lost or wasted and calculated waste production in different regions.[a]

Region	Fruits and vegetable production (million tonnes)	Lost or wasted (%)	Waste production (calculated)
Europe	205	46	94.3
North America, Oceania	190	52	98.8
Industrialised Asia	635	36	228.6
Sub-Saharan Africa	195	52	101.4
North Africa West and Central Asia	145	56	81.2
South and Southeast Asia	305	52	75.4
Latin America	190	55	104.5

[a]Adapted from ref. 9.

of waste production, it is estimated that the citrus fruit juice industry generates 15 million tons of waste per year.[8] In medium- and high-income countries, food is wasted to a great extent: it is thrown away even if it is still suitable for human consumption. Significant food loss and waste do, however, also occur early in the food supply chain. In low-income countries, food is mainly lost during the early and middle stages of the food supply chain; much less food is wasted at the consumer level.[9]

Fruit and vegetable waste can be used to generate products that may be of great interest to the food, pharmaceutical or cosmetics industries. However, knowledge of the chemical composition of fruit and vegetables, as well as of the novel products derived from them and their potential effects on human health, is necessary to make better use of them.

7.4 Chemical Composition of Fruit and Vegetables

Fruit and vegetables are rich in water and usually have low fat and low protein contents. However, an important component of much fruit and many vegetables is DF. They also contain numerous bioactive components (phytochemicals) that have beneficial effects on human health (Table 7.2).[10] These bioactive compounds are substances from various chemical families that have widely varied structures and properties, such as polyphenols (which include anthocyanin pigments, tannins, flavonols, isoflavones, lignans and stilbenoids), aromatic acid derivatives, glucosinolates, sulphur derivatives from Alliaceae, terpenoids and carotenoids, among many others.[11]

Another component of fruit and vegetables for which health benefits have also been reported is DF. DF acts as a normalising agent of bowel movements and prevents cardiovascular diseases. The occurrence of non-infectious diseases such as coronary diseases has been related to low DF intake. In addition, a relationship has also been reported between DF intake and the risk of colon cancer, as well as reduced blood serum cholesterol levels and the prevention of hyperglycaemia in diabetic patients.[16] Because of these beneficial effects, in recent years, various products have been developed with a high-DF content.

7.4.1 DF and Polyphenol Interaction

DF and antioxidants are generally addressed separately as two distinct groups of food constituents in both chemical and nutrition studies. However, a substantial proportion of the antioxidants contained in fruit and vegetables are linked to DF,[17] and some of the postulated benefits of DF intake can be attributed to these associated compounds. These compounds, the so-called non-extractable polyphenols (NEPP), are not bioaccessible in the human small intestine, but they may exert significant health effects when they reach the colon.[18]

Numerous studies have reported interactions between phenolic compounds and carbohydrates, showing that such interactions play an important

Table 7.2 Chemical composition (%) and phytochemical content of various different fruits and vegetables used in the juice industry.[a]

	Water	Protein	Total lipid	Dietary fibre	Ash	Carbohydrate	Phytochemicals
Apple	85.56	0.26	0.17	2.40	0.30	11.31	Polyphenols, flavonols, catechins, procyanidins, folates
Grape	80.5	0.72	0.16	0.9	0.48	17.7	Polyphenols, resveratrol, anthocyanins, catechins, procyanidins, caffeic acid derivatives, flavonols
Lime	88.26	0.70	0.20	2.80	–	8.04	Phenolic acids, flavonoids, flavanones
Mandarin	85.17	0.81	0.31	1.80	0.38	–	Phenolic acids, flavonoids, flavanones
Mango	83.46	0.82	0.38	1.60	0.5	–	Polyphenols, carotenoids
Orange	87.14	0.70	0.21	2.40	0.5	–	Flavonols, cryptoxanthin, terpenes (limonene), folates
Pineapple	87.24	0.55	0.13	1.4	–	–	
Celery	95.43	0.69	0.17	1.6	–	–	Flavonones, apigenin and luteolin derivatives
Carrot	88.29	0.93	0.24	2.8	0.71	–	Carotenoids (α- and β-carotene), cumarines, caffeic acid derivatives, folates
Tomato	94.52	0.88	0.20	1.20	–	–	Lycopene, caffeic acid derivatives, chalcones, flavonols

[a]From ref. 11–15.

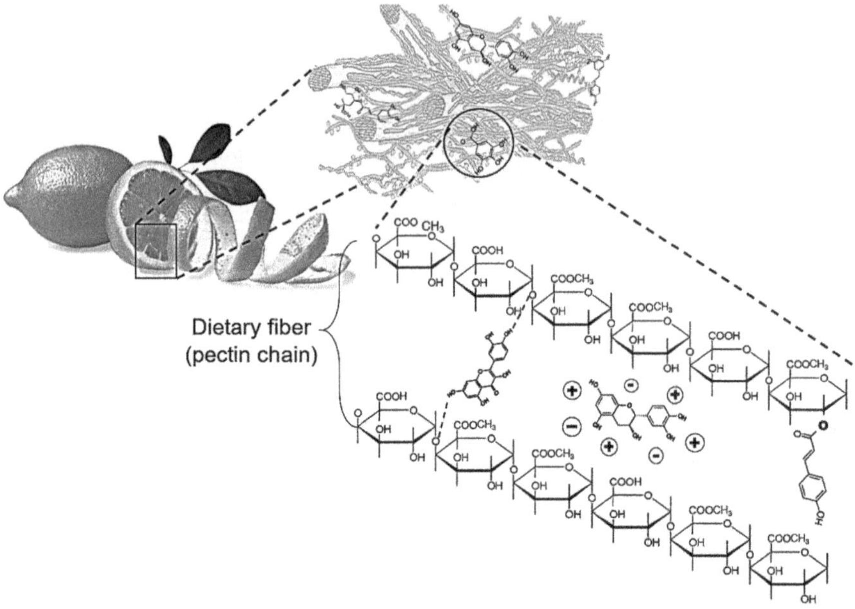

Figure 7.1 Schematic representation of entrapped phenolic compounds in a fruit matrix and their interactions with DF. (Adapted from ref. 20).

role in the human body. In particular, the interaction of phenolic compounds with DF has been highlighted.[19] Within the fruit matrix, phenolic compounds can be found in their free form or associated with the cell wall. It has been reported that simple phenolic compounds such as benzoic acid or benzaldehydes and their derivatives are generally covalently attached to polysaccharides in the cell wall, forming ester bonds. In addition to covalent bonds, there are also other types of interactions between polyphenols and DF, such as the formation of hydrogen bonds (Figure 7.1).

Given the large number of studies carried out on the association of antioxidants with DF and the related health benefits, the term "antioxidant DF" appeared, considering DF to be antioxidant DF if 1 g of it has an antioxidant capacity against free radicals of 50–100 mg equivalent of vitamin E and a DF content of over 50% of dry matter from the natural constituents of the material.[21] These materials represent an important amount of NEPP. Therefore, they can be considered NEPP-rich DF or NEPP-rich materials. As mentioned, the antioxidant compounds attached to DF do not dissolve and are not absorbed in the small intestine; they arrive unaltered in the large intestine, where they are degraded to a greater or lesser degree during colonic fermentation together with polysaccharides and other constituents of the DF (Figure 7.2). While antioxidant vitamins are absorbed in the small intestine, most polyphenols and a significant fraction of carotenoids carried by DF are released into the colon together with carbohydrates and DF, producing bioavailable metabolites,[17] creating an antioxidant environment (eliminating free radicals) and balancing the pro-oxidant effects of the diet.[22]

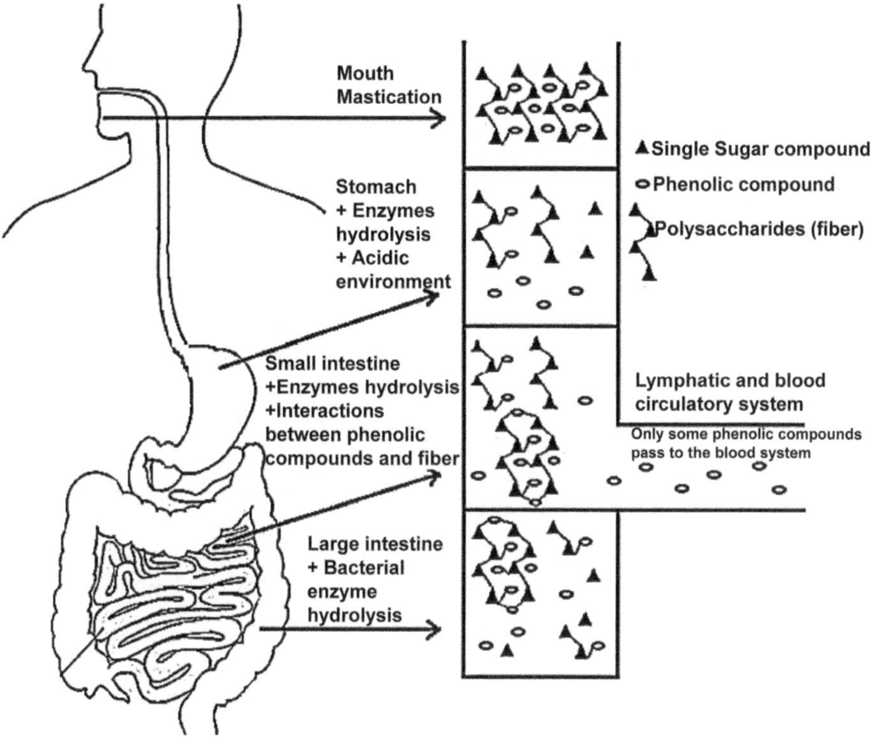

Figure 7.2 General human bioabsorption of phenolic compounds from foods also containing DF, such as fruit. (Adapted from ref. 23).

The polyphenols associated with DF can be classified into extractable polyphenols (EPP) and NEPP, with this classification referring to the process by which these compounds can be detached. EPP are those that can be detached from the DF with a solvent, while NEPP are those that require a more powerful extraction method (using acids, hydroxides, enzymes and heat) to detach them from the food matrix.[24] In different fruits, the content of these fractions varies depending on the type of fruit (Table 7.3) and the nature of the compounds constituting both the DF and the phenolic compounds.

7.5 Fruit and Vegetables as Sources of NEPP-rich Materials and Effects of Processing

7.5.1 Fruit

Some types of fruit are suitable sources from which to prepare NEPP-rich materials. This is the case of some tropical fruits, and also of *Fucus* algae and DF obtained from by-products of the wine[32] and juice industries. It has similarly been found that some fruit by-products are excellent sources of DF

Table 7.3 Extractable and non-extractable phenolic contents of some different fruits.

	Total phenolic content (mg of GAE/g of DW)[a]		
Fruit	Extractable	Non-extractable	Ref.
Jujube	1.26 ± 0.09	2.89 ± 0.10	25
Leaves of blueberry	2.26 ± 0.34	3.73 ± 0.56	26
Pigmented maize	7.65 ± 2.21	223.09 ± 27.85	27
Blueberry	117.58 ± 7.17	23.15 ± 2.04	28
Green apple	9.18 ± 1.79	5.70 ± 0.90	29
Red apple	8.04 ± 0.27	4.47 ± 1.30	29
Orange peel	28.51 ± 1.89	157.01 ± 23.8	30
Lime peel	21.61 ± 0.53	21.15 ± 0.74	14
Mandarin peel	18.26 ± 0.29	45.15 ± 0.81	15
Mango	144.34 ± 1.37	129.95 ± 8.41	31
Papaya	93.18 ± 8.66	118.98 ± 3.92	31
Pineapple	56.83 ± 0.60	50.79 ± 0.65	31

[a]GAE = gallic acid equivalent, DW=dry weight

Table 7.4 Chemical compositions and antioxidant capacities of mandarin and lime waste.

Composition	Mandarin[15]	Lime[14]
Soluble dietary fibre	8.84 ± 0.09	20.15 ± 0.36
Insoluble dietary fibre	34.73 ± 1.32	46.73 ± 0.32
Total dietary fibre	43.57 ± 1.42	66.88 ± 0.98
Total phenols		
Extractable	18.26 ± 0.29	21.61 ± 0.53
Non-extractable	45.15 ± 0.81	21.15 ± 0.74
Antioxidant capacity		
ABTS		
Extractable	342.21 ± 2.38	569.04 ± 4.37
Non-extractable	427.31 ± 12.62	839.69 ± 19.41
ORAC		
Extractable	261.83 ± 15.26	535.64 ± 5.267
Non-extractable	639.76 ± 14.75	960.68 ± 22.96

with high antioxidant capacities.[33] These products can be used as ingredients in food, since they can prevent lipid oxidation in meat and fish products, thereby maintaining nutritional quality and prolonging shelf life.[34] An example of a fruit by-product is the waste generated in the citrus juice industry, where only 50% of total fruit is used,[35] leaving 15×10^6 tons of waste per year[36] and causing a serious environmental problem.

Previously, characterisation of citrus juice and waste focused only on oranges, lemons and grapefruit. However, limes and mandarins have recently attracted attention due to the increase in consumption of these juices and the consequent increase in the waste derived from these fruits. Considering the DF content, as well as the high levels of EPP and NEPP, and the antioxidant capacity of mandarin[15] and lime[14] waste from the juice industry, it could be considered as antioxidant DF (Table 7.4).

The mandarin showed more total DF[15] than orange, lemon or mandarin peel.[37] It also showed a high Trolox equivalent antioxidant capacity, being higher in NEPP, using both the 2,2′-azino-bis(3-ethylbenzothiazoline-6-sulfonic acid (ABTS) and the oxygen radical antioxidant capacity (ORAC) assays. Lime waste has a higher total DF content as well as a higher antioxidant capacity than mandarin. However, in comparison to mandarin, lime waste presents less non-extractable total phenols.

To use this type of waste, it is necessary to pre-process it according to the use that will be made of it. The characteristics of the raw material (waste) and the type of process both affect the chemical composition of the resulting product and can improve its sensory attributes and shelf life, while lowering or enhancing its nutritional or nutraceutical value.[38–41] Processing treatments (chemical, enzymatic, mechanical/grinding, thermal or thermo-mechanical/extrusion and controlled instantaneous decompression) directly or indirectly change the physicochemical properties of DF, potentially improving its functionality.[42] Generally, for this type of waste, drying is the most commonly used process, after which the residue is ground to yield flour, which is used in food products such as pasta, soups or baked goods to improve their shelf life and sensory or nutritional characteristics.

In studies carried out at different drying temperatures on mandarin waste from the juice industry, the amount of non-extractable total phenols and flavonoids was found to be higher as the drying temperature increased[15] (Table 7.5). The results reported by Esparza *et al.* agree with those reported by Seok-Moon *et al.*[43] and Chen *et al.*,[44] who also found a significant increase in these compounds when satsuma and orange waste was dried at 150 °C compared to at 100 °C. This increase may be due to the availability of precursors of phenolic compounds by non-enzymatic inter-conversion between phenolic molecules.[45] Also, an increment of total phenolic compounds was found when grapefruit peel was dried at 150 °C compared to at 90 °C.

In the case of lime waste,[14] a higher soluble DF content was observed in the samples dried at 90 °C and 120 °C than in those dried at 60 °C. The opposite effect was observed for the insoluble DF. As in the case of mandarin,[15] the contents of extractable and non-extractable total phenols and non-extractable flavonoids were all higher in the mandarin residues dried at 90 °C and 120 °C than in those dried at 60 °C. In the case of extractable flavonoids, only the contents in the samples dried at 120 °C were higher.

It was further observed that, when using the ABTS technique, the antioxidant capacities of the extractable and non-extractable total phenols, which were also evaluated during this drying process, were higher as the drying temperature increased. This was the case for both mandarin and lime residues. In the case of the antioxidant capacity determined by the ORAC method, the opposite effect was observed. In agreement with the chemical principles of these techniques, this indicates that at a higher drying temperatures, the reducing capacity of the NEPP-rich material increases, while its capacity to donate hydrogen atoms decreases.

Table 7.5 Effect of drying on chemical compositions and antioxidant capacities of mandarin and lime waste.

	Mandarin[15]			Lime[14]		
Composition	60 °C	90 °C	120 °C	60 °C	90 °C	120 °C
Soluble dietary fibre	8.5 ± 0.11	7.8 ± 0.35	7.09 ± 0.59	10.1 ± 1.23	12.5 ± 1.13	14.79 ± 1.0
Insoluble dietary fibre	35.6 ± 2.45	35.8 ± 2.84	35.9 ± 3.23	45.8 ± 2.7	44.0 ± 2.7	42.33 ± 2.7
Total dietary fibre	44.0 ± 2.53	43.55 ± 2.81	43.0 ± 3.08	55.9 ± 1.1	56.5 ± 2.0	57.1 ± 2.8
Total phenols						
Extractable	17.34 ± 1.0	19.48 ± 0.62	21.69 ± 0.9	15.67 ± 0.37	17.41 ± 0.88	26.47 ± 2.0
Non-extractable	47.47 ± 1.0	49.0 ± 2.2	74.5 ± 0.7	21.59 ± 1.76	20.36 ± 1.12	36.81 ± 0.88
Flavonoids						
Extractable	10.2 ± 0.36	11.0 ± 0.56	12.87 ± 1.0	8.67 ± 0.67	8.84 ± 0.58	11.69 ± 0.39
Non-extractable	11.96 ± 0.23	15.68 ± 0.16	21.56 ± 0.24	14.98 ± 0.41	19.37 ± 1.18	19.77 ± 0.57
Antioxidant capacity						
ABTS						
Extractable	310.4 ± 15	447.0 ± 14	574.7 ± 19	540.5 ± 19	554.2 ± 32	678.0 ± 5
Non-extractable	307.5 ± 19	364.7 ± 16	549.5 ± 19	570.8 ± 41	770.1 ± 70	883.6 ± 79
ORAC						
Extractable	296.0 ± 14	240.67 ± 17.53	196.0 ± 8	223.5 ± 10	253.6 ± 10	201 ± 3
Non-extractable	670.3 ± 5.05	619.1 ± 16	589.6 ± 14	544.8 ± 10	529.0 ± 34	468.9 ± 27

The drying process, which is the most commonly used in the food industry, is acceptable for citrus fruits with a high phenolic content, since these compounds become more available than in the non-processed fruit. Increasing the drying temperature also makes the concentration of phenolic compounds greater, since the physicochemical interactions between them and DF make them even more available.[46]

7.5.2 Vegetables

Recently, researchers have studied applications of processed vegetable by-products, especially because of their contents of bioactive compounds, in order to reduce the waste produced and thus the potential environmental pollution.

Usually, vegetables are subjected to different process, physically by cutting or slicing them, and different heat treatments, either dry or moist, including blanching, boiling, baking, frying, drying, *etc.* All these operations are known to degrade or increase the quantities of different phytochemicals.[47,48]

Perhaps one of the processes that affects the phytochemical content of vegetables most is blanching through heat treatment in water. One of the purposes of blanching is to inactivate enzymes, so that oxidative ones will not oxidise the antioxidants present in the commodity. The hot water will decrease the amount of water-soluble and heat-sensitive components in the tissue.

It is well known that from 20% to 25% of vitamin C is lost during the blanching process,[49] due to leaching from the tissue into the cooking water. Carotenoids and anthocyanins are decreased as the blanching temperature and time increase.

It is common for vegetables to be blanched before any further treatment. The effects of the blanching step (99 °C for 2 minutes and 15 seconds) on the antioxidant capacities of kale, spinach and swamp cabbage have been studied, and they were reduced significantly; in fact, these effects were observed even after just 1 minute of blanching.[50]

After blanching (90 °C for 1 minute) peel obtained as a by-product from preparing ready-to-eat carrots, the β-carotene, total phenolic and *in vitro* potential antioxidant activity were all decreased, while pectin and total DF were increased. Thus, carrot peel could be used to produce antioxidant DF powder.[51]

During heat treatment, the decreases in the different nutrients and phytochemicals in vegetables depend on the temperature and time.

The content of phenolic compounds varies during processing, storage and cooking depending on the type of commodity. Thermal treatment (blanching, baking or canning) seems to increase the extractability of phenolics. Degradation of total phenolics may occur during fresh and frozen storage because they are water soluble or sensitive to oxidation. In stored canned samples, total phenolic decreases may be because these compounds leach out from the tissue into the canning medium.[38]

Meanwhile, since carotenoids are lipid soluble, they do not leach out into water media during processing; however, they can be degraded depending on temperature, light, acidity and the presence of oxygen. Carotenoids and lycopene seem to be somewhat stable during processing, apart from some oxidation and isomerisation. Thermal treatment might release some carotenoids from cellular matrices, increasing their concentrations in samples. DF content is not really affected by processing, unless the peel is removed from the commodity or the moisture content is changed.[39] Table 7.6 shows some of the effects on various phytochemicals of the different post-harvest treatments to which some vegetables are subjected.

It was concluded that cooking (boiling or frying) usually increases the neutral DF, acidic DF, cellulose, hemicellulose and lignin content.[67]

Food preparation (mixing or cooking) can enhance the extractability and bioavailability of β-carotene from the food matrix by either softening or disrupting the plant cell walls and breaking the carotene–protein complexes located in the chloroplasts in fresh green vegetables, while in red vegetables the amount does not change or may be diminished.[68,69]

Cooking (conventional open-pan or in a pressure cooker) increases soluble DF and decreases insoluble DF in vegetables, especially carrots and cabbage, compared to fresh.[70] In different cabbage cultivars, the amount of DF depended on the cultivar, seasonal harvesting time, processing heat treatment and storage.[71]

In a study of the effect of boiling and microwave heating, several vegetables[72] were cooked in boiling water for 30 minutes or on a ceramic plate covered by a plastic lid in a microwave. It was observed that radical-scavenging capacity increased in asparagus, carrots, aubergines and green peppers, perhaps due to the thermal destruction of the cell walls and subcellular compartments with the release of more compounds. Indeed, it has been reported that during thermal chemical reactions, more potent radical-scavenging antioxidants are released,[73] and also that flavonol glucoside is degraded, forming compounds with greater activity.[74] In contrast, a significant reduction in radical-scavenging activity was observed in broccoli, cabbage, radishes, kidney beans, onions, spinach and tomatoes after these treatments. It is well known that ascorbic acid leaches out of vegetable tissue into the cooking water, making it recommendable to use as little water as possible while boiling.

In broccoli, the effect of boiling time on the cooked tissue and cooking water was studied.[72] The radical-scavenging activity was higher in the cooking water than in the cooked tissue, while the total activity of the cooked tissue and cooking water together was higher than for fresh broccoli. In radishes, radical scavenging was lower in the cooked tissue than in the cooking water and total activity was lower in the cooked tissue and cooking water than in fresh radish. In aubergines, the radical-scavenging activity of the cooked tissue and cooking water together usually increases with boiling time, thus boiling optimises it. The radical-scavenging activity and

Table 7.6 Effect of processing on the phytochemical contents of vegetables.

Vegetable	Process	Effect on phytochemical	Reference
Fresh cut spinach	Modified atmosphere	There was no difference in total flavonoid content and this did not change during storage, but the antioxidant capacity decreased	52
Cauliflower, peas, beans, potatoes and corn	Chemical and thermic treatment (changing the pH)	The dietary fibre content in the water used to cook these vegetables was higher at pH 10 compared to pH 4. Cauliflower and peas > beans > potatoes > corn	53
Carrots	Drying	Lycopene and β-carotene were not affected after drying at below 70 °C. High carotenoid retention using microwave vacuum drying at short periods	54
Peas, spinach and carrots	Frozen	β-carotene was diminished by more than 50% after frozen storage	55
Green beans and broccoli	Frozen	β-carotene did not present significant decrements upon processing and storage	55
White cauliflower	Blanching (steam)	Polyphenols, carotenoids, flavonoid and vitamin C presented lowest reductions	56
Carrots	Blanching (steam and microwave)	Increase in the carotenoid content on a dry weight basis	57
Broccoli, carrots and green beans	Blanching and frozen	There was no significant difference in the phenol content when comparing blanched samples to unblanched samples	58
Peas, carrots, spinach, cabbage, cauliflower and yellow and white turnips	Heat treatment	Total phenolic content decreased in boiling, frying or microwave cooking	59
Broccoli, cauliflower and white cabbage	Heat treatment	In microwave cooking, only vitamin C, β-carotene, total phenolic content and antioxidant activities were lost	60
White and golden corn	Frozen	After freezing, the total carotenoid contents were increased in both type of corn, while canning did not affect carotenoid contents	61

(*continued*)

Table 7.6 (*continued*)

Vegetable	Process	Effect on phytochemical	Reference
Artichoke baby	Boiling (covered pressure)	Higher total phenolics, caffeic acid, chlorogenic acid, cynarin contents and free radical scavenging activity. Decrement of total dietary fibre	62
Purple skin aubergine	Heat treatment (boiling, steaming and microwave)	After cooking, the antioxidant activity and total phenolic content were significantly increased compared to raw ones	63
Spears of green asparagus	Heat treatment	Higher values were observed in the antioxidant activity, total phenols, quercetin, rutin, β-carotene and lutein/zeaxanthin	64
Tomatoes	Drying (sun, oven, vacuum oven and frying)	Significant losses of total phenolic content, L-ascorbic acid, glutathione, cysteine and cupric ion, reducing antioxidant capacity in all drying methods	65 and 66

ascorbic acid content varies depending on the texture of the vegetal and its polyphenol profile.

Brassica vegetables (broccoli, cauliflower, cabbage and choy sum) were cooked by boiling, steaming and microwaving and their antioxidant contents were determined. The values for antioxidant contents were in the order steamed > boiled > microwaved, and they all diminished as the cooking time increased. The greatest antioxidant losses were observed in microwaved cabbage and spinach.[75] A comparative study of the effect of boiling (5 minutes), steaming (7.5 minutes) and microwave cooking (1000 W for 1 or 1.5 minutes) on broccoli, spinach, squash, leek, pepper, peas and green beans reported the highest amount of total phenolics in pepper and the greatest antioxidant activity in broccoli.[76] All the vegetables except squash were rich in total phenolics and had high antioxidant activity when using the 1,1-difenil-2-picrilhidrazil radical (DPPH) method. Cooking had no detrimental effect on the total antioxidant activity or total phenolic content of the vegetables, except for total phenolics in squash, peas and leeks. The authors recommended the use of moderate heat treatments to improve antioxidant concentrations.

It is well known that carrots are a good source of DF, carotenoids and phenolic compounds, and their by-products still contain these bioactive compounds. It has been reported that 20% of β-carotene in carrots is degraded during drying at 90 °C.[77] Drying at between 70 °C and 90 °C presented the highest degradation of carotenoids in carrots, and for total phenolics, the greatest degradation was at between 40 °C and 60 °C. Thus, antioxidant activity is affected by temperature (40–90 °C) and time.[78]

The hydration properties of DF powder produced from carrot peel were not affected by drying at between 60 °C and 80 °C. In terms of the antioxidant activity, thermal degradation during both blanching and drying caused decreases in the contents of β-carotene and phenolic compounds, hence leading to the loss of antioxidant activity of the final product.[51]

Frozen sweet corn kernels have more total phenolics, total carotenoids and antioxidant activity (DPPH) than fresh ones. After fresh corn is boiled, microwaved or stir-fried, increments in the total phenolics, total carotenoids and antioxidant activity are observed; the opposite happens when these processes are applied to frozen corn.[79]

In tomatoes, thermal processing (2, 15 and 30 minutes at 88 °C) increases the total antioxidant activity and lycopene bioaccessibility, while the total phenolics and flavonoid content did not change significantly compared to the fresh commodity.[80]

In tomato juice and after thermal processing of baked tomatoes, tomato sauce and tomato soup (at 180 °C, 200 °C and 220 °C at durations of between 5 and 50 minutes), the vitamin C content decreased and increases were found in both total phenolics and water-soluble antioxidant activity with the ferric reducing antioxidant power (FRAP), Trolox equivalent antioxidant capacity (TEAC) and photochemiluminescence (PCL) methods.[81] Neither the amounts of total water-insoluble antioxidants (lycopene and β-carotene) nor the water-soluble fraction (phenolic compounds, vitamin C and flavonoids) of tomato puree processed at high pressure were affected compared to fresh tomatoes, whereas the water-soluble fraction increased upon cold storage compared to controls.[82] Processed canned tomato puree (92 °C for 10 minutes, canned and heated at 100 °C for 10 minutes and 50 minutes) compared to a canned puree from the United States Department of Agriculture (USDA) database indicates that temperature and time during tomato processing affect carotenoid content, polyphenol degradation and vitamin C. Meanwhile, in relation to the lyophilisation of tomato puree, the carotenoid content was lower than in fresh tomatoes, while the total polyphenol content and vitamin C were not affected.[83]

The content of EPP and NEPP is greater in fruit than in vegetables. However, in both cases, the amount of NEPP is higher than the EPP content.[24] It is clear that in both fruit and vegetables, processing causes several changes in the content of EPP and NEPP and, at the same time, these changes affect the antioxidant capacity, mainly due to heat treatments.

We should mention that according to the data presented in this section, heat treatment increases the content of EPP and NEPP due to a combination of the release of compounds from the food matrix and the concentration of compounds due to the decreased water content. However, this affects NEPP more. The change in the contents of these compounds also affects the antioxidant capacity depending on the type of fruit or vegetable under study. Besides depending on the type of fruit or vegetable, antioxidant capacity also depends on the method used to determine it. For example, when using the ABTS radical, although the antioxidant capacities of the EPP and NEPP extracted from lime and mandarin residues were higher after drying at 120 °C, in the case of the ORAC method, these capacities decrease as the drying temperature increases.

Although it is true that some processes increase the bioavailability of several phytochemicals and in this way their beneficial effects to human health are enhanced, special care should be taken not to generalise this information, since these effects will depend on the material under study (the fruit or vegetable, its variety and the geographic region where it is grown), in addition to the type of processing and how long it is applied.

7.6 Uses and Commercialisation of NEPP-rich Materials

DF alone or together with antioxidants has been used in food products such as beverages, fresh dairy products, frozen dairy products, meat, pasta, soups, baked goods and confectionery to improve shelf life and sensory or nutritional characteristics.[42,84]

NEPP-rich materials from vegetables can be used to replace fat, retain water or oil or to maintain shape after cooking. Fruit DF is beneficial due to its blends of soluble and insoluble DF, as well as its antioxidant content. Algae or seed DF is good for holding water content, emulsifying, thickening or forming gel. The advantage of using DF from whole cereals, legumes, vegetables and fruit is that some antioxidants are also present.[21,85] The waste materials, such as peel, seeds and stones, from fruit and vegetables can be used as sources of DF, phytochemicals and antioxidant capacity; some of them exhibit antimicrobial activity.[86] Tomato peel DF has been recommended to replace volume and for thickening or texturising in low-fat foods; it is a rich source of DF and an antioxidant enhancer.[87]

7.6.1 In Dairy Products

DF is used to improve consistency or texture and the feeling in the mouth, lowering syneresis;[88–90] in ice cream it is used to help develop crystallisation and resist melting.[91,92] In yogurt, Hervet-Hernandez *et al.*[93] studied the effect of polyphenols from grape pomace, while Shinde *et al.*[94] used apple skin in the viability of probiotic bacteria. Grape extracts can be used to make cheese with a high polyphenol content.[95]

7.6.2 In Cereal Products

Grape marc polyphenols have been exploited to enrich pasta[96] and biscuits.[97] To enhance the antioxidant capacity of biscuits, mango peel powder was added.[98] The use of dried vegetables in snacks[99] and adding tomato lycopene were also studied.[100]

7.6.3 In Baked Products

DF is used to improve loaf volume, ensuring the crumb is soft and the loaf is firm and prolonging its freshness; the degree of these changes depends on the DF source and the particle size, along with the process of elaboration adopted.[101–103]

7.6.4 In Meat Products

DF has been used to improve nutritional value (fat replacement) and sensory characteristics (especially tenderness).[104–106] A turkey spread was formulated by adding DF (inulin and oligofructose)[107] and antioxidants (vitamins A and E) to reduce fat and salt. The antioxidant properties of extracts of *Euryale ferox* seed shells and their preservation effects on pork sausages were studied.[108] It has been reported that fruit peels are good NEPP-rich materials and enhance the development of thermotolerant lactic bacteria in cooked sausages during storage.[109] Furthermore, tomato peel was also used to dry fermented sausages[110] and added to beef hamburgers.[111]

DF can be used in restructured fishery products (such as surimi) because of its water-binding and gelling characteristics, using sliced or chopped muscle to obtain a high-quality product.[112]

7.6.5 In Pharmaceuticals

Dry artichoke extracts obtained from their by-products are commercialised as drugs against liver diseases due to their high free radical-scavenging activity and their capacity to inhibit lipid peroxidation.[113] Also, potato peel has been recommended as an NEPP-rich material due to the compounds in it that are nutritionally and pharmacologically interesting.[114]

7.7 Conclusions

Fruit and vegetables can be subjected to several processes before commercialisation or used as raw materials to make other products. It has been reported that, according to the treatment used, the content of extractable and non-extractable phenolic compounds, as well as the DF content, can either decrease or increase. Likewise, such treatments affect the antioxidant properties of the phytochemicals present in these foods. It is crucial to take into account these effects during processing and to choose the treatment and conditions in which the content of phytochemicals is reduced as little as possible or even where their content is enhanced. Moreover, it is important to recover and use residues that are generated as by-products of the processing of fruit and vegetables, since they have important contents of phenolic compounds that can add value to products manufactured using these residues.

References

1. G. Castellano, *MATCH Commun. Math. Comput. Chem.*, 2012, **67**, 231.
2. X. Dong-Ping, L. Ya, M. Xiao, Z. Tong, Z. Yue, Z. Jie, Z. Jiao-Jiao and L. Hua-Bin, *Int. J. Mol. Sci.*, 2017, **18**, 96.
3. JFPT, *J. Food Process. Technol.*, Homepage, Consulted on Mar 14, 2017, https://www.omicsonline.org/food-processing-technology.php.
4. J. Hopkins, *Foodspan. Lesson 9: Processing: Farm to Factory 1*, 2016, Consulted on Mar 14, 2017, www.food.panlearning.org.
5. K. E. Charlton, P. Patrick, L. Dowling, K. Khulani and E. Jensen, *South Afr. J. Clin. Nutr.*, 2004, **17**(2), 56.
6. C. Zeng, *Nutr. Food Sci.*, 2013, **43**(5), 438.
7. FAOSat, 2012, Consulted on Mar 15, 2017, http://faostat.fao.org/site/567/DesktopDefault.aspx?PageID=567#ancor, http://www.fao.org/faostat/en/#data/CC.
8. M. A. Martín, J. A. Siles, A. F. Chica and A. Martín, Biomethanization of orange peel waste, *Bioresour. Technol.*, 2010, **101**(23), 8993.
9. J. Gustavsson, Ch. Cederberg, U. Sonesson, R. Van Otterdijk and A. Meybeck, *Global food losses and food waste – Extent, causes and prevention*, FAO, Rome, 2011.
10. M. Cámara, M. Sánchez de Cortés and M. E. Torija, *Nutrición y Salud. Frutas y verduras, fuentes de Salud*, Comunidad de Madrid, 2008, in line: http://www.publicaciones-isp.org/productos/t034.

11. C. Santos-Buelga and F. A. Tomás-Barberán, *Sustancias fitoquímicas de frutas y hortalizas, su posible papel beneficioso para la salud*, Tarragona, España, Ediciones de Horticultura, Fruit and Veg Project, 2001.
12. USDA, 2017, Consulted on Mar 23, 2017, https://ndb.nal.usda.gov/ndb/.
13. J. M. Morillas-Ruiz and J. M. Delgado-Alarcón, *Nutr. Clin. Diet. Hosp.*, 2012, **32**(2), 8.
14. F. J. Esparza-Martínez, R. Miranda-López and S. H. Guzmán-Maldonado, *Ind. Crops Prod.*, 2016, **84**, 1.
15. F. J. Esparza-Martínez, R. Miranda-López and S. H. Guzmán-Maldonado, *Plant Foods Hum. Nutr.*, 2016, **71**(3), 294.
16. A. Nawirska and M. Kwasniewska, *Food Chem.*, 2005, **91**(2), 221.
17. F. Saura-Calixto, J. Serrano and I. Goñi, *Food Chem.*, 2007, **101**(2), 492.
18. M. P. Gonthier, C. Remesy, A. Scalbert, V. Cheynier, J. M. Souquet, K. Poutanen and A. M. Aura, *Biomed. Pharmacother.*, 2006, **60**(9), 536.
19. L. Jakobek, *Food Chem.*, 2015, **175**, 556.
20. A. E. Quirós-Sauceda, H. Palafox, R. M. Robles-Sánchez and G. A. González-Aguilar, *BIOtecnia*, 2011, **13**(3), 3.
21. F. Saura-Calixto, *J. Agric. Food Chem.*, 1998, **46**(10), 4303.
22. M. V. Selma, J. C. Espín and F. A. Tomás-Barberán, *J. Agric. Food Chem.*, 2009, **57**(15), 6485.
23. H. Palafox-Carlos, J. F. Ayala-Zavala and G. A. González-Aguilar, *J. Food Sci.*, 2011, **76**(1), R6.
24. S. Arranz and F. Saura-Calixto, *J. Cereal Sci.*, 2010, **51**(3), 313.
25. D. Li-Juan, G. Qing-Han, J. Xiao-Long, M. Yu-Jie, X. Fang-Yi and W. Min, *J. Agric. Food Chem.*, 2013, **61**(48), 11840.
26. A. Cheng, W. Wang, Z. Gong and L. Liu, *Czech J. Food Sci.*, 2013, **31**(3), 275.
27. G. A. Camelo-Méndez and L. A. Bello-Pérez, *J. Chem., Biol. Phys. Sci.*, 2014, **4**(5), 6.
28. A. Cheng, C. Han, X. Fang, J. Sun, X. Chen and F. Wan, *J. Sci. Food Agric.*, 2014, **96**(10), 3393.
29. L. Condezo-Hoyos, I. P. Mohanty and G. D. Noratto, *Food Chem.*, 2014, **161**, 208.
30. A. Delpino-Rius, J. Eras, F. Vilaró, M. A. Cubero, M. Balcells and R. Canela-Garayoa, *Food Chem.*, 2015, **172**, 575.
31. G. Velderrain-Rodríguez, A. Quirós-Sauceda, G. Mercado-Mercado, J. F. Ayala-Zavala, H. Astiazarán-García, R. M. Robles-Sánchez, A. Wall-Medrano, S. Sayago-Ayerdi and G. A. González-Aguilar, *Food Sci. Technol.*, 2016, **36**(2), 188.
32. M. J. Pozuelo, A. Agis-Torres, D. Hervert-Hernández, M. E. López-Oliva, E. Muñoz-Martínez, R. Rotger and I. Goñi, *J. Food Sci.*, 2012, 77(2), H59.
33. C. Ajila, M. Aalami, K. Leelavathi and U. Rao, *Innovative Food Sci. Emerging Technol.*, 2010, **11**(1), 219.
34. I. Sánchez-Alonso, A. Jiménez-Escrig, F. Saura-Calixto and A. J. Borderías, *Food Chem.*, 2007, **101**(1), 372.

35. Y. Lario, E. Sendra, J. García-Pérez, C. Fuentes, E. Sayas-Barberá, J. Fernández-López and J. A. Pérez-Álvarez, *Innovative Food Sci. Emerging Technol.*, 2009, **5**, 113.
36. F. R. Marin, C. Soler-Rivas, O. Benavente-García, J. Castillo and J. A. Pérez-Álvarez, *Food Chem.*, 2007, **100**(2), 736.
37. N. Ghanem, D. Mihoubi, N. Kechaou and N. B. Mihoubi, *Ind. Crops Prod.*, 2012, **40**, 167.
38. J. C. Rickman, C. M. Bruhn and D. M. Barrett, *J. Sci. Food Agric.*, 2007, **87**(6), 930.
39. J. C. Rickman, C. M. Bruhn and D. M. Barrett, *J. Sci. Food Agric.*, 2007, **87**(7), 1185.
40. K. D. Sharma, S. Karki, N. S. Thakur and S. Attri, *J. Food Technol.*, 2012, **49**(1), 22.
41. D. J. Williams, D. Edwards, I. Hamerning, L. Jin and A. P. James, *Food Res. Int.*, 2013, **52**(1), 327.
42. D. Dhingra, M. Michael, H. Rajput and R. T. Patil, *Food Sci. Technol.*, 2012, **49**(3), 255.
43. J. Seok-Moon, K. So-Young, K. Dong-Ryul, J. Seong-Chun, K. C. Nam, D. U. Ahn and L. Seung-Cheol, *J. Agric. Food Chem.*, 2004, **52**(11), 3389.
44. M. L. Chen, D. J. Yang and S. C. Liu, *Int. J. Food Sci. Technol.*, 2011, **46**(6), 1179.
45. F. Que, L. Mao, X. Fang and T. Wu, *Int. J. Food Sci. Technol.*, 2008, **43**(7), 1195.
46. A. E. Quirós-Sauceda, J. F. Ayala-Zavala, S. G. Sáyago-Ayerdi, R. Vélez-de La Rocha, A. Sañudo-Barajas and G. A. González-Aguilar, *J. Appl. Bot. Food Qual.*, 2014, **87**, 227.
47. J. Volden, G. B. Bengtsson and T. Wicklund, *Food Chem.*, 2009, **112**(4), 967.
48. A. Rawson, A. Patras, B. K. Tiwari, F. Noci, T. Koutchma and N. Brunton, *Food Res. Int.*, 2011, **44**(7), 1875.
49. O. Fennema, *Food Technol.*, 1977, **31**(12), 32.
50. A. Ismail, Z. M. Marjan and C. W. Foong, *Food Chem.*, 2004, **87**(4), 581.
51. P. Chantaro, S. Devahastin and N. Chiewchan, *LWT–Food Sci. Technol.*, 2008, **41**(10), 1987.
52. M. I. Gil, F. Ferreres and F. A. Tomas-Barberan, *J. Agric. Food Chem.*, 1999, **47**(6), 2213.
53. L. M. Brandt, M. A. Jeltema, M. E. Zabik and B. D. Jeltema, *J. Food Sci.*, 1984, **49**(3), 900.
54. M. Regier, E. Mayer-Miebach, D. Behsnilian, E. Neff and H. P. Schumann, *Drying Technol.*, 2005, **23**(4), 989.
55. A. Bouzari, D. Holstege and D. M. Barrett, *J. Agric. Food Chem.*, 2015, **63**(3), 957.
56. F. A. Ahmed and R. F. M. Ali, *BioMed Res. Int.*, 2013, 367819, 9 pages.
57. G. K. Sharma, A. D. Semwal and S. S. Arya, *J. Food Sci. Technol.*, 2000, **37**(2), 196.

58. A. Patras, B. K. Tiwari and N. P. Brunton, *LWT–Food Sci. Technol.*, 2011, **44**(1), 299.
59. B. Sultana, F. Anwar and S. Iqbal, *Int. J. Food Sci. Technol.*, 2008, **43**(3), 560.
60. M. H. A. Shams El-Din, M. M. Abdel-Kader, S. K. Makhlouf and O. S. S. Mohamed, *World Appl. Sci. J.*, 2013, **26**(6), 697.
61. C. E. Scott and A. L. Eldridge, *J. Food Compos. Anal.*, 2005, **18**(6), 551.
62. M. Lutz, C. Henriquez and M. Escobar, *J. Food Compos. Anal.*, 2011, **24**(1), 49.
63. A. Chumyam, K. Whangchai, J. Jungklang, B. Faiyue and K. Saengnil, *ScienceAsia*, 2013, **39**, 246.
64. S. Fanasca, Y. Rouphael, E. Venneira, E. Azzini, A. Durazzo and G. Maiani, *Int. J. Food Sci. Technol.*, 2009, **44**(5), 1017.
65. O. A. Gümüsay, A. A. Borazan, N. Ercal and O. Demirkol, *Food Chem.*, 2015, **173**, 156.
66. E. Sahlin, G. P. Savage and C. E. Lister, *J. Food Compos. Anal.*, 2004, **17**(5), 635.
67. J. Herranz, C. Vidal-Valverde and E. Rojas-Hidalgo, *J. Food Sci.*, 1981, **46**(6), 1927.
68. J. M. Dietz, S. S. Kantha and J. W. Erdman, *Plant Foods Hum. Nutr.*, 1988, **38**(4), 333.
69. S. Bernhardt and E. Schlich, *J. Food Eng.*, 2006, **77**(2), 327.
70. F. Khanum, M. S. Swamy, K. R. Sudarshanashna, K. Santhanam and K. R. Viswanathan, *Plant Foods Hum. Nutr.*, 2000, **55**(3), 207.
71. M. S. Wennberg, G. M. Engqvist and E. M. G.-L. Nyman, *J. Food Sci.*, 2003, **68**(5), 1615–1621.
72. T. Yamaguchi, T. Mizobuchi, R. Kajikawa, H. Kawashima, F. Miyabe, J. Terao, H. Takamura and T. Matoba, *Food Sci. Technol. Res.*, 2001, **7**(3), 250.
73. H. Maeda, T. Katsuki, T. Akaike and R. Yasutake, *Jpn. J. Cancer Res.*, 1992, **83**(9), 923.
74. D. P. Makris and J. T. Rossiter, *J. Agric. Food Chem.*, 2000, **48**(9), 3830.
75. S. Wachtel-Galor, K. Wing-Wong and I. F. F. Benzie, *Food Chem.*, 2008, **110**(3), 706.
76. N. Turkmen, F. Sari and Y. S. Velioglu, *Food Chem.*, 2005, **93**(4), 713–718.
77. E. Mayer-Miebach, D. Behsnilian, M. Regier and H. P. Schuchmann, *Food Res. Int.*, 2005, **38**(8–9), 1103.
78. D. Uurrea, V. Eim, C. Rosello and S. Simal, *Rev. Aliment. Hoy*, 2012, **21**(27), 68.
79. J. Song, Ch. Liu, D. Li and L. Meng, *Czech J. Food Sci.*, 2013, **31**(6), 607.
80. V. Dewanto, X. Wu, K. K. Adom and R. H. Liu, *J. Agric. Food Chem.*, 2002, **50**(10), 3010.
81. S. Gahler, K. Otto and V. Bohm, *J. Agric. Food Chem.*, 2003, **51**(27), 7962.
82. A. Fernandez-Garcia, P. Butz and B. Tauscher, *J. Food Sci.*, 2001, **66**(7), 1033.

83. S. George, F. Tourniaire, H. Gautier, P. Goupy, E. Rock and C. Caris-Veyrat, *Food Chem.*, 2011, **124**(4), 1603.
84. M. Elleuch, D. Bedigian, O. Roiseux, S. Besbes, C. Blecker and H. Attia, *Food Chem.*, 2011, **124**(2), 411.
85. F. Esposito, G. Arlotti, A. M. Bonifati, A. Napolitano, D. Vitale and V. Fogliano, *Food Res. Int.*, 2005, **38**(10), 1167.
86. N. S. Sonia, C. Mini and P. R. Geethalekshmi, *Agric. Rev.*, 2016, **37**(1), 35.
87. I. Navarro-Gonzalez, V. Garcia-Valverde, J. Garcia-Alonso and M. J. Periago, *Food Res. Int.*, 2011, **44**(5), 1528.
88. F. J. Garcia-Perez, Y. Lario, J. Fernandez-Lopez, E. Sayas, J. A. Perez-Alvarez and E. Sendra, *Color Res. Appl.*, 2005, **30**(6), 457.
89. C. S. Brennan and C. M. Tudorica, *Int. J. Food Sci. Technol.*, 2008, **43**(5), 824.
90. T. Sanz, A. Salvador, A. Jiménez and S. Fiszman, *Eur. Food Res. Technol.*, 2008, **227**(5), 1515.
91. A. Regand and H. D. Goff, *Food Hydrocolloids*, 2003, **17**(1), 95.
92. C. Soukoulis, D. Lebesi and C. Tzia, *Food Chem.*, 2009, **115**(2), 665.
93. D. Hervet-Hernandez, C. Pintado, R. Roter and I. Goni, *Int. J. Food Microbiol.*, 2009, **136**(1), 119.
94. T. S. Shinde, J. D. Brooks and D. Sun-Waterhouse, *Int. J. Food Sci. Technol.*, 2015, **50**(6), 1303.
95. D. F. Da Silva, P. T. Matumoto-Pintro, L. Bazinet, C. Couillard and M. Britten, *J. Dairy Sci.*, 2015, **98**(3), 1552.
96. V. Marinelli, L. Padalino, D. Nardiello, M. A. Del Nobile and A. Conte, *J. Chem.*, 2015, 734578, 8 pages.
97. A. Pasqualone, A. M. Bianco, V. M. Paradiso, C. Summo, G. Gambacorta and F. Caponio, *Food Res. Int.*, 2014, **65**, 385.
98. C. M. Ajila, K. Leelavathi and U. J. S. Prasada Rao, *J. Cereal Sci.*, 2008, **48**(2), 319.
99. L. Huang and M. Zhang, *Drying Technol.*, 2012, **30**(5), 448.
100. Z. Dehghan-Shoar, A. K. Hardacre and C. S. Brennan, *Food Chem.*, 2010, **123**(4), 1117.
101. J. Wang, C. M. Rosell and C. B. De Barber, *Food Chem.*, 2002, **79**(2), 221.
102. A. Sangnark and A. Noomhorm, *Food Res. Int.*, 2004, **37**(1), 66.
103. C. M. Rosell and E. Santos, *J. Food Eng.*, 2010, **98**(2), 273.
104. M. L. Garcia, R. Dominguez, M. D. Galvez, C. Casas and M. D. Selgas, *Meat Sci.*, 2002, **60**(3), 227.
105. V. S. Eim, S. Simal, C. Rossello and A. Femenia, *Meat Sci.*, 2008, **80**(2), 173–182.
106. S. G. Sayago-Ayerdi, A. Brenes and I. Goñi, *LWT–Food Sci. Technol.*, 2009, **42**(5), 971.
107. F. B. Garrido, Z. A. Gonzalez, M. C. Dondero and E. R. Wittig de Penna, *Rev. Chil. Nutri.*, 2010, **37**(3), 360.
108. C. Zhang, *J. Food Process. Preserv.*, 2014, **39**(6), 1172.
109. J. Diaz-Vela, A. Totosaus and M. L. Perez-Chabela, *J. Food Process. Preserv.*, 2015, **39**(6), 2630.

110. M. M. Calvo, M. L. Garcia and M. D. Selgas, *Meat Sci.*, 2008, **80**(2), 167.
111. M. I. García, M. M. Calvo and M. D. Selgas, *Meat Sci.*, 2009, **83**(1), 45.
112. A. J. Borderias, I. Sanchez-Alonso and M. Perez-Mateos, *Trends Food Sci. Technol.*, 2005, **16**(10), 458.
113. R. Llorach, J. C. Espin, F. A. Tomas-Barberan and F. Ferreres, *J. Agric. Food Chem.*, 2002, **50**(12), 3458.
114. A. Schieber and M. D. Aranda Saldaña, *Food*, 2009, **3**(2), 23.

CHAPTER 8

Carotenoids: Concept, Dietary Sources and Structure–Activity Relationships for Antioxidant and Health-related Properties

L. R. B. MARIUTTI[†a], R. C. CHISTÉ[†b] AND A. Z. MERCADANTE*[c]

[a]University of Campinas (UNICAMP), Department of Food and Nutrition, Rua Monteiro Lobato, 80, 13083-862, Campinas, São Paulo, Brazil; [b]Federal University of Pará (UFPA), Institute of Technology (ITEC), Faculty of Food Engineering (FEA), Rua Augusto Corrêa, 01, Guamá, 66075-110, Belém, Pará, Brazil; [c]University of Campinas (UNICAMP), Department of Food Science, Rua Monteiro Lobato, 80, 13083-862, Campinas, São Paulo, Brazil
*E-mail: azm@unicamp.br, azmercadante@gmail.com

8.1 Introduction

The group of carotenoids comprises more than 800 compounds sharing in common a skeleton formed by two isoprenoid unities linked in such a way that the molecule is linear and has inverted symmetry in the center and several conjugated double bonds (c.d.b.) in the chain. This basic skeleton of

[†]Both authors contributed equally to this chapter.

Food Chemistry, Function and Analysis No. 5
Non-extractable Polyphenols and Carotenoids: Importance in Human Nutrition and Health
Edited by Fulgencio Saura-Calixto and Jara Pérez-Jiménez

Published by the Royal Society of Chemistry, www.rsc.org

40 carbon atoms can be modified by hydrogenation, dehydrogenation, cyclization, shortening or chain extension, isomerization, introduction of substituents or combinations of these processes.

Carotenoids are divided into two major groups: carotenes that have only carbon and hydrogen in their structure; and xanthophylls that present different oxygen-containing chemical groups, such as alcohol, ketone, aldehyde and epoxide (Figure 8.1). In addition, in most fruits, xanthophylls are generally esterified with fatty acids[1] (Figure 8.1). Regardless of the presence of oxygen-containing groups, almost all carotenoids are hydrophobic. Carotenoid nomenclature, semi-systematic naming and carbon numbering follow the IUPAC (Commission on Nomenclature of Organic Chemistry) and IUPAC-IUB (Commission on Biochemical Nomenclature) rules,[2] but the trivial names are the most used ones.

The c.d.b. and chemical groups conjugated to these double bonds that are able to share non-binding electron pairs are the parts of the structure responsible for the absorption of light in the visible region, which is known as the chromophore. As the number of c.d.b. increases, the energy required to promote electron transitions of the valence layer decreases and the absorption wavelength increases. In this way, carotenoids with five or fewer c.d.b., such as phytofluene and phytoene, exhibit absorption in the UV region and are colorless. Increasing the number of c.d.b. and consequently the maximum absorption wavelength (λ_{max}) changes the color from light yellow (ζ-carotene, 7 c.d.b.), orange (β-carotene, 11 c.d.b.) to red (astaxanthin, 13 c.d.b.). Besides the number of c.d.b. and auxochrome groups, the spatial arrangement of the chromophore components influences energy absorption. For example, despite the fact that β-carotene and lycopene have 11 c.d.b., the λ_{max} of β-carotene is lower than that of lycopene because the two double bonds of the β-ring are not in the same plane as those from the polyene chain and thus they require more energy for the electronic transition. On the other hand, structures with different molecular weights, but with the same chromophore, show identical UV-visible spectra (*e.g.* β-carotene and zeaxanthin).

8.2 Structure and Physical–Chemical Properties

The *trans*-conformation of all double bonds (all-*E*) of carotenoids is more stable than the *cis* (*Z*) one, and therefore the all-*E* is in general the most abundant carotenoid isomer in nature. Due to the steric hindrance generated by the presence of the methyl groups at the C-19, C-20, C-19′ and C-20′ positions, the 9*Z*, 13*Z* and 15*Z* isomers are the most common ones and are found in larger amounts than *Z* isomers at other positions. There are a few exceptions; for example, bixin, the major carotenoid in annatto seeds, has *cis*-conformation at C-9′ (Figure 8.1) and 5*Z*-lycopene is found in human plasma. The *Z* isomers present a slightly lower λ_{max} value than the corresponding all-*E* isomers when in the same solvent.[3]

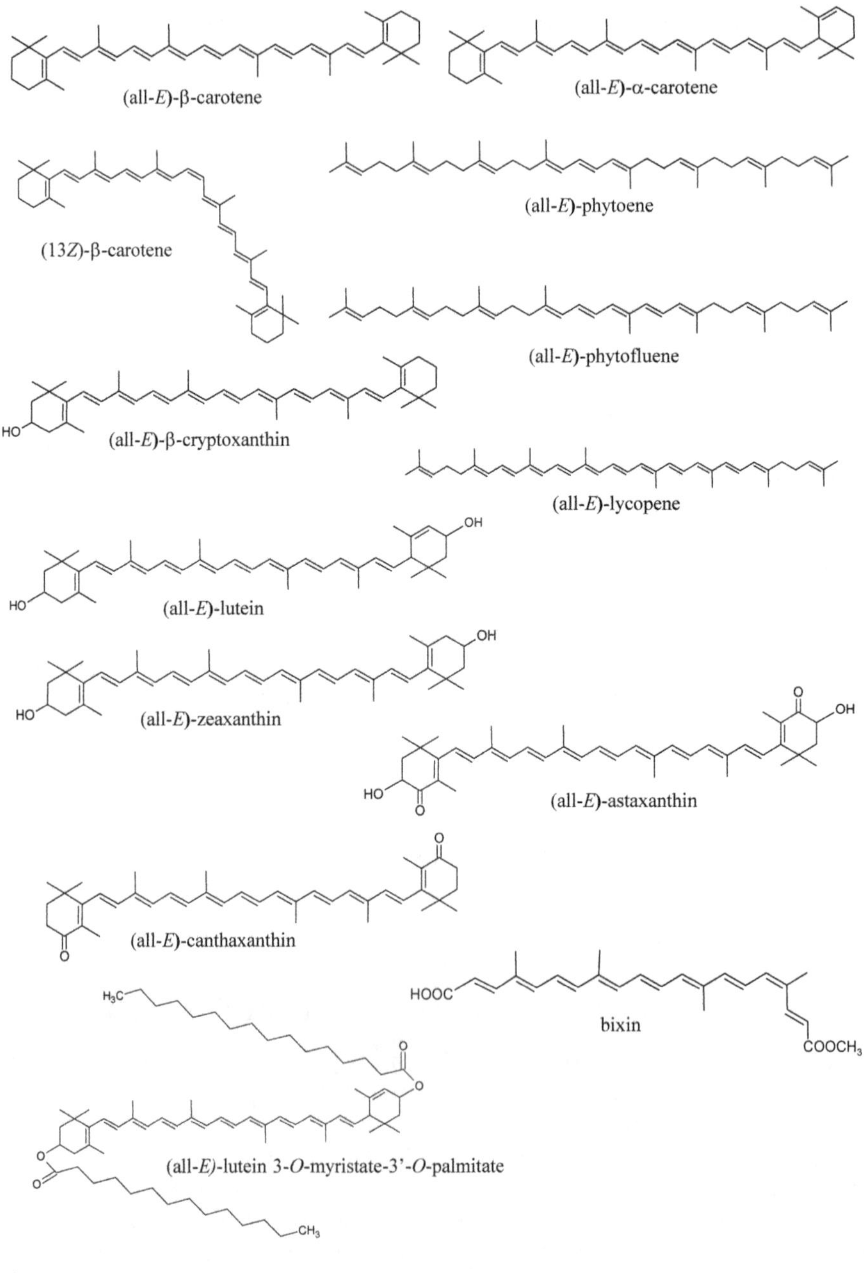

Figure 8.1 Structures of carotenes and xanthophylls.

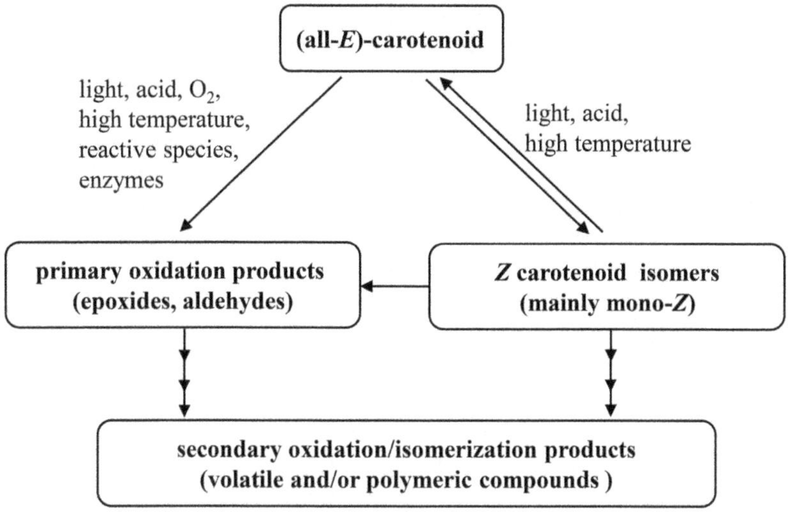

Figure 8.2 General scheme of carotenoid degradation. Adapted from Borsarelli and Mercadante.[5]

Although carotenoids are naturally stable in plant or animal matrices, these pigments can easily undergo degradation under heat, light, acid and enzymes exposure, among others factors, when tissue is damaged, with consequent modification of color and biological activity. Differences in degradation are related to the three-dimensional shape of the carotenoid structure, which determines hydrophobicity, crystalline state, ease of crystal formation, type of organization (multilayers or aggregates), as well as location of biosynthesis or storage in the cells.[4] Carotenoid degradation includes simultaneous transformations: reversible $E \leftrightarrow Z$ isomerization, irreversible rearrangement of epoxide to furanoid and irreversible degradation forming oxidation products such as epoxides and apocarotenoids, which can further furnish volatile compounds.[5–7] These global changes, observed in natural and non-natural systems, are shown in Figure 8.2.

For example, oxidation of lycopene with atmospheric oxygen catalyzed by a metalloporphyrin was evaluated in a model mimicking the active center of cytochrome P450 enzymes, which are present in the liver, the organ where carotenoids can be stored and eventually metabolized.[6] In this system, three types of products derived from lycopene were detected: *(Z)*-isomers, monoepoxides and apo-lycopenals. A mechanism was proposed with *(Z)*-isomers being firstly formed and then oxidized into lycopene monoxides. The epoxide groups underwent oxidative cleavage, giving rise to apo-lycopenals of all molecular weights, and finally the longest-chain apo-lycopenals were further cleaved into shorter apo-lycopenals and apo-lycopenones.[6]

Although in another completely different scenario, similar chemical transformations were also observed after heating a food system mimicking fruit juice at 90 °C for 4 hours. Formation of mono-*Z*-isomers (*e.g.* 9*Z*, 13*Z* and 15Z-β-carotene), rearrangement of 5,6-epoxide to 5,8-furanoid (from violaxanthin to auroxanthin) and formation of oxidation products such as 5,6-epoxy-β-cryptoxanthin and 12′-apo-β-carotenal occurred.[7] However, only a small fraction of the degraded carotenoids could be related to the formation of isomers and/or oxidation products, with a 36% loss mainly attributed to the generation of low-molecular-weight volatile compounds.[7–9]

Another interesting example was verified in different tomato varieties submitted to heat treatment in water or water : oil (8 : 2) at 100 °C for 30 minutes. Under these treatments, lycopene, prolycopene, δ-carotene and γ-carotene were practically not isomerized, while *Z*-isomers of β-carotene increased by 21% and those of lutein increased by 27%. A possible explanation for this fact is related to the linear structure of lycopene that facilitates formation of aggregates or crystals, while the presence of two bulky β- or ε-ionone rings (*e.g.* of β-carotene) does not easily allow aggregation into an ordered and stable system.[4]

Despite the apparent consensus about the general mechanism of carotenoid degradation and formed products, the preferential reaction and the amount of each formed product depends on both intrinsic factors, such as the structure and physical state of the carotenoid and its location within the cellular organelles, and extrinsic factors, such as medium/food composition, temperature, light intensity, concentration of oxygen and metals, among others.[5] Thus, a comparison of data on the extent of carotenoid degradation is a difficult task, since different conditions are commonly employed in systems with different compositions. Another difficulty is that different kinetic models, such as first order, zero order and bi-exponential, have been assigned to describe the degradation of carotenoids.[5]

8.3 Antioxidant Properties of Carotenoids Involving Reactive Oxygen/Nitrogen Species and Structure–Activity Relationships

8.3.1 Oxidative Stress and Reactive Oxygen Species and Reactive Nitrogen Species Generation

Oxidation is a key reaction that is fundamental to aerobic life and cellular metabolism. In this sense, pro-oxidant species are naturally produced by the physiological system and they exert fundamental biological functions. Reactive oxygen species (ROS), as the name implies, are derived from molecular oxygen with redox activity and high reactivity, while reactive nitrogen species (RNS) are those derived from nitric oxide ($^{\bullet}NO$).[10,11] In the living organism, ROS and RNS are involved in energy production, phagocytosis, cell growth regulation, intercellular signaling and synthesis of important biological substances. However, whenever there is an imbalance between the generation of pro-oxidant reactive species and the action of antioxidant systems (endogenous and exogenous), the cellular components, such as lipids, proteins, DNA and even tissue, can be damaged (oxidative stress), resulting in several diseases, such as diabetes, cancer, allergies, inflammation, neurodegeneration and cardiovascular diseases.[12] Although the ROS and RNS produced by phagocytes represent an extremely important mechanism of the body's defense, their excess may cause or aggravate damage to inflammatory sites.

ROS and RNS are produced within cell compartments as a consequence of human metabolism (*e.g.* exercise and mitochondrial electron transport), environmental pollutants, changed atmospheric conditions (hyperoxia), environmental stimuli (sunlight ionizing radiation) and lifestyle stressors (excessive alcohol consumption and cigarette smoking).[13]

In food matrices, the occurrence of ROS/RNS is unavoidable due to their biological nature, which may cause loss of nutritional value and food quality due to physical and chemical changes during processing, storage or commercialization. ROS are the main agents responsible for the oxidation reactions in foods, as they can react with lipids, proteins, carbohydrates and vitamins, producing undesirable volatile compounds, degrading essential fatty acids, amino acids and vitamins and producing carcinogenic compounds.[14] Furthermore, ROS/RNS can change the functional properties of proteins, lipids and carbohydrates due to the formation of oxidized dimers and trimers, making the food product less acceptable or even unacceptable to consumers.[15,16]

The general scheme for the interrelation between the formation of ROS and RNS can be seen in Figure 8.3. The production of ROS in the organism is generally initiated by the formation of superoxide anion radicals ($O_2^{\bullet-}$) that

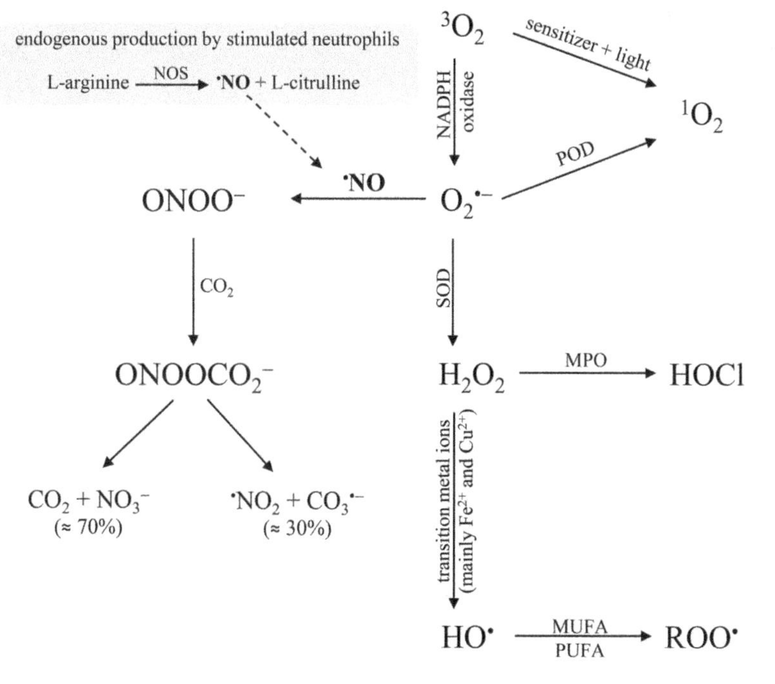

Figure 8.3 General scheme for the interrelation between the formation of ROS and RNS in the organism. POD: peroxidase; SOD: superoxide dismutase; MPO: myeloperoxidase; PUFA: polyunsaturated fatty acids; MUFA: monounsaturated fatty acids; NOS: nitric oxide synthase.

are produced *in vivo* through one-electron reduction of molecular oxygen ($^3O_2$) during some physiological stress (*e.g.* inflammation), which is quickly converted to hydrogen peroxide (H_2O_2), either spontaneously or by enzyme catalysis (by superoxide dismutase).[17] Despite its low reactivity, H_2O_2 interacts with transition metal ions (mainly Fe^{2+} and Cu^{2+}) generating hydroxyl radicals ($HO^•$) through the Fenton reaction. $HO^•$ is one of the most oxidant ROS, causing severe damage to biological targets in its proximity, and it initiates lipid peroxidation due to its high reactivity with polyunsaturated fatty acids and monounsaturated fatty acids, resulting in the production of peroxyl radicals ($ROO^•$) and other cytotoxic products.[18–20] In addition, the enzyme myeloperoxidase (H_2O_2 oxidoreductase), which is specifically found in leukocytes, including neutrophils and eosinophils, catalyzes the reaction between chloride ions (Cl^-) and H_2O_2 with the formation of hypochlorous acid (HOCl), a powerful bactericidal agent.[21] In addition, singlet oxygen ($^1O_2$), a highly reactive form of oxygen, can be produced *in vivo* by peroxidase enzymes during reactions catalyzed by lipoxygenase and from photosensitization reactions through the energy transfer between the excited triplet state of sensitizers (*e.g.* chlorophylls, riboflavin) and $^3O_2$.[22–24]

Nitric oxide ($^•NO$) is a RNS that is produced in different cell types by a family of isoenzymes named nitric oxide synthases (NOS), by converting L-arginine to L-citrulline and is involved in several physiological functions.[25,26] The $^•NO$ produced during neutrophil activity is responsible for immunity; however, its overproduction is involved in diverse pathogenesis (*e.g.* inflammation).[27,28] In addition, by the reaction with $O_2^{•-}$, $^•NO$ can generate the peroxynitrite anion ($ONOO^-$),[29] which is a very effective oxidant with reactivity similar to $HO^•$[30] and is responsible for mediating $^•NO$ toxicity that causes damage to cell biomolecules, leading to cell death.[31] In contrast to $O_2^{•-}$, $^•NO$ readily diffuses through the cytoplasm and plasma membranes. As such, $ONOO^-$ formation occurs close to the site of $O_2^{•-}$ production.[26] The $ONOO^-$ reaction with physiological CO_2 is one of the fastest known reactions, resulting in the formation of the nitrosoperoxycarbonate anion ($ONOOCO_2^-$). Its decomposition leads to the formation of $CO_2 + NO_3^-$ (about 70%) and free radicals as $^•NO_2$ and carbonate radical anions ($CO_3^{•-}$) (about 30%).[11,26]

In the physiological system, the overproduction of pro-oxidant species is counteracted by the endogenous antioxidant system produced by the body, such as glutathione and several enzymes (superoxide dismutase, catalase and glutathione peroxidase). In addition to the endogenous antioxidant system, the exogenous antioxidants, which are compounds absorbed from the diet, are of paramount importance to decrease or inhibit the deleterious effects of pro-oxidant species. In this sense, carotenoids, one of the main class of exogenous compounds with antioxidant properties, have been studied for their role against *in vivo* oxidative stress and oxidative-related processes in foods and have received great attention from the scientific community. The antioxidant properties of carotenoids against the oxidizing effects of ROS and RNS have been extensively studied in the oxidative deterioration of food products, as well as in the therapeutic approaches against several human diseases, such as atherosclerosis, age-related macular degeneration, cataracts, chronic inflammation, neurodegenerative diseases and cancer.[32–35]

8.3.2 *In vitro* Antioxidant Properties of Carotenoids

Carotenoids are able to scavenge ROS/RNS and inhibit the formation of oxidant species. As previously discussed, the structure of carotenoids with different end-groups and c.d.b. makes these compounds very reactive and these characteristics are responsible for their associated antioxidant properties.

The mechanisms by which carotenoids can exert their action can be divided into three main pathways: single-electron transfer (SET) (eqn (8.1)), hydrogen atom transfer (HAT) (eqn (8.2)) or radical addition (eqn (8.3)).[36,37] Knowledge of these reactions is very important in order to understand and select the proper methods to be used to measure the antioxidant capacity of any bioactive compound.[38]

$$R^{\bullet} + CAR \rightarrow R^{-} + CAR^{\bullet+} \tag{8.1}$$

$$R^{\bullet} + CAR \rightarrow RH + CAR^{\bullet} \tag{8.2}$$

$$R^{\bullet} + CAR \rightarrow [R\text{-}CAR^{\bullet}] \tag{8.3}$$

where, $R^{\bullet}$ is the oxidant radical specie and CAR is the carotenoid molecule. In these reactions, the newly formed radicals, after antioxidant action, are more stable and less reactive than $R^{\bullet}$.

There are several *in vitro* antioxidant capacity assays available in the literature, which differ from each other in terms of reaction means (aqueous buffers, organic solvents), probes, substrates, reaction conditions and, consequently, antioxidant mechanisms. Some examples of assays involving the SET reaction are 2,2′-azinobis-(3-ethylbenzothiazoline-6-sulfonic acid) (ABTS), ferric ion-reducing antioxidant parameter (FRAP), diphenyl-1-picrylhydrazyl assay (DPPH), copper (II) reduction capacity and total phenolic assay by Folin–Ciocalteu reagent. Examples of assays involving the HAT reaction are oxygen radical absorbance capacity (ORAC), total radical trapping antioxidant parameter (TRAP) and crocin bleaching/inhibition of linoleic acid oxidation assay.[39] Currently, it is known that the antioxidant properties of the carotenoids are closely related to their chemical structures, including aspects such as the number of c.d.b., type of end-groups and oxygen-containing substituents.[36,37] The structure–activity relationship of carotenoids is well established when the capacity to scavenge the synthetic and non-physiological radical cation $ABTS^{\bullet+}$[41,43] is considered, and also for physiologically relevant ROS, such as $ROO^{\bullet}$[37] and $^1O_2$[22,40] (Table 8.1).

Table 8.1 *In vitro* chemical antioxidant capacity of the physiologically relevant carotenoids based on their ability to scavenge 2,2′-azinobis-(3-ethylbenzothiazoline-6-sulfonic acid) (ABTS) and peroxyl radicals (ROO•) and to quench singlet oxygen ($^1O_2$).

Carotenoid	$ABTS^{•+}$-scavenging capacity[a]	$ROO^•$-scavenging capacity[b]	$^1O_2$-quenching capacity: Chemical quenching kr (M^{-1} s^{-1})[c]	Physical quenching kq (M^{-1} s^{-1})[d]
Lycopene	2.9	8.67	23×10^5	31×10^9
β-Carotene	1.9	3.24	9×10^5	14×10^9
α-Carotene	1.3	2.62	16×10^5	19×10^9
β-Cryptoxanthin	2.0	2.85	–	6×10^9
Zeaxanthin	1.4	3.52	–	10×10^9
Lutein	1.5	1.90	18×10^5	8×10^9
Astaxanthin	0.03	6.50	–	24×10^9

[a]Trolox-equivalent antioxidant capacity values were expressed as mM. Model system: hexane/acetone (9 : 1, v/v).[41]
[b]Values expressed as α-tocopherol relative. Model system: 4,4-difluoro-5-(4-phenyl-1,3-butadienyl)-4-bora-3a,4a-diaza-*s*-indacene-3-undecanoic acid (C_{11}–BODIPY$^{581/591}$, as the probe), azobisisobutyronitrile (AIBN) as ROO• generator, both in DMSO/MTBE (10 : 1, v/v).[37]
[c]Model system: surfactant sodium bis(2-ethylhexyl) sulfosuccinate/hexane/H_2O reverse micelle; $^1O_2$ generator: sensitization of Bengal rose; detection: time-resolved phosphorescence.[150]
[d]Model system: ethanol/chloroform/H_2O (50 : 50 : 1, v/v/v) at 37 °C; $^1O_2$ generator: thermodissociation of disodium 3,3′-(1,4-naphthalene) bispropionate ($NDPO_2$); detection: infrared chemiluminescence.[22]

Based on the ability of carotenoids to scavenge $ABTS^{•+}$, a tocopherol equivalent antioxidant capacity (TEAC) value was assigned to each compound and a wide range of *in vitro* antioxidant activities have already been reported. According to the literature, the ability of carotenes to scavenge $ABTS^{•+}$ increases with the extension of the chromophore and maximum overlap of the carbon–carbon double bond molecular orbitals.[41] For example, among the carotenes, lycopene (ψ,ψ-carotene; TEAC value = 2.9 mM) is more effective than β-carotene (β,β-carotene; 1.9 mM), which in turn is a more efficient $ABTS^{•+}$ scavenger than α-carotene (β,ε-carotene; 1.3 mM).[41,42] A later study reported the same behavior, but with no difference between α- and β-carotene activities, as determined by the αTEAC assay.[43] As discussed in Section 8.2, both β-carotene and lycopene have 11 c.d.b.; however, β-carotene has two β-rings and a steric hindrance between the methyl substituent at C-5 of the ring and the hydrogen atom at C-8 of the chain that induces a conformation in which the rings and the C-5,6 and C-5′,6′ double bonds are twisted out of plane.[41] Such a conformation (orbital overlap) reduces the contribution of the double bonds of the ring to the chromophore. As α-carotene has 10 c.d.b., it shows lower antioxidant activity as compared to lycopene and β-carotene. Furthermore, the ability of carotenoids to scavenge $ABTS^{•+}$ seems to be modulated by the presence of functional groups in the terminal rings, as in the case of xanthophylls. On the one hand, the presence of a carbonyl group (CO, keto group) decreases the scavenging capacity of astaxanthin (TEAC value = 0.03 mM) and canthaxanthin (0.02 mM),[41] because the carbonyl groups at the 4 and 4′ positions in the rings reduce the unpaired electron density across the 11 c.d.b., resulting in a decreased propensity for electron donation.[42] On the other hand, in comparison to β-carotene antioxidant activity (TEAC value = 1.9 mM), the presence of a single OH group in the β-ring, as in β-cryptoxanthin (β,β-carotene-3-ol; 2 mM) seems not to decrease the scavenging capacity; however, OH groups at both the C-3 and C-3′ positions, as in zeaxanthin (β,β-carotene-3,3′-diol; 1.4 mM), significantly reduces the antioxidant activity.[41,43] Thus, the requirement for SET mechanisms in the case of $ABTS^{\bullet+}$ may explain the high reactivity of $ABTS^{\bullet+}$ with carotenes with electron-dense structures and its very low reactivity with ketoxanthophylls, which exhibit high electron delocalization across the polyene chain.[42]

Conversely, the ability of carotenoids to absorb light in the visible region of the electromagnetic spectrum seems to interfere in the determination of their scavenging capacity against $DPPH^{\bullet}$ – another stable, synthetic and non-physiological radical. In fact, considering the spectrum-overlapping characteristic observed between the mixture of carotenoids and $DPPH^{\bullet}$, the scavenging capacity of carotenoids against $DPPH^{\bullet}$ seems not to be a reliable approach to determining the antioxidant potential of individual carotenoids or carotenoid-rich extracts, which may also induce erroneous interpretations. Müller *et al.*[43] assessed the antioxidant capacities of 19 carotenoids (6 carotenes and 13 xanthophylls) by various methods, including DPPH assay, to achieve a wide range of technical principles, and surprisingly they reported that none of the compounds exhibited antioxidant activity against $DPPH^{\bullet}$. These authors reported that a dark purple–brown color was produced after mixing a carotenoid solution with $DPPH^{\bullet}$ solution, which interfered with the maximum absorbance of $DPPH^{\bullet}$ (λ_{max} = 515 nm).

Regarding physiologically relevant ROS, the structure–activity relationships of carotenoids as scavengers of $ROO^{\bullet}$ are also well documented.[37,43,44] However, results are conflicting probably due to the use of different reaction systems, such as aqueous buffers (not appropriate for carotenoid analysis), different probes (fluorescein – hydrophilic; or C_{11}-BODIPY$^{581/591}$ – lipophilic) and different radical generators (2,2′-azobis-2,4-dimethyl valeronitrile – lipophilic; or 2,2′-azobis[2-methylpropionamidine] dihydrochloride – hydrophilic). Rodrigues *et al.*[37] established the structure–activity relationships for seven carotenes and eight xanthophylls and concluded that the opening of the β-ionone ring and the increase of chromophore extension were the major factors leading to the increase of the $ROO^{\bullet}$-scavenging capacity of carotenoids, with lycopene being the most efficient scavenger of $ROO^{\bullet}$ (chemical assay), followed by astaxanthin, γ-carotene and bixin. Lycopene (linear chain) was 1.4- and 2.7-times more efficient than γ-carotene (one β-ring) and β-carotene (two β-rings), respectively.[37] Another study reported that the reaction rate constant (k) between β-carotene and $ROO^{\bullet}$ is 0.41 μM minute^{-1}, while this value increases to 0.58 μM minute^{-1} for lycopene.[44] The high reactivity of lycopene with $ROO^{\bullet}$ may be explained by the addition of $ROO^{\bullet}$ to its linear long system, mainly at the allylic hydrogen at C-4/C-4′, since the hydrogen abstraction at other positions is energetically not favored.[45] As has also been described for the ABTS radical, the addition of oxygenated functional groups (OH and CO) in the terminal ring and whether or not it is part of the chromophore also influence the scavenging capacity against $ROO^{\bullet}$. Taking into consideration the same chromophore, as in the case of β-carotene, β-cryptoxanthin and zeaxanthin, the addition of one OH group to the terminal ring (β-cryptoxanthin) did not influence the scavenging capacity; however, the addition of two OH groups (zeaxanthin) increased the antioxidant capacity by 10% in relation to β-carotene (3.24 α-tocopherol relative).[37] In contrast, these authors did not observe any influence when two OH groups were added to the α-carotene structure (10 c.d.b., one β-ring and one ε-ring), which originates lutein with the same chromophore. The increase in the scavenging capacity against $ROO^{\bullet}$ with the increase of the chromophore extension was observed in the case of astaxanthin, which is originated by the addition of two OH and two CO groups to the β-carotene structure. Astaxanthin was two-times more efficient than β-carotene, and if only the addition of the CO groups is considered, comparing astaxanthin and zeaxanthin, the scavenging capacity of astaxanthin increased by about 1.8-times, since these groups are part of the molecule chromophore.[37] The increase of the c.d.b. system thermodynamically favors the antioxidant mechanisms (eqns (8.1)–(8.3)), since the resulting reaction products are more stable.[36] The increase in the scavenging capacity of astaxanthin, as compared to zeaxanthin, may be attributed to the increase in the extension of the chromophore due to the two free electron pairs of the oxygen, which might contribute to the electron resonance system of c.d.b.[22] Moreover, the balance between the keto and enol forms of astaxanthin possibly activates the OH groups, which results in the formation of an *ortho*-dihydroxy-conjugate polyene system that acts as a chain-breaking antioxidant (similar to α-tocopherol).[46]

Carotenoids are known to be the most efficient quenchers of $^1O_2$ and the triplet state of sensitizers, since they play a very important role in photosynthesis as light-harvesting pigments as well as protecting biomolecules against the oxidizing effects of $^1O_2$. Carotenoid molecules quench $^1O_2$ mainly by physical quenching (energy transfer mechanism): in a photosensitized reaction, photons are absorbed by a sensitizer (S), promoting it to its excited form ($^1S^*$) that, by intersystem crossing, furnishes a long-lived, energy-rich state (triplet state) of the sensitizer ($^3S^*$) (eqn (8.4)). Following the reaction, the $^3S^*$ can directly react with molecular oxygen ($^3O_2$), generating $^1O_2$ (eqn (8.5)), and both reactive species ($^3S^*$ and $^1O_2$) can be efficiently quenched by carotenoids due to an efficient energy transfer process, resulting in the carotenoid triplet state ($^3CAR^*$) (eqns (8.6) and (8.7)). Finally, $^3CAR^*$ returns harmlessly to its ground state (CAR), releasing heat (eqn (8.8)), and the same molecule is now able to repeat the quenching process several times. At the same time, but at a low rate, carotenoids can also protect biological targets from $^1O_2$ *via* chemical quenching, resulting in carotenoid oxidation products ($CARO_2$) (eqn (8.9)).[23,24]

$$S \rightarrow {}^1S^* \rightarrow {}^3S^* \quad (8.4)$$

$${}^3S^* + {}^3O_2 \rightarrow S + {}^1O_2 \quad (8.5)$$

$${}^3S^* + CAR \rightarrow S + {}^3CAR^* \quad (8.6)$$

$${}^1O_2 + CAR \rightarrow {}^3O_2 + {}^3CAR^* \quad (8.7)$$

$${}^3CAR^* \rightarrow CAR + heat \quad (8.8)$$

$${}^1O_2 + CAR \rightarrow CARO_2 \quad (8.9)$$

The quenching ability of carotenoids against $^1O_2$ is based on the number of c.d.b., the presence of functional groups and possibly on the oxidation potential.[22] Lycopene (11 c.d.b., linear chain) has been considered the most efficient $^1O_2$ quencher with the highest physical quenching rate constant ($kq = 31 \times 10^9$ M^{-1} s^{-1}).[22] When γ-carotene (11 c.d.b. and one β-ring) and β-carotene (11 c.d.b. and two β-rings) structures are compared to lycopene, the presence of the β-ionone ring decreases the quenching ability of carotenoids, since the linear chain provides a lower energy level.[22] The highest occupied molecular orbital–lowest unoccupied molecular orbital gap influences the ability of carotenoids to quench $^1O_2$, which is related to the ionization energy and the difference between the energies of its singlet and triplet states.[47,48] If the difference of energy is low, the $^1O_2$ quenching ability of carotenoid favors the physical mechanism.[48] Furthermore, the rate constant for the physical quenching of bixin (11 c.d.b. with carboxylate and methyl ester groups instead of two β-rings) ($kq = 14 \times 10^9$ M^{-1} s^{-1}) is the same observed for β-carotene, while crocin (7 c.d.b. and no β-rings) is less efficient ($kq = 1.1 \times 10^9$ M^{-1} s^{-1}), which indicates that the quenching ability is mainly based on the number of c.d.b.[22] As expected, the quenching ability increases by replacing the H in positions 4 and 4′ of the β-ring by carbonyl groups (C=O), as in the case of astaxanthin and cantaxanthin (13 c.d.b.), and this fact can be related to the two free electron pairs that contribute to the electron resonance system of c.d.b.[22] Similar behavior is also observed for the $ROO^•$ scavenging capacity, as shown above. Nevertheless, in the case of mono- or di-hydroxylation, the quenching ability seems to decrease due to chemical reactions that may take place between $^1O_2$ and the OH groups.[22]

The structure–activity relationships of carotenoids against the other ROS and RNS are not well established in the literature; however, a few interesting studies described the antioxidant capacity of carotenoid standards, or extracts with high contents of carotenoids, against such reactive species.[49–52] For example, bixin, the main carotenoid found in annatto seeds, was reported to possess high *in vitro* scavenging capacity against $O_2^{•-}$, H_2O_2, HOCl, $^1O_2$, $^•NO$ and $ONOO^-$ with IC_{50} values at low μg mL^{-1} ranges.[53] Among the tested ROS and RNS, these authors reported high scavenging capacity against HOCl (IC_{50} = 0.3 μg mL^{-1}) and $ONOO^-$ (0.4–1.0 μg mL^{-1}).[53] The antioxidant capacity, antitumor promoting and anticarcinogenesis effects of astaxanthin and lutein and their reaction products with $ONOO^-$, nitroastaxanthin and nitrolutein, were investigated and all the compounds also showed excellent $^1O_2$ quenching activity. Mortensen *et al.*[54] determined the mechanisms and reaction rates of radical scavenging by lycopene, lutein, zeaxanthin, astaxanthin and canthaxanthin against nitrogen dioxide ($NO_2^•$), thiyl ($RS^•$) and sulphonyl ($RSO_2^•$). They reported that all the tested carotenoids reacted with $NO_2^•$ *via* electron transfer reactions to generate carotenoid radical cations ($Car^{•+}$), while $RS^•$ reacts *via* radical addition processes to generate carotenoid-thiyl radical adducts $[RS\text{-}Car]^•$ and $RSO_2^•$ undergoes both radical addition $[RSO_2\text{-}Car]^•$ and electron abstraction $[Car^{•+}]$.[54]

Regarding the antioxidant capacity of carotenoids against ROS in *in vitro* cellular assays, some studies in the literature that describe the structure–activity relationship can be found and they are usually related to $ROO^•$. The results reinforce the notion that lycopene exhibits the highest scavenging efficiency due to its linear polyene chain structure. In a study that investigated the carotenoids usually detected in human blood plasma as potential inhibitors of $ROO^•$-induced hemolysis of human erythrocytes, lycopene is the most efficient one (IC_{50} = 0.24 μM), followed by β-carotene (0.32 μM), lutein (0.38 μM) and zeaxanthin (0.43 μM).[55] When *tert*-butyl-hydroperoxide is the lipid peroxidation initiator, $ROO^•$ and alkoxyl radicals ($RO^•$) are generated and lycopene also proved to be the most efficient carotenoid with the lowest IC_{50} value (2.2 μM), while β-cryptoxanthin, zeaxanthin, lutein and β-carotene also presented antioxidant capacity, although they did not achieve 50% inhibition up to highest tested physiological concentration (3 μM).[56]

8.4 Dietary Sources and Intake of Carotenes and Xanthophylls

The most important sources of carotenoids in the human diet are vegetables and fruits, as their contribution from food products of animal origin, such as dairy products, eggs, some fishes and seafood, is less important. Other sources are colorants, either synthetic and natural carotenoids or carotenoid-rich extracts, which are widely used for the coloration of processed food products. Despite the fact that 70–90 different carotenoids are found in foods, α-carotene, β-carotene, lycopene, phytoene, phytofluene, β-cryptoxanthin, lutein and zeaxanthin account for more than 95% of the total carotenoids typically detected in the human plasma.[57,58] Tables 8.2 and 8.3 shows some food sources of these carotenoids.

Many intrinsic and extrinsic factors influence the carotenoid contents in foods, such as variety/cultivar/genotype/breed, edaphoclimatic and ripening conditions, feed composition, pre- and post-harvest handling, food processing and storage.[59] For example, acerola fruits of cv. Olivier showed about two- to three-times higher β-carotene contents than cv. Waldy Cati 30, both harvested at the same orchard in Brazil[60] (Table 8.2). Differences of up to eight times can be observed in the lycopene contents of different tomato varieties, although

Table 8.2 Food sources of provitamin A carotenoids.

Carotenoids	Foods	Variety and origin	Contents (mg/100 g fresh weight)	References
β-Carotene	Carrot (*Daucus carota* L.)	cv. Nantes (Brazil)	6.2	71
		cv. not described (Spain)	6.6	72
		Type orange (Japan)	5.2	69
	Bok choy (*Brassica rapa* L. ssp. chinensis)	cv. Amur (Switzerland)	3.1 (bed, spring) 3.6 (pot, spring) 3.8 (field, summer) 5.4 (tunnel, summer)	61
	Kale (*Brassica oleracea* L. var. acephala)	cv. not described (Portugal)	2.6–6.4	73
	Tossa jute (*Corchorus olitorius* L.)	cv. not described (Japan)	12.0	69
	Pumpkin (*Cucurbita moschata* L.)	Genotype 12 (Brazil)	38.7 (all-*E*), 8.4 (Σ*Z*)	66
		cv. Menina Brasileira (Brazil)	6.7	65
	Spinach (*Spinacia oleracea* L.)	cv. not described (Spain)	3.3	72
	Acerola (*Malpighia punicifolia* L.)	cv. Waldy Cati 30 (Brazil)	0.3–0.6	60
		cv. Olivier (Brazil)	0.9–1.7	60
		Type sweet (Japan)	1.2	74
	Banana (*Musa* spp.)	cv. Fe'i (Solomon Islands)	5.9	70
	Buriti (*Mauritia vinifera* Mart.)	cv. not described (Brazil)	37.2 (all-*E*), 8.7 (Σ*Z*)	67
	Mango (*Mangifera indica* L.)	cv. Keitt (Brazil)	0.7–1.5	75
		cv. Tommy Atkins (Brazil)	0.6	75
		Not described (Indonesia)	0.1–1.2	76
		cv. Irwin (Japan)	2.0	74
	Papaya (*Carica papaya* L.)	cv. Sun rise (Japan)	2.0	74
α-Carotene	Carrot (*Daucus carota* L.)	cv. Nantes (Brazil)	3.5	71
		cv. not described (Spain)	2.9	72
		Type orange (Japan)	2.3	69
	Pumpkin (*Cucurbita moschata* L.)	Genotype 12 (Brazil)	18.6	66
		cv. Menina Brasileira (Brazil)	2.7	65
	Banana (*Musa* spp.)	cv. Fe'i (Solomon Islands)	2.4	70

(*continued*)

Table 8.2 (*continued*)

Carotenoids	Foods	Variety and origin	Contents (mg/100 g fresh weight)	References
β-Crypto-xanthin	Orange (*Citrus sinensis* L.)	var. Navel Lane late (Portugal)	0.1–0.2	73
		var. Valencia (Corsica)	0.2	62
		var. Valencia (Costa Rica)	0.06	62
	Mandarin (*Citrus reticulata* L.)	cv. Saga mandarin (Japan)	1.4	74
		cv. Youko (Japan)	1.2	74
		cv. Ponkan (Japan)	1.1	74
		cv. Ponkan (Brazil)	0.9 (all-*E*), 0.3 (Σ*Z*)	83
		cv. Hansen (Corsica)	1.5	62
		cv. Hansen (New Caledonia)	1.6	62
	Tangor (*Citrus reticulata* × *Citrus sinensis*)	cv. Harumi (Japan)	1.3	74
		cv. Murcott (Brazil)	0.9 (all-*E*), 0.3 (Σ*Z*)	83
	Papaya (*Carica papaya* L.)	cv. Golden (Brazil)	0.3 (all-*E*)	81
		cv. Sun rise (Japan)	3.2	74
	Loquat (*Eriobotrya japonica* L.)	cv. Mizumo IAC 1567-411 (Brazil)	0.7 (all-*E*)	82

these are cultivated in different environments (Table 8.3). The influence of season and type of cultivation (bed, pot, tunnel, field) was evaluated in a green leafy grown in Switzerland; β-carotene production tended to be higher in summer while no differences in lutein contents were observed between spring and summer cultivations.[61] In the same line, geographic location differently affected the carotenoids in mandarin cv. Hansen: five-fold differences were observed in terms of the concentrations of colorless carotenes, while β-cryptoxanthin levels remained similar.[62]

Manipulation of growing conditions and choice of adequate genotypes can be strategies to improve carotenoid concentration. For example, the HarvestPlus program applies conventional breeding for biofortification to enhance provitamin A contents in different crops;[63] in the case of pumpkins,[64] all five biofortified genotypes achieved more than five-times the provitamin A content of native pumpkins.[65,66] Despite the seasonal and regional availability, another strategy would be the propagation of foods rich in carotenoids from native biodiversity, such as buriti from Brazil,[67] sastra from Panama,[68] tossa jute from Japan[69] and banana from Solomon Islands[70] (Table 8.2).

β-Carotene is present in most fruits and vegetables;[60,61,65–67,69–76] however, α-carotene is provided by a limited number of foods.[65,66,69–72] The major contributors to the dietary daily intake of these carotenes are pumpkins, carrots and lettuce in Brazil,[77] carrots in Spain[78] and Luxembourg[79] and carrots and pumpkins in Australia.[80]

Table 8.3 Food sources of some non-provitamin A carotenoids.

Carotenoids	Foods	Variety and origin	Contents (mg/100 g fresh weight)	References
Lycopene	Papaya (*Carica papaya* L.)	cv. Golden (Brazil)	1.3 (all-*E*), 0.1 (Σ*Z*)	81
		cv. not described (Indonesia)	4.3–7.6	76
		cv. Sun rise (Japan)	1.5	74
	Tomato (*Lycopersicon esculentum* Mill)	cv. Carmen (Brazil)	3.5	71
		var. Lido (Portugal)	8	84
		var. "comum" (Spain)	2.1	72
		Type raf (Spain)	3.7	85
		var. cherry (Spain)	0.8	85
		cv. Daniela (Spain)	1.9	85
		Type Pomodorino Vesuviano (Italy)	7.9 (all-*E*), 0.7 (Σ*Z*)	86
	Watermelon (*Citrullus lamatus* Thunb)	cv. Crimson Sweet (Brazil)	3.5	71
		cv. Crimson Sweet (USA)	3.7	87
		cv. Dixie Lee (USA)	6.9	87
	Guava (*Psidium guajava* L.)	cv. not described (Japan)	2.3	74
Phytoene	Carrot (*Daucus carota* L.)	cv. not described (Luxembourg)	1.4	79
	Red pepper (*Capsicum annuum* L.)	Type red (Luxembourg)	1.7	79
	Tomato (*Solanum lycopersicum* Mill)	cv. not described (Luxembourg)	1.4	79
	Acerola (*Malpighia punicifolia* L.)	Type sweet (Japan)	11.5	74
	Apricot (*Prunus armeniaca* L.)	cv. not described (Luxembourg)	7.2	79
	Mandarin (*Citrus reticulata* L.)	cv. Saga mandarin (Japan)	0.4	74
		cv. Youko (Japan)	2.5	74
		cv. Hansen (Corsica)	0.5	62
		cv. Hansen (New Caledonia)	0.1	62
	Tangor (*Citrus reticulata* × *Citrus sinensis*)	cv. Harumi (Japan)	0.9	74
Phytofluene	Carrot (*Daucus carota* L.)	cv. not described (Luxembourg)	0.6	79
	Red pepper (*Capsicum annuum* L.)	Type red (Luxembourg)	0.5	79
	Tomato (*Lycopersicon esculentum* Mill)	cv. not described (Luxembourg)	0.4	79
	Apricot (*Prunus armeniaca* L.)	cv. not described (Luxembourg)	2.4	79

(*continued*)

Table 8.3 *(continued)*

Carotenoids	Foods	Variety and origin	Contents (mg/100 g fresh weight)	References
	Mandarin (*Citrus reticulata* L.)	cv. Hansen (Corsica)	0.5	62
		cv. Hansen (New Caledonia)	0.1	62
Lutein	Kale (*Brassica oleracea* L. var. acephala)	cv. not described (Portugal)	3.7–7.2	73
	Brassica rapa L. ssp. chinensis	cv. Amur (Switzerland)	6.3 (bed, spring) 7.1 (pot, spring) 9.4 (field, summer) 8.6 (tunnel, summer)	61
	Tossa jute (*Corchorus olitorius* L.)	cv. not described (Japan)	13.6	69
Zeaxanthin	Sweet pepper (*Capsicum annuum* L.)	Type orange (Japan)	10.2	69
	Sastra/achachairu (*Garcinia intermedia* [Pittier] Hammel)	cv. not described (Panama)	8.5	68
	Membrillo (*Gustavia superba* [Kunth] O. Berg)	cv. not described (Panama)	3.8	68

β-Cryptoxanthin is mainly found in fruits, such as papaya[74,81] and some cultivars of loquat[82] and citrus, mandarins[62,74,83] and tangor,[74,83] showing higher levels than oranges.[62,73] Considering the daily intake in all the four countries mentioned above, the major contributors of β-cryptoxanthin to the diet are citrus fruits, oranges and tangerines.[77–80]

The major sources of lycopene are tomato,[71,72,84–86] papaya,[74,76,81] red guava[74] and watermelon,[71,87] with tomato and its products being the major contributors to lycopene daily intake in Australia, Brazil, Spain and Luxembourg.[77,79,80,88] Watermelon also plays an important role in lycopene daily intake in Australia and Brazil.

Phytoene and phytofluene are widely distributed in fruits and vegetables; however, these carotenes have been less studied.[89] The contents of phytoene are usually higher than those of phytofluene (*e.g.* in carrot, red pepper, tomato and apricot);[79] however, similar amounts of both carotenes are found in mandarin cv. Hansen.[62] The major contributors to the daily intake of phytoene and phytofluene in Luxembourg are tomatoes (raw and sauce), carrots and citrus fruits;[79] however, data on daily intake of these colorless carotenes were not reported for Australia, Brazil and Spain. In fact, as far as we are concerned, the daily intakes of phytoene and phytofluene are only available for Luxembourg.

Table 8.4 Daily intake per capita (mg) of carotenoids in different countries.[a,c]

Carotenoids	Australia	Brazil	Luxembourg	Spain	United States
α- + β-Carotene	7.7	1.1	7.6	1.6	2.3
Lycopene	3.3	0.7	1.8	3.0	5.1
Phytoene	ND	ND	2.0	ND	ND
Phytofluene	ND	ND	0.7	ND	ND
β-Cryptoxanthin	0.2	0.1	1.4	0.2	0.2
Lutein	0.8[b]	0.8	1.5	0.7	1.4[b]
Zeaxanthin	ND	0.01	0.3	0.01	ND

[a]ND: not determined.
[b]Value is the sum of lutein and zeaxanthin contributions.
[c]Data sources: Ref. 77–80, 88 and 90.

Lutein is widely distributed among fruits and vegetables, but its best sources are green leafy vegetables,[61,69,73] such as those from *Brassica* genus[61,73] (Table 8.3). Kale, oranges and curly lettuce are the main contributors to the daily intake of lutein in Brazil,[77] spinach, Swiss chard and tomatoes (fresh and sauce) in Spain[88] and green leafy vegetables in Luxembourg.[79] In Australia, broccoli, green beans and oranges are the major contributors to the daily intake of lutein and zeaxanthin.[80] Zeaxanthin is found in very few foods, usually in low levels, except in some types of pepper[69] and exotic fruits.[68] The major contributors to the daily intake are oranges in Brazil,[77] bell peppers in Luxembourg[79] and peppers (green and red), spinach and oranges in Spain.[88]

The daily dietary intake per person of provitamin A carotenoids (α-carotene, β-carotene, β-cryptoxanthin) is much higher in Australia[80] and Luxembourg[79] when compared to the United States,[90] Spain[78] and Brazil[77] (Table 8.4). Also, the daily dietary intake per person of total carotenoids is much higher in Luxembourg (15.3 mg) and Australia (12.0 mg) compared to the United States (9.0 mg), Spain (5.6 mg) and Brazil (2.7 mg). The higher carotenoid intake level in Luxembourg can be explained, at least in part, by the inclusion of colorless carotenes, which was not taken in account in the countries. Other important sources of differences in total carotenoid daily intake are related to different dietary assessments, the age range of the subjects among the countries and the foods assessed for the intake (only the contributions of fresh fruits and vegetables were considered for the intake of carotenoids in Brazil, while processed foods were considered in the other above-mentioned studies). It is important to remember that these studies do not account for the foods consumed only regionally and not distributed across the country (*e.g.* most fruits from Amazonia except açai pulp, such as buriti, which does not easily reach other more populated states in Brazil).

Finally, it is important to highlight that not only the carotenoid composition and daily intake in different foods should be considered, but also the bioavailability of carotenoids from specific sources; however, this aspect is not included in food composition tables, nor in the scope of studies mentioned above.

8.5 Health-related Properties of Carotenoids

Dietary intake of fruits and vegetables is positively correlated to plasma total carotenoid levels.[91] However, carotenoids are found in human plasma always in low concentrations and only a few are normally detected, as previously mentioned.[57,58,92] Higher amounts of carotenoids are accumulated in the liver, adrenal gland and reproductive tissues, with levels being even ten-times greater than in lipid tissues.[92]

The main physiological role of carotenoids in humans is their provitamin A activity. Vitamin A is an essential nutrient for humans and its deficiency is responsible for visual anomalies such as night blindness and xerophthalmia, besides compromising the immune system and being implicated in an increased incidence or severity of respiratory infections, gastrointestinal infections and even measles.[93] Vitamin A deficiency is a public health issue affecting preschool-age children and pregnant women in developing countries, especially in Africa and South-East Asia. The WHO Global Database on Vitamin A Deficiency (http://www.who.int/vmnis/database/vitamina/en/ – accessed July 26, 2017)[94] worldwide estimates are that night blindness affects 5.2 million preschool-age children and 9.8 million pregnant women, while low serum retinol (below 0.70 μmol L^{-1}) affects about 190 million preschool-age children and 19.1 million pregnant women.

Conversion of β-carotene to retinol by β,β-carotene 15,15′-oxygenase (BCO1), a dioxygenase that incorporates only oxygen from O_2 into retinal,[95] and reductase enzymes (Figure 8.4) occurs mainly in the liver and intestine,[93,96] but fat, lungs and kidneys are also capable of converting β-carotene into retinoids.[93] BCO1 was also found in the skin, eyes and other tissues.[93] Retinol is taken up by intestinal cells, esterified to retinyl ester by lecithin : retinol acyltransferase for incorporation into chylomicrons and finally absorbed *via* the lymphatic blood system or the portal circulation.[96]

After β-carotene ingestion, both β-carotene and its metabolite, retinol, can be found in plasma since only half of the amount ingested through the diet is converted to retinol.[93,96] Only carotenoids possessing a non-substituted β-ring terminal group can be converted to vitamin A (retinol), such as β-carotene, α-carotene, β-cryptoxanthin and *Z* (*cis*) isomers of β-carotene (Figure 8.1). In fact, the provitamin A activity of carotenoids other than (all-*E*) β-carotene, which possess two unsubstituted β-ionone rings, are around 10–50%.[96]

To guide dietary recommendations of provitamin A carotenoid intake, conversion factors from β-carotene to vitamin A (Table 8.5) can be applied to estimate the provitamin A activity correspondent to the ingested carotenoid content from food. The US Institute of Medicine (IOM)[97] introduced the term "retinol activity equivalent" in substitution for "retinol equivalent" used by Food and Agriculture Organization/World Health Organization (FAO/WHO),[98] due to the evidence of lower rates of carotenoid conversion to vitamin A in humans. However, it is important to point out that both bioavailability and conversion of carotenoids to vitamin A are affected by factors related to the

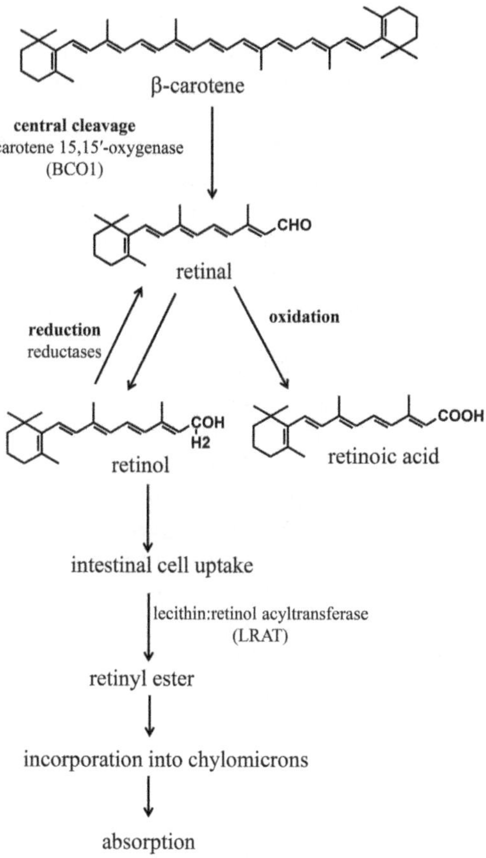

Figure 8.4 Conversion of β-carotene into vitamin A and its absorption.

Table 8.5 Conversion factors for provitamin A carotenoids to vitamin A.[a]

Unit	Conversion factor/vitamin A equivalency
1 µg RAE	1 µg retinol 2 µg β-carotene in supplements 12 µg β-carotene in mixed foods 24 µg other provitamin A carotenoids[b] in mixed foods 0.5 µg RE
1 µg RE	1 µg retinol 2 µg β-carotene in supplements 6 µg β-carotene in mixed foods 12 µg other provitamin A carotenoids[b] in mixed foods 2 µg RAE

[a]RAE: retinol activity equivalent; RE: retinol activity equivalent.
[b]Other provitamin A carotenoids = α-carotene and β-cryptoxanthin.

ingested food, such as matrix, processing, co-ingestion with fat or fiber, preformed vitamin A and other carotenoids, and also to individual humans, such as deficiency or excess of some nutrients, including vitamin A, genetic polymorphisms related to carotenoid metabolism and gut conditions, among others.[99]

Although ingestion of β-carotene is considered a safe source of vitamin A, since there is an inverse relationship between oral doses of β-carotene and intestinal conversion to vitamin A,[100] high doses of supplemental β-carotene (20–30 mg day^{-1}) alone[101] or in combination with preformed vitamin A (25 000 IU)[102] were found to increase the risk of lung cancer and development of cardiovascular diseases in smokers and people exposed to asbestos during the ATBC (Alpha-Tocopherol, Beta-Carotene Cancer Prevention Study)[101] and CARET (Carotene and Retinol Efficacy Trial)[102] intervention studies. Depletion of the retinoic acid in the tissues reduced retinoid signaling, and RAR-β gene expression was observed in ferrets exposed to tobacco smoke and receiving high doses of supplemental β-carotene, equivalent to 30 mg day^{-1} in humans, possibly due to β-carotene and its metabolites, especially excentric cleavage oxidative metabolites formed due to oxidation induced by ROS from cigarette smoke, suggesting that these alterations could be related to the enhanced risk of lung cancer.[103,104] On the other hand, animals consuming lower doses of supplemental β-carotene possibly produced lower amounts of excentric cleavage oxidative metabolites, which could have exerted beneficial antiproliferative effects, and metabolized some retinoid acid instead, contributing to the vitamin A supply.[103,104]

The Physicians' Health Study I (PHS I) did not show any effects regarding β-carotene supplementation (50 mg of supplemental β-carotene every other day for 12 years) on cancer and heart disease incidence in a healthy population.[105] However, a subgroup with the lowest baseline levels of β-carotene in the blood presented some potential decreases in terms of total and prostate cancer incidence,[105] a fact that is similar to that observed in the Linxian General Population Nutrition Intervention Trial (NIT) with daily supplementation with β-carotene (15 mg) combined with vitamin E (30 mg) and selenium (50 μg) to a malnourished population from Linxian, China, which lowered the mortality rate from cancer, especially stomach cancer, after 6 years of treatment.[106] Furthermore, 10 years after ceasing the supplementation, the participants from NIT who received β-carotene, vitamin E and selenium supplementation showed lower mortality (5%) and gastric cancer incidence (11%), especially those younger than 55 years of age.[107]

The results of several epidemiological studies suggest that carotenoids and foods rich in carotenoids are associated with a decreased risk of prostate cancer, but they are not always consistent.[108,109] The interest in studying lycopene as an anti-prostate cancer agent came from the positive results in small clinical studies and *in vitro* studies measuring biomarkers of prostate cancer that highlighted lycopene as a promising agent for the prevention of this type of cancer.[110] Several mechanisms were proposed to explain the anticarcinogenic action of lycopene, such as inhibition of cancerous cell proliferation at the G0–G1 cell cycle phase, protection of biomolecules

(DNA, lipoproteins) against oxidation,[111] activation of phase II drug metabolism and improvement of junction communication.[112] Meta-analysis studies that investigated the association between dietary intake of lycopene or tomato/tomato products and the risk of prostate cancer found that high lycopene levels in plasma or high lycopene ingestion were inversely associated with prostate cancer risk.[111,113–115] Considering the dose response for dietary lycopene, a decrease of 1–3% in the risk of prostate cancer was found for each additional consumption of 1–2 mg of lycopene per day; however, no association was found with a reduced risk of advanced prostate cancer.[113,115] Similar trends were observed for high α-carotene intake and blood levels association with decreased risk of prostate cancer, but not for β-carotene.[113]

Up to now, most of the trials were conducted testing the effects of complex foods containing lycopene, such as tomatoes and tomato products, rather than the isolated effects of lycopene consumption.[108] The estimate of lycopene intake from food sources is mainly obtained from food diaries and food frequency questionnaires and calculated based on lycopene contents available from food composition databases. However, these values are not totally accurate since tomatoes and tomato composition vary between varieties and brands, among other factors. Thus, it would be more logical to measure lycopene plasma concentration to try to establish a relationship between dietary lycopene intake and the risk of prostate cancer in epidemiological studies. Some studies show a positive correlation between food products rich in lycopene and serum lycopene concentration.[116–118] However, dietary intake of lycopene is not always strongly correlated with plasma concentrations (r = 0.16–0.47), possibly because of the differences in lycopene bioavailability due to individuals (age, genetic factors, co-ingestion with other food components, *etc.*) or food sources (*e.g.* cooked *versus* raw tomatoes).[110] In this sense, higher preventive effects against prostate cancer were shown for cooked tomato products than raw tomatoes.[111] This effect could be attributed to the high concentration of lycopene in tomato products such as tomato sauces and pulps, but also to the higher bioavailability of lycopene in thermally processed foods, which in many cases also contain vegetable oil in the formulation.[119] In addition, the possible protective effects of other bioactive compounds present in tomatoes and tomato products on prostate cancer risk cannot be disregarded.[120] In fact, tomato powder was more effective than lycopene in the inhibition of carcinogenesis in rodent models, indicating that other phytochemicals could be implicated in the etiology of prostate cancer prevention.[121,122]

A diet rich in fruits and vegetables is also associated with a reduced risk of cardiovascular disease, and an inverse relationship between lycopene concentrations in blood and tissues and the risk of cardiovascular disease has been observed in epidemiological studies,[123–126] such as in the reduction of the development of atherosclerosis,[123] myocardial infarction risk,[124] risk of acute coronary events or stroke[125] and increased low-density lipoprotein cholesterol levels.[127] A systematic review comparing the efficacy of supplemental lycopene with consumption of tomato products on the risk factors of cardiovascular disease showed that, in general, consumption of tomato products resulted in greater improvements in biomarkers of oxidative stress, inflammation, endothelial function and lipid metabolism than supplemental lycopene, while the opposite was observed for blood pressure control.[128]

Lutein, zeaxanthin and *meso*-zeaxanthin accumulate in the eye macula, which is the central part of the human retina responsible for visual acuity, as well as being responsible for the yellowish color of this ocular region.[129–131] Lutein and zeaxanthin are not synthetized by humans; therefore, they must be obtained from the diet, mainly from green leafy vegetables, while *meso*-zeaxanthin is probably formed at the macula by metabolic transformations of dietary carotenoids.[129]

Age-related macular degeneration (AMD) is a degenerative disease that compromises the macula, leading to loss of central vision, and is the major cause of irreversible blindness in the elderly in developed countries.[129–131] The capacity of lutein to scavenge ROS and RNS and its action as a blue light filter (~450 nm), preventing the formation of reactive species in the macula, are the primary mechanisms of protection against AMD.[130,131] In fact, metabolites of lutein and zeaxanthin, such as *meso*-zeaxanthin, 3′-epilutein and 3′-oxolutein, a direct oxidation product of lutein present in monkey retinas and human eye tissues,[131,132] were detected in primate retinas, lens and uveal tracts,[131–133] corroborating that xanthophyll oxidation is one of the protection mechanisms against ROS and RNS oxidation.[134]

The Age-Related Eye Disease Study (AREDS), conducted by the National Eye Institute (NEI), established that intake of high daily doses of supplemental vitamins C (500 mg day^{-1}) and E (400 IU day^{-1}), β-carotene (15 mg day^{-1}), zinc (8 mg day^{-1}) and copper (2 mg day^{-1}) (AREDS formulation) can slow down the progress of AMD.[130,135] However, in the same period of the AREDS study, the results of the ATBC[101] and CARET[102] studies were published, and concerns about supplementation of smokers, who constitute nearly half of the population at risk of AMD, with high doses of β-carotene were raised.[136] In addition, β-carotene is not detectable in the retina, and results of the Eye Disease Case–Control study showed that lutein and zeaxanthin would be more appropriate physiologically for supplementation considering the risk of AMD progression.[137] Thus, other formulations containing ω-3 fatty acids, lutein (10 mg day^{-1}), zeaxanthin (2 mg day^{-1}) and a low level of zinc were tested in the AREDS2 follow-up study.[136] Smokers and former smokers received a formulation without β-carotene due to the risk of lung cancer. No benefits were observed due to the incorporation of ω-3 fatty acids or a mixture of lutein and zeaxanthin (5 : 1) to the original formulation. However, the subgroups receiving the AREDS2 formulation without β-carotene, especially those subjects with low lutein and zeaxanthin ingestion through the diet, showed 18–25% lower risks of AMD than the non-supplemented subjects or subgroups receiving the AREDS formulation containing only β-carotene.[129,130,136] Nowadays, the AREDS2 formulation is the standard recommendation for reducing the risk of advanced AMD in patients slowing progression to severe visual loss.[130]

Lutein also accumulates in the human brain, both in infants and the elderly, and in macaque monkeys.[138–140] There is evidence that lutein accumulation in neural tissues, such as retinal and brain tissues, is related to the fact that the BCO2 enzyme cannot cleave this carotenoid.[141] Lutein content in both the macula and the brain tissues has been correlated with cognitive function in adults,[138,139,142,143] and there is also evidence of this from a clinical trial with lutein supplementation.[144] Since polyunsaturated fatty acids in the brain are susceptible to ROS and RNS oxidative attack, it is highly possible that one of the mechanisms by which lutein protects the brain functions is by scavenging these reactive species; however, these mechanisms are not yet completely known.[145]

There is also evidence that carotenoid intake and plasma concentrations are positively related to bone health in men and women.[146,147] However, high intakes of vitamin A were associated with vitamin D insufficiency and loss of bone density, which are risk factors for osteoporosis.[148,149]

Colorless carotenoids, such as phytoene and phytofluene, probably exert health benefits; however, these carotenoids are usually not considered in food databases and epidemiologic studies. Since they are present in foods with high contents of lycopene, like tomatoes and tomato products, it is possible that part of the beneficial effects attributed to this carotenoid could be due to the presence of colorless carotenoids.[89]

Acknowledgements

AZM acknowledges the financial support of the São Paulo Research Foundation (FAPESP) (grant 2013/07914-8) and CNPq (Conselho Nacional de Desenvolvimento Científico e Tecnológico, Brazil, grants 308484/2014-2 and 402463/2016-1) and RCC acknowledges the financial support of CNPq through grant 403121/2016-7.

The authors declare no competing financial interest.

References

1. A. Z. Mercadante, D. B. Rodrigues, F. C. Petry and L. R. B. Mariutti, *Food Res. Int.*, 2017, **99**, 830–850.
2. Report of IUPAC Commission on Nomenclature of Organic Chemistry and IUPAC-IUB Commission on Biochemical Nomenclature - Nomenclature of Carotenoids (Rules approved 1974), *Pure Appl. Chem.*, 1975, **41**, 405.
3. G. Britton, *Carotenoids*, Birkhäuser, Basel, 1995.
4. M. Nguyen, D. Francis and S. Schwartz, *J. Sci. Food Agric.*, 2001, **81**, 910–917.
5. C. D. Borsarelli and A. Z. Mercadante, *Thermal and Photochemical Degradation of Carotenoids*, CRC Press, Boca Raton, 2009.
6. C. Caris-Veyrat, A. Schmid, M. Carail and V. Böhm, *J. Agric. Food Chem.*, 2003, **51**, 7318–7325.

7. L. Q. Zepka and A. Z. Mercadante, *Food Chem.*, 2009, **117**, 28–34.
8. L. Q. Zepka, C. D. Borsarelli, M. A. A. P. da Silva and A. Z. Mercadante, *J. Agric. Food Chem.*, 2009, **57**, 7841–7845.
9. L. Q. Zepka, D. S. Garruti, K. L. Sampaio, A. Z. Mercadante and M. A. A. P. Da Silva, *Food Res. Int.*, 2014, **56**, 108–114.
10. A. Gomes, E. Fernandes and J. L. F. C. Lima, *J. Biochem. Biophys. Methods*, 2005, **65**, 45–80.
11. A. Gomes, E. Fernandes and J. L. F. C. Lima, *J. Fluoresc.*, 2006, **16**, 119–139.
12. M. Valko, D. Leibfritz, J. Moncol, M. T. Cronin, M. Mazur and J. Telser, *Int. J. Biochem. Cell Biol.*, 2007, **39**, 44–84.
13. E. Birben, U. M. Sahiner, C. Sackesen, S. Erzurum and O. Kalayci, *World Allergy Organ. J.*, 2012, **5**, 9–19.
14. E. Choe and D. B. Min, *Crit. Rev. Food Sci. Nutr.*, 2006, **46**, 1–22.
15. J. Y. Lee, S. Min, E. O. Choe and D. B. Min, *J. Food Sci.*, 2003, **68**, 1933–1937.
16. E. Choe and D. B. Min, *J. Food Sci.*, 2005, **70**, R142–R159.
17. R. C. Chisté, M. Freitas, A. Z. Mercadante and E. Fernandes, *Curr. Med. Chem.*, 2015, **22**, 4234–4256.
18. K. H. Cheeseman and T. F. Slater, *Br. Med. Bull.*, 1993, **49**, 481–493.
19. H. Jaeschke, *Proc. Soc. Exp. Biol. Med.*, 1995, **209**, 104–111.
20. S. X. Chen and P. Schopfer, *Eur. J. Biochem.*, 1999, **260**, 726–735.
21. Y. Xia and J. L. Zweier, *Anal. Biochem.*, 1997, **245**, 93–96.
22. P. Di Mascio, S. Kaiser and H. Sies, *Arch. Biochem. Biophys.*, 1989, **274**, 532–538.
23. M. A. Montenegro, O. Rios Ade, A. Z. Mercadante, M. A. Nazareno and C. D. Borsarelli, *J. Agric. Food Chem.*, 2004, **52**, 367–373.
24. A. O. Rios, A. Z. Mercadante and C. D. Borsarelli, *Dyes Pigm.*, 2007, **74**, 561–565.
25. G. L. Squadrito and W. A. Pryor, *Free Radical Biol. Med.*, 1998, **25**, 392–403.
26. M. Freitas, J. L. F. C. Lima and E. Fernandes, *Anal. Chim. Acta*, 2009, **649**, 8–23.
27. P. Kostka, *Anal. Chem.*, 1995, **67**, 411R–416R.
28. T. Nagano, *Luminescence*, 1999, **14**, 283–290.
29. N. Miyasaka and Y. Hirata, *Life Sci.*, 1997, **61**, 2073–2081.
30. R. E. Shackelford, W. K. Kaufmann and R. S. Paules, *Free Radical Biol. Med.*, 2000, **28**, 1387–1404.
31. K. I. Priyadarsini, S. M. Khopde, S. S. Kumar and H. Mohan, *J. Agric. Food Chem.*, 2002, **50**, 2200–2206.
32. N. I. Krinsky and E. J. Johnson, *Mol. Aspects Med.*, 2005, **26**, 459–516.
33. A. V. Rao and L. G. Rao, *Pharmacol. Res.*, 2007, **55**, 207–216.
34. J. Fiedor and K. Burda, *Nutrients*, 2014, **6**, 466–488.
35. A. Kaulmann and T. Bohn, *Nutr. Res.*, 2014, **34**, 907–929.
36. A. El-Agamey, G. M. Lowe, D. J. McGarvey, A. Mortensen, D. M. Phillip, T. G. Truscott and A. J. Young, *Arch. Biochem. Biophys.*, 2004, **430**, 37–48.

37. E. Rodrigues, L. R. Mariutti, R. C. Chisté and A. Z. Mercadante, *Food Chem.*, 2012, **135**, 2103–2111.
38. R. L. Prior, X. Wu and K. Schaich, *J. Agric. Food Chem.*, 2005, **53**, 4290–4302.
39. D. Huang, B. Ou and R. L. Prior, *J. Agric. Food Chem.*, 2005, **53**, 1841–1856.
40. D. Baltschun, S. Beutner, K. Briviba, H.-D. Martin, J. Paust, M. Peters, S. Röver, H. Sies, W. Stahl, A. Steigel and F. Stenhorst, *Liebigs Ann./Recl.*, 1997, **1997**, 1887–1893.
41. N. J. Miller, J. Sampson, L. P. Candeias, P. M. Bramley and C. A. Rice-Evans, *FEBS Lett.*, 1996, **384**, 240–242.
42. C. A. Rice-Evans, J. Sampson, P. M. Bramley and D. E. Holloway, *Free Radical Res.*, 1997, **26**, 381–398.
43. L. Müller, K. Fröhlich and V. Böhm, *Food Chem.*, 2011, **129**, 139–148.
44. A. A. Woodall, S. W. Lee, R. J. Weesie, M. J. Jackson and G. Britton, *Biochim. Biophys. Acta*, 1997, **1336**, 33–42.
45. M. G. Simic, *Methods Enzymol.*, 1992, **213**, 444–453.
46. Y. M. Naguib, *J. Agric. Food Chem.*, 2000, **48**, 1150–1154.
47. Á. Vázquez-Mayagoitia, R. Vargas, J. A. Nichols, P. Fuentealba and J. Garza, *Chem. Phys. Lett.*, 2006, **419**, 207–212.
48. D. L. Monego, M. B. da Rosa and P. C. do Nascimento, *Food Chem.*, 2017, **217**, 37–44.
49. R. C. Chisté, M. Freitas, A. Z. Mercadante and E. Fernandes, *Food Chem.*, 2012, **135**, 1740–1749.
50. A. B. Ribeiro, R. C. Chisté, M. Freitas, A. F. da Silva, J. V. Visentainer and E. Fernandes, *Food Chem.*, 2014, **165**, 140–148.
51. A. Berto, A. B. Ribeiro, E. Sentandreu, N. E. de Souza, A. Z. Mercadante, R. C. Chisté and E. Fernandes, *Food Funct.*, 2015, **6**, 3081–3090.
52. É. O. Barizão, J. V. Visentainer, V. de Cinque Almeida, D. Ribeiro, R. C. Chisté and E. Fernandes, *Food Res. Int.*, 2016, **86**, 24–33.
53. R. C. Chisté, A. Z. Mercadante, A. Gomes, E. Fernandes, J. L. Lima and N. Bragagnolo, *Food Chem.*, 2011, **127**, 419–426.
54. A. Mortensen, L. H. Skibsted, J. Sampson, C. Rice-Evans and S. A. Everett, *FEBS Lett.*, 1997, **418**, 91–97.
55. R. C. Chisté, M. Freitas, A. Z. Mercadante and E. Fernandes, *J. Food Sci.*, 2014, **79**, H1841–H1847.
56. R. C. Chisté, M. Freitas, A. Z. Mercadante and E. Fernandes, *Life Sci.*, 2014, **99**, 52–60.
57. B. Olmedilla, F. Granado, S. Southon, A. J. Wright, I. Blanco, E. Gil-Martinez, H. Berg, B. Corridan, A. M. Roussel, M. Chopra and D. I. Thurnham, *Br. J. Nutr.*, 2001, **85**, 227–238.
58. F. Khachik, C. J. Spangler, J. C. Smith Jr, L. M. Canfield, A. Steck and H. Pfander, *Anal. Chem.*, 1997, **69**, 1873–1881.
59. A. Z. Mercadante, *Carotenoids in Foods: Sources and Stability During Processing and Storage*, CRC Press, Boca Raton, 2008.
60. V. V. De Rosso and A. Z. Mercadante, *Food Res. Int.*, 2005, **38**, 1073–1077.

61. C. Reif, E. Arrigoni, F. Berger, D. Baumgartner and L. Nyström, *LWT–Food Sci. Technol.*, 2013, **53**, 378–381.
62. C. Dhuique-Mayer, A.-L. Fanciullino, C. Dubois and P. Ollitrault, *J. Agric. Food Chem.*, 2009, **57**, 9160–9168.
63. H. E. Bouis and A. Saltzman, *Global Food Secur.*, 2017, **12**, 49–58.
64. A. Saltzman, E. Birol, H. E. Bouis, E. Boy, F. F. De Moura, Y. Islam and W. H. Pfeiffer, *Global Food Secur.*, 2013, **2**, 9–17.
65. C. H. Azevedo-Meleiro and D. B. Rodriguez-Amaya, *J. Agric. Food Chem.*, 2007, **55**, 4027–4033.
66. E. M. G. Ribeiro, C. Chitchumroonchokchai, L. M. J. de Carvalho, F. F. de Moura, J. L. V. de Carvalho and M. L. Failla, *Food Res. Int.*, 2015, **77**, 620–626.
67. V. V. De Rosso and A. Z. Mercadante, *J. Agric. Food Chem.*, 2007, **55**, 5062–5072.
68. E. Murillo, A. J. Meléndez-Martínez and F. Portugal, *Food Chem.*, 2010, **122**, 167–172.
69. K. Aizawa and T. Inakuma, *Food Sci. Technol. Res.*, 2007, **13**, 247–252.
70. L. Englberger, G. Lyons, W. Foley, J. Daniells, B. Aalbersberg, U. Dolodolotawake, C. Watoto, E. Iramu, B. Taki, F. Wehi, P. Warito and M. Taylor, *J. Food Compos. Anal.*, 2010, **23**, 624–632.
71. P. Y. Niizu and D. B. Rodriguez-Amaya, *J. Food Compos. Anal.*, 2005, **18**, 739–749.
72. F. Granado, B. Olmedilla, I. Blanco and E. Rojas-Hidalgo, *J. Agric. Food Chem.*, 1992, **40**, 2135–2140.
73. M. G. Dias, M. F. G. F. C. Camões and L. Oliveira, *Food Chem.*, 2009, **113**, 808–815.
74. M. Yano, M. Kato, Y. Ikoma, A. Kawasaki, Y. Fukazawa, M. Sugiura, H. Matsumoto, Y. Oohara, A. Nagao and K. Ogawa, *Food Sci. Technol. Res.*, 2005, **11**, 13–18.
75. A. Z. Mercadante and D. B. Rodriguez-Amaya, *J. Agric. Food Chem.*, 1998, **46**, 128–130.
76. B. Setiawan, A. Sulaeman, D. W. Giraud and J. A. Driskell, *J. Food Compos. Anal.*, 2001, **14**, 169–176.
77. L. Vargas-Murga, V. V. de Rosso, A. Z. Mercadante and B. Olmedilla-Alonso, *J. Food Compos. Anal.*, 2016, **50**, 88–96.
78. B. Beltran-de-Miguel, R. Estevez-Santiago and B. Olmedilla-Alonso, *Int. J. Food Sci. Nutr.*, 2015, **66**, 706–712.
79. E. Biehler, A. a. Alkerwi, L. Hoffmann, E. Krause, M. Guillaume, M.-L. Lair and T. Bohn, *J. Food Compos. Anal.*, 2012, **25**, 56–65.
80. F. Manzi, V. Flood, K. Webb and P. Mitchell, *Public Health Nutr.*, 2002, **5**, 347–352.
81. G. P. M. Barreto, J. P. Fabi, V. V. De Rosso, B. R. Cordenunsi, F. M. Lajolo, J. R. O. do Nascimento and A. Z. Mercadante, *J. Food Compos. Anal.*, 2011, **24**, 620–624.
82. A. F. Faria, P. N. Hasegawa, E. A. Chagas, R. Pio, E. Purgatto and A. Z. Mercadante, *J. Food Compos. Anal.*, 2009, **22**, 196–203.

83. F. C. Petry and A. Z. Mercadante, *Food Funct.*, 2017, **8**, 3951–3963.
84. M. G. Dias, M. F. Camoes and L. Oliveira, *Food Chem.*, 2008, **109**, 815–824.
85. A. G. Frenich, M. E. Torres, A. B. Vega, J. L. Vidal and P. P. Bolanos, *J. Agric. Food Chem.*, 2005, **53**, 7371–7376.
86. M. Fattore, D. Montesano, E. Pagano, R. Teta, F. Borrelli, A. Mangoni, S. Seccia and S. Albrizio, *J. Food Compos. Anal.*, 2016, **53**, 61–68.
87. P. Perkins-Veazie, J. K. Collins, S. D. Pair and W. Roberts, *J. Sci. Food Agric.*, 2001, **81**, 983–987.
88. R. Estevez-Santiago, B. Beltran-de-Miguel and B. Olmedilla-Alonso, *Int. J. Food Sci. Nutr.*, 2016, **67**, 305–313.
89. A. J. Melendez-Martinez, P. Mapelli-Brahm, A. Benitez-Gonzalez and C. M. Stinco, *Arch. Biochem. Biophys.*, 2015, **572**, 188–200.
90. U.S. Department of Agriculture and Agricultural Research Service, *Nutrient Intakes from Food and Beverages: Mean Amounts Consumed per Individual, by Gender and Age, What We Eat in America, NHANES 2013-2014*, www.ars.usda.gov/nea/bhnrc/fsrg, accessed August, 2017.
91. W. K. Al-Delaimy, N. Slimani, P. Ferrari, T. Key, E. Spencer, I. Johansson, G. Johansson, I. Mattisson, E. Wirfalt, S. Sieri, A. Agudo, E. Celentano, D. Palli, C. Sacerdote, R. Tumino, M. Dorronsoro, M. C. Ocke, H. B. Bueno-De-Mesquita, K. Overvad, M. D. Chirlaque, A. Trichopoulou, A. Naska, A. Tjonneland, A. Olsen, E. Lund, G. Skeie, E. Ardanaz, E. Kesse, M. C. Boutron-Ruault, F. Clavel-Chapelon, S. Bingham, A. A. Welch, C. Martinez-Garcia, G. Nagel, J. Linseisen, J. R. Quiros, P. H. Peeters, C. H. van Gils, H. Boeing, A. L. van Kappel, J. P. Steghens and E. Riboli, *Eur. J. Clin. Nutr.*, 2005, **59**, 1397–1408.
92. K. Canene-Adams and J. W. Erdman Jr, in *Carotenoids: Nutrition and Health*, ed. G. Britton, S. Liaaen-Jensen and H. Pfander, Birkhäuser Basel, Basel, Editon edn, 2009, vol. 5, pp. 115–148.
93. G. Tang and R. M. Russell, in *Carotenoids: Nutrition and Health*, ed. G. Britton, S. Liaaen-Jensen and H. Pfander, Birkhäuser, Basel, Editon edn, 2009, vol. 5, pp. 149–172.
94. World Health Organization, *WHO Global Database on Vitamin A Deficiency*, http://www.who.int/vmnis/database/vitamina/en, accessed 07/26, 2017.
95. C. D. Seña, K. M. Riedl, S. Narayanasamy, R. W. Curley Jr, S. J. Schwartz and E. H. Harrison, *J. Biol. Chem.*, 2014, **289**, 13661–13666.
96. E. H. Harrison, *Biochim. Biophys. Acta*, 2012, **1821**, 70–77.
97. NAS-IOM, *Dietary Reference Intakes for Vitamin A, Vitamin K, Arsenic, Boron, Chromium, Copper, Iodine, Iron, Manganese, Molybdenum, Nickel, Silicon, Vanadium, and Zinc*, National Academic Press, 2001.
98. NAS-NRC, *Recommended Dietary Allowances*, 10th edn, National Academy of Science, 1989.
99. M. J. Haskell, *Am. J. Clin. Nutr.*, 2012, **96**, 1193S–1203S.
100. J. A. Novotny, D. J. Harrison, R. Pawlosky, V. P. Flanagan, E. H. Harrison and A. C. Kurilich, *J. Nutr.*, 2010, **140**, 915–918.

101. The Alpha-Tocopherol Beta Carotene Cancer Prevention Study Group, *N. Engl. J. Med.*, 1994, **330**, 1029–1035.
102. G. S. Omenn, G. E. Goodman, M. D. Thornquist, J. Balmes, M. R. Cullen, A. Glass, J. P. Keogh, F. L. Meyskens Jr, B. Valanis, J. H. Williams Jr, S. Barnhart, M. G. Cherniack, C. A. Brodkin and S. Hammar, *J. Natl. Cancer Inst.*, 1996, **88**, 1550–1559.
103. X. D. Wang, C. Liu, R. T. Bronson, D. E. Smith, N. I. Krinsky and M. Russell, *J. Natl. Cancer Inst.*, 1999, **91**, 60–66.
104. M. R. Russell, *Pure Appl. Chem.*, 2002, **74**, 1461–1467.
105. C. H. Hennekens, J. E. Buring, J. E. Manson, M. Stampfer, B. Rosner, N. R. Cook, C. Belanger, F. LaMotte, J. M. Gaziano, P. M. Ridker, W. Willett and R. Peto, *N. Engl. J. Med.*, 1996, **334**, 1145–1149.
106. W. J. Blot, J. Y. Li, P. R. Taylor, W. Guo, S. Dawsey, G. Q. Wang, C. S. Yang, S. F. Zheng, M. Gail and G. Y. Li, *et al.*, *J. Natl. Cancer Inst.*, 1993, **85**, 1483–1492.
107. Y. L. Qiao, S. M. Dawsey, F. Kamangar, J. H. Fan, C. C. Abnet, X. D. Sun, L. L. Johnson, M. H. Gail, Z. W. Dong, B. Yu, S. D. Mark and P. R. Taylor, *J. Natl. Cancer Inst.*, 2009, **101**, 507–518.
108. C. L. Rock, in *Carotenoids: Nutrition and Health*, ed. G. Britton, S. Liaaen-Jensen and H. Pfander, Birkhäuser Basel, Basel, Editon edn, 2009, vol. 5, pp. 269–286.
109. D. Ilic and M. Misso, *Maturitas*, 2012, **72**, 269–276.
110. C. W. Hadley, E. C. Miller, S. J. Schwartz and S. K. Clinton, *Exp. Biol. Med.*, 2002, **227**, 869–880.
111. M. Etminan, B. Takkouche and F. Caamano-Isorna, *Cancer Epidemiol., Biomarkers Prev.*, 2004, **13**, 340–345.
112. A. Herzog and K. Wertz, *Agro Food Ind. Hi-Tech*, 2005, **16**, 13–15.
113. Y. Wang, R. Cui, Y. Xiao, J. Fang and Q. Xu, *PLoS One*, 2015, **10**, e0137427.
114. P. Chen, W. Zhang, X. Wang, K. Zhao, D. S. Negi, L. Zhuo, M. Qi, X. Wang and X. Zhang, *Medicine*, 2015, **94**, e1260.
115. J. L. Rowles 3rd, K. M. Ranard, J. W. Smith, R. An and J. W. Erdman Jr, *Prostate Cancer Prostatic Dis.*, 2017, **20**, 361–377.
116. V. Ganji and M. R. Kafai, H. Third National and S. Nutrition Examination, *J. Nutr.*, 2005, **135**, 567–572.
117. C. M. Allen, S. J. Schwartz, N. E. Craft, E. L. Giovannucci, V. L. De Groff and S. K. Clinton, *Nutr. Cancer*, 2003, **47**, 48–56.
118. R. Re, G. D. Mishra, C. W. Thane and C. J. Bates, *Eur. J. Clin. Nutr.*, 2003, **57**, 1545–1554.
119. F. J. Barba, L. R. B. Mariutti, N. Bragagnolo, A. Z. Mercadante, G. V. Barbosa-Cánovas and V. Orlien, *Trends Food Sci. Technol.*, 2017, **67**, 195–206.
120. H. L. Tan, J. M. Thomas-Ahner, E. M. Grainger, L. Wan, D. M. Francis, S. J. Schwartz, J. W. Erdman Jr and S. K. Clinton, *Cancer Metastasis Rev.*, 2010, **29**, 553–568.
121. T. W. Boileau, Z. Liao, S. Kim, S. Lemeshow, J. W. Erdman Jr and S. K. Clinton, *J. Natl. Cancer Inst.*, 2003, **95**, 1578–1586.
122. K. Canene-Adams, B. L. Lindshield, S. Wang, E. H. Jeffery, S. K. Clinton and J. W. Erdman Jr, *Cancer Res.*, 2007, **67**, 836–843.

123. K. Klipstein-Grobusch, L. J. Launer, J. M. Geleijnse, H. Boeing, A. Hofman and J. C. Witteman, *Atherosclerosis*, 2000, **148**, 49–56.
124. L. Kohlmeier, J. D. Kark, E. Gomez-Gracia, B. C. Martin, S. E. Steck, A. F. M. Kardinaal, J. Ringstad, M. Thamm, V. Masaev, R. Riemersma, J. M. Martin-Moreno, J. K. Huttunen and F. J. Kok, *Am. J. Epidemiol.*, 1997, **146**, 618–626.
125. T. H. Rissanen, S. Voutilainen, K. Nyyssonen, T. A. Lakka, J. Sivenius, R. Salonen, G. A. Kaplan and J. T. Salonen, *Br. J. Nutr.*, 2001, **85**, 749–754.
126. H. D. Sesso, J. E. Buring, E. P. Norkus and J. M. Gaziano, *Am. J. Clin. Nutr.*, 2004, **79**, 47–53.
127. K. Ried and P. Fakler, *Maturitas*, 2011, **68**, 299–310.
128. B. Burton-Freeman and H. D. Sesso, *Adv. Nutr.*, 2014, **5**, 457–485.
129. P. S. Bernstein, B. Li, P. P. Vachali, A. Gorusupudi, R. Shyam, B. S. Henriksen and J. M. Nolan, *Prog. Retinal Eye Res.*, 2016, **50**, 34–66.
130. A. Gorusupudi, K. Nelson and P. S. Bernstein, *Adv. Nutr.*, 2017, **8**, 40–53.
131. J. Mares, *Annu. Rev. Nutr.*, 2016, **36**, 571–602.
132. P. S. Bernstein, F. Khachik, L. S. Carvalho, G. J. Muir, D. Y. Zhao and N. B. Katz, *Exp. Eye Res.*, 2001, **72**, 215–223.
133. R. A. Bone, J. T. Landrum, L. Fernandez and S. L. Tarsis, *Invest. Ophthalmol. Visual Sci.*, 1988, **29**, 843–849.
134. F. Khachik, F. F. de Moura, D. Y. Zhao, C. P. Aebischer and P. S. Bernstein, *Invest. Ophthalmol. Visual Sci.*, 2002, **43**, 3383–3392.
135. Age-Related Eye Disease Study Research Group, *Arch. Ophthalmol.*, 2001, **119**, 1439–1452.
136. The Age-Related Eye Disease Study 2 Research Group, E. Y. Chew, T. E. Clemons, J. P. SanGiovanni, R. P. Danis, F. L. Ferris, M. J. Elman, A. Antoszyk, A. Ruby, D. Orth, S. B. Bressler, G. Fish, B. Hubbard, M. Klein, S. Chandra, B. Blodi, A. Domalpally, T. Friberg, W. Wong, P. Rosenfeld, E. Agron, C. Toth, P. Bernstein and R. Sperduto, *JAMA Ophthalmol.*, 2014, **132**, 142–149.
137. J. M. Seddon, U. A. Ajani, R. D. Sperduto, R. Hiller, N. Blair, T. C. Burton, M. D. Farber, E. S. Gragoudas, J. Haller and D. T. Miller, *et al.*, *JAMA*, 1994, **272**, 1413–1420.
138. E. J. Johnson, R. Vishwanathan, M. A. Johnson, D. B. Hausman, A. Davey, T. M. Scott, R. C. Green, L. S. Miller, M. Gearing, J. Woodard, P. T. Nelson, H. Y. Chung, W. Schalch, J. Wittwer and L. W. Poon, *J. Aging Res.*, 2013, **2013**, 951786.
139. R. Vishwanathan, A. Iannaccone, T. M. Scott, S. B. Kritchevsky, B. J. Jennings, G. Carboni, G. Forma, S. Satterfield, T. Harris, K. C. Johnson, W. Schalch, L. M. Renzi, C. Rosano and E. J. Johnson, *Age Ageing*, 2014, **43**, 271–275.
140. E. J. Johnson, *Nutr. Rev.*, 2014, **72**, 605–612.
141. B. Li, P. P. Vachali, A. Gorusupudi, Z. Shen, H. Sharifzadeh, B. M. Besch, K. Nelson, M. M. Horvath, J. M. Frederick, W. Baehr and P. S. Bernstein, *Proc. Natl. Acad. Sci. U. S. A.*, 2014, **111**, 10173–10178.
142. J. Feeney, C. Finucane, G. M. Savva, H. Cronin, S. Beatty, J. M. Nolan and R. A. Kenny, *Neurobiol. Aging*, 2013, **34**, 2449–2456.

143. L. M. Renzi, M. J. Dengler, A. Puente, L. S. Miller and B. R. Hammond Jr, *Neurobiol. Aging*, 2014, **35**, 1695–1699.
144. E. R. Bovier, L. M. Renzi and B. R. Hammond, *PLoS One*, 2014, **9**, e108178.
145. J. W. Erdman Jr, J. W. Smith, M. J. Kuchan, E. S. Mohn, E. J. Johnson, S. S. Rubakhin, L. Wang, J. V. Sweedler and M. Neuringer, *Foods*, 2015, **4**, 547–564.
146. R. P. G. Hayhoe, M. A. H. Lentjes, A. A. Mulligan, R. N. Luben, K. T. Khaw and A. A. Welch, *Br. J. Nutr.*, 2017, **117**, 1439–1453.
147. S. Sahni, M. T. Hannan, J. Blumberg, L. A. Cupples, D. P. Kiel and K. L. Tucker, *J. Bone Miner. Res.*, 2009, **24**, 1086–1094.
148. J. M. Mata-Granados, J. R. Cuenca-Acevedo, M. D. Luque de Castro, M. F. Holick and J. M. Quesada-Gomez, *Arch. Osteoporosis*, 2013, **8**, 124.
149. H. H. Conaway, P. Henning and U. H. Lerner, *Endocr. Rev.*, 2013, **34**, 766–797.
150. M. A. Montenegro, M. A. Nazareno, E. N. Durantini and C. D. Borsarelli, *Photochem. Photobiol.*, 2002, **75**, 353–361.

CHAPTER 9

Metabolic Fate of Bioaccessible and Non-bioaccessible Carotenoids

TORSTEN BOHN

Luxembourg Institute of Health (LIH), Population Health Department, rue 1 A-B, Thomas Edison, L-1445 Strassen, Luxembourg
*E-mail: torsten.bohn@gmx.ch

9.1 Introduction

Carotenoids are among the most abundant lipophilic secondary plant compounds consumed in the diet. Carotenoids include mostly C-40 tetraterpenoids, though also C-30 tri-[1] and C-50 pentaterpenoids[2] have been reported, at least in various bacteria, though due to their minute intake, they are not of dietary relevance. However, their dietary variety is increased when also counting the many apo-carotenoids (*i.e.* carotenoid metabolites derived from oxidative cleavage of carotenoids, which may occur in bacteria, fungi and plants, but also in animals).[3] Major plant-derived apo-carotenoids include, for example, bixin, crocin, picrocrocin, abscisic acid, strigolactone and mycorradicin.[4]

Though several hundreds of carotenoids have been described in nature, it is estimated that only a few dozen or so predominate in our daily diet.[5,6] Their dietary intake as well as circulating blood levels have been related to

Food Chemistry, Function and Analysis No. 5
Non-extractable Polyphenols and Carotenoids: Importance in Human Nutrition and Health
Edited by Fulgencio Saura-Calixto and Jara Pérez-Jiménez

Published by the Royal Society of Chemistry, www.rsc.org

the reduced risk of several chronic diseases, including certain cardiometabolic conditions such as the metabolic syndrome[7] and certain types of cancer,[8,9] but also allergies[10] and asthma.[11] A carotenoid health index based on total plasma carotenoids has been proposed by Donaldson, with <1 μM total carotenoids constituting a very high health risk[12] regarding total mortality (lower threshold), and less so for metabolic syndrome and cancer (higher threshold).

Some compounds such as beta- and alpha-carotene as well as beta-cryptoxanthin possess a beta-ionone ring, and these compounds can be cleaved by the human body into vitamin A active compounds such as retinal. In addition, lutein and zeaxanthin, belonging to the oxygen-carrying xanthophylls (as opposed to the oxygen-free carotenes), have been related to improved vision aspects in subjects with age-related macular degeneration (AMD),[13] the major cause of blindness in the elderly. While in the past, most health benefits have been attributed to the radical-scavenging properties[14] and light-absorbing properties[15] of the native carotenoids, more recent reports have rather emphasized the potential of carotenoids and their metabolites to interact with nuclear transcription factors, such as with NF-κB and Nrf2,[16,17] influencing gene expression following their binding to nuclear receptors, which then regulate many inflammatory agents such as cytokines (IL-6, IL-1β) and antioxidant-related enzymes, including superoxide dismutase (SOD), catalyze (CAT) and heme-oxygenase (HO-1).

These latter mechanisms have therefore been highlighted as relating to reduced inflammation and strengthening the body's own antioxidant defense system. However, it has also been suggested that concentration-dependent effects could occur, with lower physiological concentrations fostering these health-beneficial effects, while higher concentrations may bear the risk of pro-inflammatory, pro-oxidant and thus adverse health effects.[18,19] Such negative effects were found in several supplementation trials in which beta-carotene was given concomitantly with vitamin E and A, as in the ATBC[20] and CARET trials,[21] respectively, and were found to increase, not decrease, the lung cancer rate, possibly due to pro-carciongenic intermediate formation following reaction with cytochrome oxidases. Thus, these results have not only questioned the dose-related effects, but also the nature of the bioactive constituents (*i.e.* native carotenoids *vs.* their metabolites and/or degradation products).

In fact, carotenoids have only low bioavailability – the amount that is absorbed and used for physiological functions and or storage (Table 9.1) – of approximately between 5% and 30%. The majority of carotenoid absorption is thought to take place in the small intestine. Absorption in the stomach, due to their lipophilic nature requiring micellization, is thought to be non-existent, while absorption from the colon remains an open question, but has never been shown.[18] Most bioavailability trials show absorption occurring between 2 and 8 hours following their intake, as measured by their appearance in the chylomicron-rich fraction of the plasma.[22–24] The low bioavailability of carotenoids is mostly due to the low solubility of these highly

Table 9.1 Concentration of carotenoids in human blood plasma/serum and various tissues.

Carotenoid	Tissue	Concentration (nmol g^{-1} or nmol mL^{-1})[a]	Reference
Beta-carotene	Plasma, male/female	0.36 ± 0.01/0.54 ± 0.01	56
Lycopene	Plasma, male/female	0.74 ± 0.01/0.71 ± 0.01	56
Lutein	Plasma, male/female	0.38 ± 0.01/0.44 ± 0.01	56
Zeaxanthin	Plasma, male/female	0.09 ± 0.01/0.09 ± 0.01	56
Alpha-carotene	Plasma, male/female	0.12 ± 0.01/0.20 ± 0.01	56
Beta-cryptoxanthin	Plasma, male/female	0.23 ± 0.01/0.34 ± 0.01	56
Total carotenoids	Plasma, male/female	1.94 ± 0.02/2.35 ± 0.03	56
Zeta-carotene	Plasma, females	0.11 ± 0.09	57
Phytoene	Plasma, females	0.04 ± 0.02	57
Phytofluene	Plasma, females	0.17 ± 0.07	57
Lutein + zeaxanthin	Retina	65–225 pmol	58
Beta-carotene	Adipose tissue	0.32 ± 0.10	Reviewed in ref. 6
Lycopene	Adipose tissue	0.36 ± 0.23	Reviewed in ref. 6
Lutein/zeaxanthin	Adipose tissue	1.19 ± 0.56	Reviewed in ref. 6
Beta-carotene	Liver	5.9 ± 6.3	Reviewed in ref. 6
Lycopene	Liver	8.4 ± 11.5	Reviewed in ref. 6
Lutein/zeaxanthin	Liver	2.2 ± 1.6	Reviewed in ref. 6
Total carotenoids	Skin	0–0.7	59
Total carotenoids	Lung	1.95 ± 2.82	53
Beta-carotene	Lung	0.35 ± 0.44	53
Alpha-carotene	Lung	0.23 ± 0.27	53
Beta-cryptoxanthin	Lung	0.42 ± 0.75	53
Lycopene	Lung	0.57 ± 1.11	53
Lutein	Lung	0.48 ± 0.66	53
Total carotenoids	Kidney	3.05 ± 4.21	53
Beta-carotene	Kidney	0.55 ± 0.73	53
Alpha-carotene	Kidney	0.30 ± 0.40	53
Beta-cryptoxanthin	Kidney	0.45 ± 1.04	53
Lycopene	Kidney	0.62 ± 0.62	53
Lutein	Kidney	1.21 ± 2.83	53
Beta-carotene	Prostate	0.54 ± 0.09 (SEM)	60
Lycopene	Prostate	0.80 ± 0.08 (SEM)	60
Lutein	Various brain tissues[b]	0.02–0.08	61
Zeaxanthin	Various brain tissues	0.01–0.03	61
Beta-cryptoxanthin	Various brain tissues	<0.01	61
Beta-carotene	Various brain tissues	0.01–0.03	61
Total carotenoids	Adrenals	9.4 ± 7.8 (SEM)	62
Beta-cryptoxanthin	Adrenals	0.66 (0.01–2.90)	62
Lycopene	Adrenals	1.90 (0.19–5.60)	62
Alpha-carotene	Adrenals	1.22 (0.11–7.52)	62
Beta-carotene	Adrenals	5.60 (0.68–31.83)	62

[a]Unless otherwise stated, mean ± SD.
[b]Infants: prefrontal cortex, frontal cortex, hippocampus, auditory cortex and occipital cortex.

apolar constituents (log *p* values ~8–12[25]) in the gut and limited availability for absorption (*i.e.* poor bioaccessibility).[26] In addition, following cellular uptake into the enterocyte, carotenoids may be further metabolized by beta-carotene oxygenases 1/2 (BCO1/2) to produce several apo-carotenoids.[27–29] These may then be re-esterified[30,31] prior to their further sequestration by chylomicrons, as is known for retinol, and later experience tissue distribution *via* low-density lipoproteins/high-density lipoproteins in particular.

However, little information exists regarding potential metabolites, their concentrations in various tissues and also the fate of the large fraction of intestinally non-absorbed carotenoids, as these may be further fermented by the microbiota, similarly to, for example, polyphenols, though this remains unknown.[18] In addition, recent studies on naturally occurring apo-carotenoids such as on norbixin or bixin from annatto seeds or crocetin from safflower suggest that these compounds appear to be highly bioavailable[32,33] and may also be very bioactive, as has been shown for abscisic acid, a plant hormone derived from carotenoids, which was shown to improve glucose tolerance[34] and inflammation-related aspects in inflammatory bowel diseases (IBD).[35] For several of these apo-carotenoids, it is also not known whether they can be produced in the human body or if they are entirely derived from consuming plant-based foods.

This is of importance, as it appears that several of the carotenoid metabolites may be bioactive – in fact, more than the native compounds, a fact that may be ascribed to their better cytosolic solubility and potential to interact with transcription factors, as well as their higher electrophilicity, resulting in better binding to, for example, cysteine residues of NF-κB[36] and Nrf2.[37] In addition, other metabolites, such as those of lycopene, have been reported to interact with retinoic acid receptor (RAR)[38,39] and retinoid X receptor (RXR) nuclear receptors,[40] thus mimicking the immune-related functions of vitamin A active compounds. However, the further fate of these more polar carotenoid derivatives remains largely speculative.

This review tries to summarize our current knowledge regarding the metabolic fate of both bioaccessible and non-bioaccessible carotenoids and to point out the boundaries of our knowledge regarding the metabolism of these mostly colorful dietary constituents and their relation to health aspects.

9.2 Carotenoid Concentrations: from Diet to Tissue Levels

The most frequently consumed carotenoids in the diet include beta-carotene, lycopene, lutein, alpha-carotene, beta-cryptoxanthin, phytofluene, phytoene, neoxanthin, violaxanthin and zeta-carotene, though the sequence of intake may change according to dietary habits. For example, in the USA and Italy, due to the intake of tomato products including ketchup, lycopene appears to constitute the most frequently consumed carotenoid, with an intake of up to approximately 10.5 and 7.5 mg day^{-1} in those countries, respectively.[41,42] Total carotenoid dietary intake ranges typically between 10 and 20 mg day^{-1}.[43–46] In plant-based food items, concentrations vary widely, but can reach up to 20 mg/100 g in some cases, such as for spinach and tomatoes,[47] though other leafy vegetables such as cabbages,[48] as well as carrots,[49] sweet potatoes[50] and a few animal foods such as eggs and salmon[47] constitute important and frequently consumed dietary carotenoid sources.

As not all carotenoids and their food sources have comparable bioavailability, human tissue and plasma levels do not necessarily reflect total dietary intake (Table 9.1), though significant and reasonably strong correlations between various tissue carotenoid concentrations have been reported, such as for plasma with skin and buccal cells[51] as well as dermal levels,[52] and also between various organs such as kidney, liver and lung,[53] suggesting that plasma may serve as a reasonable indicator for comparing and perhaps assessing carotenoid status. At least for several tissues, only limited discrimination occurs between the concentration in the circulatory system and the tissues (Table 9.1), though certain tissues such as the liver and retina appear to store carotenoids at higher concentrations (see Chapter 8).

It is also interesting to note that, compared to other frequently consumed phytochemicals such as polyphenols, which are consumed in much higher amounts (up to 1 g day^{-1}), plasma concentrations are at least equally high, even when taking into account polyphenol metabolites such as glucuronides/sulfates.[54] In fact, carotenoids appear to constitute the most predominant phytochemicals in human plasma, owing to their reasonably long half-life, due to their more limited metabolism compared to, for example, polyphenols. In a human depletion study by Burri *et al.*[55] with 18–42 year-old females, half-lives of 27 days (lycopene), 35 days (beta-carotene) and 76 days (lutein) were found, likely reflecting losses from deeper body pools such as the liver and adipose tissue, perhaps suggesting increased metabolism of the carotenes compared to the xanthophylls. Similar half-lives were corroborated in other studies, as reviewed by Bohn *et al.*[6]

9.3 Carotenoid Bioaccessibility During Gastrointestinal Digestion

9.3.1 Overview of Factors Influencing Bioaccessibility

In order to come into contact with the small intestinal epithelium, where carotenoid absorption takes place, these very lipophilic molecules need to be solubilized in the aqueous environment (*i.e.* packaged into mixed micelles of approximately 5–10 nm in diameter).[25,63] These are formed during intestinal digestion and consist of a mixture of partially digested lipids (free fatty acids, mono- and di-glycerides), bile salts, phospholipids and other liposoluble micro-constituents from the diet, such as cholesterol or vitamin E.[64,65] Following their formation, mixed micelles can then diffuse through the mucus layer[66] to the unstirred water layer of the epithelium, where they interact with the cell membrane, following liberation of carotenoids and their cellular uptake *via* either transporters or passive diffusion (see Chapter 4). The factors fostering adhesion to the epithelium and uptake include adhesion (governed by surface charges) and mobility in particular, with negatively charged particles resulting in generally higher transport rates.[67] This solubilization and becoming available for further absorption is also termed bioaccessibility.

Thus, micellization is a crucial step and a prerequisite for carotenoid bioavailability. Though only constituting the first important step of bioavailability after matrix release, several studies have reported a good correlation of carotenoid bioaccessibility with carotenoid bioavailability measures in human subjects,[68,69] also highlighting the pivotal role of carotenoid bioaccessibility (reviewed by Biehler and Bohn[70]).

Consequently, all factors that impinge on bioaccessibility are likely factors that can alter carotenoid bioavailability. These factors have often being summarized by the mnemonic term "SLAMENGHI," comprising carotenoid species, their molecular/chemical linkage, amount of carotenoids, dietary matrix, effectors of absorption/bioconversion, nutrient status of the host, genetic factors, other host factors such as diseases, and their interactions.[71] Regarding the metabolic fate of carotenoids, a first important consideration is therefore their bioaccessibility. While an extensive review of dietary and host-related factors is beyond the scope of this book chapter, and the reader is referred to other more comprehensive reviews,[5,6,26,72] the most important considerations that determine carotenoid bioaccessibility will be briefly highlighted in the following.

9.3.2 Matrix-related Factors, Bioaccessibility and Carotenoid Fate

In order to become bioaccessible, carotenoids first have to be released from the food matrix. In the following, they then have to be packed into mixed micelles. Consequently, all factors that influence these two steps determine whether carotenoids will be absorbed or, alternatively, passed on to the colon. The major factors influencing carotenoid release from the food matrix and their transfer to mixed micelles are shown in Tables 9.2 and 9.3, respectively, and include the type of carotenoid, type of matrix, food processing aspects, preparatory aspects and the amount of dietary fat and fiber present in the carotenoid containing meal. Food processing in particular appears to be a double-edged sword, as there is a critical balance between heat application that may macerate the matrix and contribute to the release of carotenoids and the negative influence *via* facilitating air and oxygen access, which together with heat may result in carotenoid oxidative degradation.

However, when summarizing the most crucial factors, it can be stated that softer food matrices with macerized cells wall, those containing carotenoids in rather oily droplets compared to crystalline forms, low amounts of dietary fiber, sufficient presence of dietary lipids and small particle size are the main factors enhancing carotenoid absorption[5,73,74] and reducing the amounts of unabsorbed carotenoids that are passed on to the colon. Additional dietary

Table 9.2 Aspects of food processing influencing carotenoid release from the matrix and therefore bioaccessibility and bioavailability.

Type of processing	Food matrix	Carotenoid(s) investigated	Type of study	Effect on bioavailability marker	Explanation of effects	Reference
Heating	Water fruit juice, milk fruit juice, soymilk fruit juice	Total and individual carotenoids	*In vitro* digestion	Reduced bioaccessibility following heat treatment	Unclear	99
Heating, homogenization	Tomato juice	Lycopene	Human postprandial trial	Higher bioavailability from heat-induced *cis*-isomers	*Cis*-isomer formation, higher micellization	100
Heating, homogenization	Carrot juice	Beta-carotene	*In vitro* digestion	Up to 20-times higher bioaccessibility of carotenes from juice than from raw carrots	Cell wall maceration, cell rupture	101
Cooking + pureeing	Carrots: raw *vs.* cooked + pureed	Beta-carotene	Human study with ileostomists	>20% higher absorption from processed carrots	Cell wall maceration	102
High-pressure homogenization	Tomato-based systems	Beta-carotene, lycopene	*In vitro* digestion	Decreased bioaccessibility (50→30%) of beta-carotene with high-pressure homogenization when insoluble fiber was present	Forming more stable fiber network during high-pressure homogenization	103
High-pressure homogenization	Water fruit juice, milk fruit juice, soymilk fruit juice	Total and individual carotenoids	*In vitro* digestion	No consistent negative effects on juices in general	Slight changes due to fiber network rearrangements and cell maceration	99

(*continued*)

Table 9.2 (*continued*)

Type of processing	Food matrix	Carotenoid(s) investigated	Type of study	Effect on bioavailability marker	Explanation of effects	Reference
High-intensity pulsed electric fields	Water fruit juice, milk fruit juice, soymilk fruit juice	Total and individual carotenoids	*In vitro* digestion	No clear negative effects on juices in general	Slight changes due to fiber network rearrangements and cell maceration	99
Particle size	Carrots and tomatoes	Beta-carotene, lycopene	*In vitro* digestion	Beta-carotene and lycopene bioaccessibility increased with particles <125 μm	Destruction of cell wall as major barrier of carotenoid release	104
Grinding, particle reduction	Different leafy vegetables	Beta-carotene	*In vitro* digestion	Several-fold improved bioaccessibility of ground leafy vegetables	Enhanced surface and improved release kinetics	105

Table 9.3 Aspects of food matrix and co-consumed dietary constituents influencing carotenoid micellization and therefore bioaccessibility and bioavailability, showing exemplary studies.[a]

Type of dietary constituents	Food matrix	Carotenoid(s) investigated	Type of study	Effect on bioavailability marker	Explanation of effects	Reference
Presence as crystals *vs.* dissolved in lipids	Mango, papaya, tomato, carrot	Beta-carotene, lycopene, lutein, beta-cryptoxanthin	Bioaccessibility following *in vitro* digestion	Fractional liberation and bioaccessibility of beta-carotene several times higher from mango/papaya than carrot/tomato	Different physical states, with crystalline forms taking more time to dissolve	106
Dietary fiber	Supplements given with or without either pectin, guar, alginate, cellulose or wheat bran	Beta-carotene, lycopene, alpha-carotene, canthaxanthin, lutein	Human postprandial	Decreased plasma AUC (24 h), 33–43% for beta-carotene, 40–74% reduction for lycopene, no effect on other carotenoids	Binding of fiber to bile salts and phospholipids, effect on lipase, desquamination of mucosal cells	107
Dietary fiber	Alginates, carboxymethylcelluloses, and methylcelluloses, pectins added to purified carotenoids in oil	Lutein, beta-carotene	*In vitro* digestion and cellular uptake (Caco-2 model)	Up to 31% reduced micellization and up to 45% cellular uptake	Increased viscosity, bile salt binding	108
Dietary lipids	Avocado fruit/oil	Alpha-carotene, beta-carotene, lutein	Human postprandial	Increase: 7.2-, 15.3- and 5.1-fold of TRL-AUC, respectively *vs.* no added lipids	Enhanced micellization of carotenoids with lipids	24
Dietary lipids	Yellow peppers with medium-chain triglycerides, long-chain triglycerides or indigestible lipids	Beta-carotene	*In vitro* digestion	Higher bioaccessibility, especially with long-chain FA	Enhanced micellization of carotenoids with lipids	109

(*continued*)

Table 9.3 (*continued*)

Type of dietary constituents	Food matrix	Carotenoid(s) investigated	Type of study	Effect on bioavailability marker	Explanation of effects	Reference
Free and esterified plant sterols (2.2 g)	Beta-carotene isotopic marker in milk	Beta-carotene	Human trial, plasma appearance	50% reduced absorption with plant sterols	Competition for micellization	80
Proteins	Artificial micelle systems digested with milk cream or canola oil	Beta-carotene	*In vitro* digestion	Higher bioaccessibility of carotenoids from milk cream than canola oil	Degradation products of proteins aiding in emulsification	84
Calcium	Tomato paste	Lycopene	Human postprandial	83% reduced plasma-AUC of lycopene with 500 mg dose	Binding of calcium with bile salts and FA	76
DM: calcium, magnesium	Spinach, carrot juice, tomato juice, apricot juice, individual carotenoids	Beta-carotene, lycopene, lutein, neoxanthin	*In vitro* digestion and cellular uptake (Caco-2 model)	Reduction by up to 100% with equivalent full RDA intake	Binding of DM with bile salts and FA, causing precipitation of bile salts and poorly soluble soaps	75,110,111

[a]AUC: area under concentration *vs*. time curve; DM: divalent minerals; TRL: triacyl glycerol-rich lipoprotein fraction; RDA: recommended dietary allowance; FA: fatty acids.

factors that may play a role but require more studies are dietary minerals, compounds competing for micellization, such as phytosterols and other lipophilic vitamins, and proteins. Dietary minerals may reduce carotenoid solubilization in the gut due to binding of released fatty acids and bile salts,[75] though the two human studies conducted thus far remain somewhat contradictory.[76,77] Other lipophilic constituents such as cholesterol and phytosterols may also compete for micellization and therefore reduce carotenoid bioaccessibility. This was shown for high concentrations of cholesterol *in vitro*,[78,79] phytosterols limiting beta-carotene[80] but not beta-cryptoxanthin availability in humans,[81] and also according to a recent meta-analysis where plant sterol and plant stanol intake lowered carotene and, less so, xanthophyll concentrations in human blood plasma,[82] and vitamin E lowered canthaxanthin absorption in rats.[83] Proteins may produce peptides with emulsifying properties during digestion that could aid in the transition of carotenoids from lipid droplets to mixed micelles,[84,85] and their emulsifying properties have been employed for encapsulating carotenoids to enhance both shelf-life stability and also bioavailability aspects.[86] However, no human trials investigating the effects of proteins on carotenoid absorption have been reported.

Also, due to a typically lower aqueous solubility and the higher tendency to aggregate (especially lycopene) and to be present in the matrix in crystalline form, carotenes tend to be of lower bioaccessibility compared to xanthophylls (reviewed by Desmarchelier and Borel[5] and Bohn[87]), and this appears to cause their generally lower fractional absorption and bioavailability expressed as percentage of ingested dose,[24,88,89] resulting in a larger proportion of carotenes *vs.* xanthophylls reaching the colon. For xanthophylls, lower bioaccessibility and bioavailability of the densely packed H-aggregates compared to more loosely packed J-aggregates (present in the absence of hydrogen bonds between molecules such as at lower pH or when esterified) have been proposed.[90] In addition, *cis*-isomers of carotenoids appear to be of higher bioaccessibility, most likely due to their more bent structures and shorter apparent molecular lengths, which improve micellization[91,92] and often bioavailability.[24,93–95] For this reason, the rather apolar phytoene and phytofluene, predominantly present in their 15-*cis* form and as variable *cis*-isomers,[96] respectively, were found to be highly bioaccessible[91] and also bioavailable (almost 60%), as shown in a postprandial study with isotopically labeled phytoene,[97] though the nature of the isomer(s) given was not revealed. A high bioavailability was also shown for a *cis*-product of lycopene, tetra-*cis* lycopene or prolycopene, which showed an almost eight to ten-times higher area under the curve (AUC) than all-*trans* lycopene from tangerines compared to red tomato sauces[93,98]

9.3.3 Degradation in the Gastrointestinal Tract

Most carotenoids are present in the all-*trans* form in fruits and vegetables and their products, though through heat, light and oxygen exposure, *trans–cis* isomerization as well as oxidation to certain apo-carotenoids (mostly apo-carotenals) can occur, as reviewed previously.[27,112] In a recent study by

Kopec *et al.*,[113] it was investigated whether there were digestive losses of beta-carotene, lutein and lycopene dissolved in sunflower oil, as studied by a gastrointestinal (GI) *in vitro* digestion system from a simulated meal containing mostly sucrose and additional phospholipids. While *trans–cis* isomerization was found to be insignificant, losses of lutein, lycopene and beta-carotene during GI digestion with digestive enzymes (pepsin, pancreatin) were 40%, 20% and 40%, respectively. Metmyoglobin further increased losses of lycopene and beta-carotene to approximately 30% and 80%, respectively, possibly due to oxidative degradation as triggered by iron. It is possible that the formed microcrystals of lycopene protected it from further oxidation and thus losses were lower compared to beta-carotene. The absence of digestion enzymes even enhanced degradation, perhaps as proteins digested with enzymes conferred some protection by forming a protective layer. Similar losses of 20% and 30% for lycopene and beta-carotene, respectively, during simulated digestion in a dynamic model were reported earlier by Blanquet-Diot *et al.* based on tomato-rich meals.[114] In another study, losses of beta-carotene from digested spinach of up to 70% after the small intestinal phase were reported,[115] based on digested spinach (Table 9.3), while lutein losses were lower (25%), perhaps due to higher micellization and thus protection form degradation. Losses of carotenoid standards dissolved in acetone/oil were also considerably higher compared to carotenoids in juice, demonstrating that the matrix could have protective effects, preventing carotenoid oxidation. Other studies reported virtually no detectable losses of lutein during simulated digestion.[114,116] It is likely that differences in test meals (matrix *vs.* solvents), time and pH of digestion or even mechanical factors such as shaking speed did contribute to the observed differences. The absence of significant isomerization was in line with previous studies with lutein,[117] beta-carotene[118] and lycopene[92,119] *in vitro*, and also *in vivo* for beta-carotene.[69] In the study by Kopec *et al.*,[113] no further oxidation products were detected, which was different from an earlier report by the same authors employing artificial emulsifiers[120] and the presence of iron, where several epoxides and apo-carotenals were detected (Table 9.4). Thus, though losses of carotenoids appear to be quite variable, mostly depending on the original matrix and co-digested food constituents, typically 50% or more are expected to reach the site of absorption. Indeed, it can be postulated that a matrix that releases carotenoids rapidly during digestion ensures rapid transition of the lipid droplets to mixed micelles and is able to bind iron (otherwise triggering degradation), which would be associated with only minimal carotenoid degradation.

Exceptions regarding digestive stability exist for carotenoids that are unstable under acidic conditions, as these are more prone to structural changes during the gastric phase, where the pH could drop to as low as 2 (absence of meal). This has been shown for the epoxycarotenoids violaxanthin and neoxanthin, which undergo, at least partly, epoxide–furanoid transition during digestion.[75] Nevertheless, their bioaccessibility appeared to be high, at around 49% and 30% in the latter study, respectively. In another

Table 9.4 Degradation products of carotenoids formed during simulated digestion.

Degradation product	Parent compounds	Factors fostering degradation	Place of degradation	Type of study	Reference
Neochrome	Neoxanthin	Low pH	Likely gastric phase	*In vitro* digestion	75 and 127
Auroxanthin Luteoxanthin	Violaxanthin	Low pH	Likely gastric phase	*In vitro* digestion	75 and 128
(*E*)-beta-apo-13-carotenone (*E*)-beta-apo-15-carotenal(retinal) (*E*)-beta-apo-14′-carotenal (*E*)-beta-apo-12′-carotenal (*E*)-beta-apo-10′-carotenal (*E*)-beta-apo-8′-carotenal 5,6-Seco-beta-carotene-5,6-dione (*E*)-beta-carotene-5,8,5′,8′-diepoxide (*E*)- and (*Z*)-beta-carotene-5,6,5′,6′-diepoxide (*E*)- and (*Z*)-beta-carotene-5,6-epoxide (*E*)-beta-carotene-5,6,5′,8′-diepoxide (*E*)- and (*Z*)-beta-carotene-5,8-epoxide (*Z*)-β-carotene	Beta-carotene	Oxidizing agent (iron)	Gastric phase	Simplified gastric digestion model, pH 4	120 and 129

study, recovery of violaxanthin from raw spinach was below 10% in an *in vitro* digestion system. However, the pH of both models was adjusted to 2, which does not reflect a postprandial pH in the stomach, when the pH could rise to as high as pH 5.[121] However, following epoxide–furanoid transition, violaxanthin would react to form auroxanthin and luteoxanthin, and neoxanthin to from neochrome; epimers of these compounds can likewise be formed. These compounds have been shown to be taken up by Caco-2 cells as a model of the intestinal epithelium, following *in vitro* digestion with a gastric pH of 3.[111,122] However, concentrations in the human bloodstream[123] were shown to be very low, possibly supporting the fragility of the original epoxy-carotenoids. In another human study with three subjects, none of the original compounds or breakdown products of violaxanthin and neoxanthin could be found in the plasma following 50 mg bolus doses.[124] It is also possible that these compounds react to form further unknown metabolites or are shuffled out of the epithelial cell back into the gut lumen.

Taken together, losses from test meals have been reported to average around 20–30% for most carotenoids. The absence of *cis*-isomer formation during digestion suggests that only minor isomerization occurs or that isomers quickly further degrade, which appears less likely as *cis*-isomers appear to be generally as stable during digestion as the all-*trans* form.[114] Several metabolites have been reported during GI digestion, as summarized in Table 9.4. Whether these can be further absorbed is unclear. Studies that have investigated more polar carotenoid metabolites from plants indicate that shorter, more polar apo-carotenoids can be absorbed. For example, in a human postprandial study with crocetin from saffron from crocus flowers,[32] plasma AUC following doses of 7.5, 15 and 22.5 mg were 670, 1130 and 1840 ng hour^{-1} mL^{-1} (100, 200 and 250 ng mL^{-1} peak concentrations), respectively, apparently not saturable at these doses, and bioavailability levels were comparable to native carotenoids (*e.g.* 250 ng hour^{-1} mL^{-1} following the consumption of 19 mg lycopene from tomato soup[76]), though plasma appearance was slower for lycopene. Also, doses of 16 mg of bixin and 0.5 mg of norbixin were well bioavailable, reaching 12 and 58 ng mL^{-1} plasma peak concentrations.[33] In a very recent study, various apo-carotenals were show to be taken up by Caco-2 cells, also emphasizing that these appear to be absorbable.[125] Likewise, in a mouse study, apo-10′-lycopenoic acid, a tentative *in vivo* metabolite of lycopene, was shown to be bioavailable in mice and stored in the liver, while acting on SIRT1 and decreasing hepatic fat accumulation.[126]

9.4 Host-related Factors Governing Carotenoid Digestion, Cellular Uptake and Absorption

9.4.1 Carotenoids during GI Digestion

Host-related factors predominantly include the available intestinal surface area for absorption, which can be diminished *via* various diseases such as IBD[132] or surgically removed short bowel syndrome following, for example, cancer surgery and factors that influence the amount of digestion enzymes and bile salts needed for carotenoid absorption, such as pancreatitis.[133] Also, older age may contribute to decreased intestinal surface[134] and result in lower circulating carotenoid concentrations. Recently, these factors have been reviewed and found to contribute to inter-individual variation of plasma carotenoid concentrations.[6]

Again, during intestinal digestion, lipase and bile salts appear to constitute the most crucial agents fostering carotenoid solubilization, and several genetic aspects (*i.e.* single-nucleotide polymorphisms [SNP]) have shown to be associated with altered carotenoid bioavailability as measured by plasma or plasma-triacylglycerol-rich lipoprotein (TRL) fractions. These included *PNLIP* (encoding pancreatic lipase) and *CLPS* (encoding colipase), which were shown to be related to lycopene[23] and cholesterol absorption,[135] respectively. Several *in vitro* studies have shown that without either bile or pancreatin, carotenoid bioaccessibility was severely compromised. While some studies emphasized that the absence of pancreatin had a more drastic effect,[78] others emphasized that the absence of bile more drastically reduced carotenoid bioaccessibility,[122,136] possibly differing due to specific interactions with the matrix or digestion conditions.

Not much is known regarding the importance of gastric lipase, which presumably digests about 25% of ingested lipids,[121] thus contributing to carotenoid transfer from lipid droplets to mixed micelles, due to the fact that this enzyme has not been commercially available. Also in the gastric phase, low pH (due to an empty stomach) again may foster the degradation of epoxy-carotenoids (violaxanthin, neoxanthin) to their furanoid forms. Pepsin is not known to have any significant influence on carotenoid bioaccessibility, as most carotenoid-containing food items are rather low in total protein content, though in one study, higher pepsin concentrations enhanced lycopene bioaccessibility from a tomato puree *in vitro*,[137] perhaps due to the formation of emulsifying peptides.

Esters from xanthophylls, such as those present in several fruits (mango and papaya, rich in, for example, beta-cryptoxanthin esters) and also some leafy vegetables (mangold, rich in lutein esters) are even more apolar than the non-esterified carotenoids and likewise rely on micellization for their absorption. It remains unclear to what extent cleavage by carboxyl-ester lipase (also termed cholesterol esterase) or also triacylglycerol lipase occurs, which are both secreted by the pancreas,[138] with the latter also present in intestinal cells[139] and possibly active at the brush border, as reviewed by Reboul.[30] *In vitro*, cholesterol esterase does not seem to constitute an effective enzyme for cleaving xanthophyll esters,[140] while *in vivo* cleavage appears complete, perhaps suggesting the involvement of other enzymes. However, as native esters of carotenoids in the plasma appear only at very low concentrations,[141] it is assumed that esters are cleaved prior to absorption.[142] On the other hand, esters have been found to be taken up by Caco-2 cell models.[140] It is thus possible that they are further cleaved intracellularly or are not further absorbed and/or re-secreted in the GI tract, though more research is needed here. It is also worth noting that there are indications that xanthophylls, upon absorption, may be re-esterified, perhaps when present at high concentrations, as shown earlier with lutein in humans.[141]

9.4.2 Carotenoid Cellular Uptake

It is likely that at lower, rather physiological concentrations, uptake *via* transporters predominates, while higher (supplemental) concentrations may favor passive diffusion. In fact, a saturable absorption curve (for doses from 10 to 120 mg) has been determined for lycopene, for example,[143,144] supporting the notion that at these concentrations the transporter uptake predominates. Meanwhile, several transporters have been identified or suggested to participate in carotenoid cellular uptake from the small intestine, including scavenger receptor class B type 1 (SRB1), cluster of differentiation 36 (CD36), Niemann-Pick C1-like 1 (NPC1L1) and ATP binding cassette subfamily A member 1 (ABCA1) for further uptake to the basolateral site.

Several gene association studies were conducted by Borel *et al.* and other researchers, and several SNP of transport enzymes were found to be associated with newly absorbed carotenoids, as determined by their concentration in the plasma-TRL fraction: *SCARB1* (encoding for SRB1; transport of lutein/zeaxanthin, beta-cryptoxanthin, lycopene, beta-carotene, alpha-carotene likely), *CD36* (lycopene, alpha-carotene, lutein/zeaxanthin, beta-cryptoxanthin transport likely), *ABCA1* (beta-carotene, lycopene, lutein/zeaxanthin transport likely) and *NPC1L1* (lycopene, lutein/zeaxanthin transport likely), as reviewed recently.[6] Despite the fact that the selectivity of some of these transporters remains to be further elucidated, it is not quite clear whether at this step of absorption a huge discrimination between the various carotenoids exists. Not considering micellization, compared to all-*trans*-beta-carotene, beta-carotene *cis*-forms were shown to be poorly taken up, at least in Caco-2 cell experiments,[145] similarly to lutein and lycopene. The higher cellular uptake of beta-carotene compared to lycopene in such cell models was found earlier,[146] also observed by O'Sullivan *et al.*[147] when comparing beta-carotene and lutein Caco-2 cellular uptake, though secretion was not effected systematically. However, differences in cellular uptake/absorption efficacy may be lower between other carotenoids, such as between lycopene and astaxanthin cellular uptake into HT-29 cells.[148] In another study by O'Sullivan *et al.*, astaxanthin and also lutein showed high cellular uptake into Caco-2 cells.[149] Differences in cell lines, passage number, time of incubation, dosing and preparation of artificial micelles may contribute to such obtained differences.

As several of these transporters also accept other substrates, these may, at least at higher concentrations, compete for one another. For example, SRB1, CD36 and ABCA1 are all also cholesterol transporters, while NPC1L1 has also shown some selectivity for vitamins E, K and D, CD36 for vitamins D and K and SRB1 for vitamins E, K and D, as reviewed by Reboul and Borel[150] and Yamanashi *et al.*[151] However, no human postprandial studies have investigated the interactions of high doses of vitamins with carotenoid absorption, though it has been observed in an intervention that high doses of plant sterols and sterol esters reduced beta-carotene absorption in humans by approximately 50%.[80] As these sterols (2.2 g day^{-1}) are normally poorly absorbable, however, it is also more likely that their negative influence originates from competition for micelle incorporation (Table 9.2).

In summary, carotenoid cellular uptake is at least in part occurring *via* transporters, and thus higher doses are expected and have been shown to result in lower fractional absorption, also questioning the intake of high doses of individual carotenoids at the same time, such as in high-dose supplements.

9.4.3 Further Carotenoid Metabolism in the Enterocytes and Other Tissues

Once taken up by the enterocyte, carotenoids may be further cleaved, either centrally by BCO1 or eccentrically by BCO2. While BCO1 is present in the cytosol, BCO2 is present in the mitochondria.[152] While beta-carotene appears to be a particular target for BCO1, resulting in the production of provitamin A active retinal, BCO2 has shown to be able to produce apo-carotenals from beta-carotene, at least apo-10′-beta-apo-carotenal, as shown in HepG2 liver cells[153] and in mice studies.[154] BCO2 has further been suggested to cleave lycopene into apo-10′-lycopenal,[38,155] but also apo-8′- and apo-12′-lycopenal[156] were found in the livers of rats. While BCO1 thus appears to favor full-length provitamin A carotenoids (beta-carotene, alpha-carotene and beta-cryptoxanthin, though not lycopene, lutein and 9-*cis*-beta-carotene[157]), BCO2 favors the cleavage of xanthophylls.[158] However, cleavage is far from being complete and species dependent, and has been estimated for xanthophylls and BCO2 to be 10–40-times weaker than in mice.[159] It is also noteworthy that longer apo-carotenoids are themselves substrates for BCO1/2, as shown for beta-apo-4′-carotenal, beta-apo-8′-carotenal, beta-apo-10′-carotenal and beta-apo-12′-carotenal in chickens and rats, as reviewed by Kim *et al.*[3]

The further fate of these metabolites is not entirely clear. While retinal can react to form retinol and is then likely to be re-esterified by LRAT prior to further absorption, this is not the case for uncleaved carotenes and xanthophylls. BCO2 metabolites are potentially further metabolized *via* hydroxylation, as some hydroxylated compounds were detected earlier *in vivo* (Figure 9.1), though their origin (plant *vs. in vivo*) could not be clearly differentiated.[57,160,161] It also is possible that these compounds can then be further conjugated to form sulfated and glucuronidated metabolites, as glucuronidation has been shown for retinol.[162] This was demonstrated following the administration of isotopically labeled all-*trans* beta-carotene in rats, upon which the majority of radioactively labeled compounds were studied in bile, and 73–96% were water-soluble compounds that yielded retinol or retinoic acid following cleavage by glucuronidase.[163] Retinoyl-beta-glucuronide (RAG) was also found in the plasma of humans, though at rather low concentrations (*i.e.* ~2.4 ng mL^{-1}).[164] Further glucuronidated metabolites such as 9-*cis*-4-oxo RAG were identified in subjects treated with 9-*cis* retinoic acid.[165] These findings would also be in line with excretion of beta-carotene *via* urine, which was detected *via* isotopic studies (6.5% of the administered dose),[166] though the nature of the compounds was not determined at that time. However, it is apparent that BCO1 and BCO2 are among the most important carotenoid metabolism steps in humans.

In addition to further metabolism, carotenoids or their metabolites may also be re-secreted back into the intestinal lumen. This may happen *via* cellular abrasion of the enterocytes or *via* shuttling carotenoids back *via* transporters. Whether, how and to what extent this happens is unclear. As highlighted in a review by Reboul,[30] SRB1 has been shown to be able to act as a transporter in both directions, and vitamin E[167] and D[168] efflux to the apical side was shown for Caco-2 cells, and a similar phenomenon could also be assumed for carotenoids. However, other transporters, including other ATP binding cassette subfamily proteins such as ABCG5,[169] ABCB1[23] and ABCG8,[170] may also play a role, as suggested for lutein and ABCG5 for beta-carotene[171] in human trials investigating genetic polymorphisms.

In summary, the further metabolism following a potential cleavage by BCO1/2 remains unclear, but is likely to include further reactions with phase II enzymes, such as glucuronidation and perhaps sulfation, though a number of other oxidation products such as the formation of keto-compounds[172,173] and penta-cyclic compounds[57] have also been suggested (Figure 9.1).

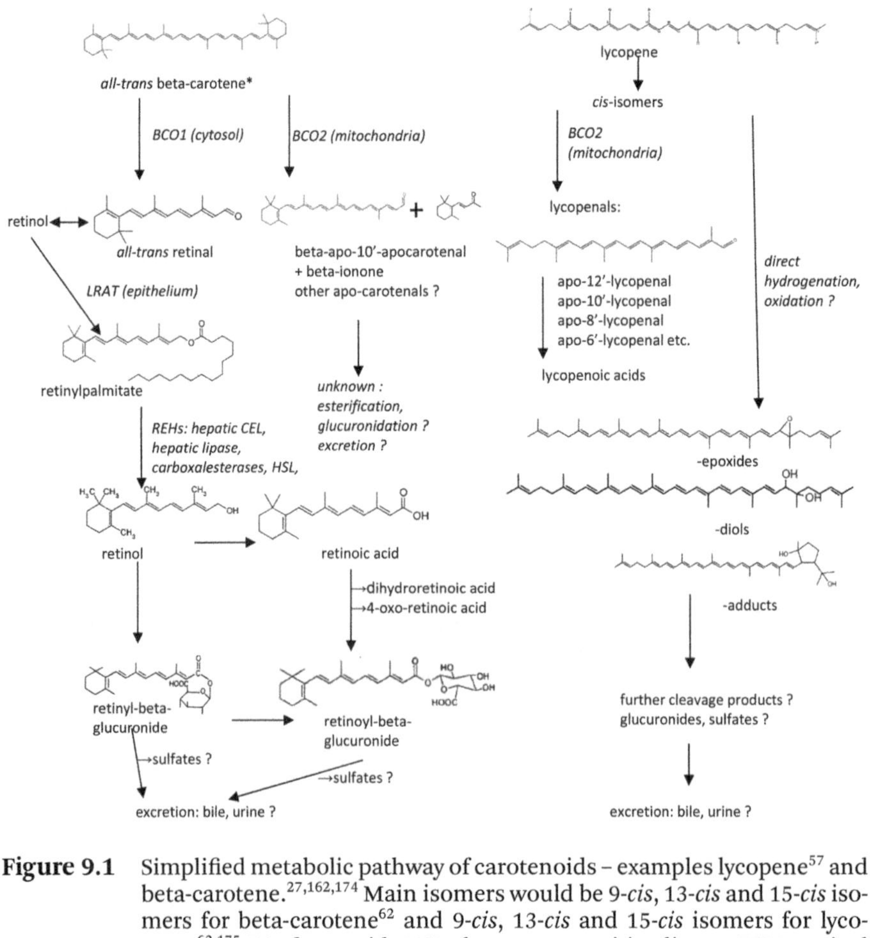

Figure 9.1 Simplified metabolic pathway of carotenoids – examples lycopene[57] and beta-carotene.[27,162,174] Main isomers would be 9-*cis*, 13-*cis* and 15-*cis* isomers for beta-carotene[62] and 9-*cis*, 13-*cis* and 15-*cis* isomers for lycopene.[62,175] FA: fatty acid; HSL: hormone-sensitive lipase. REH: retinyl ester hydrolysis enzymes. [a]*Cis*–*trans*-isomerization not considered here.

9.5 Colonic Fate of Carotenoids

All carotenoids that are not absorbed in the small intestine will eventually reach the colon, either in bioaccessible form (emulsified) or precipitated. In addition to non-absorbed carotenoids, carotenoids are also re-excreted *via* bile and the pancreas back into the intestinal lumen (*i.e.* following their original absorption), as shown by, for example, isotopic studies in salmon.[176] As absorption of carotenoids is only approximately 10–40%, the remaining 60–90% should reach the colon, not counting potential endogenous losses of carotenoids or metabolites. Several *in vitro* studies have shown that carotenoids, following colonic fermentation, are not completely recovered, but that they are degraded (*i.e.* fermented) by the microbiota. For example, in a study with fermented plum and cabbage carotenoids, only 4–25% of carotenoids were recovered following colonic fermentation.[177] In another study by Goni *et al.*, colonic recovery of lutein, lycopene and beta-carotene were 19/19%, 17/3% and 24/21%, respectively, for digested fruits/vegetables.[178] In another study by those authors, losses of pure beta-carotene standard were as high as 98%,[130] though no metabolites were reported (Table 9.5).

However, the nature of these microbiota-fermented carotenoids is not known. Unlike for polyphenols, which are known to undergo numerous reactions in the gut, including ring fission, deglycosylation, hydrolysis, deglucuronidation and demethylation, no literature exists on carotenoid products in the large intestine. Though it is possible that more polar degradation products are produced, this is speculative at present. Some carotenoid metabolites originating from the body, however, are also expected, as retinoyl-glucuronide excreted *via* the bile following beta-carotene administration may be de-glucuronidated.

Furthermore, it remains unknown whether carotenoids can be absorbed from the colon. In an earlier human trial, it was found that maximum carotenoid concentrations in colonic cells trailed behind that of blood plasma by 5–7 days, suggesting insignificant direct absorption from the luminal content,[179] but rather absorption *via* the small intestine followed by carotenoid distribution *via* the circulatory system. In an insightful rat study by Oshima *et al.*,[180] rats received intracolonic doses of lycopene, and low amounts were detected in serum and higher amounts in the liver, suggesting that colonic uptake does play a role. However, mixed micelles are present in the colon, and there is indirect evidence of absorption of other lipophilic compounds (*e.g.* of vitamin K with long side chains [phylloquinone, log P: 10.9, https://www.phenomenex.com/Compound?id=Vitamin+K]), as these were found in the livers of humans,[181] though their availability is possibly low as deduced following colorectal administration in rats.[182] Finally, as BCO1 has been detected in colonic cells, at least in the mouse,[183] carotenoid metabolism is also expected to occur in colonic epithelial cells. In summary, more research is warranted on potential colonic degradation products or possible colonic uptake of native carotenoids or their products, which may be low, but perhaps not zero.

Table 9.5 Losses during carotenoid digestion in the gastrointestinal tract, including the colon when studied.[a]

Carotenoid(s) studied	System	Meal	Losses/degradation	Other observations	Reference
Small intestine					
Beta-carotene	*In vitro* (static)	Carotenoids in sunflower oil, sugar and phospholipids	40% losses after intestinal phase	No oxidation products, no isomers observed	113
Lutein	*In vitro* (static)	Carotenoids in sunflower oil, sugar and phospholipids	40% losses after intestinal phase	No oxidation products, no isomers observed	113
Lycopene	*In vitro* (static)	Carotenoids in sunflower oil, sugar and phospholipids	20% losses after intestinal phase	No oxidation products, no isomers observed	113
Zeaxanthin	*In vitro* (TIM)	Tomatoes and beadlets	No losses	No *trans–cis* isomerization during digestion	114
Lutein	*In vitro* (TIM)	Tomatoes and beadlets	No losses	No *trans–cis* isomerization during digestion	114
Beta-carotene	*In vitro* (TIM)	Tomatoes and beadlets	30% losses after intestinal phase	No *trans–cis* isomerization during digestion	114
Lycopene	*In vitro* (TIM)	Tomatoes and beadlets	20% losses after intestinal phase	No *trans–cis* isomerization during digestion	114
Colon					
Total carotenoids (beta-carotene + lutein)	*In vitro* (static)	*S. oleracea*, *S. americanum* and *C. aconitifolius* carotenoids	92%, 96% and 23% losses during colonic fermentation	No degradation products investigated	130
Lutein	*In vitro* (static)	*S. oleracea*, *S. americanum* and *C. aconitifolius* carotenoids	92.5%, 95.5% and 23.0% losses during colonic fermentation	No degradation products investigated	130
Beta-carotene	*In vitro* (static)	*S. oleracea*, *S. americanum* and *C. aconitifolius* carotenoids	98.5%, 97.0% and 23.0% losses during colonic fermentation	No degradation products investigated	130
Total carotenoids (beta-carotene, lutein)	*In vitro* (static)	Various plum and cabbage varieties	4–25% of after colonic fermentation	No degradation products investigated	131

[a]TIM: TNO gastrointestinal model (dynamic multi-compartmental model).

9.6 Further Metabolism of Carotenoids Following Cellular Uptake

9.6.1 Conversion into Vitamin A

Clearly, the fraction of provitamin A carotenoids that can be absorbed and then be cleaved and turned into vitamin A active compounds, a term also called "bioconversion," is very variable, depending, as emphasized above, on the food matrix and factors influencing the absorption of beta-carotene, and finally host factors such as BCO1 cleavage activity. In a study by Leung *et al.*,[184] it was shown that bioconversion differed between variants of BCO1, with a double mutant of the SNP R267S:rs12934922 and A379V:rs7501331 showing 57% reduced catalytic cleavage activity. In a human clinical study by Lietz *et al.*,[185] three polymorphisms of the *BCO1* gene (rs6420424, rs11645428 and rs6564851) hampered BCO1 catalytic activity by 59%, 51% and 48%, respectively. The expression of BCO1 appears to be regulated by the intestine-specific homeobox transcription factor (ISX), as retinoic acids appear to stimulate its expression, which in turn reduces expression of BCO1, limiting vitamin A formation in the intestinal cell.[186] Thus, the fate of beta-carotene (*i.e.* its conversion to retinal) is influenced by genetic factors, which is possibly the cause of some people being rather high converters of beta-carotene to vitamin A, while others (~45% in Westernized countries) are low converters[187]. However, for more detailed reviews on carotenoids as sources of vitamin A, the reader is referred to other articles[188–190]

9.6.2 Intracellular Transport and Sequestration of Carotenoids into Chylomicrons

Following carotenoid cellular uptake, potential cleavage and/or further re-esterification, carotenoids are secreted *via* chylomicrons into the lymphatic system. This appears to be the case for all compounds with a log $P > 5$ (which would include all major dietary carotenoids), below which compounds can be further transported *via* the portal vein.[191] As apo-carotenoids also have a log P value of approximately 5 or slightly higher (http://foodb.ca/compounds/FDB022576), these are also assumed to be sequestered *via* the portal pathway, at least to some extent, though they have not been detected so far. In a study by Kopec *et al.* with orally administered ^{13}C-beta-carotene, ^{13}C-retinol was observed in whole plasma but not in the TRL fraction, while no non-vitamin A active ^{13}C-beta-apo-carotenals, ^{13}C-beta-apo-carotenols or ^{13}C-beta-apo-carotenoic acids were observed in the plasma or TRL samples. Thus, at least for beta-carotene metabolism, apo-carotenoids are not secreted *via* chylomicrons to a measurable extent.[192]

The details of chylomicron sequestration in the enterocyte are not completely elucidated, but possibly include binding of carotenoids to some intracellular transport proteins (due to their otherwise low cytosolic solubility) and their carrying to the endoplasmic reticulum, followed by the Golgi apparatus for sequestration into chylomicrons. Regarding the binding proteins, potential candidates include fatty acid binding protein (FABP), and also BCO1 itself, as reviewed previously.[6] A human postprandial study found an association between a SNP in FABP and lycopene in the plasma TRL fraction,[193] supporting a potential role for its participation in intracellular transport.

However, carotenoids are eventually assembled, by the aid of apolipoprotein A1 (APOA1),[194] into chylomicrons. They are then released *via* exocytosis to the lacteals and the lymphatic system. Another route has been speculated upon, which includes secretion of carotenoids to the basolateral site directly *via* APOA1, by lipid transfer to high-density lipoprotein (HDL) from the membrane protein ABCA1. This is supported by a study showing direct transport of xanthophylls *via* ABCA1 to APOA1 in hamsters.[195]

It is interesting to note that after consumption of a carotenoid-rich meal, carotenoids appear to be stored for a certain time in the enterocytes and are sequestered following the consumption of a later meal rich in lipids, which appears to foster the formation and secretion of carotenoid-loaded chylomicrons,[22,24] and thus follow-up meals could influence the kinetics, if not the bioavailability, of carotenoids.

9.7 Carotenoid Biodistribution, Target Tissues and Excretion

Following their sequestration into chylomicrons and transport in the lymph, carotenoids reach the bloodstream and are transported to the liver. Chylomicrons are partly stripped of their triglycerides along the way to react with chylomicron remnants, not affecting carotenoids. While the liver appears partly to constitute a storage organ for carotenoids, and more so for vitamin A, carotenoids are also redistributed into various lipoproteins. While xanthophylls appear to be distributed about evenly between HDL and low-density lipoprotein (LDL), carotenes appear to be associated primarily with the LDL fraction.[196]

Carotenoids are then released *via* the lipoproteins to different tissues. Carotenoids have been detected in many different tissues and, based on the concentrations encountered (Table 9.1), there does not appear to exist a dominating primary target tissue, although the adipose tissue appears to be an important if not the major storage organ for carotenoids, and liver concentrations are typically higher than those of other organs. It is plausible that tissues expressing a larger number of especially LDL receptors are candidates for storing higher concentrations of carotenoids, such as adipose tissue,[197] prostate tissue[60] or adrenal tissue,[62] which appears to be the case (Table 9.1), at least for adipose tissue and adrenal tissue. As many tissues contain both BCO1 and BCO2,[27] similar metabolites as in the enterocytes may be formed, though few data are available on tissue metabolites. Because of the relation of carotenoids to AMD, the macula constitutes an important target tissue for lutein, zeaxanthin and *meso*-zeaxanthin (a stereoisomer of zeaxanthin produced *in vivo* from lutein and present to some extent in seafood) – for further information, please consult further reviews.[198] Also, a non-enzymatic oxidation product of lutein, 3′-oxolutein, has been related to the human retina.[199]

Following administration of isotopically labeled phytoene (^{13}C-labeled), Moran *et al.*[97] developed a compartment model (Figure 9.2) highlighting major body pools of phytoene. A similar model was developed for ^{13}C-labeled lycopene,[97] also highlighting ongoing *trans* → *cis* isomerization in the human body. These studies also demonstrated that the largest pools were slow-turnover pools, possibly representing various body tissues of slow exchanges such as the adipose tissue, followed by fast-turnover pools, which may include liver metabolism.

Hardly any data are available regarding carotenoid excretion pathways. As Zile *et al.* found that the majority of vitamin A active compounds following administration of isotopic ATRA to rats[163] were present in the water-soluble fraction in the bile as retinol and retinoic acid glucuronides (in all-*trans* and 13-*cis* forms), it can be assumed that, at least for beta-carotene, biliary

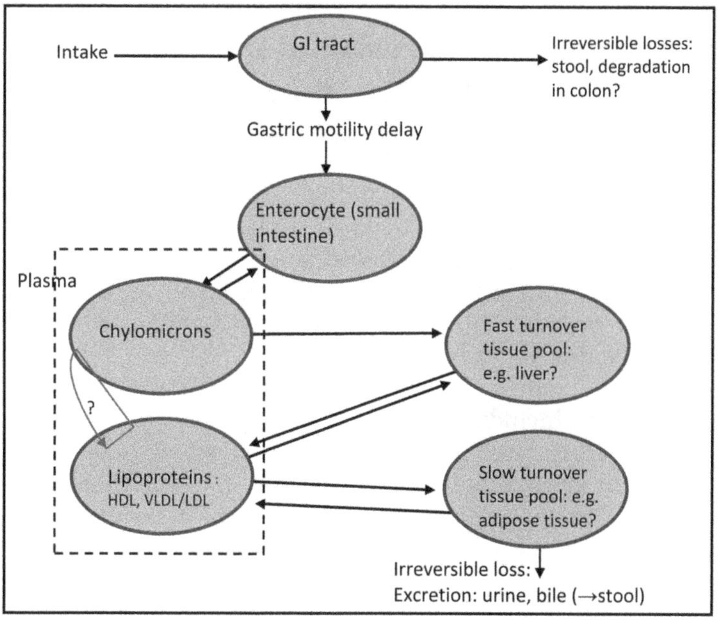

Figure 9.2 Human compartmental kinetic model of non-provitamin A carotenoids based on measurements in plasma following dosings as carried out by Diwadkar-Navsariwala[144] and stable isotope administration by Moran *et al.*[97,202] for lycopene and phytofluene. Arrows represent the direction of exchange between the different pools and ovals are compartments (kinetically homogenous pools). Gastric motility delay likely represents GI transit time. More details on beta-carotene compartment models can be found in the thesis by Park.[203]

excretion as glucuronides is one main pathway of excretion. As urinary excretion was also demonstrated following the radioactive administration of ^{14}C-beta-carotene,[166] it could be speculated that glucuronides or other polar metabolites such as apo-carotenoids are partly secreted in the urine. In the study of Ho *et al.*,[166] approximately 6.5% (range 1.7–15.8%) of the orally administered dose was excreted in urine within 30 days. Interestingly, mono- and di-glucuronides of crocetin were also found in plasma of rats following oral administration at higher concentrations than the free molecule, also suggesting that glucuronidation could be a fate of apo-carotenoids.[200]

By contrast, the cumulative dose excreted in feces ranged between 28% and 100% (mean 56%) within 14 days.[166] Similar results were found by Hickenbottom *et al.*[201] following administration of a small dose of radiolabeled beta-carotene, demonstrating cumulative recoveries in urine and stool of 4% and 45%, respectively. As the latter constituted mostly non-absorbed beta-carotene detected within the first 2 days of test meal intake, and only 1% in addition was recovered later due to endogenous losses, urinary excretion appeared to be the major route of excretion in this trial.

9.8 Interrelation of Carotenoid Degradation Products, Metabolites and Health Aspects

9.8.1 Health Benefits Associated with Native Carotenoids

It ought to be emphasized that our current gaps in knowledge regarding further carotenoid metabolism coincide with our lacking a grasp on carotenoid health-beneficial aspects, and this failure to characterize the true bioactive constituents of carotenoid-rich food items, if any, will impede any further meaningful intervention trials aimed at ameliorating ailments related to low intake and circulating carotenoids.

While direct antioxidant effects of the native carotenoids clearly have been shown *in vitro*, including quenching of singlet oxygen,[204,205] protection against UV light as reviewed previously[206] and quenching of peroxyl radicals originating, for example, from lipid oxidation,[207] it is unclear whether these effects can be translated as such to the *in vivo* situation. However, oxidative stress measured by the dichlorofluorescein method in cell culture (Caco-2) following exposure to H_2O_2 and astaxanthin at 10 μM reduced oxidative stress, indicating that such high but intestinally reachable concentrations of native carotenoids can have beneficial effects.[208] As the cleavage rate into derivatives (*via* BCO1/2) is slow, it could be assumed that carotenoids may also protect the cell membranes from oxidative stress *in vivo*. In this respect, carotenes are likely to be situated in the more apolar area of the fluid mosaic model of the cell membrane (*i.e.* in the middle part[209]), while xanthophylls may span the membrane, occupying the outer domains with the polar parts of their molecules.[210]

9.8.2 Health Benefits Associated with Carotenoid *In Vivo* Metabolites

As most studies did not investigate carotenoid degradation products and their concentrations in tissues following *in vivo* studies, it is difficult to ascribe any observed effects of carotenoids to either the native compounds or their metabolites. Several studies have pointed out that carotenoid metabolites (*e.g.* the resulting apo-carotenals) have a higher affinity to interact with transcription factors in the cytosol due to their higher solubility and higher electrophilicity. In a study by Linnewiel *et al.*,[37] several apo-carotenals obtained from the oxidation of lycopene with potassium permanganate were tested in several transfected cell lines (MCF-7 breast cancer cells, T47D mammary cancer cells and LNCaP prostate cancer cells) regarding their interaction with Nrf2. It was found that: (1) aldehydes, not acids, were most active; (2) the activity of the compounds depended on the relative position of the methyl group 3 to the terminal aldehyde; and (3) the optimal length of the dialdehyde derivative was 12 carbons in the main chain. In another study by the same authors,[36] the authors investigated the effect of synthetic apo-carotenals on T47D, HOS and MC3T3-E1 osteoblast-like cells transfected with an NF-κB reporter gene. Again, while native lycopene was found to be rather ineffective, apo-carotenals strongly influenced NF-κB activation, possibly by binding to cysteine residues of NF-κB *via* the reaction of α,β-unsaturated carbonyls with nucleophilic protein thiols to a Michael adduct. Similarly, in a rat study with lutein,[211] photo-oxidative degradation products reduced lipopolysaccharide (LPS)-triggered inflammation, including NF-κB downstream products such as malondialdehyde, prostaglandin E2, TNF-α and IL-6, more strongly than lutein in serum.

In several studies, it was suggested that lycopene metabolites could interact with RXR and RAR, which are important for regulating immune system functions, among others. In a recent study by Aydemir *et al.*,[39] it was shown that lycopene fed to mice regulated genes affected by vitamin A deficiency, bringing their expression back to levels seen in rats without vitamin A deficiency. This was ascribed – though it was not directly detected – to lycopene metabolites such as to apo-15′-lycopenoid acid,[212] highlighting the potency of these metabolites, suggesting that these have vitamin A-like properties.

In a recent review by Bonet *et al.*,[213] it has been proposed that apo-carotenals of beta-carotene and possibly other carotenoids influence peroxisome proliferator-activated receptors (by acting as agonists or antagonists) in adipose tissue, thereby regulating proliferation of adipose tissue. In support of this, several human studies have found that obese subjects typically have lower levels of circulating carotenoids,[214,215] though it is not clear whether this is due to rather low intake of carotenoids, increased storage of carotenoids in adipose tissue or increased turnover in other tissues.

Finally, several plant-derived apo-carotenoids have been shown to be bioactive in humans. For example, the plant hormone abscisic acid has been shown to ameliorate experimental IBD in an animal model,[216] to act as a cytokine in humans, positively influencing systemic sclerosis markers *in vitro*,[35] and to improve glucose tolerance in humans in small quantities (0.5–1 µg kg bodyweight^{-1}),[34] possibly *via* enhancing insulin production and GLUT4-related glucose cellular uptake.

In summary, though more confirmation is needed, many studies point out that carotenoid metabolites (such as those produced by BCO1/2) can be very bioactive and may be more potent than their native precursors, effecting many inflammation, oxidative stress and cell proliferation pathways *via* interacting with transcription factors and nuclear receptors. Thus, understanding the metabolism of carotenoids is key to our understanding of their health-related benefits or risks, and more research in this domain is needed. At present, research regarding carotenoid metabolites is limited, at least in part due to methodological limitations. Firstly, not many carotenoid metabolites are commercially available, and even if they are available, they are quite expensive. The same is true for isotopically labeled carotenoids for following the metabolism of orally administered tracers. Secondly, their sensitive and selective analysis requires sophisticated and likewise expensive analytical instrumentation (*i.e.* chromatographic separation techniques coupled to mass-spectrometry, such as atmospheric-pressure-ionization based liquid chromatography-mass spectrometry ([APCI-]LC-MS-MS), which many laboratories do not have access to, even though this technique has become more available in recent years).

9.9 Conclusions

The metabolic fate of carotenoids depends on several factors (Figure 9.3). When carotenoids can be released from the matrix and are bioaccessible, they may enter the body pool of carotenoids, while their colonic fate remains unknown. Their release from the diet is dependent on both the food matrix and food constituents, foremost the presence of lipids, and host factors such as the availability of sufficient bile and pancreatin. Following their cellular uptake, which depends on the expression of active transporters (SRB1, CD36, NPC1L1 and ABCA1) related to SNPs, provitamin A carotenoids may then be preferably cleaved by BCO1 into retinal, while non-provitamin A carotenoids may be preferentially cleaved by BCO2 into several apo-carotenals, if at all.

In what follows, carotenoids and their metabolites and retinyl esters (produced from retinol) will be transported to the liver, where they may be stored or further distributed in the body by HDL and very-low-density lipoprotein/LDL. Apo-carotenoids may likewise be transported to target tissues, though their transport is not understood and the transport of the free form in the bloodstream appears limited. Though no single specific target tissue of carotenoids exists, the liver appears to constitute a major storage tissue, especially for vitamin A, while a large part of carotenoids is stored in adipose tissue. The macula of the retina is a target for lutein and zeaxanthin. Further metabolism in tissues expressing BCO1/2 appears likely. The resulting apo-carotenoids are expected to interact with transcription factors in the cytosol of the cell, including NF-κB, Nrf2 and some metabolites such as retinal, and some lycopeneoids may interact with the nuclear receptors RAR and RXR. In the process that follows this, it is likely that apo-carotenoids

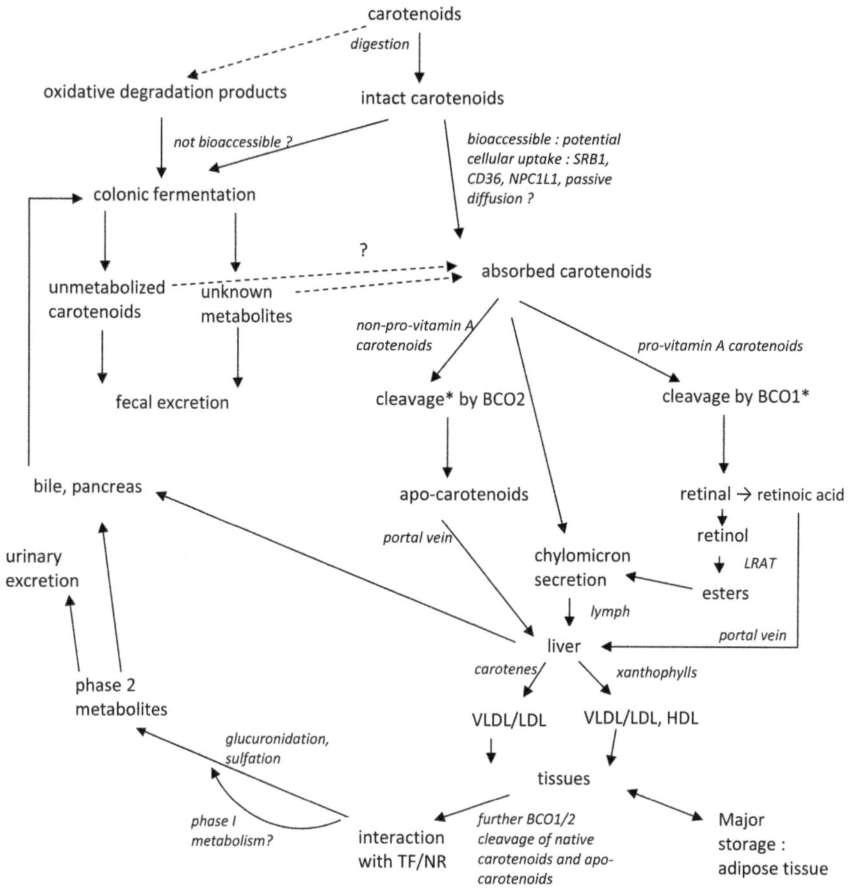

Figure 9.3 Simplified overview on the metabolic fate of dietary-derived carotenoids in the human body. TF: transcription factor; NR: nuclear receptors. [a]Preferred cleavage pathway, though provitamin A may also be cleaved by BCO2 and non-provitamin A carotenoids by BCO1.

are eventually metabolized by phase II enzymes (sulfotransferases and UDP-glucuronosyltransferases). The resulting sulfates and glucuronides are more water soluble and can be excreted *via* the kidneys, while native carotenoids may be secreted back into the gut *via* bile.

Conflict of Interest

The author declares no conflict of interest.

Acknowledgements

The insights obtained from the EU-COST actions POSITIVE (FA-1403) and EUROCAROTEN (CA-15136) are greatly appreciated.

References

1. J. Paniagua-Michel, J. Olmos-Soto and M. A. Ruiz, *Methods Mol. Biol.*, 2012, **892**, 1–12.
2. Y. Yang, R. Yatsunami and A. Ando, *et al.*, *J. Bacteriol.*, 2015, **197**, 1614–1623.
3. Y. S. Kim and D. K. Oh, *Appl. Microbiol. Biotechnol.*, 2010, **88**, 807–816.
4. J. C. Beltran and C. Stange, *Subcell. Biochem.*, 2016, **79**, 239–272.
5. C. Desmarchelier and P. Borel, *Trends Food Sci. Technol.*, 2017, **69**, 270–280.
6. T. Bohn, C. Desmarchelier and L. O. Dragsted, *et al.*, *Mol. Nutr. Food Res.*, 2017, **61**, DOI: 10.1002/mnfr.201600685.
7. E. T. Leermakers, S. K. Darweesh and C. P. Baena, *et al.*, *Am. J. Clin. Nutr.*, 2016, **103**, 481–494.
8. J. Chen, W. Jiang and L. Shao, *et al.*, *Int. J. Food Sci. Nutr.*, 2016, **67**, 744–753.
9. N. Yu, X. Su and Z. Wang, *et al.*, *Nutrients*, 2015, **7**, 9309–9324.
10. I. Kompauer, J. Heinrich and G. Wolfram, *et al.*, *Public Health Nutr.*, 2006, **9**, 472–479.
11. L. G. Wood, M. L. Garg and R. J. Blake, *et al.*, *J. Am. Coll. Nutr.*, 2005, **24**, 448–455.
12. M. S. Donaldson, *Nutrients*, 2011, **3**, 1003–1022.
13. L. Ma, R. Liu and J. H. Du, *et al.*, *Nutrients*, 2016, **8**, 426.
14. N. I. Krinsky and E. J. Johnson, *Mol. Aspects Med.*, 2005, **26**, 459–516.
15. W. Stahl and H. Sies, *Am. J. Clin. Nutr.*, 2012, **96**, 1179s–1184s.
16. A. Kaulmann and T. Bohn, *Nutr. Res.*, 2014, **34**, 907–929.
17. Y. Sharoni, K. Linnewiel-Hermoni and M. Khanin, *et al.*, *Mol. Nutr. Food Res.*, 2012, **56**, 259–269.
18. T. Bohn, *Int. J. Vitam. Nutr. Res.*, 2016, **10**, 1–5.
19. G. Bjelakovic, D. Nikolova and L. L. Gluud, *et al.*, *Cochrane Database Syst. Rev.*, 2012, **14**, Cd007176.
20. D. Albanes, O. P. Heinonen and J. K. Huttunen, *et al.*, *Am. J. Clin. Nutr.*, 1995, **62**, 1427s–1430s.
21. G. S. Omenn, G. E. Goodman and M. D. Thornquist, *et al.*, *N. Engl. J. Med.*, 1996, **334**, 1150–1155.
22. T. Bohn, M. Blackwood and D. Francis, *et al.*, *Nutr. Cancer*, 2013, **65**, 919–929.
23. P. Borel, C. Desmarchelier and M. Nowicki, *et al.*, *Free Radical Biol. Med.*, 2015, **83**, 238–244.
24. N. Z. Unlu, T. Bohn and S. K. Clinton, *et al.*, *J. Nutr.*, 2005, **135**, 431–436.
25. C. Sy, B. Gleize and O. Dangles, *et al.*, *Mol. Nutr. Food Res.*, 2012, **56**, 1385–1397.
26. T. Bohn, *Curr. Nutr. Food Sci.*, 2008, **4**, 53–72.
27. A. Eroglu and E. H. Harrison, *J. Lipid Res.*, 2013, **54**, 1719–1730.
28. A. Eroglu, D. P. Hruszkewycz and C. dela Sena, *et al.*, *J. Biol. Chem.*, 2012, **287**, 15886–15895.

29. G. P. Lobo, J. Amengual and G. Palczewski, *et al.*, *Biochim. Biophys. Acta*, 2012, **1821**, 78–87.
30. E. Reboul, *Nutrients*, 2013, **5**, 3563–3581.
31. E. H. Harrison, *Biochim. Biophys. Acta*, 2012, **1821**, 70–77.
32. N. Umigai, K. Murakami and M. V. Ulit, *et al.*, *Phytomedicine*, 2011, **18**, 575–578.
33. L. W. Levy, E. Regalado and S. Navarrete, *et al.*, *Analyst*, 1997, **122**, 977–980.
34. M. Magnone, P. Ameri and A. Salis, *et al.*, *FASEB J.*, 2015, **29**, 4783–4793.
35. S. Bruzzone, F. Battaglia and E. Mannino, *et al.*, *Biochem. Biophys. Res. Commun.*, 2012, **422**, 70–74.
36. K. Linnewiel-Hermoni, Y. Motro and Y. Miller, *et al.*, *Free Radical Biol. Med.*, 2014, **75**, 105–120.
37. K. Linnewiel, H. Ernst and C. Caris-Veyrat, *et al.*, *Free Radical Biol. Med.*, 2009, **47**, 659–667.
38. X. D. Wang, *Am. J. Clin. Nutr.*, 2012, **96**, 1214s–1222s.
39. G. Aydemir, Y. Kasiri and E. M. Bartok, *et al.*, *Mol. Nutr. Food Res.*, 2016, **60**, 2413–2420.
40. G. Aydemir, Y. Kasiri and E. Birta, *et al.*, *Mol. Nutr. Food Res.*, 2013, **57**, 739–747.
41. M. Lucarini, S. Lanzi, L. D'Evoli, L. D'Evoli and A. Aguzzi, *et al.*, *Int. J. Vitam. Nutr. Res.*, 2006, **76**, 103–109.
42. M. Porrini and P. Riso, *J. Nutr.*, 2005, **135**, 2042s–2045s.
43. E. Biehler, A. Alkerwi and L. Hoffmann, *et al.*, *J. Food Compos. Anal.*, 2012, **25**, 56–65.
44. M. E. O'Neill, Y. Carroll and B. Corridan, *et al.*, *Br. J. Nutr.*, 2001, **85**, 499–507.
45. M. M. Murphy, L. M. Barraj and D. Herman, *et al.*, *J. Acad. Nutr. Diet.*, 2012, **112**, 222–229.
46. E. N. Story, R. E. Kopec and S. J. Schwartz, *et al.*, *Annu. Rev. Food Sci. Technol.*, 2010, **1**, 189–210.
47. USDA, *National Nutrient Database for Standard Reference Release 28*, 2015, http://ndb.nal.usda.gov/ndb/foods/show, accessed 05.11.2015: USDA.
48. A. Kaulmann, M. C. Jonville and Y. J. Schneider, *et al.*, *Food Chem.*, 2014, **155**, 240–250.
49. C. Stange-Klein and M. Rodriguez-Conception, Carotenoids in carrots, in *Pigments in Fruits and Vegetables*, ed. C. Chen, Springer, Heidelberg, 2015.
50. F. Gurmu, S. Hussein and M. Laing, *Int. J. Vitam. Nutr. Res.*, 2014, **84**, 65–78.
51. Y. M. Peng, Y. S. Peng and Y. Lin, *et al.*, *Nutr. Cancer*, 1995, **23**, 233–246.
52. S. Scarmo, B. Cartmel and H. Lin, *et al.*, *Arch. Biochem. Biophys.*, 2010, **504**, 34–39.
53. H. H. Schmitz, C. L. Poor and R. B. Wellman, *et al.*, *J. Nutr.*, 1991, **121**, 1613–1621.

54. C. Manach, G. Williamson and C. Morand, *et al.*, *Am. J. Clin. Nutr.*, 2005, **81**, 230S–242S.
55. B. J. Burri, T. R. Neidlinger and A. J. Clifford, *J. Nutr.*, 2001, **131**, 2096–2100.
56. W. K. Al-Delaimy, P. Ferrari and N. Slimani, *et al.*, *Eur. J. Clin. Nutr.*, 2005, **59**, 1387–1396.
57. F. Khachik, C. J. Spangler and J. C. Smith Jr, *et al.*, *Anal. Chem.*, 1997, **69**, 1873–1881.
58. G. J. Handelman, E. A. Dratz and C. C. Reay, *et al.*, *Invest. Ophthalmol. Visual Sci.*, 1988, **29**, 850–855.
59. S. Alaluf, U. Heinrich and W. Stahl, *et al.*, *J. Nutr.*, 2002, **132**, 399–403.
60. S. K. Clinton, C. Emenhiser and S. J. Schwartz, *et al.*, *Cancer Epidemiol., Biomarkers Prev.*, 1996, **5**, 823–833.
61. R. Vishwanathan, M. J. Kuchan and S. Sen, *et al.*, *J. Pediatr. Gastroenterol. Nutr.*, 2014, **59**, 659–665.
62. W. Stahl, W. Schwarz and A. R. Sundquist, *et al.*, *Arch. Biochem. Biophys.*, 1992, **294**, 173–177.
63. R. S. Parker, *Eur. J. Clin. Nutr.*, 1997, **51**(suppl. 1), S86–S90.
64. O. Hernell, J. E. Staggers and M. C. Carey, *Biochemistry*, 1990, **29**, 2041–2056.
65. P. Borel, P. Grolier and M. Armand, *et al.*, *J. Lipid Res.*, 1996, **37**, 250–261.
66. R. E. Kopec and M. Failla, *J. Food Compost. Anal.*, 2017, DOI: 10.1016/j.jfca.2017.1006.1008.
67. W. Xu, P. Ling and T. Zhang, *J. Drug Delivery*, 2013, **2013**, 340315.
68. E. Reboul, M. Richelle and E. Perrot, *et al.*, *J. Agric. Food Chem.*, 2006, **54**, 8749–8755.
69. V. Tyssandier, E. Reboul and J. F. Dumas, *et al.*, *Am. J. Physiol.: Gastrointest. Liver Physiol.*, 2003, **284**, G913–G923.
70. E. Biehler and T. Bohn, *Curr. Nutr. Food Sci.*, 2010, **6**, 44–69.
71. C. E. West and J. J. Castenmiller, *Int. J. Vitam. Nutr. Res.*, 1998, **68**, 371–377.
72. P. Borel, *Clin. Chem. Lab. Med.*, 2003, **41**, 979–994.
73. H. Palafox-Carlos, J. F. Ayala-Zavala and G. A. Gonzalez-Aguilar, *J. Food Sci.*, 2011, **76**, R6–R15.
74. T. Bohn, Provitamin A carotenoids - occurrence, intake, and bioavailability, in *Vitamin a and Carotenoids: Chemistry, Analysis, Function and Effects (Food and Nutritional Components in Focus)*, ed. V. Preedy, RSC Publishing, London, 2012.
75. E. Biehler, L. Hoffmann and E. Krause, *et al.*, *J. Nutr.*, 2011, **141**, 1769–1776.
76. P. Borel, C. Desmarchelier and U. Dumont, *et al.*, *Br. J. Nutr.*, 2016, **116**, 2091–2096.
77. J. Corte-Real, C. Guignard and M. Gantenbein, *et al.*, *Br. J. Nutr.*, 2017, **117**, 1560–1569.
78. E. Biehler, A. Kaulmann and L. Hoffmann, *et al.*, *Food Chem.*, 2011, **125**, 1328–1334.

79. D. M. Fahy, Y. C. O'Callaghan and N. M. O'Brien, *Food Addit. Contam.*, 2004, **21**, 42–51.
80. M. Richelle, M. Enslen and C. Hager, *et al.*, *Am. J. Clin. Nutr.*, 2004, **80**, 171–177.
81. F. Granado-Lorencio, E. Donoso-Navarro and L. M. Sanchez-Siles, *et al.*, *J. Agric. Food Chem.*, 2011, **59**, 11819–11824.
82. S. Baumgartner, R. T. Ras and E. A. Trautwein, *et al.*, *Eur. J. Nutr.*, 2017, **56**, 909–923.
83. S. H. Hageman, L. She and H. C. Furr, *et al.*, *Lipids*, 1999, **34**, 627–631.
84. J. Corte-Real, E. Richling and L. Hoffmann, *et al.*, *Nutr. Res.*, 2014, **34**, 1101–1110.
85. J. F. Zayas, *Functionality of Proteins in Food*, Springer, Heidelberg, 1997.
86. C. Soukoulis and T. Bohn, *Crit. Rev. Food Sci. Nutr.*, 2018, **58**, 1–36.
87. T. Bohn, *Curr. Nutr. Food Sci*, 2008, **4**, 240–258.
88. R. E. Kopec, J. L. Cooperstone and R. M. Schweiggert, *et al.*, *J. Nutr.*, 2014, **144**, 1158–1166.
89. K. H. van het Hof, I. A. Brouwer and C. E. West, *et al.*, *Am. J. Clin. Nutr.*, 1999, **70**, 261–268.
90. R. M. Schweiggert and R. Carle, *Crit. Rev. Food Sci. Nutr.*, 2017, **57**, 1807–1830.
91. P. Mapelli-Brahm, J. Corte-Real and A. J. Melendez-Martinez, *et al.*, *Food Chem.*, 2017, **229**, 304–311.
92. M. L. Failla, C. Chitchumroonchokchai and B. K. Ishida, *J. Nutr.*, 2008, **138**, 482–486.
93. N. Z. Unlu, T. Bohn and D. Francis, *et al.*, *J. Agric. Food Chem.*, 2007, **55**, 1597–1603.
94. A. C. Boileau, N. R. Merchen and K. Wasson, *et al.*, *J. Nutr.*, 1999, **129**, 1176–1181.
95. T. W. Boileau, A. C. Boileau and J. W. Erdman Jr, *Exp. Biol. Med. (Maywood)*, 2002, **227**, 914–919.
96. A. J. Melendez-Martinez, P. Mapelli-Brahm and A. Benitez-Gonzalez, *et al.*, *Arch. Biochem. Biophys.*, 2015, **572**, 188–200.
97. N. E. Moran, J. A. Novotny and M. J. Cichon, *et al.*, *J. Nutr.*, 2016, **146**, 368–376.
98. J. L. Cooperstone, R. A. Ralston and K. M. Riedl, *et al.*, *Mol. Nutr. Food Res.*, 2015, **59**, 658–669.
99. M. J. Rodriguez-Roque, B. de Ancos and R. Sanchez-Vega, *et al.*, *Food Funct.*, 2016, 7, 380–389.
100. N. Z. Unlu, T. Bohn and D. M. Francis, *et al.*, *Br. J. Nutr.*, 2007, **98**, 140–146.
101. E. A. Tydeman, M. L. Parker and M. S. Wickham, *et al.*, *J. Agric. Food Chem.*, 2010, **58**, 9847–9854.
102. O. Livny, R. Reifen and I. Levy, *et al.*, *Eur. J. Nutr.*, 2003, **42**, 338–345.
103. P. Palmero, A. Panozzo and I. Colle, *et al.*, *Food Chem.*, 2016, **199**, 423–432.
104. K. R. Moelants, L. Lemmens and M. Vandebroeck, *et al.*, *J. Agric. Food Chem.*, 2012, **60**, 11995–12003.

105. W. Sriwichai *Déterminants de la bioaccessibilité des caroténoides et tocophérols de légumes feuilles: comparaison variétale et influence du prodéde*, GAIA, 2016.
106. R. M. Schweiggert, D. Mezger and F. Schimpf, *et al.*, *Food Chem.*, 2012, **135**, 2736–2742.
107. J. Riedl, J. Linseisen and J. Hoffmann, *et al.*, *J. Nutr.*, 1999, **129**, 2170–2176.
108. L. Yonekura and A. Nagao, *Biosci., Biotechnol., Biochem.*, 2009, **73**, 196–199.
109. X. Liu, J. Bi and H. Xiao, *et al.*, *J. Agric. Food Chem.*, 2015, **63**, 8534–8543.
110. J. Corte-Real, M. Bertucci and C. Soukoulis, *et al.*, *Food Funct.*, 2017, **8**, 1008–1019.
111. J. Corte-Real, M. Iddir and C. Soukoulis, *et al.*, *Food Chem.*, 2016, **197**, 546–553.
112. D. B. Rodriguez-Amaya, Effects of processing and storage, in *Food Carotenoids*, John Wiley & Sons, Ltd, 2015, pp. 132–173.
113. R. E. Kopec, B. Gleize and P. Borel, *et al.*, *Food Funct.*, 2017, **8**, 1494–1503.
114. S. Blanquet-Diot, M. Soufi and M. Rambeau, *et al.*, *J. Nutr.*, 2009, **139**, 876–883.
115. J. Courraud, J. Berger and J. P. Cristol, *et al.*, *Food Chem.*, 2013, **136**, 871–877.
116. C. Chitchumroonchokchai, S. J. Schwartz and M. L. Failla, *J. Nutr.*, 2004, **134**, 2280–2286.
117. F. Granado-Lorencio, B. Olmedilla-Alonso and C. Herrero-Barbudo, *et al.*, *J. Agric. Food Chem.*, 2007, **55**, 6387–6394.
118. M. G. Ferruzzi, J. L. Lumpkin and S. J. Schwartz, *et al.*, *J. Agric. Food Chem.*, 2006, **54**, 2780–2785.
119. M. Richelle, B. Sanchez and I. Tavazzi, *et al.*, *Br. J. Nutr.*, 2010, **103**, 1800–1807.
120. C. Sy, O. Dangles and P. Borel, *et al.*, *Free Radical Biol. Med.*, 2013, **63**, 195–206.
121. M. Alminger, A. M. Aura and T. Bohn, *et al.*, *Compr. Rev. Food Sci. Food Saf.*, 2014, **13**, 413–436.
122. J. Corte-Real, C. Desmarchelier and P. Borel, *et al.*, *Food Chem.*, 2017, **239**, 751–759.
123. A. Asai, L. Yonekura and A. Nagao, *Br. J. Nutr.*, 2008, **100**, 273–277.
124. A. B. Barua and J. A. Olson, *J. Nutr.*, 2001, **131**, 3212–3215.
125. B. O. Durojaye, K. M. Riedl and R. W. Curley, *et al.*, *FASEB J.*, 2017, **31**, 635.18.
126. J. Chung, K. Koo and F. Lian, *et al.*, *J. Nutr.*, 2012, **142**, 405–410.
127. A. Asai, M. Terasaki and A. Nagao, *J. Nutr.*, 2004, **134**, 2237–2243.
128. F. Khachik, G. R. Beecher and N. F. Whittaker, *J. Agric. Food Chem.*, 1986, **34**, 603–616.
129. C. Sy, O. Dangles and P. Borel, *et al.*, *Arch. Biochem. Biophys.*, 2015, **572**, 89–100.

130. J. Serrano, I. Goni and F. Saura-Calixto, *J. Agric. Food Chem.*, 2005, **53**, 2936–2940.
131. A. Kaulmann, C. M. Andre and Y. J. Schneider, *et al.*, *Food Chem.*, 2015, **197**, 325–332.
132. M. S. Ward, D. Y. Zhao and P. S. Bernstein, *J. Ocul. Biol. Dis. Inf.*, 2008, **1**, 12–18.
133. P. Layer and J. Keller, *J. Clin. Gastroenterol.*, 1999, **28**, 3–10.
134. J. Meier and A. Sturm, *Dig. Dis.*, 2009, **27**, 240–245.
135. S. D'Silva, X. Xiao and M. E. Lowe, *J. Lipid Res.*, 2007, **48**, 2478–2484.
136. D. A. Garrett, M. L. Failla and R. J. Sarama, *J. Agric. Food Chem.*, 1999, **47**, 4301–4309.
137. M. J. Periago, S. Bravo and F. J. Garcia-Alonso, *et al.*, *J. Agric. Food Chem.*, 2013, **61**, 3859–3867.
138. D. Y. Hui and P. N. Howles, *J. Lipid Res.*, 2002, **43**, 2017–2030.
139. T. Imai, M. Imoto and H. Sakamoto, *et al.*, *Drug Metab. Dispos.*, 2005, **33**, 1185–1190.
140. C. Chitchumroonchokchai and M. L. Failla, *J. Nutr.*, 2006, **136**, 588–594.
141. F. Granado, B. Olmedilla and E. Gil-Martinez, *et al.*, *Br. J. Nutr.*, 1998, **80**, 445–449.
142. D. E. Breithaupt, A. Bamedi and U. Wirt, *Comp. Biochem. Physiol., Part B: Biochem. Mol. Biol.*, 2002, **132**, 721–728.
143. D. M. Gustin, K. A. Rodvold and J. A. Sosman, *et al.*, *Cancer Epidemiol., Biomarkers Prev.*, 2004, **13**, 850–860.
144. V. Diwadkar-Navsariwala, J. A. Novotny and D. M. Gustin, *et al.*, *J. Lipid Res.*, 2003, **44**, 1927–1939.
145. A. During, M. M. Hussain and D. W. Morel, *et al.*, *J. Lipid Res.*, 2002, **43**, 1086–1095.
146. A. Kaulmann, T. Serchi and J. Renaut, *et al.*, *Br. J. Nutr.*, 2012, **108**, 963–973.
147. L. O'Sullivan, S. A. Aisling and N. M. O'Brien, *Int. J. Vitam. Nutr. Res.*, 2009, **79**, 337–347.
148. H. S. Ribeiro, J. M. Guerrero and K. Briviba, *et al.*, *J. Agric. Food Chem.*, 2006, **54**, 9366–9369.
149. L. O'Sullivan, L. Ryan and N. O'Brien, *Br. J. Nutr.*, 2007, **98**, 38–44.
150. E. Reboul and P. Borel, *Prog. Lipid Res.*, 2011, **50**, 388–402.
151. Y. Yamanashi, T. Takada and R. Kurauchi, *et al.*, *J. Atheroscler. Thromb.*, 2017, **24**, 347–359.
152. G. Palczewski, J. Amengual and C. L. Hoppel, *et al.*, *FASEB J.*, 2014, **28**, 4457–4469.
153. J. Amengual, G. P. Lobo and M. Golczak, *et al.*, *FASEB J.*, 2011, **25**, 948–959.
154. J. Amengual, M. A. Widjaja-Adhi and S. Rodriguez-Santiago, *et al.*, *J. Biol. Chem.*, 2013, **288**, 34081–34096.
155. K. Q. Hu, C. Liu and H. Ernst, *et al.*, *J. Biol. Chem.*, 2006, **281**, 19327–19338.
156. M. Gajic, S. Zaripheh and F. Sun, *et al.*, *J. Nutr.*, 2006, **136**, 1552–1557.
157. C. dela Sena, S. Narayanasamy and K. M. Riedl, *et al.*, *J. Biol. Chem.*, 2013, **288**, 37094–37103.

158. L. Wu, X. Guo and W. Wang, *et al.*, *Exp. Biol. Med. (Maywood)*, 2016, **241**, 1879–1887.
159. B. Li, P. P. Vachali and A. Gorusupudi, *et al.*, *Proc. Natl. Acad. Sci. U. S. A.*, 2014, **111**, 10173–10178.
160. F. Khachik, L. Carvalho and P. S. Bernstein, *et al.*, *Exp. Biol. Med. (Maywood)*, 2002, **227**, 845–851.
161. F. Khachik, F. F. de Moura and D. Y. Zhao, *et al.*, *Invest. Ophthalmol. Visual Sci.*, 2002, **43**, 3383–3392.
162. A. B. Barua and N. Sidell, *J. Nutr.*, 2004, **134**, 286s–289s.
163. M. H. Zile, R. C. Inhorn and H. F. DeLuca, *J. Biol. Chem.*, 1982, **257**, 3537–3543.
164. A. B. Barua and J. A. Olson, *Am. J. Clin. Nutr.*, 1986, **43**, 481–485.
165. J. O. Sass, E. Masgrau and J. H. Saurat, *et al.*, *Drug Metab. Dispos.*, 1995, **23**, 887–891.
166. C. C. Ho, F. F. de Moura and S. H. Kim, *et al.*, *J. Nutr.*, 2009, **139**, 1480–1486.
167. E. Reboul, A. Klein and F. Bietrix, *et al.*, *J. Biol. Chem.*, 2006, **281**, 4739–4745.
168. E. Reboul, A. Goncalves and C. Comera, *et al.*, *Mol. Nutr. Food Res.*, 2011, **55**, 691–702.
169. K. L. Herron, M. M. McGrane and D. Waters, *et al.*, *J. Nutr.*, 2006, **136**, 1161–1165.
170. P. Borel, F. S. de Edelenyi and S. Vincent-Baudry, *et al.*, *Ann. Med.*, 2011, **43**, 47–59.
171. P. Borel, C. Desmarchelier and M. Nowicki, *et al.*, *J. Nutr.*, 2015, **145**, 1740–1747.
172. S. Picone, A. Ritieni and A. Fabiano, *et al.*, *Clin. Biochem.*, 2012, **45**, 1558–1563.
173. E. Kotake-Nara and A. Nagao, *Mar. Drugs*, 2011, **9**, 1024–1037.
174. D. N. D'Ambrosio, R. D. Clugston and W. S. Blaner, *Nutrients*, 2011, **3**, 63–103.
175. C. W. Hadley, S. K. Clinton and S. J. Schwartz, *J. Nutr.*, 2003, **133**, 727–732.
176. O. J. Torrissen and K. Ingebrigtsen, *Aquaculture*, 1992, **108**, 381–385.
177. A. Kaulmann, C. M. Andre and Y. J. Schneider, *et al.*, *Food Chem.*, 2016, **197**, 325–332.
178. I. Goni, J. Serrano and F. Saura-Calixto, *J. Agric. Food Chem.*, 2006, **54**, 5382–5387.
179. P. P. Nair, A. Lohani and E. P. Norkus, *et al.*, *Cancer Epidemiol., Biomarkers Prev.*, 1996, **5**, 913–916.
180. S. Oshima, T. Inakuma and T. Narisawa, *J. Nutr. Sci. Vitaminol. (Tokyo)*, 1999, **45**, 129–134.
181. M. J. Shearer, A. Bach and M. Kohlmeier, *J. Nutr.*, 1996, **126**, 1181s–1186s.
182. M. M. Groenen-van Dooren, J. E. Ronden and B. A. Soute, *et al.*, *Biochem. Pharmacol.*, 1995, **50**, 797–801.

183. A. Lindqvist, Y. G. He and S. Andersson, *J. Histochem. Cytochem.*, 2005, **53**, 1403–1412.
184. W. C. Leung, S. Hessel and C. Meplan, *et al.*, *FASEB J.*, 2009, **23**, 1041–1053.
185. G. Lietz, A. Oxley and W. Leung, *et al.*, *J. Nutr.*, 2012, **142**, 161s–165s.
186. G. P. Lobo, J. Amengual and D. Baus, *et al.*, *J. Biol. Chem.*, 2013, **288**, 9017–9027.
187. S. J. Hickenbottom, J. R. Follett and Y. Lin, *et al.*, *Am. J. Clin. Nutr.*, 2002, **75**, 900–907.
188. G. Lietz, A. Oxley and C. Boesch-Saadatmandi, *et al.*, *Mol. Nutr. Food Res.*, 2012, **56**, 241–250.
189. A. M. Priyadarshani, *Crit. Rev. Food Sci. Nutr.*, 2017, **57**, 1710–1717.
190. S. A. Tanumihardjo, N. Palacios and K. V. Pixley, *Int. J. Vitam. Nutr. Res.*, 2010, **80**, 336–350.
191. W. N. A. Charman and V. J. Stella, *Int. J. Pharm.*, 1986, **33**, 165–172.
192. R. E. Kopec, C. Caris-Veyrat and M. Nowicki, *et al.*, *FASEB J.*, 2017, **31**, 32–35.
193. P. Borel, M. Moussa and E. Reboul, *et al.*, *J. Nutr.*, 2007, **137**, 2653–2659.
194. S. Dash, C. Xiao and C. Morgantini, *et al.*, *Annu. Rev. Nutr.*, 2015, **35**, 265–294.
195. E. J. Niesor, E. Chaput and J. L. Mary, *et al.*, *Lipids*, 2014, **49**, 1233–1243.
196. H. C. Furr, *J. Nutr.*, 2004, **134**, 281s–285s.
197. H. Y. Chung, A. L. Ferreira and S. Epstein, *et al.*, *Am. J. Clin. Nutr.*, 2009, **90**, 533–539.
198. P. S. Bernstein, B. Li and P. P. Vachali, *et al.*, *Prog. Retinal Eye Res.*, 2016, **50**, 34–66.
199. P. Bhosale and P. S. Bernstein, *Anal. Biochem.*, 2005, **345**, 296–301.
200. A. Asai, T. Nakano and M. Takahashi, *et al.*, *J. Agric. Food Chem.*, 2005, **53**, 7302–7306.
201. S. J. Hickenbottom, S. L. Lemke and S. R. Dueker, *et al.*, *Eur. J. Nutr.*, 2002, **41**, 141–147.
202. N. E. Moran, M. J. Cichon and K. M. Riedl, *et al.*, *Am. J. Clin. Nutr.*, 2015, **102**, 1436–1449.
203. H. Park, Model-based compartmental analysis of the kinetics of retinol and beta-carotene in humans: statistical considerations in designing and building models for retinol plus expanded models for beta-carotene, PhD, Pennsylvania State University, 2011.
204. F. Bohm, R. Edge and T. G. Truscott, *Acta Biochim. Pol.*, 2012, **59**, 27–30.
205. T. Heymann, P. Heinz and M. A. Glomb, *J. Agric. Food Chem.*, 2015, **63**, 3279–3287.
206. E. Fernandez-Garcia, *Food Funct.*, 2014, **5**, 1994–2003.
207. A. A. Woodall, G. Britton and M. J. Jackson, *Biochim. Biophys. Acta*, 1997, **1336**, 575–586.
208. Z. Li, X. Dong and H. Liu, *et al.*, *Mol. Vis.*, 2013, **19**, 1656–1666.
209. J. Fiedor and K. Burda, *Nutrients*, 2014, **6**, 466–488.

210. F. J. Pashkow, D. G. Watumull and C. L. Campbell, *Am. J. Cardiol.*, 2008, **101**, 58d–68d.
211. B. Nidhi, G. Sharavana and T. R. Ramaprasad, *et al.*, *Food Funct.*, 2015, **6**, 450–460.
212. C. Caris-Veyrat, A. L. Garcia and E. Reynaud, *et al.*, *Int. J. Vitam. Nutr. Res.*, 2016, in press.
213. M. L. Bonet, J. A. Canas and J. Ribot, *et al.*, *Subcell. Biochem.*, 2016, **79**, 377–414.
214. Y. Kitamura, K. Tanaka and C. Kiyohara, *et al.*, *Int. J. Epidemiol.*, 1997, **26**, 307–314.
215. L. Wang, J. M. Gaziano and E. P. Norkus, *et al.*, *Am. J. Clin. Nutr.*, 2008, **88**, 747–754.
216. A. J. Guri, R. Hontecillas and H. Si, *et al.*, *Clin. Nutr.*, 2007, **26**, 107–116.

CHAPTER 10

Models for Studying Polyphenols and Carotenoids Digestion, Bioaccessibility and Colonic Fermentation

SENEM KAMILOGLU AND ESRA CAPANOGLU*

Istanbul Technical University, Department of Food Engineering, Maslak, Istanbul, 34469, Turkey
*E-mail: capanogl@itu.edu.tr

10.1 Introduction

Epidemiological studies have shown that a high intake of fruits and vegetables has been associated with a reduced risk of chronic diseases, including cardiovascular diseases, diabetes and cancer. These potential beneficial effects have been attributed to the presence of bioactive compounds, such as polyphenols and carotenoids.[1,2] As a part of our diet, polyphenols and carotenoids are ingested as complex mixtures immersed in a food matrix, which undergo a digestion process. It is important to determine how the digestion process affects polyphenols and carotenoids, as this will affect their bioaccessibility and hence their possible beneficial effects.[3]

Bioaccessibility is defined as the amount of a food component that is released from a complex food matrix in the lumen of the gastrointestinal tract and, therefore, could potentially be available for absorption into the

Food Chemistry, Function and Analysis No. 5
Non-extractable Polyphenols and Carotenoids: Importance in Human Nutrition and Health
Edited by Fulgencio Saura-Calixto and Jara Pérez-Jiménez

Published by the Royal Society of Chemistry, www.rsc.org

body. Only polyphenols and carotenoids released from the food matrix by the action of digestive enzymes and bacterial microflora are bioaccessible in the gut and therefore potentially bioavailable.[4] Methods for determining the bioavailability of bioactive compounds involve human (*in vivo*) or simulated experiments performed in a laboratory (*in vitro*). *In vivo* methods provide direct data of bioavailability and have been used for a variety of nutrients.[5] In general, *in vivo* bioavailability studies refer to the consumption of certain doses of a nutrient and following changes in its concentration in the blood plasma over time. The major disadvantages of *in vivo* studies are the variability in the physiological state of individuals and the possible interaction of the nutrient with other components in the diet.[6] *In vitro* methods simulating digestion are commonly used to study the gastrointestinal behavior of bioactive compounds and have the advantages of being more rapid, less expensive and less labor intensive and do not have ethical restrictions.[7]

It is estimated that only 5–10% of the total intake of dietary polyphenols, mainly those with monomeric and dimeric structures and belonging to the so-called extractable polyphenols (EPP), may be directly absorbed in the small intestine, generally after deconjugation reactions such as deglycosylation.[8] In contrast, non-extractable polyphenols (NEPP) are not released by digestive enzymes and are not bioaccessible in the small intestine. Thus, *in vitro* bioaccessibility models may allow for identifying the EPP and NEPP of individual foods or whole diets.[4] The remaining 90–95% reach the colon, where they are further metabolized to compounds with different physiological significance by the enzymatic action of the colonic bacteria.[9] In the particular case of NEPP, they may be directly fermented[10] or may need previous release from the food matrix by enzymes present in the colonic microbiota.[11] Considering this fact, *in vitro* colon models have been developed in order to discover new polyphenol metabolites with health-promoting potential from both EPP and NEPP. On the other hand, carotenoids are primarily absorbed in the small intestine, and colonic metabolites have not been reported so far.[12]

The aim of the present chapter is to highlight the frequently employed models for studying polyphenol and carotenoid digestion, bioaccessibility and colonic fermentation. In this context, *in vitro* gastrointestinal digestion, cell culture, intestinal tissue and colonic fermentation models are discussed.

10.2 Gastrointestinal Digestion Models

The methods that mimic the gastrointestinal digestion process under laboratory conditions are known as *in vitro* gastrointestinal digestion models. Generally, *in vitro* gastrointestinal digestion models are divided into two broad categories: static models, where the products of digestion generally stay immobile and physical processes such as shearing, mixing and hydration are not simulated; and dynamic models, which try to integrate physical and mechanical processes and temporal changes in luminal conditions to mimic parameters *in vivo*.[5,6] Both static and dynamic models have already been used to measure the bioaccessibility of several polyphenols[13–17] and carotenoids[18–20] from various food sources. In the following sections, the features of these two models are discussed.

10.2.1 Static Models

In static *in vitro* digestion models, the stomach and the upper part of the small intestine are simulated by addition of enzymes and salts, adjustment of pH and incubation at the physiological condition of 37 °C. During the gastric phase, samples are incubated at gastric pH in the presence of gastric enzymes such as pepsin. Then, by increasing the pH and addition of pancreatic enzymes and bile salts, the upper part of the small intestine is simulated. Some models also contain an initial oral digestion phase, which mimics mastication in the presence of amylase.[21]

The simplicity of static models allows for the analysis of multiple samples simultaneously. However, unlike dynamic models, static models do not consider the dynamic physiological responses to the introduction of a food bolus, such as pH increases and the following decreases in the stomach and enzyme secretions in response to the food bolus being introduced. Adaptations of the static models have been carried out for the investigation of various phytochemicals, such as ultracentrifugation and/or filtration, in order to study the micellar phase of lipophilic constituents. While this treatment is normally not done in order to study polyphenol bioaccessibility, additional steps such as dialysis have occasionally been applied.[12]

The diversity of static *in vitro* gastrointestinal digestion model conditions has limited the ability to compare results across different studies. For instance, a broad range of enzymes from different sources (*e.g.* porcine, rabbit or human origin) has been used. Furthermore, differences in pH, minerals, ionic strengths and incubation periods, which alter enzyme activity and other factors, may also significantly affect the results. To overcome this issue, a standardized and practical static digestion model based on physiologically relevant conditions was proposed. In this international consensus paper, a frameset of parameters including the oral, gastric and small intestinal digestion are outlined (Figure 10.1) and their relevance is discussed in relation to the available *in vivo* data. At each step of digestion, a constant final ratio of food to enzymes, salt, bile acids, *etc.*, of 50 : 50 (w/v) is targeted. This harmonized, static, *in vitro* gastrointestinal digestion model should help with the generation of more comparable data in the future.[7]

Up to the moment, static digestion models have been the most common ones used to assess the metabolic fate of NEPP. Thus, a study in the Spanish diet showed that 42% of polyphenols consumed by this population were not bioaccessible in the small intestine; this corresponds to NEPP, since EPP are bioaccessible in the small intestine, although only a small fraction is absorbable.[4] Regarding individual foods, it was reported that only 1% of anthocyanins in black carrots were bioaccessible in the small intestine, while most of them remained associated with the cell wall, as proven by confocal

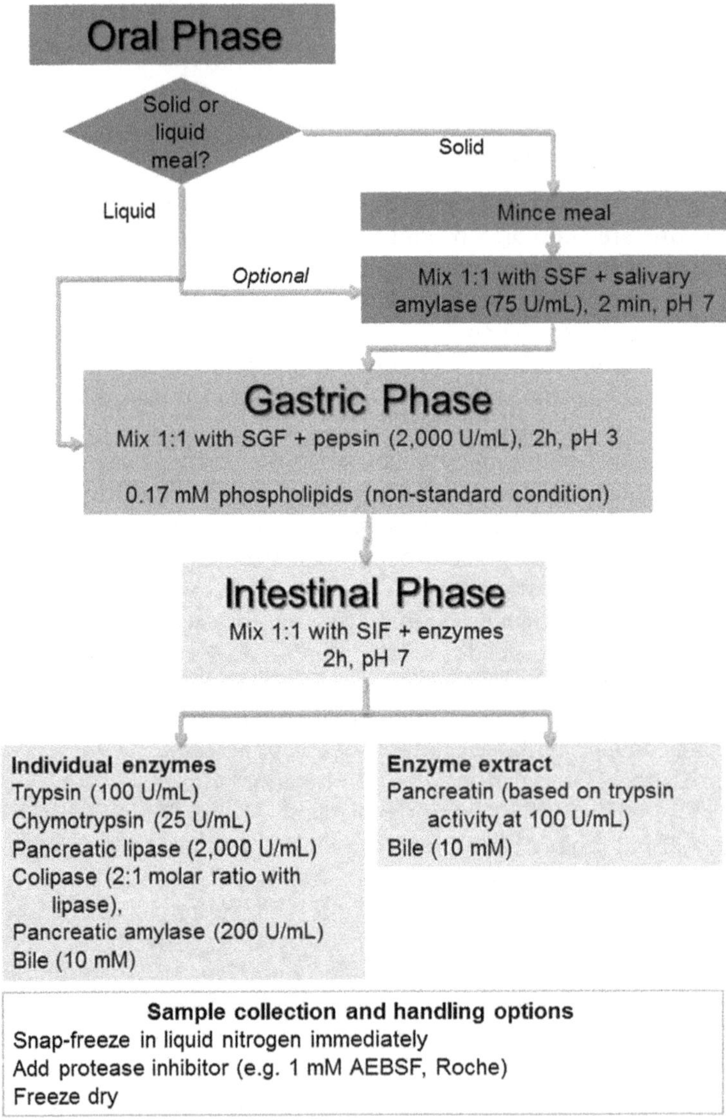

Figure 10.1 Overview and flow diagram of the standardized *in vitro* digestion method. Enzyme activities are in U mL^{-1} of the final digestion mixture at each corresponding digestion phase. AEBSF: 4-(2-aminoethyl) benzenesulfonyl fluoride hydrochloride; SSF: simulated salivary fluid; SGF: simulated gastric fluid; SIF: simulated intestinal fluid. (Reproduced from ref. 7 with permission from The Royal Society of Chemistry.)

microscopy measurements.[22] In addition, a recent study on fish enriched with grape pomace reported that the higher the grape pomace dietary content, the higher the proportion of polyphenols reaching the large intestine due to increased interactions with the food matrix and, therefore, of the NEPP fraction.[23]

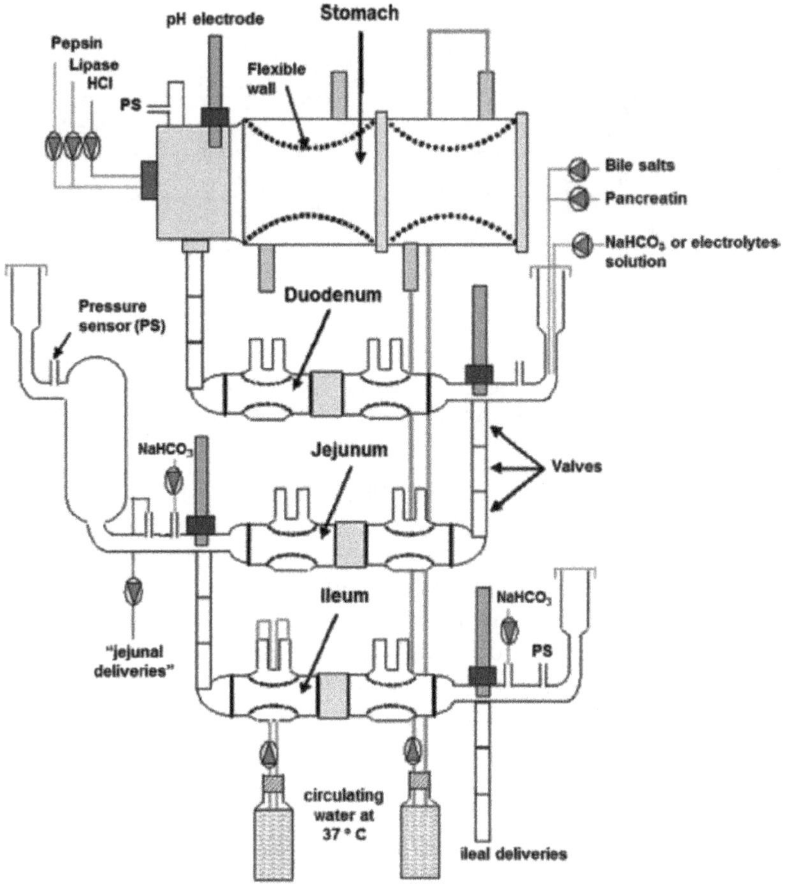

Figure 10.2 Schematic representation of TIM-1. (Reproduced from, *Adv. Colloid Interface Sci.*, **243**, L. de Souza Simões, D. A. Madalena, A. C. Pinheiro, J. A. Teixeira, A. A. Vicente and Ó. L. Ramos, Micro- and nano bio-based delivery systems for food applications: *In vitro* behavior, 23,[25] Copyright 2017, with permission from Elsevier.)

10.2.2 Dynamic Models

The dynamic *in vitro* digestion models have the same basic principles of static models; however, they better mimic the conditions in the gastrointestinal tract. In general, dynamic models have advantages over static models, including continuous removal of nutrients, simulation of peristalsis and more controlled, gradual changes in pH and enzyme levels.[24]

Different dynamic models have been developed to gain detailed insights into changes occurring during digestion. Among them, the TNO gastrointestinal digestion model (TIM-1) is composed of four serial sections simulating the stomach, duodenum, jejunum and ileum (Figure 10.2).[25] Each compartment consists of two glass jackets lined with flexible walls. The TIM-1 system allows the simulation of the gastric emptying rate, peristaltic movements,

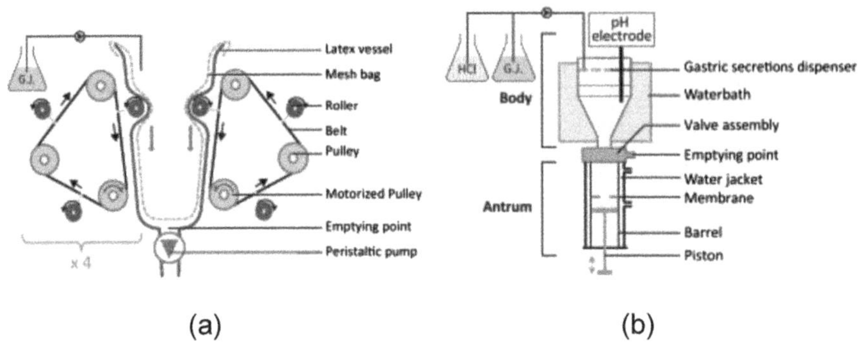

Figure 10.3 Schematic representation of (a) human gastric simulator and (b) dynamic gastric model. (Adapted from *Trends Biotechnol.*, **30**, A. Guerra, L. Etienne-Mesmin, V. Livrelli, S. Denis, S. Blanquet-Diot and M. Alric, Relevance and challenges in modeling human gastric and small intestinal digestion, 591,[27] Copyright 2012, with permission from Elsevier.)

transit time through the small intestine and gradual pH changes in the different compartments.[26] An adapted model of TIM-1, a computer-controlled gastrointestinal model called Tiny-TIM, is a simplified and downscaled TIM-1 for rapid screening. The main features of this system are the same as for TIM-1; however, instead of four compartments, the Tiny-TIM model consists of two compartments that mimic the stomach and the small intestine.[12]

Some other dynamic models that simulate the stomach conditions are the human gastric simulator (HGS) and dynamic gastric model (DGM) (Figure 10.3).[27] The HGS and DGM models were developed at the Institute of Food Research and the University of California-Davis, respectively.[28,29] The HGS is composed of a latex chamber surrounded by a mechanical driving system to effectively mimic the frequency and intensity of the peristaltic movements in the stomach. The DGM is composed of two consecutive compartments that simulate physical mixing, transit and breakdown forces (including flow, shear and hydration), pH gradients and gastric secretions.[12]

Regarding NEPP, dynamic models have been used in combination with colonic fermentation models (TIM-2), as described below.

10.3 Cell Culture Models

Cell cultures are useful model systems for studying the absorption of polyphenols and carotenoids through the intestinal epithelium. The benefit of using cell culture models is their consistency and reproducibility. In addition, they are high-throughput models that allow for the screening of a large number of samples. Several studies have investigated the transport kinetics of polyphenols[30–33] and carotenoids.[34–36] In the following sections, different intestinal cell culture models, including Caco-2, HT-29, co-cultures and some other intestinal cell lines, are discussed. Application of these cell cultures to the evaluation of the absorption of NEPP, particularly in the large intestine, remains to be done.

10.3.1 Caco-2 Cell Line

The Caco-2 (Cancer coli-2) cell line, which was established by Fogh and co-workers in 1977 from a human colon adenocarcinoma, was originally used to screen the cytotoxic effects of anti-tumor drugs and to study drug resistance mechanisms.[37] Caco-2 cells exhibit the characteristics of mature enterocytes upon spontaneous differentiation in culture conditions. The cell surface facing the top medium develops a brush border that resembles the luminal membrane of the intestinal epithelium, whereas the cell surface facing the bottom medium develops into the basolateral membrane.[31,38] Although Caco-2 cells originate from the colon, they express the morphological and functional characteristics of small bowel enterocytes. The Caco-2 monolayer hosts multiple transporters, receptors and metabolic enzymes.[39]

To simulate the conditions in the intestine *in vivo*, transport experiments are performed using filter-based inserts, where Caco-2 cells are seeded and allowed to grow and differentiate to monolayers for approximately 21 days post-seeding (Figure 10.4).[40] Prior to carrying out a transport experiment, the integrity of the Caco-2 monolayer should be checked by measuring the transepithelial electrical resistance (TEER) or by examining the permeability of paracellular markers such as Lucifer yellow, mannitol, inulin, Dextran or PEG 4000.[39] TEER is a non-invasive method that measures the impedance between the lumen and basolateral tissue. A constant direct current applied by two electrodes, one connected with the lumen side and the other one with the basolateral side, is used to measure the TEER. The related cell resistance is measured by applying Ohm's law.[41] According to the literature, an acceptable TEER value for Caco-2 cell monolayers should be from 200 to 1000 Ohm cm^{-2}.[42] Since its original isolation, the Caco-2 cell line has been propagated in several laboratories around the world, producing Caco-2 cells of different numbers of passages in culture. From early to late passages, the expressions of typical differentiation markers of intestinal enterocytes were shown to increase. Accordingly, the TEER value has also been reported to increase in later passages of the cell monolayer.[43]

Some clonal cell lines have been obtained from Caco-2 to decrease the heterogeneity of the Caco-2 parental cell line and to improve the performance and stability of this cellular model. Among them, the TC7 clone displays similar features to Caco-2 such as the presence of a brush border membrane and microvilli and the formation of tight junctions. In addition, the TC7 clone also appears to be similar to Caco-2 cells on the basis of biochemical attributes and permeability characteristics.[44] Considering all of this, the TC7 clone is often used to simulate polyphenol[45–47] and carotenoid[48,49] transport. On the other hand, it was reported that the TC7 clone may not be a suitable model for intestinal absorption of highly lipophilic and poorly absorbed compounds and compounds where transporter-mediated routes and/or first-pass metabolism are involved.[50] Therefore, the validity of using the TC7

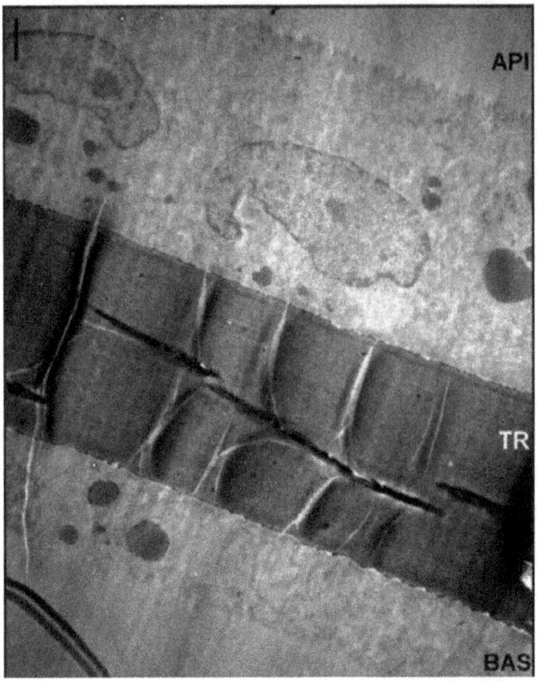

Figure 10.4 Transmission electron microscopy image of Caco-2 cells after 21 days in culture. API: apical side; TR: Transwell™ membrane; BAS: basolateral side. (Adapted from *Eur. J. Pharm. Biopharm.*, **119**, A. Lechanteur, A. Almeida and B. Sarmento, Elucidation of the impact of cell culture conditions of Caco-2 cell monolayer on barrier integrity and intestinal permeability, 137,[40] Copyright 2017, with permission from Elsevier.)

clone in the study of intestinal transport of bioactive compounds should be reconsidered.

10.3.2 HT29 Cell Line

The human colon adenocarcinoma cell line HT29 was isolated by Fogh and Trempe in 1964 and initially used to investigate various aspects of human cancers.[51] More recently, the HT29 cell line has been used as a model for absorption, secretion and transport by intestinal cells. Under standard culture conditions, these cells grow as a non-polarized, undifferentiated monolayer. However, altering culture conditions or treating the cells with different inducers results in a differentiated and polarized morphology, characterized by a redistribution of membrane antigens and the development of an apical brush border membrane.[52]

Under appropriate conditions, the cell differentiation of HT29 is quite similar to that observed in Caco-2 cells. On the other hand, these two cell cultures have some differences. For instance, the differentiation period of the HT29 cell line is longer (30 days) than Caco-2 (15–20 days) and the enzyme activity levels in the HT29 cell line are lower than the levels in Caco-2.[53]

However, the major difference between HT29 and Caco-2 is that HT29 cells can secrete mucin.[54] A stable, homogeneous, differentiated, mucin-secreting clone (HT29-MTX) has been developed by stepwise adaptation to methotrexate (MTX).[55] HT29-MTX cells co-cultured with parental Caco-2 cells are increasingly being used in transport studies of polyphenols and carotenoids, as is discussed in the next section. Other mucus-secreting clones of the HT29 cell line have also been developed, such as HT29-B6 and HT29-FU.[56]

10.3.3 Co-culture Models

Current trends in the study of bioactive compounds involve the use of co-culture systems to more closely simulate *in vivo* conditions. The most common models for intestinal absorption of polyphenols and carotenoids is co-culturing Caco-2 cells with TC7 or HT29-MTX clones. As mentioned earlier, the use of TC7 monocultures for studying intestinal transport may not be a very proper approach. However, optimizing the ratio of Caco-2 : TC7 populations in the co-culture setup could provide a more representative and homogeneous model.[56] Co-culture of Caco-2 and HT29-MTX ensures the generation of a model including the two cell types that are well represented in normal epithelium. Furthermore, the mucus layer that is formed on top of the epithelial cells due to the HT29-MTX clone is critical for the prediction of intestinal permeability, as it acts as a barrier against the absorption of particular compounds, especially those that are lipophilic (*e.g.* carotenoids).[53] In addition, the lack of a mucus layer usually causes an overestimation of the permeability of highly diffusible small compounds. Considering the above, it can be concluded that the co-culture models of Caco-2 and TC7 or Caco-2 and HT29-MTX may be useful methods for the investigation of the transport of bioactive compounds over the intestinal epithelium.

10.4 Intestinal Tissue Models

Some extended models including everted sac, Ussing chamber, intestinal segments and 3D cultures are used to mimic the intestinal tract in more detail. In the following sections, these *in vitro* intestinal tissue models are discussed. These kind of systems have not been yet used to assess the metabolic fate of NEPP.

10.4.1 Everted Sac

The everted sac technique was initially developed to measure the transport of glucose and amino acids across the rat and hamster intestine.[58] In the everted sac model, the intestinal section is everted and tied after filling the sac with buffer, which is then placed in buffer with the compound of interest.[59] Under optimum conditions, the tissue is viable for up to 2 hours, and the system is mostly used for studying drug absorption.[60–62] This technique is also used to study the absorption of polyphenols[63,64] and carotenoids[65] by measuring the amount of compound transferred to the serosal side and in the epithelial cells. The disadvantage of the everted sac method is that the absorption rate of the compounds may be unrealistically slow compared to the *in vivo* situation, as the compound of interest needs to pass the muscle layers and the fluid is stationary.[51]

10.4.2 Ussing Chamber

The Ussing chamber model was first used to study the transport of nutrients, ions and drugs across several epithelial tissues.[66] One of the most studied epithelia is the intestine, where a small section of intestinal mucosa is situated between two chambers with buffer solution, preserving the epithelial polarity.[67] One of the advantages of this model is that tissues can be used from various sections of the intestine (duodenum to colon), from different species of specific microbiota, genetic and/or disease backgrounds and from humans or animals that have been subjected *in vivo* to food ingredients or drug compounds. On the other hand, the major disadvantages of this method are that it needs careful and laborious preparation, it has a relatively low throughput and it shows short-term survival in culture.[57] Despite these drawbacks, this model has a good resemblance to the *in vivo* situation and therefore is used to study the intestinal absorption of several polyphenols.[68–70]

10.4.3 3D Cultures

Recently, there has been an increased interest in the development of 3D cell culture models that mimic the morphological and functional features of their *in vivo* parental tissues.[71] 3D culture models are prepared from cell lines, primary cells and/or organ cultures using various methods, including spontaneous aggregation in a suspension culture, implantation onto 3D scaffolds (*e.g.* collagen, extracts of the extracellular matrix or synthetic materials) and culture in Transwell® systems or in rotating culture bioreactors.[72] These models have the advantage of bridging the gap between animal models and/or *ex vivo* tissue studies and cell culture-based research. However, they also have some disadvantages, including the increased time and cost of experimental setup, the need for model optimization, reproducibility and validation when different cell types are incorporated and the level of expertise required to design tractable experimental setups.[57]

10.5 Colonic Fermentation Models

Phytochemicals that reach the colon are subjected to microbial degradation, and those degradation products derived from the ingested compounds may play a role in their beneficial health effects. For this reason, while studying the digestion of bioactive compounds, inclusion of a colonic fermentation stage is of great importance. To this end, dynamic gut models such as the

Simulator of the Human Intestinal Microbial Ecosystem (SHIME®), the TNO Intestinal Model 2 (TIM-2) and the SIMulator Gastro-Intestinal tract (SIMGI®) have been developed in which gut microbiota are cultured in one or multiple connected, pH-controlled vessels representing different parts of the human colon.

10.5.1 SHIME®

The SHIME® is a multi-compartment dynamic simulator of the entire human gastrointestinal tract incorporating the stomach, small intestine and ascending, transverse and descending colon (Figure 10.5).[73] The system consists of double-jacketed glass vessels that are connected through peristaltic pumps. A nutritional medium is added to the gastric compartment, whereas pancreatic and bile solutions are inserted into the small intestinal compartment. Following the gastric and small intestinal compartments, colonic digestion is initiated. The colon compartments are continuously stirred with magnetic stir bars. Depending on the group of interest, the retention time in colon vary from 24 to 72 hours. While the entire system runs at 37 °C, the pH of the colon compartments is maintained between 5.6 and 5.9 in the ascending colon, between 6.1 and 6.4 in the transverse colon and between 6.6 and 6.9 in the descending colon.[74,75] Careful control of the environmental parameters in the reactors allows for obtaining complex and stable microbial communities, which are highly similar in both structure and function to the microbial communities in the different regions of the human colon. A typical SHIME® experiment consists of four stages: a stabilization period (2 weeks) to allow for adaptation of the microbial community (high quality of the inoculum) to the environmental conditions in the respective colon regions; a basal period (2 weeks) in which the system reactor is operated under nominal conditions

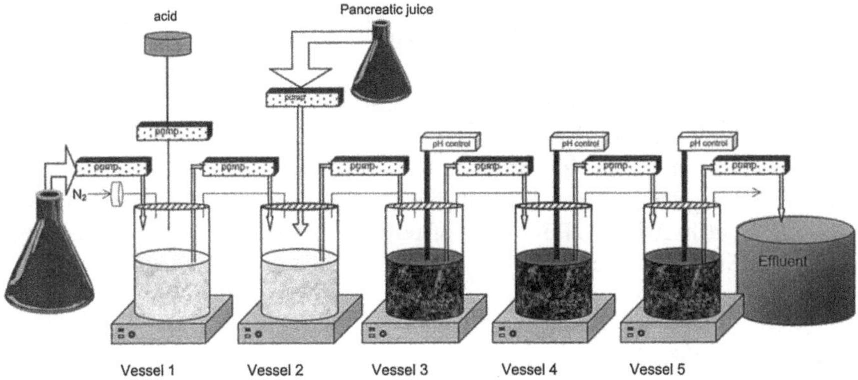

Figure 10.5 Schematic representation of SHIME®. Vessels 1, 2, 3, 4 and 5 are stomach, small intestine, ascending colon, transverse colon and descending colon, respectively. (Reproduced from *Am. J. Clin. Nutr.* (2000; **130**: 2599–2606), American Society for Nutrition.[73])

and basal parameters are measured; a treatment period (2/3 weeks) in which the effect of a specific compound on the gastrointestinal microbial community is tested; and a washout period (2 weeks) to determine for how long the changes induced by the tested substance can still be measured in the absence of the substance itself.[76]

The presence of different metabotypes has been suggested to differentiate a bioactive metabolite-producing phenotype from a non-producing phenotype.[77] On this basis, SHIME® is inoculated with the fecal microbiome from one individual to maintain the microbial metabolic phenotype toward specific polyphenols.[78,79] So far, the SHIME® model has been used to study the colonic fermentation of polyphenols from flax seed,[80] black tea and red wine/grape juice,[81,82] cranberry and grape seed,[83] pomegranate[84] and aronia.[85] On the other hand, as far as is known, no previous study has used this model to investigate carotenoid fermentation in the colon.

10.5.2 TIM-2

The TNO *in vitro* model of the colon, TIM-2, is composed of four interconnected glass compartments with a flexible membrane that contracts and creates peristaltic movements that mix the luminal contents and move them through the system (Figure 10.6).[86] Peristaltic movements are obtained by altering the pressure of the water between the glass jacket and the membrane. These peristaltic movements of the flexible membrane ensure a better mixing and movement of the components through the entire system than would be achieved by stirring or shaking, where phase separation of fluids and solids may take place. The system works at 37 °C and a pH of 5.8 (the pH of the proximal colon), which are measured by a temperature sensor and a pH electrode, respectively. Sodium hydroxide is used to control the pH, which may be reduced by the acids that are produced by the microbiota. The dialysis liquid is pumped to maintain the appropriate electrolyte and metabolite concentrations. A level sensor is used to control the volume and nitrogen is flushed to keep the model anaerobic.[87] The TIM-2 model has been used for several applications, including the fermentation of polyphenols from tea, citrus or soy flavonoid supplements.[75,88]

This system was used to compare the colonic fermentation of different diets, observing that at least 66% of NEPP reaching the colon was released from the food matrix by the action of colonic microbiota, which was measured as antioxidant capacity.[89]

10.5.3 SIMGI®

The SIMGI® consists of five compartments simulating the stomach, small intestine and the ascending, transverse and descending colon regions, which are interconnected with pipes and peristaltic valve pumps that transfer the contents between the consecutive compartments (Figure 10.7). The stomach

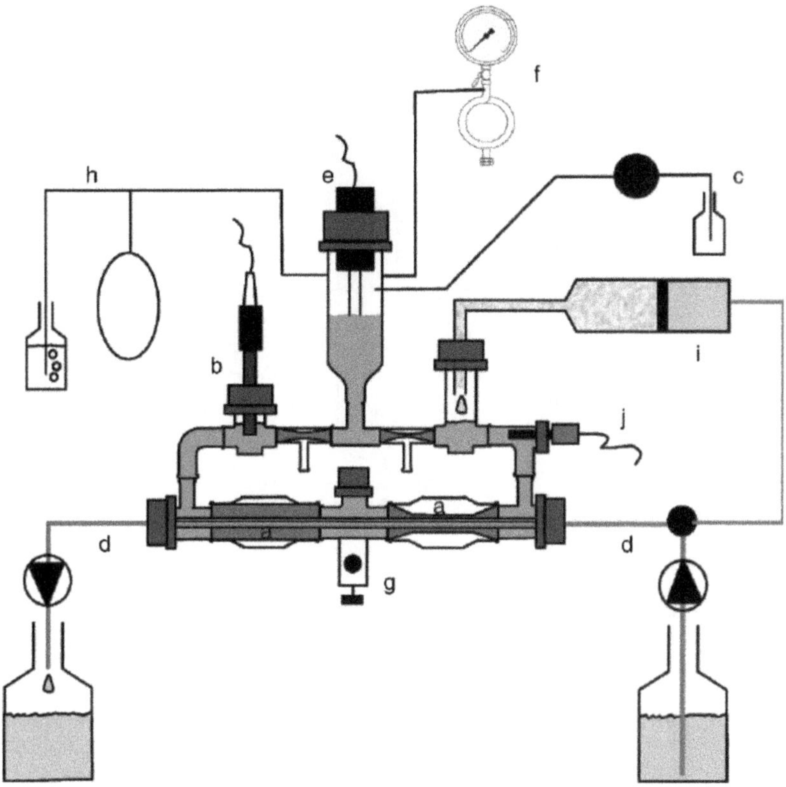

Figure 10.6 Schematic representation of TIM-2. (a) Peristaltic compartments containing fecal matter; (b) pH electrode; (c) alkali pump; (d) dialysis liquid circuit with hollow fiber membrane; (e) level sensor; (f) N_2 gas inlet; (g) sampling port; (h) gas outlet; (i) "ileal efflux" container; (j) temperature sensor. (Reproduced from Maathuis *et al.*[86] with permission from Taylor & Francis.)

is composed of two transparent and rigid methacrylate plastic modules covering a reservoir of flexible silicone walls where the gastric contents are mixed by peristaltic movements. The other four compartments consist of double-jacketed glass reactors that are continuously stirred with a magnetic stirrer. The entire system works at 37 °C and the pH values in the colonic units are kept at 5.6, 6.3 and 6.8 for the ascending, transverse and descending colon, respectively.[90] In terms of operating conditions in the colon, the SIMGI® model is quite similar to the SHIME® model.

The SIMGI® model has only recently been developed and hence it has a short history in terms of evaluating the digestion of polyphenols, whereas no previous study has investigated the digestion of carotenoids using this model. So far, digestion of polyphenols from red wine[91] and pure polyphenol compounds (chlorogenic acid, caffeic acid, ferulic acid and rutin)[92] have been studied using the SIMGI® model.

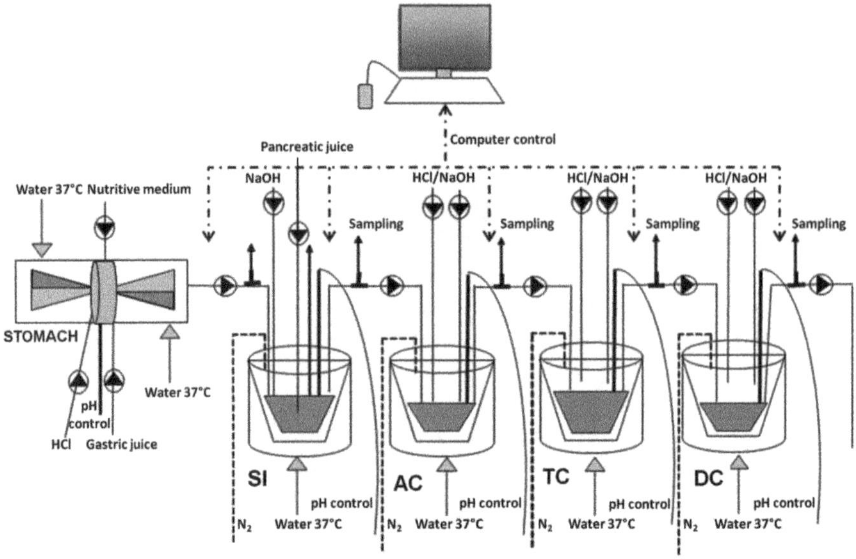

Figure 10.7 Schematic representation of SIMGI®. SI: small intestine; AC: ascending colon; TC: transverse colon; DC: descending colon. (Reproduced from *LWT - Food Science and Technology*, **61**, E. Barroso, C. Cueva, C. Peláez, M. Carmen, Martínez-Cuesta and T. Requena, Development of human colonic microbiota in the computer-controlled dynamic SIMulator of the GastroIntestinal tract SIMGI, 283–289,[90] Copyright 2015, with permission from Elsevier.)

10.5.4 Other Fermentation Models

In addition to the fermentation models discussed above, the VTT (Technical Research Center of Finland) one-compartment fermentation model is also used to study the *in vitro* colonic microbial metabolism of polyphenols.[93–95] The application of this model includes a strict anaerobic environment in order to provide an active microbial community. Furthermore, to ensure reproducibility between the experiments, fecal samples from several donors are pooled. Other than that, several other in-house-developed colonic fermentation models are also used to study the digestion of polyphenols[96,97] and carotenoids.[98] By using one of these systems, it was observed that the presence of dietary fiber enhanced the colonic fermentation of NEPP.[97] Also, an *in vitro* study on the water-insoluble fraction of cocoa (composed mostly of NEPP) showed that the colonic fermentation of this product generated significant increases in the concentrations of bifidobacteria and lactobacilli.[99]

10.6 Conclusions

Several models have been developed to evaluate the digestion and absorption of polyphenols and carotenoids. These models allow researchers to understand the uptake of bioactive compounds from food products. However, only a few of the models described in the literature have been validated.

Hence, more validation studies are needed to compare the outcomes of *in vitro* studies with *in vivo* results. On the other hand, the models presented in this chapter cannot directly predict the amount of bioactive compound that will be absorbed and utilized under *in vivo* conditions. Still, much insightful information can be obtained from applying digestion models in order to study the digestion, bioaccessibility and colonic fermentation of polyphenols and carotenoids. Studies on this topic should systematically consider the presence of NEPP in foods, since they have usually neglected the presence of this class of bioactive compounds. Nevertheless, the few studies dealing with the *in vitro* metabolic fate of these compounds have shown that they are not bioaccessible in the small intestine, but may become bioaccessible in the large intestine by the action of the colonic microbiota.

References

1. K. B. Pandey and S. I. Rizvi, *Oxid. Med. Cell. Longevity*, 2009, **2**, 270.
2. A. V. Rao and L. G. Rao, *Pharmacol. Res.*, 2007, **55**, 207.
3. M. J. Bermudez-Soto, F. A. Tomas-Barberan and M. T. Garcia-Conesa, *Food Chem.*, 2007, **102**, 865.
4. F. Saura-Calixto, J. Serrano and I. Goni, *Food Chem.*, 2007, **101**, 492.
5. M. Hamdan, F. J. Moyano and D. Schuhardt, *J. Sci. Food Agric.*, 2009, **89**, 1195.
6. J. Parada and J. M. Aguilera, *J. Food Sci.*, 2007, **72**, R21.
7. M. Minekus, M. Alminger, P. Alvito, S. Ballance, T. Bohn, C. Bourlieu, F. Carriere, R. Boutrou, M. Corredig, D. Dupont, C. Dufour, L. Egger, M. Golding, S. Karakaya, B. Kirkhus, S. Le Feunteun, U. Lesmes, A. Macierzanka, A. Mackie, S. Marze, D. J. McClements, O. Menard, I. Recio, C. N. Santos, R. P. Singh, G. E. Vegarud, M. S. J. Wickham, W. Weitschies and A. Brodkorb, *Food Funct.*, 2014, **5**, 1113.
8. C. Manach, G. Williamson, C. Morand, A. Scalbert and C. Remesy, *J. Am. Coll. Nutr.*, 2005, **81**, 230S.
9. E. Bowey, H. Adlercreutz and I. Rowland, *Food Chem. Toxicol.*, 2003, **41**, 631.
10. M. L. Mateos-Martín, J. Pérez-Jiménez, E. Fuguet and J. L. Torres, *Br. J. Nutr.*, 2012, **108**, 290.
11. M. F. Andreasen, P. A. Kroon, G. Williamson and M. T. Garcia-Conesa, *Free Radical Biol. Med.*, 2001, **31**, 304.
12. M. Alminger, A. M. Aura, T. Bohn, C. Dufour, S. El, A. Gomes, S. Karakaya, M. Martínez-Cuesta, G. McDougall and T. Requena, *Compr. Rev. Food Sci. Food Saf.*, 2014, **13**, 413.
13. S. Kamiloglu, A. A. Pasli, B. Ozcelik, J. Van Camp and E. Capanoglu, *Food Chem.*, 2015, **186**, 74.
14. S. Kamiloglu and E. Capanoglu, *Int. J. Food Sci. Technol.*, 2013, **48**, 2621.
15. J. Bouayed, L. Hoffmann and T. Bohn, *Food Chem.*, 2011, **128**, 14.
16. G. J. McDougall, P. Dobson, P. Smith, A. Blake and D. Stewart, *J. Agric. Food Chem.*, 2005, **53**, 5896.
17. D. Tagliazucchi, E. Verzelloni, D. Bertolini and A. Conte, *Food Chem.*, 2010, **120**, 599.

18. S. Kamiloglu, M. Demirci, S. Selen, G. Toydemir, D. Boyacioglu and E. Capanoglu, *J. Sci. Food Agric.*, 2014, **94**, 2225.
19. C. Nimalaratne, P. Savard, S. F. Gauthier, A. Schieber and J. Wu, *J. Agric. Food Chem.*, 2015, **63**, 2956.
20. S. Blanquet-Diot, M. Soufi, M. Rambeau, E. Rock and M. Alric, *J. Nutr.*, 2009, **139**, 876.
21. A. Bengtsson, M. Larsson Alminger and U. Svanberg, *J. Agric. Food Chem.*, 2009, **57**, 9693.
22. A. Padayachee, G. Netzel, M. Netzel, L. Day, D. Mikkelsen and M. J. Gidley, *Food Funct.*, 2013, **4**, 906.
23. A. Solari-Godiño, J. Pérez-Jiménez, F. Saura-Calixto, A. J. Borderías and H. M. Moreno, *Food Res. Int.*, 2017, **102**, 639.
24. C. Svelander, In vitro *bioaccessibility of carotenes: Influence of microstructure in tomato and carrot as modified by processing*, Chalmers University of Technology, 2011.
25. L. de Souza Simões, D. A. Madalena, A. C. Pinheiro, J. A. Teixeira, A. A. Vicente and Ó. L. Ramos, *Adv. Colloid Interface Sci.*, 2017, **243**, 23.
26. M. Minekus, P. Marteau and R. Havenaar, *ATLA, Altern. Lab. Anim.*, 1995, **23**, 197.
27. A. Guerra, L. Etienne-Mesmin, V. Livrelli, S. Denis, S. Blanquet-Diot and M. Alric, *Trends Biotechnol.*, 2012, **30**, 591.
28. M. Vardakou, A. Mercuri, S. A. Barker, D. Q. Craig, R. M. Faulks and M. S. Wickham, *AAPS PharmSciTech*, 2011, **12**, 620.
29. F. Kong and R. P. Singh, *J. Food Sci.*, 2010, **75**, E627.
30. S. Kamiloglu, C. Grootaert, E. Capanoglu, C. Ozkan, G. Smagghe, K. Raes and J. Camp, *Mol. Nutr. Food Res.*, 2017, **61**, 1600455.
31. W. Yi, C. C. Akoh, J. Fischer and G. Krewer, *J. Agric. Food Chem.*, 2006, **54**, 5651.
32. X. J. Tian, X. W. Yang, X. Yang and K. Wang, *Int. J. Pharm.*, 2009, **367**, 58.
33. S. Deprez, I. Mila, J. F. Huneau, D. Tome and A. Scalbert, *Antioxid. Redox Signaling*, 2001, **3**, 957.
34. A. Kaulmann, C. M. André, Y.-J. Schneider, L. Hoffmann and T. Bohn, *Food Chem.*, 2016, **197**, 325.
35. M. G. Ferruzzi, M. L. Failla and S. J. Schwartz, *J. Agric. Food Chem.*, 2001, **49**, 2082.
36. E. Reboul, A. Lydia, C. Mikail, O. Ghiringhelli, M. André, H. Portugal, D. Jourdheuil-Rahmani, M.-J. Amiot, D. Lairon and P. Borel, *Biochem. J.*, 2005, **387**, 455.
37. F. Zucco, A. Batto, G. Bises, J. Chambaz, A. Chiusolo, R. Consalvo, H. Cross, G. Negro, I. de Angelis and G. Fabre, *Atla-Nottingham*, 2005, **33**, 603.
38. T. Zhang, C. Lv, L. Chen, G. Bai, G. Zhao and C. Xu, *Food Res. Int.*, 2014, **62**, 183.
39. H. Sun, E. C. Chow, S. Liu, Y. Du and K. S. Pang, *Expert Opin. Drug Metab. Toxicol.*, 2008, **4**, 395.

40. A. Lechanteur, A. Almeida and B. Sarmento, *Eur. J. Pharm. Biopharm.*, 2017, **119**, 137.
41. E. Finotti, R. Gezzi, F. Nobili, I. Garaguso and M. Friedman, *RSC Adv.*, 2015, **5**, 22490.
42. H. Sabboh-Jourdan, F. Valla, I. Epriliati and M. J. Gidley, *Eur. J. Nutr.*, 2011, **50**, 31.
43. Y. Sambuy, I. De Angelis, G. Ranaldi, M. Scarino, A. Stammati and F. Zucco, *Cell Biol. Toxicol.*, 2005, **21**, 1.
44. P. V. Balimane and S. Chong, *Drug Discovery Today*, 2005, **10**, 335.
45. R. Barrington, G. Williamson, R. N. Bennett, B. D. Davis, J. S. Brodbelt and P. A. Kroon, *J. Funct. Foods*, 2009, **1**, 74.
46. G. Pereira-Caro, R. Mateos, S. Saha, A. Madrona, J. L. Espartero, L. Bravo and P. A. Kroon, *J. Agric. Food Chem.*, 2010, **58**, 11501.
47. A. Soler, M. P. Romero, A. Macia, S. Saha, C. S. Furniss, P. A. Kroon and M. J. Motilva, *Food Chem.*, 2010, **119**, 703.
48. C. Dhuique-Mayer, P. Borel, E. Reboul, B. Caporiccio, P. Besancon and M. J. Amiot, *Br. J. Nutr.*, 2007, **97**, 883.
49. E. Biehler, L. Hoffmann, E. Krause and T. Bohn, *J. Nutr.*, 2011, **141**, 1769.
50. L. Turco, T. Catone, F. Caloni, E. Di Consiglio, E. Testai and A. Stammati, *Toxicol. In vitro*, 2011, **25**, 13.
51. J. Fogh and G. Trempe, *Human Tumor Cells In vitro*, Springer, 1975, pp. 115–159.
52. A. Zweibaum, M. Pinto, G. Chevalier, E. Dussaulx, N. Triadou, B. Lacroix, K. Haffen, J. L. Brun and M. Rousset, *J. Cell. Physiol.*, 1985, **122**, 21.
53. A. Zweibaum, M. Laburthe, E. Grasset and D. Louvard, *Compr. Physiol.*, 2011, **Sup 19**, 223.
54. J. Maoret, J. Font, C. Augeron, P. Codogno, C. Bauvy, M. Aubery and C. Laboisse, *Biochem. J.*, 1989, **258**, 793.
55. T. Lesuffleur, A. Barbat, E. Dussaulx and A. Zweibaum, *Cancer Res.*, 1990, **50**, 6334.
56. G. B. Gonzales, J. Van Camp, H. Vissenaekens, K. Raes, G. Smagghe and C. Grootaert, *Compr. Rev. Food Sci. Food Saf.*, 2015, **14**, 741.
57. I. Behrens, P. Stenberg, P. Artursson and T. Kissel, *Pharm. Res.*, 2001, **18**, 1138.
58. T. H. Wilson and G. Wiseman, *J. Physiol.*, 1954, **123**, 116.
59. T. Mariappan and S. Singh, *Mol. Pharm.*, 2004, **1**, 363.
60. S. A. Acra and F. K. Ghishan, *J. Parenter. Enteral Nutr.*, 1991, **15**, 93S.
61. G. Roeselers, M. Ponomarenko, S. Lukovac and H. M. Wortelboer, *Best Pract. Res., Clin. Gastroenterol.*, 2013, **27**, 101.
62. M. A. Alam, F. I. Al-Jenoobi and A. M. Al-mohizea, *J. Pharm. Pharmacol.*, 2012, **64**, 326.
63. J. M. Gee, M. S. DuPont, A. J. Day, G. W. Plumb, G. Williamson and I. T. Johnson, *J. Nutr.*, 2000, **130**, 2765.
64. N. M. A. Hassimotto, M. I. Genovese and F. M. Lajolo, *Nutr. Res.*, 2008, **28**, 198.

65. S. Veda and K. Srinivasan, *J. Funct. Foods*, 2009, **1**, 394.
66. H. H. Ussing and K. Zerahn, *Acta Physiol.*, 1951, **23**, 110.
67. L. L. Clarke, *Am. J. Physiol.: Gastrointest. Liver Physiol.*, 2009, **296**, G1151.
68. D. Scherbl, S. Muentnich and E. Richling, *Food Res. Int.*, 2014, **63**, 456.
69. H. Deußer, D. Rogoll, W. Scheppach, A. Volk, R. Melcher and E. Richling, *Biotechnol. J.*, 2013, **8**, 363.
70. T. Erk, J. Hauser, G. Williamson, M. Renouf, H. Steiling, F. Dionisi and E. Richling, *BioFactors*, 2014, **40**, 103.
71. R. Salerno-Goncalves, A. Fasano and M. B. Sztein, *Gastroenterology*, 2011, **141**, e18.
72. K. L. Schmeichel and M. J. Bissell, *J. Cell Sci.*, 2003, **116**, 2377.
73. P. De Boever, B. Deplancke and W. Verstraete, *J. Nutr.*, 2000, **130**, 2599.
74. K. Molly, M. V. Woestyne and W. Verstraete, *Appl. Microbiol. Biotechnol.*, 1993, **39**, 254.
75. K. Verhoeckx, P. Cotter, I. López-Expósito, C. Kleiveland, T. Lea, A. Mackie, T. Requena, D. Swiatecka and H. Wichers, *The Impact of Food Bioactives on Health*, Springer, 2015.
76. S. Kamiloglu, *Bioavailability and bioactivity of black carrot polyphenols using* in vitro *digestion models combined with a co-culture model of intestinal and endothelial cell lines*, Ghent University, 2016.
77. S. Bolca, T. Van de Wiele and S. Possemiers, *Curr. Opin. Biotechnol.*, 2013, **24**, 220.
78. S. Possemiers, S. Bolca, C. Grootaert, A. Heyerick, K. Decroos, W. Dhooge, D. De Keukeleire, S. Rabot, W. Verstraete and T. Van de Wiele, *J. Nutr.*, 2006, **136**, 1862.
79. K. Decroos, E. Eeckhaut, S. Possemiers and W. Verstraete, *J. Nutr.*, 2006, **136**, 946.
80. E. Eeckhaut, K. Struijs, S. Possemiers, J. P. Vincken, D. D. Keukeleire and W. Verstraete, *J. Agric. Food Chem.*, 2008, **56**, 4806.
81. G. Gross, D. M. Jacobs, S. Peters, S. Possemiers, J. van Duynhoven, E. E. Vaughan and T. Van de Wiele, *J. Agric. Food Chem.*, 2010, **58**, 10236.
82. F. Van Dorsten, S. Peters, G. Gross, V. Gomez-Roldan, M. Klinkenberg, R. De Vos, E. Vaughan, J. Van Duynhoven, S. Possemiers and T. Van de Wiele, *J. Agric. Food Chem.*, 2012, **60**, 11331.
83. F. Sánchez-Patán, E. Barroso, T. Van de Wiele, A. Jiménez-Girón, P. J. Martín-Alvarez, M. V. Moreno-Arribas, M. C. Martínez-Cuesta, C. Peláez, T. Requena and B. Bartolomé, *Food Chem.*, 2015, **183**, 273.
84. R. Garcia-Villalba, H. Vissenaekens, J. Pitart, M. Romo-Vaquero, J. C. Espín, C. Grootaert, M. V. Selma, K. Raes, G. Smagghe and S. Possemiers, *J. Agric. Food Chem.*, 2017, **65**, 5480.
85. T. Wu, C. Grootaert, S. Voorspoels, G. Jacobs, J. Pitart, S. Kamiloglu, S. Possemiers, M. Heinonen, N. Kardum and M. Glibetic, *J. Funct. Foods*, 2017, **38**, 128.
86. A. Maathuis, A. Hoffman, A. Evans, L. Sanders and K. Venema, *J. Am. Coll. Nutr.*, 2009, **28**, 657.

87. M. Minekus, M. Smeets-Peeters, A. Bernalier, S. Marol-Bonnin, R. Havenaar, P. Marteau, M. Alric and G. Fonty, *Appl. Microbiol. Biotechnol.*, 1999, **53**, 108.
88. K. Gao, A. Xu, C. Krul, K. Venema, Y. Liu, Y. Niu, J. Lu, L. Bensoussan, N. P. Seeram and D. Heber, *J. Nutr.*, 2006, **136**, 52.
89. M. Tabernero, K. Venema, A. J. Maathuis and F. D. Saura-Calixto, *J. Agric. Food Chem.*, 2011, **59**, 8968.
90. E. Barroso, C. Cueva, C. Peláez, M. C. Martínez-Cuesta and T. Requena, *LWT–Food Sci. Technol.*, 2015, **61**, 283.
91. C. Cueva, A. Jiménez-Girón, I. Muñoz-González, A. Esteban-Fernández, I. Gil-Sánchez, M. Dueñas, P. Martín-Álvarez, M. Pozo-Bayón, B. Bartolomé and M. Moreno-Arribas, *Food Res. Int.*, 2015, **72**, 149.
92. S. S. Ekbatan, L. Sleno, K. Sabally, J. Khairallah, B. Azadi, L. Rodes, S. Prakash, D. J. Donnelly and S. Kubow, *Food Chem.*, 2016, **204**, 453.
93. V. M. Koistinen, E. Nordlund, K. Katina, I. Mattila, K. Poutanen, K. Hanhineva and A. M. Aura, *J. Agric. Food Chem.*, 2017, **65**, 1854.
94. A. M. Aura, K. O'leary, G. Williamson, M. Ojala, M. Bailey, R. Puupponen-Pimiä, A. Nuutila, K. M. Oksman-Caldentey and K. Poutanen, *J. Agric. Food Chem.*, 2002, **50**, 1725.
95. A. M. Aura, P. Martin-Lopez, K. A. O'Leary, G. Williamson, K. M. Oksman-Caldentey, K. Poutanen and C. Santos-Buelga, *Eur. J. Nutr.*, 2005, **44**, 133.
96. A. R. Rechner, M. A. Smith, G. Kuhnle, G. R. Gibson, E. S. Debnam, S. K. S. Srai, K. P. Moore and C. A. Rice-Evans, *Free Radical Biol. Med.*, 2004, **36**, 212.
97. F. Saura-Calixto, J. Pérez-Jiménez, S. Touriño, J. Serrano, E. Fuguet, J. L. Torres and I. Goñi, *Mol. Nutr. Food Res.*, 2010, **54**, 939.
98. J. Serrano, I. Goni and F. Saura-Calixto, *J. Agric. Food Chem.*, 2005, **53**, 2936.
99. V. Fogliano, M. L. Corollaro, P. Vitaglione, A. Napolitano, R. Ferracane, F. Travaglia, M. Arlorio, A. Costabile, A. Klinder and G. Gibson, *Mol. Nutr. Food Res.*, 2011, **55**, S44.

CHAPTER 11

Metabolic Fate of Extractable and Non-extractable Polyphenols

I. A. LUDWIG[a], L. RUBIÓ[a,b], J. I. MOSELE[c,d] AND M. J. MOTILVA*[a]

[a]Food Technology Department, Universitat de Lleida-AGROTECNIO Center, Lleida, Alcalde Rovira Roure 191, 25198 Lleida, Spain; [b]Functional Nutrition, Oxidation and Cardiovascular Diseases Group (NFOC-Salut), Unit of Lipids and Atherosclerosis Research (URLA), Centro de Investigación Biomédica en Red de Diabetes y Enfermedades Metabólicas Asociadas (CIBERDEM), Hospital Universitari Sant Joan, IISPV, Technological Center of Nutrition and Health (CTNS), Faculty of Medicine and Health Sciences, Universitat Rovira i Virgili, Sant Llorenç, 21, 43201, Reus, Spain; [c]Universidad de Buenos Aires, Facultad de Farmacia y Bioquímica, Departamento de Química Analítica y Fisicoquímica, Cátedra de Fisicoquímica, Buenos Aires, Argentina; [d]CONICET-Universidad de Buenos Aires, Instituto de Bioquímica y Medicina Molecular (IBIMOL), Buenos Aires, Argentina
*E-mail: motilva@tecal.udl.es

11.1 Introduction

Plants produce a great variety of secondary metabolites, with the phenolic compounds, usually called (poly)phenols, being the most widely distributed in dietary sources. Due to their abundance in certain plant foods, they are

Food Chemistry, Function and Analysis No. 5
Non-extractable Polyphenols and Carotenoids: Importance in Human Nutrition and Health
Edited by Fulgencio Saura-Calixto and Jara Pérez-Jiménez

Published by the Royal Society of Chemistry, www.rsc.org

important candidates for the beneficial effects associated with vegetable- and fruit-rich diets. Indeed, an increased consumption of fruit and vegetables has been related, with different levels of evidence, to the prevention of several chronic diseases.[1,2] Among the constituents responsible for these beneficial effects, (poly)phenols are considered to play an important role. However, most studies are based on underestimated (poly)phenol contents, because they include only those compounds that are found in the supernatants derived from aqueous–organic extractions from foods (*i.e.* extractable [poly]phenols [EPP]), which are usually considered as total (poly)phenols.

However, it has been reported that a significant fraction of (poly)phenols remains in the residues of the sample extraction: the so-called non-extractable (poly)phenols (NEPP) or macromolecular antioxidants.[3,4] The NEPP have three specific features: (i) they are either polymeric (poly)phenols or single (poly) phenols linked to macromolecular food constituents (polysaccharides or proteins); (ii) they are not extracted by common aqueous–organic solvents; and (iii) they are accessible and bioavailable only in the large intestine after gastrointestinal digestion. In contrast, soluble (poly)phenols or EPP are mainly low-molecular-weight structures that are soluble in aqueous–organic mixtures and are potentially bioavailable in the stomach and small intestine.[5,6]

Besides, the NEPP can be divided into two fractions: hydrolysable polyphenols (HPP), which are low-molecular-weight phenolic compounds strongly associated with polysaccharides or proteins and hydrolysable tannins (gallotannins and ellagitannins); and proanthocyanidins (PA), which are high-molecular-weight structures. Despite NEPP having been commonly ignored in chemical analyses, when plant foods are consumed, both EPP and NEPP are ingested.

To understand the possible health effects associated with the intake of any bioactive compound, including NEPP, it is necessary to know the different events that occur in the human gut, including their bioaccessibility, possible transformation by colonic microbiota and subsequent absorption of the derived catabolites in the colon. All these aspects have been studied in depth over recent years for EPP,[7] but studies on NEPP are scarce. Despite that, the available work that focuses on NEPP suggests that they are subjected to extensive transformation in the gut, giving rise to several absorbable metabolites that have been reported to have different health effects.

The present chapter begins by describing all of the metabolic aspects related to the NEPP. It then examines the bioavailability of NEPP in *in vivo* studies and the last section reviews the *in vitro* gastrointestinal digestion and colonic fermentation studies with NEPP, which are predictive models that provide very valuable information regarding the behaviour of NEPP during digestion in the gastrointestinal tract.

11.2 Occurrence of EPP and NEPP in Foodstuffs

The different classes of NEPP are widespread among the different families of plant foods (cereals, fruit, vegetables, nuts and legumes), although some of them are particularly characteristic of a particular food family (Table 11.1).

Table 11.1 Main sources of extractable and non-extractable (poly)phenols in the Spanish diet.

Classes of (poly)phenols		Main foodstuff sources in the Spanish diet	Amount (mg 100 g^{-1} dry weight)[a]
Extractable (poly)phenols	Phenolic acids, flavonoids (...)	Fruit	306
		Vegetables	196
		Nuts	130
		Legumes	58
		Cereals	57
Non-extractable (poly)phenols	*Hydrolysable polyphenols (HPP)* Phenolic acids (ferulic acid, caffeic acid, gallic acid, *p*-hydroxybenzoic *etc.*), hydrolysable tannins, (−)-epicatechin, catechin	Fruit	367
		Vegetables	326
		Legumes	256
		Cereals	210
		Nuts	136
	Proanthocyanidins	Fruit	514
		Legumes (...)	312
		Nuts	199

[a]Amounts extracted from Arranz *et al.*[8]

11.3 Metabolic Fate of NEPP

11.3.1 Intake and Bioaccessibility of NEEP

Only two studies (in Spain[9] and Finland[10]) have determined total (poly)phenol intake, including EPP and NEPP, with values ranging from 2591 to 3016 mg day^{-1} and 863 mg day^{-1}, respectively. NEPP contribute about 79% of the total (poly)phenol intake in Spain with fruit being the highest contributor to the NEPP intake, followed by cereals, vegetables, legumes and nuts.[8] Another recent study analysed NEPP in the 24 most widely consumed fruit and vegetables in four European countries (France, Germany, The Netherlands and Spain), corroborating that NEPP are the major contributors (57%) to the total (poly)phenol content of fruit and vegetables.[11] These results constitute evidence of the appreciable contribution of NEPP to the total (poly)phenol intake.

(Poly)phenols released from the food matrix during the digestive process (named bioaccessible [poly]phenols) are potentially bioavailable or susceptible to absorption through the gut barrier. It is estimated that about 48% of total polyphenols from solid vegetable foods in the diet are bioaccessible in the small intestine.[9] By contrast, (poly)phenols from beverages are completely bioaccessible because they pass directly into the intestinal fluids, and they are the largest contributors of bioaccessible (poly)phenols in the small intestine. However, only a small part of these compounds can be absorbed through the intestinal mucosa and therefore metabolised. Numerous studies have shown that the bioavailability of the bioaccessible polyphenols in the small intestine is very low, with values between 5% and 10% being reported.[12]

There are abundant data in the literature on the content and composition of EPP, but only one study has evaluated the intestinal bioaccessibility of

NEPP with an *in vitro* gastrointestinal model combining enzyme digestion and colonic fermentation.[9] This study reported that about 50% of hydrolysable phenolics in the NEPP fraction become bioaccessible in the small intestine and, in the case of PA, most of them reach the colon nearly intact. Similarly, other studies have shown that PA are only partially depolymerised in the small intestine.[13,14]

Therefore, in addition to (poly)phenols associated with the indigestible food fraction, a major part of the (poly)phenols potentially absorbable in the small intestine may also reach the colon because of their low bioavailability. If we assume that only 5–10% of dietary (poly)phenols (including beverage [poly]phenols) are absorbed in the small intestine, this means that the main fraction of the dietary (poly)phenols reaches the colon. This fraction is made up of (poly)phenols solubilised in the digesta plus those associated with the indigestible fraction. However, a minor fraction of dietary (poly)phenols (10%) is detected in faeces, which suggests that 90% of them may be bioaccessible in the gut as intact molecules and/or as colonic fermentation metabolites. A general scheme of the metabolic fate of NEPP and EPP is shown in Figure 11.1.

11.3.2 Colonic Transformation of NEPP

In the colon, NEPP released from the food matrix and polymeric NEPP undergo different microbial transformations. They become accessible either by fermentation of the molecules to which they are associated (carbohydrates or proteins) by colonic microbiota or by the action of some intestinal enzymes able to break covalent bonds, such as esterases.[15] It has been estimated that 95% of PA arriving in the colon are released from the food matrix by these two mechanisms.[9] Once they are released from the food matrix, they undergo colonic fermentation, which produces potentially absorbable metabolites.

Work focusing strictly on the fermentation of dietary NEPP in *in vivo* studies is still scarce and mostly qualitative, but it has already provided clear evidence of their colonic catabolism. For instance, supplementing rats with an NEPP concentrate resulted in the presence of phenolic metabolites in the urine and faeces at concentrations at least ten-fold higher than in a non-supplemented group.[14] Many studies of the metabolic fate of whole foods, which evidently contain NEPP, have provided data regarding the compounds that correspond to NEPP microbial metabolites.

11.3.2.1 Metabolites Derived from Hydrolysable (Poly)phenols

Although free phenolic acids released from the food matrix may be absorbed directly, they may also undergo additional transformation resulting from interaction with the colonic microbiota. For instance, simple phenolic acids, such as ferulic acid (FA), are transformed into dihydroferulic acid, which is later metabolised to other phenolic acids common to some products of the PA colonic metabolism.

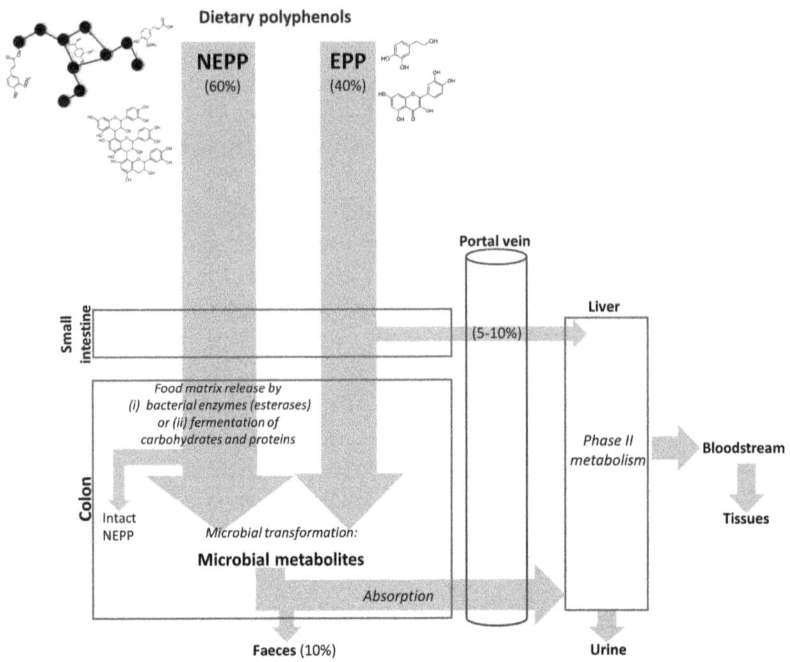

Figure 11.1 General scheme of the metabolic fate of NEPP and EPP. The former, representing hydrolysable phenolic compounds associated with a polysaccharide chain and polymeric NEPA, pass through the small intestine without any transformation, together with those EPP that are not absorbed in the small intestine. Once in the colon, and mainly through the action of the microbiota, NEPP are released and new compounds form. Most of these microbial metabolites are absorbed though the portal vein and some of them reach the liver, giving rise to phase II metabolites. Once formed, these metabolites pass into the bloodstream as a first step towards reaching the target tissues and finally are excreted in urine. A minor part (10%) is associated with the fermentation residue and is excreted in faeces.

11.3.2.2 Metabolites Derived from PA

These include the monomeric constituents of non-extractable proanthocyanidins (NEPA) polymers, mainly (epi)catechin, and a wide variety of compounds derived from the breakdown of the basic skeleton of flavanols and further degradation, which gives rise first to valerolactones, then to several phenolic acids corresponding to successive degradations (hydroxyphenylpropionic acids, hydroxyphenylacetic acids, *etc.*) and then to small molecules such as hippuric acid.

11.3.2.3 Metabolites Derived from Hydrolysable Tannins

Ellagic acid released from ellagitannins in the gastrointestinal tract is transformed by the microbiota into a particular class of compounds, the urolithins, while gallotannins are hydrolysed into gallic acid and further transformed into catechol and hydroxybenzoic acid.

The presence of these microbial metabolites from NEPP has been determined in caecal contents and faeces, but their determination is most common in urine and blood samples in both animal studies and human clinical trials (see Section 11.4). Most of these studies concluded that the NEPP colonic metabolites are absorbed and therefore become bioavailable, and can have different biological effects on the human body, as well as possible local effects on the colon.

11.3.3 Bioavailability of NEPP

The metabolites described above do not differ from those described as derived from EPP colonic fermentation,[7,16] as the colonic fermentation of both EPP and NEPP gives rise to the same compounds. However, the specific characteristics of NEPP (high molecular weight and/or association with other macromolecules present in the food matrix) generate two specific features in their absorption as compared with EPP. On the one hand, the absorption of NEPP is retarded in relation to that of EPP, as shown, for instance, by the delayed peak (C_{max}) in plasma FA levels after the intake of wheat bran (an important NEPP source) in comparison with that of pure FA.[17] Such data indicate that the circulating metabolites derived from NEPP remain for longer in the human body than those derived from EPP. This fact suggests that the dietary NEPP may continually provide the colonic microbiota with a significant amount of fermentable substrates that, due to their specific nature, need more time to be released and fermented than EPP.

On the other hand, since NEPP do not reach the colon alone, but in combination with other fermentable dietary components, such as proteins and polysaccharides, their fermentation rate may be affected. Indeed, some studies in rats have shown that the presence of dietary fibre enhances the fermentation of (poly)phenols.[18,19] Moreover, in the work by Saura-Calixto *et al.*,[20] the *in vitro* fermentability of a concentrate of polymeric NEPP was 23%, while that of a concentrate of NEPP associated with dietary fibre was 50%, probably due to the increase in the bacterial activity. Therefore, the colonic fermentation of NEPP would be accompanied by a synergism with the other components of the food to which they are associated, an aspect that is not present in the fermentation of EPP, which are free in the food matrix.

11.4 Recent Advances in *In vivo* Studies with NEPP-rich Foods

Protective effects against chronic diseases, such as cardiovascular disease, neurodegeneration and cancer, have been associated with (poly)phenolic compounds.[1] To exert any of these effects, (poly)phenols must be bioavailable and reach the target tissues. For many decades, the bioavailability of (poly)phenols has been underestimated since only the parent compounds present in food have been taken into consideration. (Poly)phenols, however, are extensively metabolised during their passage through the gastrointestinal tract and are transformed into a plethora of lower-molecular-weight compounds that can be more easily absorbed. One of the main sites for these reactions is the colon, and this is especially true for NEPP, since they reach the large intestine almost intact. As mentioned above, most studies into the bioavailability of hydrolysable phenolics and tannins and PA determine the phenolic contents in food by liquid/liquid and liquid/solid extraction with organic solvents and disregard the NEPP content. This, together with the fact that after the *in vivo* gastrointestinal digestion an important part of the NEPP and EPP fractions reaches the colon, means that it is almost impossible to obtain numeric data on the bioavailability of NEPP. Besides, as mentioned previously, NEPP metabolites do not differ from those described as being derived from EPP colonic fermentation, as the colonic fermentation of both EPP and NEPP gives rise to the same compounds.

11.4.1 Proanthocyanidins

PA comprise a group of polyphenols made up of flavan-3-ol subunits with high structural diversity and a wide ranging degree of polymerisation (DP). The most abundant type of PA are the procyanidins, consisting of (epi)catechin units, while less abundant propelargonidins and prodelphinidins contain (epi)afzelechin and (epi)gallocatechin units, respectively. Depending on the type of interflavanic bond between the subunits, PA are classified into B-type PA with C4–C8 or C4–C6 linkages or A-type PA with an additional ether type bond (C2–O–C7). Typical sources of B-type PA include fruit like grapes, pears and apples, as well as legumes, cocoa, sorghum and such beverages as wine, cider and beer, while A-type PA are less abundant and are mainly found in nuts, plums, cranberries, avocados and cinnamon.[21] Polymeric and oligomeric PA represent the bulk of dietary PA, most of which being present in cocoa, berries, grapes and sorghum.

The bioavailability of PA is, to a great extent, influenced by their DP. No mammalian enzymes have been reported with the ability to degrade the high-molecular-weight molecules into forms that are small enough to be readily absorbed in the small intestine.[22] It has been speculated that PA could be depolymerised into bioavailable monomers under the acidic conditions in the stomach.[23] However, *in vivo* studies failed to corroborate this hypothesis[24,25] and only dimers have been reported as being absorbed intact in the upper gastrointestinal tract, but to a very limited extent compared with the monomeric flavan-3-ols. The majority of dimers, however, pass to the colon together with oligomers and polymers and are extensively metabolised by the colonic microbiota. The proposed catabolic reactions occurring in the colon have been reviewed by Monagas *et al.*[26] and involve the release of flavan-3-ol subunits as a first step. In the case of galloylated subunits, the gallic acid is cleaved by microbial esterase and further decarboxylated to pyrogallol. The C ring of the flavan-3-ol subunit is opened, giving rise to diphenylpropan-2-ol, which is later converted into 5-(3′,4′,5′-trihydroxyphenyl)-γ-valerolacton in the case of (epi)gallocatechin and 5-(3′,4′-dihydroxyphenyl)-γ-valerolacton for (epi)catechin. The lacton ring later breaks, giving rise to 5-(3′,4′-dihydroxyphenyl)-valeric acid and 4-hydroxy-5-(3′,4′-dihydroxyphenyl)-valeric acid, which are subsequently biotransformed to hydroxylated phenylpropionic and benzoic acids by a successive loss of carbon atoms from the side chain *via* β-oxidation. The possibility of the formation of hydroxylated phenylacetic acids *via* α-oxidation has been postulated,[27] although the cleavage of the upper unit of dimeric PA to form phenylacetic acid has been proposed as the main catabolic pathway[27,28] leading to phenylacetic acid derivatives.

Most studies into the bioavailability of PA have focused on the bioavailability of dimeric PA.[25,29,30] However, it is not clear to what extent polymeric PA are degraded into the monomeric flavan-3-ol units, and further to phenolic acids in the colon, since the carbon–carbon bonds between subunits have been shown to be resistant to degradation by the microflora.[31,32] Nonetheless, more recent *in vivo* studies with rats fed grape seed extract have shown that at least oligomeric PA with DP ≤ 6 are degraded by the colonic microflora to monomeric flavan-3-ols, and subsequently to valerolactons and phenolic acids as the digesta moves through the lower gastrointestinal tract.[33,34] In this sense, significantly higher levels of phenolic acids excreted in urine in human *in vivo* studies with cocoa and almond skin extracts, both rich in oligomeric and polymeric PAs, also point to at least a partial degradation of PA with higher DP and subsequent absorption of the catabolites.[35,36] Once absorbed, the lower-molecular-weight phenolics are subjected to phase II metabolism before being excreted in the urine. The valerolacton metabolites detected in urine after the intake of PA-rich foods include monoglucuronides and monosulphates of dihydroxy and monohydroxyphenyl-γ-valerolactone, in addition to methyl-sulphates of trihydroxyphenyl-γ-valerolactones, with sulphates being the most abundant conjugates.[35,37] Further glucuronide and sulphate conjugates of dihydroxy- and monohydroxyphenyl-valeric acids have been detected in the urine of human volunteers after the ingestion of almond skin extracts rich in A- and B-type PA.[36] Both valerolacton and valeric acid conjugates showed significantly higher excretion in urine 2–24 hours post-ingestion of almond skin extract compared to the control diet. The highest excretion values for most conjugates of mono- and di-substituted phenyl-γ-valerolactones were observed from 6 to 10 hours, whereas trihydroxyphenyl-γ-valerolactone glucuronide was mainly excreted at 10–24 hours after intake. This finding could be attributed to the fact that (epi)gallocatechin units in almond prodelphinidins may form part of the inner extension rather than terminal units, and may only be accessible to the gut microbiota after extended degradation of the terminal units. The urine excretion profile of valeric acid conjugates differed according to the hydroxylation pattern of the phenyl ring from non- to mono- and di-hydroxylated forms. For dihydroxylated derivatives, the highest values were found 6–10 hours after intake, for monohydroxylated forms at 6–24 hours and, finally, for non-hydroxylated forms at 10–24 hours after intake. These findings suggest that dehydroxylation reactions gradually occur as colonic metabolism progresses. With regard to phenolic acid conjugates, glucuronide and sulphates of monohydroxy and dihydroxyphenyl propionic and acetic acids have been detected in rat urine after the intake of grape extracts rich in oligomeric PA.[38]

To summarise, the results of the bioavailability of PA suggest that after colon microbial degradation of the polymeric structures, lower-molecular-weight valerolactons, valeric acids and phenolic acids are bioavailable and further metabolised after absorption by phase II enzymes.

11.4.2 Hydrolysable Tannins: Gallotannins and Ellagitannins

Hydrolysable tannins, which include gallotannins and ellagitannins, are probably the most complex group of tannins with molar masses ranging from 300 to 3000 Da, and even up to 20 000 Da.[39] Gallotannins, which are composed of gallic acid units linked *via* ester bonds to a central glucose molecule, are present only in some fruit, legumes and nuts,[39,40] and consequently little attention has been paid to their bioavailability. Li *et al.*[41] studied the pharmacokinetics of penta-*O*-galloyl-β-D-glucose (PGG) in mice after intra-peritoneal administration of 0.5 mg. Peak plasma concentrations of 3–4 μmol L^{-1} were observed 2 hours after PGG administration. When mice were administered orally 2 mg of PGG, the authors failed to detect PGG in the plasma. This was attributed to the fact that gallotannins can strongly bind to proteins and phospholipids such as albumin and lecithin, respectively, which may limit their absorption. However, the study did not include the analysis of PGG metabolites. Due to their simple composition of just glucose and galloyl units, the metabolism of gallotannins is restricted to the hydrolysis of the polymeric structure and release of gallic acid. In fact, it has been demonstrated that PGG is degraded into tri- and tetra-galloyl glucose in contact with Caco-2 cells by the action of esterases.[42] Also, the esterase activity of the colonic microbiota may cleave galloyl moieties in certain tannins and further degrade the free gallic acid units before absorption of these simpler compounds in the gut. In the absence of bioavailability data on gallotannins, studies on gallic acid and other galloylated phenolics, such as theaflavins, may help us to understand their metabolism. Gallic acid has been reported to be well absorbed in humans, with recoveries of 37% of the intake found in urine as free gallic acid and 4-*O*-methygallic acid.[43,44] When analysing the bioavailability of black tea theaflavins, Pereira-Caro *et al.*[45] concluded that part of the absorbed gallic acid appears to be converted into 3-*O*-methygallic acid and 4-*O*-methygallic acid by mammalian enzymes. Subsequently, the unabsorbed gallic acid released from the digesta in the colon can be further decarboxylated to pyrogallol, which then undergoes phase II metabolism, yielding substantial amounts of pyrogallol-1-sulphate and pyrogallol-2-sulphate and smaller amounts of pyrogallol-2-*O*-glucuronide. This study showed that from a total of 1169 μmol gallic acid units present in the ingested theaflavin extract, 64.4% was recovered in urine. The main metabolites detected were pyrogallol and pyrogallol sulphate and glucuronide conjugates, which made up 61% of the ingested gallic acid. Methylgallic acid represented 3.2% of the intake, while free gallic acid corresponded to just 0.2%. However, the formation of all these metabolites after gallotannin ingestion is speculative and further research is needed to confirm this hypothesis.

Ellagitannins are structurally more complex molecular structures than gallotannins and consist of at least one hexahydroxydiphenoyl (HHDP) group esterified to a sugar unit, most often glucose. On hydrolysis, HHDP groups are rearranged into ellagic acid, one of the intermediate metabolites of ellagitannins. Further, most ellagitannins also contain galloyl groups.[46] Ellagitannins are divided into different groups according to their chemical structure: (i) monomeric ellagitannins with one sugar core and at least one HHDP moiety, such as punicalagin, punicalin and pedunculagin; (ii) C-glycosidic ellagitannins with an open chain glycosidic core, such as vescalagin, casuariin and stachiurin; and (iii) oligomeric ellagitannins composed of two or more monomers linked by phenolic C–C and C–O bonds, like sanguiin H-6 and lambertianin C. Ellagitannins are present in many plants, with berries, including strawberries, raspberries and blackberries (the main source of sanguiin H-6 and lambertianin C), being the main contributors to their dietary intake, along with pomegranates (the main source of punicalagin and punicalin) and nuts (predominantly pedunculagin).[39,40] The metabolic fate of ellagitannins has been studied over recent decades in several foods and can be divided into two steps: (i) hydrolysis of the polymeric structure; and (ii) metabolism of the released ellagic acid and gallic acid subunits. As mentioned above, the composition of NEPP in food was not analysed in most of these studies. Some studies have detected ellagic acid and ellagic acid metabolites, such as dimethyl-ellagic acid-glucuronide and dimethyl-ellagic acid-sulphate, in plasma 1–3 hours after ingestion of raspberries and pomegranate, pointing to hydrolysis and absorption in the upper gastrointestinal tract.[47–49] However, recoveries of ellagic acid and its conjugate metabolites in urine make up less than 1% of the ingested ellagitannins.[48,49] The major site of absorption of ellagitannins is the colon, where bacterial esterases act upon the complex structure, releasing ellagic and gallic acids. These acids may be subjected to the same metabolic pathways as described above for gallotannins. Ellagic acid, on the other hand, is transformed into a specific class of compounds, the urolithins, dibenzopyran-6-one derivatives formed *via* cleavage and decarboxylation with a decreasing number of hydroxyl groups. Urolithins have been proposed as biomarkers of dietary ellagitannin intake.[50] According to their hydroxylation pattern, they are classified as follows: urolithin D, with four hydroxyl groups; urolithin C, with three hydroxyl groups; urolithin A, with two hydroxyl groups; and urolithin B, with one hydroxyl group. The absence of this class of compounds after the intake of ellagitannin-rich food in the ilea fluid of people who have had ileostomies (*i.e.* those who have had their colon removed surgically for medical reasons) demonstrated that the colon is the site of urolithin production.[48] After absorption, urolithins suffer extensive phase II metabolism, including methylation, glucuronidation and sulphation, before they are excreted in urine. A vast array of urolithin metabolites have been detected in biological fluids, the most important being the free and conjugated forms of urolithins A and B, such as urolithin A, urolithin A-glucuronide, urolithin A-sulphate, urolithin A-sulphoglucuronide, urolithin A-diglucuronide, urolithin B, urolithin B-glucuronide and urolithin B-sulphate.[51–54] Typical maximum plasma concentrations are achieved 24–48 hours post-ingestion, and urinary metabolites can be excreted even 90 hours after the consumption of ellagitannin-rich foods.[48,49,52,55] Espín *et al.*[56] conducted a study in which pigs were fed acorns, a rich source of ellagitannins, and it was found that large amounts of urolithin metabolites accumulated in the gall bladder. These results are indicative of enterohepatic recirculation and could explain the prolonged presence of urolithin metabolites in the circulatory system. On the other hand, Truchado *et al.*[52] suggested that the ellagitannins may be bound to the intestinal mucosa so that their transfer to the digesta fraction would be very slow, which could be another explanation for the long persistence of their metabolites (urolithins) in the body.

On the other hand, the plasma concentration of urolithins reached the nanomolar range, although large inter-individual variability has been reported.[49,50,57] Similarly, large person-to-person variation in the timing, quantity and kind of urolithins excreted has also been described in urine[48,49] and may arise from differences in the composition of the colonic microbiota. Although the specific bacteria involved in the metabolic fate of ellagitannins are still unknown, Tomás-Barberán *et al.*[58] observed three metabotypes when analysing the results from different intervention studies on ellagitannins. Metabotype A produced only urolithin A metabolites and included between 25% and 80% of the volunteers in the different trials. Metabotype B produced urolithin B in addition to urolithin A and was the second relevant group (10–50%), while 5–25% of the volunteers analysed produced either no urolithins or amounts below the quantification limit and were categorised as metabotype 0. The three metabotypes were observed independently of the volunteers' health status and demographic characteristics (age, gender and body mass index) and of the amount or type of ellagitannin food source ingested (walnuts and other nuts, strawberries, raspberries and other berries or pomegranates).

11.4.3 Hydrolysable Phenolic Compounds

Hydrolysable phenolics comprise low-molecular-weight phenolic compounds, mainly hydroxycinnamic acids (FA, caffeic acid and synapic acid) and hydroxybenzoic acids, esterified with cell wall polysaccharides. They are abundant in cereals such as wheat and barley, where they are located principally in the bran of the grains, with FA representing 70–90% of the total phenolic acid contents.[59,60] More recently, hydrolysable phenolics have also been analysed in fruit and vegetables, and considerable contents have been reported in apples, bananas, onions, carrots, lettuce, chard, red beetroot and white cabbage.[6,11]

Although also present in fruit and vegetables, bioavailability studies on hydrolysable phenolics have mostly focused on cereals with known contents of the bound phenolics. Although the phenolic–fibre linkage in whole-grain products is thought to prevent almost all the phenolics from being absorbed in the upper intestine,[61,62] some authors have suggested that hydrolysable phenolics may be partly liberated from the food matrix and absorbed in the small intestine.[9] Rondini *et al.*[17] compared the bioavailability of FA from wheat aleurone and free FA in the rat animal model by analysing FA and its glucuronide and sulphate conjugates in plasma and urine up to 24 hours post-gavage. Urinary excretion of FA and its conjugates after bran consumption was 15-fold less than that after intake of the pure molecule, the later representing 43% of the intake dose against 2.6% found in the bran group. Metabolites appeared rapidly in the plasma of both groups, indicating that absorption of FA starts in the small intestine. However, in the bran-supplemented group, free and conjugated FA reached a plateau 1 hour after intake and remained approximately constant up to 24 hours, while it decreased to zero after 4.5 hours for the pure compound group. Sulphated FA was the main metabolite, representing 72% of the total, followed by glucuronidated FA with 19% of the plasmatic form and native FA with only 9%. These results demonstrate that free FA is rapidly absorbed, while FA esterified to fibre is slowly released in the colon after the action of microbial esterases. The lower recovery in urine might be mainly explained by two facts: firstly, the authors did not analyse such colonic catabolites of FA as dihydroferulic acid and hydroxyphenylpropionic acid that were reported for colonic *in vitro* fermentations;[62,63] and secondly, urinary excretion of FA and their conjugate metabolites might have continued for 24 hours post-gavage as indicated by the presence of these compounds in the plasma from the last sampling time point. A more recent study analysed the bioavailability of hydroxycinnamates in rats fed a durum wheat aleurone fraction with approximately 80% bound FA.[64] A total of 14 phenolic derivatives were identified in urine as potential FA metabolites. These comprised eight conjugates (FA sulphate and glucuronide, dihydroferulic acid sulphate, coumaric acid sulphate, phenylpropionic- and hydroxyphenylpropionic acid sulphate, hydroxyphenylpropionic acid glucuronide and vanillic acid sulphate) and six free phenolic acids (coumaric acid, three hydroxyphenylpropionic acid isomers and two dihydroxyphenylacetic acid isomers). The higher amount of sulphate conjugates compared with glucuronides indicates that the aleurone-associated phenolics interact principally with rat sulphotransferases rather than uridine-5′-diphosphate glucuronosyltransferases. A low bioavailability of approximately 1% was reported in this study in 24 hour rat urine and was in line with previous studies in rats and humans that reported bioavailabilities of FA ranging from 2.3% to 4% in 24 hour urine.[17,65,66] A recent bioavailability study with human volunteers fed whole-grain wheat and aleurone-enriched wheat bread analysed plasma up to 24 hours and urine up to 48 hours post-ingestion and reported bioavailability values up to 8%, showing that phenolic metabolites are still excreted

24–48 hours post-ingestion.[67] The urine metabolites were similar to those found in rats but included dihydroferulic acid glucuronide, feruloylglycine and hydroxybenzoic acid not reported by Calani *et al.*[64] In plasma samples, FA conjugates peaked early at 1–2 hours post-ingestion, while dihydroferulic acid conjugates reached their maximum concentrations at 7 hours. Taking into account the above-mentioned results, the bioavailability of hydrolysable phenolics starts in the upper gastrointestinal tract. However, substantial amounts of these compounds reach the colon esterified to fibre, where they are slowly released and degraded to phenylpropanoid, phenylacetic and benzoic acid derivatives, which then undergo phase II metabolism, principally sulphation, before being finally excreted in urine.

Summarising the results of the bioavailability of condensed and hydrolysable tannins and phenolic compounds, it can be concluded that NEPP are bioavailable after gastrointestinal digestion and subsequent degradation in the colon. However, standardised methodologies of the quantification of NEPP in foods are required for an accurate determination of their real intake. In addition to this, longer biological sampling times (plasma and urine) in bioavailability studies should be considered in order to fully elucidate the real absorption of these compounds.

11.5 Recent Advances in *In vitro* Fermentation Studies with NEPP-rich Foods

Ethical concerns, invasive techniques and economic implications are some issues that hamper the monitoring of the *in situ* evolution of NEPP through the digestive tube (from the mouth to the anus) in human beings. One of the best alternatives for understanding what occurs during the digestion process is the use of *in vitro* experiments known as gastrointestinal digestion and colonic fermentation *in vitro* models.

In vitro gastrointestinal digestion is a predictive model that provides information about the behaviour of food components during digestion in the upper gastrointestinal tract. It comprises individual steps that imitate the physiological conditions of the mouth, stomach and small intestine. In order to achieve maximum similarity, the incubations are carried out in sequence (oral then gastric and then intestinal incubation) at 37 °C, adding different buffers at specific pH values and digestive enzymes.[9,62,68–71]

The objective of exposing food components to the different steps of the *in vitro* gastrointestinal digestion is normally to know their apparent bioaccessibility and bioavailability in each digestive cavity. The bioaccessible fraction is defined as those compounds that are released from the food matrix and dissolved in the digestive juices. In the *in vitro* experiments, the bioaccessible components are recovered in the supernatants obtained after centrifugation of the digested samples.[9] The fact of being dissolved, however, does not guarantee that these compounds are digested (enzymatically modified) and/or absorbed (bioavailable). The characterisation of the media containing those compounds able to cross a semipermeable membrane during the intestinal step is a more accurate way to determine the digestible and/or bioaccessible fraction.[9,71–73]

On the other hand, bioavailability refers to the released compounds that are absorbed and susceptible to reaching peripheral organs and tissues, where they can be used for different purposes before being excreted. In the *in vitro* experiments, bioavailability can be tentatively determined by considering those compounds that pass from the basal to the apical compartment in intestinal cell cultures, especially differentiated Caco-2 cells.[74]

The data currently available reveal that the physiological conditions during gastrointestinal digestion (pH and digestive enzymes) are not enough to completely disrupt the association between the NEPP and food components.[9,62,69] Therefore, it can be considered that the bioaccessibility and, in consequence, the bioavailability of NEPP in the small intestine are very limited. Consequently, they reach the large intestine broadly unchanged or, in the case of tannins, partially hydrolysed.[70,75]

During the colon passage, the NEPP undergo the most relevant modifications.[9,62,70] Hence, studying the digestion in the large intestine is of central importance to elucidating the reactions involved in their metabolic fate. *In vitro* colonic digestion is usually called colonic fermentation, a denomination that highlights the crucial role of gut bacteria. The large intestine hosts the most abundant and varied microorganisms in the human body, and 'microbiota' is the term commonly used to describe this complex ecosystem where bacteria are predominant.[76] Thus, a critical point during *in vitro* colonic fermentation is to maintain a propitious environment for microbial growth and development.

In general, *in vitro* colonic fermentation consists of the anaerobic microbial incubation of a substrate. The microbial inoculum can be obtained from: (i) bacterial cultures;[77] (ii) a simulator of the human intestinal microbial ecosystem (SHIME);[77] or (iii) slurries prepared with a specific percentage of caecum content or faeces of animal or human origin.[9,68–72] The temperature during the procedure is maintained at 37 °C and fermented samples are taken at different times in order to obtain enough evidence to describe the modification of NEPP over time.[9,62,68,70–72]

The common features of all the modalities of *in vitro* colonic fermentation are usually the temperature, the microbial inoculum, the anaerobic conditions and the drawing of samples at different times. What varies is principally the degree of sophistication of the material and/or equipment used. *In vitro* colonic fermentations can be broadly classified into static and dynamic. Static incubations include those modalities in which the microbial inoculum and samples are incubated in hermetic recipients (tubes, vessels or anaerobic chambers) with continuous shaking/stirring.[69–73] In some cases, the vessels are connected to an automatic controlled pH pump to maintain steady pH values during the experiment.[71] Basically, the dynamic *in vitro* fermentations consist of the interconnection of different automatically computed controlled compartments, including the stomach, small intestine and the three sections of the large intestine. The inoculated microbiota differentiates according to the conditions of the vessels that represent the ascending, transverse and descending colon. The samples pass through the system, imitating, as far as possible, the gastrointestinal motility and physiological conditions of each segment of the digestive tract.[62,77]

Pure compounds,[72] foods (raw or cooked)[62,68,70,73] and phenolic-rich extracts[68,70,77] have been used as substrates for microbial fermentations. These samples have been directly incubated[68,72,78] or obtained after *in vitro* gastrointestinal digestion.[9,62,68–73]

The enzymatic cocktail secreted by the microbiota is able to hydrolyse non-extractable tannins and break down the interactions between NEPP and food components that human digestive enzymes cannot. This results in an increase in free phenolic compounds in the colon media, which are more easily accessible for the microbiota in order to be further biotransformed. The metabolites generated from bacterial fermentation differ according to the type of NEPP. However, several authors have observed a similar kind of hierarchy. This means that at earlier stages of incubation, the intermediate metabolites of high molecular weight with a similar chemical structure to their parent compounds are abundant. Over time, the metabolites of lower molecular weight start to increase, indicating continuous bacterial degradation.[62,70,72,75,77,79]

Unfortunately, except for some works,[20,62,73] most studies have ignored the fate of NEPP during *in vitro* fermentations, especially because they have not been described in the substrate before and throughout the time of fermentation. However, it is possible to predict the distinctive metabolic profile by collecting information from studies in which tannins and hydrolysable phenols were fermented.

The fermentation of individual or pure compounds offers advantages over the fermentation of complex matrices since it guarantees that the detected metabolites are exclusive to the substrate used,[72] whereas the fermentation of whole food or extracts generates a broad spectrum of metabolites that could originate from different phenolic compounds.[68,70,75,77] The rate of production of microbial metabolites seems to differ between the food NEPP and free compounds, apparently because the fibre to which NEPP are attached promotes bacteria activity.[20,68,69]

11.5.1 Colonic Fermentation of Polymeric Tannins

This classification includes the PA and hydrolysable tannins, mainly ellagitannins and gallotannins. Catechin and epicatechin are the structural units of PA in foods and, depending on their number, they can form oligomeric or polymeric structures. Non-extractable PA should be first detached form the cell wall material to be hydrolysed into catechin and epicatechin. These free monomers become more accessible for fermentation, from which phenolic acids have been described as main metabolites.

The microbial degradation sequence of catechin and epicatechin starts with their heterocyclic C-ring cleavage to form the first intermediate,

1-(3′,4′-dihydroxyphenyl)-3-(2″,4″,6″-trihydroxyphenyl)-propan-2-ol (diphenylpropanol). The breakdown of the A-ring of diphenylpropanol generates 5-(3′,4-dihydroxyphenyl)-γ-valerolactone, which is subsequently metabolised into phenolic acids.[20,72,77,78,80] Valeric, propionic, acetic, cinnamic and benzoic acids with different hydroxylation patterns are among the phenolic acids described as final microbial metabolites of flavanols.[20,68,72,77,78,80]

In general, the metabolic products recovered during the fermentation of different PA-rich products were the same. The differences among the results were mainly due to the production rate and/or the compound representing the major metabolite. For example, the fermentation of dealcoholised wine extract produced 4-hydroxy-5-3′-hydroxyphenylvaleric acid, di- and mono-hydroxylated phenylacetic acid and di- and mono-hydroxylated phenylpropionic acid as the main metabolites.[77] The fermentation of two varieties of apple products resulted in the production of di- and mono-hydroxyphenylpropionic acid as well as benzoic acids.[68]

Punicalagin, found in pomegranates, is probably the most extended ellagitannin studied during *in vitro* fermentation. Ellagitannins formed by different units of ellagic acid are degraded into their monomeric constituents during the initial period of fermentation to be finally converted into urolithins (dibenzopyran-6-one derivatives) with different degrees of hydroxylation.[70,79,81,82] Those urolithins with the highest degree of hydroxylation (up to five groups) are generated during the initial stages of fermentation, and then microbial dehydroxylation produces different spectra of hydroxylated urolithins.[70,79,81,82] The type and amount of urolithins recovered at the end of fermentation is defined by the microbiota profile of the donors.[70,79,81,82] Human beings can be associated with a phenotype depending on the major type of urolithin generated: phenotype A (predominant production of dihydroxylated urolithin A), phenotype B (predominant production of monohydroxylated urolithin B and/or dihydroxylated isourilithin A) and phenotype 0 (non-urolithin producers).[58,70,81] The form of microbial transformation is similar for all types of ellagitannins, whereas the difference is observed mainly in the production rate of metabolites.[79]

Gallotannins need to be hydrolysed to their gallic acid units, which are mainly converted into protocatechuic acid that is simultaneously degraded to catechol and *p*-hydroxybenzoic acid by the human microbiota.[75]

11.5.2 Colonic Fermentation of Hydrolysable Phenolic Compounds

This group includes the hydroxycinnamic acids esterified to cell wall components.[83] FA is probably the most widely studied hydroxycinnamic acid due to its abundance in plant products, especially grains.[15,62] In cereals, most of the FA is covalently bound to oligosaccharides and polysaccharides, and this association, almost unaltered during gastrointestinal digestion, is prevalently disrupted in the colon due to the activity of microbial esterases.[15,62,83] The amount and facility of FA release during colonic fermentation seems to depend on the saccharide chain length and solubility (soluble or insoluble fibre).[15,83] The penetration of microbial enzymes into the non-absorbable material is crucial to starting the fermentation process. However, the characteristics of the food matrix determines to a large extent the degree of catabolism.[15,62] It was also observed that the colon bioaccessibility of FA enhances after the bioprocessing of bran cereals (fermentation and enzymatic treatment).[62]

During *in vitro* colonic fermentation, hydroxycinnamic acids are metabolised into phenolic acids with different patterns of hydroxylation, normally in accordance with the initial hydroxylation pattern of the precursor.[71,73] To give an example, the microbial incubation of bread prepared with wheat bran led to the generation of phenylpropionic and hydroxyphenylpropionic acids as the main microbial metabolites derived from FA. Other minor compounds detected were dihydroxyphenylpropionic, phenylacetic and benzoic acids.[62]

In addition, *in vitro* fermentation also enables us to study whether the release and/or metabolism of NEPP influences in the generation of other beneficial colonic microbial metabolites such as short-chain fatty acids (SCFA). The recovery of SCFA during *in vitro* fermentation has been similar,[69] higher[71] and lower in the presence of NEPP when compared to controls. This contradiction observed among the results may reflect differences in the microbial fermentability of cell wall saccharides associated with the NEPP, as well as their antimicrobial activity. Both the release of NEPP and the production of microbial derivatives increase their bioaccessibility in the colon considerably compared with the small intestine. In addition, they could participate in anti-inflammatory, antioxidant and anticarcinogenic episodes, maintaining or improving the local health status. This, plus the apparent stimulation of beneficial SCFA production, converts NEPP into interesting candidates for the treatment or prevention of colonic pathologies. In addition, if absorbed by colonocytes, they can act as compounds of biological significance in peripheral organs and tissues.

To conclude, it has been described that NEPP can also modulate the microbiota population and, in consequence, its activity, promoting the growth and development of bacteria described as beneficial for human health.[84] The transformation of phenolic compounds during digestion and colonic fermentation has aroused much interest within the scientific area, reinforcing the studies focused on the health implications of the biological (poly)phenol metabolites.

References

1. D. Del Rio, A. Rodriguez-Mateos, J. P. E. Spencer, M. Tognolini, G. Borges and A. Crozier, *Antioxid. Redox Signaling*, 2013, **18**, 1818.
2. A. Rodriguez-Mateos, D. Vauzour, C. G. Krueger, D. Shanmuganayagam, J. Reed, L. Calani, P. Mena, D. Del Rio and A. Crozier, *Arch. Toxicol.*, 2014, **88**, 1803.
3. L. Bravo, R. Abia and F. Saura-Calixto, *J. Agric. Food Chem.*, 1994, **42**, 1481.
4. S. Arranz, F. Saura-Calixto, S. Shaha and P. A. Kroon, *J. Agric. Food Chem.*, 2009, **57**, 7298.

5. F. Saura-Calixto, *J. Agric. Food Chem.*, 2012, **60**, 11195.
6. J. Pérez-Jiménez, M. E. Díaz-Rubio and F. Saura-Calixto, *Nutr. Res. Rev.*, 2013, **26**, 118.
7. C. Manach, G. Williamson, C. Morand, A. Scalbert and C. Rémésy, *Am. J. Clin. Nutr.*, 2005, **81**, 230S.
8. S. Arranz, J. M. Silván and F. Saura-Calixto, *Mol. Nutr. Food Res.*, 2010, **54**, 1646.
9. F. Saura-Calixto, J. Serrano and I. Goñi, *Food Chem.*, 2007, **101**, 492.
10. M. L. Ovaskainen, R. Törrönen, J. M. Koponen, H. Sinkko, J. Hellström, H. Reinivuo and P. Mattila, *J. Nutr.*, 2008, **138**, 562.
11. J. Pérez-Jiménez and F. Saura-Calixto, *Food Res. Int.*, 2015, **74**, 315.
12. M. N. Clifford, *Planta Med.*, 2004, **70**, 1103.
13. S. Touriño, J. Pérez-Jiménez, M. L. Mateos-Martín, E. Fuguet, M. P. Vinardell, M. Cascante and J. L. Torres, *J. Agric. Food Chem.*, 2011, **59**, 5955.
14. M. L. Mateos-Martín, J. Pérez-Jiménez, E. Fuguet and J. L. Torres, *Br. J. Nutr.*, 2012, **108**, 290.
15. P. A. Kroon, C. B. Faulds, P. Ryden, J. A. Robertson and G. Williamson, *J. Agric. Food Chem.*, 1997, **45**, 661.
16. A. Crozier, I. B. Jaganath and M. N. Clifford, *Nat. Prod. Rep.*, 2009, **26**, 1001.
17. L. Rondini, M. N. Peyrat-Maillard, A. Marsset-Baglieri, G. Fromentin, P. Durand, D. Tomé, M. Prost and C. Berset, *J. Agric. Food Chem.*, 2004, **52**, 4338.
18. O. Aprikian, V. Duclos, S. Guyot, C. Besson, C. Manach, A. Bernalier, C. Morand, C. Rémésy and C. Demigné, *J. Nutr.*, 2003, **133**, 1860.
19. J. Juskiewicz, J. Milala, A. Jurgonski, B. Krol and Z. Zdunczyk, *Nutrition*, 2011, **27**, 351.
20. F. Saura-Calixto, J. Pérez-Jiménez, S. Touriño, J. Serrano, E. Fuguet, J. L. Torres and I. Goñi, *Mol. Nutr. Food Res.*, 2010, **54**, 939.
21. L. Gu, M. A. Kelm, J. F. Hammerstone, G. Beecher, D. Holden, D. Haytowitz, S. Gebhardt and R. L. Prior, *J. Nutr.*, 2004, **134**, 613.
22. G. Williamson and M. N. Clifford, *Biochem. Pharmacol.*, 2017, **139**, 24.
23. J. P. E. Spencer, F. Chaidry, A. S. Pannala, S. K. Srai, E. Debnam and C. Rice-Evans, *Biochem. Biophys. Res. Commun.*, 2000, **272**, 236.
24. L. Y. Rios, R. N. Bennett, S. A. Lazarus, C. Rémésy, A. Scalbert and G. Williamson, *Am. J. Clin. Nutr.*, 2002, **76**, 1106.
25. C. Tsang, C. Auger, W. Mullen, A. Bornet, J. M. Rouanet, A. Crozier and P. L. Teissedre, *Br. J. Nutr.*, 2005, **94**, 170.
26. M. Monagas, M. Urpi-Sarda, F. Sánchez-Patán, R. Llorach, I. Garrido, C. Gómez-Cordovés, C. Andrés-Lacueva and B. Bartolomé, *Food Funct.*, 2010, **1**, 233.
27. S. Stoupi, G. Williamson, J. W. Drynan, D. Barron and M. N. Clifford, *Arch. Biochem. Biophys.*, 2010, **501**, 73.
28. M. M. Appeldorn, J. P. Vincken, A. M. Aura, P. C. H. Hollman and H. Gruppen, *J. Agric. Food Chem.*, 2009, **57**, 1084.
29. R. R. Holt, S. A. Lazarus, M. C. Sullards, D. D. Zhu, J. F. Hammerstone, C. G. Fraga, H. H. Schmitz and C. L. Keen, *Am. J. Clin. Nutr.*, 2002, **76**, 798.

30. K. Bittner, T. Kemme, K. Peters, S. Kersten, S. Dänicke and H. U. Humpf, *Mol. Nutr. Food Res.*, 2014, **58**, 2261.
31. E. Arrigoni, P. Balzer and R. Amado, in *Polyphenols in Food*, ed. R. Amadò, H. Andersson, S. Bardocz and F. Serra, Office for Official Publications of the European Communities, Luxembourg, 1998, Influence of polyphenols on the fermentability of polysaccharides, pp. 187–192.
32. N. Martin-Carron, *J. Agric. Food Chem.*, 1998, **46**, 1064.
33. K. M. Goodrich and A. P. Neilson, *J. Chromatogr. B: Anal. Technol. Biomed. Life Sci.*, 2014, **958**, 63.
34. K. M. Goodrich, A. T. Smithson, A. K. Ickes and A. P. Neilson, *J. Nutr. Biochem.*, 2015, **26**, 1007.
35. M. Urpi-Sarda, M. Monagas, N. Khan, R. Llorach, R. M. Lamuela-Raventós, O. Jáuregui, R. Estruch, M. Izquierdo-Pulido and C. Andrés-Lacueva, *J. Chromatogr. A*, 2009, **1216**, 7258.
36. R. Llorach, I. Garrido, M. Monagas, M. Urpi-Sarda, S. Tulipani, B. Bartolomé and C. Andrés-Lacueva, *J. Proteome Res.*, 2010, **9**, 5859.
37. M. Urpi-Sarda, I. Garrido, M. Monagas, C. Gómez-Cordovés, A. Medina-Remón, C. Andrés-Lacueva and B. Bartolomé, *J. Agric. Food Chem.*, 2009, **57**, 10134.
38. S. Touriño, E. Fuguet, M. P. Vinardell, M. Cascante and J. L. Torres, *J. Agric. Food Chem.*, 2009, **57**, 11418.
39. P. Mena, L. Calani, R. Bruni and D. Del Rio, in *Diet-microbe Interactions in the Gut*, ed. K. Tuohy and D. Del Rio, Academic Press, Cambridge, 2015, Bioactivity of high-molecular-weight polyphenols by gut microbiome, pp. 73–101.
40. A. Smeriglio, D. Barreca, E. Belloco and D. Trombetta, *Br. J. Pharmacol.*, 2017, **174**, 1244.
41. L. Li, A. A. Shaik, J. Zhang, K. Nhkata, Y. Zhang, C. Xing, S. H. Kim and J. Lü, *J. Pharm. Biomed. Anal.*, 2011, **54**, 545.
42. K. Cai, A. E. Hagerman, R. E. Minto and A. Bennick, *Biochem. Pharmacol.*, 2006, **71**, 1570.
43. S. Shahrzad and I. Bitsch, *J. Chromatogr. B: Biomed. Sci. Appl.*, 1998, **705**, 87.
44. S. Shahrzad, K. Aoyagi, A. Winter, A. Koyama and I. Bitsch, *J. Nutr.*, 2001, **131**, 1207.
45. G. Pereira-Caro, J. M. Moreno-Rojas, N. Brindani, D. Del Rio, M. E. J. Lean, Y. Hara and A. Crozier, *J. Agric. Food Chem.*, 2017, **65**, 5365.
46. T. Yoshida, Y. Amakura and M. Yoshimura, *Int. J. Mol. Sci.*, 2010, **11**, 79.
47. N. P. Seeram, R. Lee and D. Heber, *Clin. Chim. Acta*, 2004, **348**, 63.
48. R. González-Barrio, G. Borges, W. Mullen and A. Crozier, *J. Agric. Food Chem.*, 2010, **58**, 3933.
49. I. A. Ludwig, P. Mena, L. Calani, G. Borges, G. Pereira-Caro, L. Bresciani, D. Del Rio, M. E. J. Lean and A. Crozier, *Free Radical Biol. Med.*, 2015, **89**, 758.
50. B. Cerdá, F. A. Tomás-Barberán and J. C. Espín, *J. Agric. Food Chem.*, 2005, **53**, 227.

51. A. González-Sarrías, J. A. Giménez-Bastida, M. T. García-Conesa, M. B. Gómez-Sánchez, N. V. García-Talavera, A. Gil-Izquierdo, C. Sáncjez-Alvarez, L. O. Fontana-Compiano, J. P. Morga-Egea, F. A. Pastor-Quirante, F. Martínez-Díaz, F. A. Tomás-Barberan and J. C. Espín, *Mol. Nutr. Food Res.*, 2010, **54**, 311.
52. P. Truchado, M. Larrosa, M. T. García-Conesa, B. Cerdá, M. L. Vidal-Guevara, F. A. Tomás-Barberan and J. C. Espín, *J. Agric. Food Chem.*, 2012, **60**, 5749.
53. S. Tulipani, M. Urpi-Sarda, R. García-Villalba, M. Rabassa, P. López-Uriarte, M. Bulló, O. Jáuregui, F. A. Tomás-Barberan, J. Salas-Salvadó, J. C. Espín and C. Lacueva, *J. Agric. Food Chem.*, 2012, **60**, 8930.
54. M. A. Núñez-Sánchez, R. García-Villalba, T. Monedero-Saiz, N. V. García-Talavera, M. B. Gómez-Sánchez, C. Sánchez-Álvarez, A. M. García-Albert, F. J. Rodríguez-Gil, M. Ruiz-Marín, F. A. Pastor-Quirante, F. Martínez-Díaz, M. J. Yánez-Gascón, A. González-Sarrías, F. A. Tomás-Barberán and J. C. Espín, *Mol. Nutr. Food Res.*, 2014, **58**, 1199.
55. N. P. Seeram, S. M. Henning, Y. Zhang, M. Suchard, Z. Li and D. Heber, *J. Nutr.*, 2006, **136**, 2481.
56. J. C. Espín, R. González-Barrio, B. Cerdá, C. López-Bote, A. I. Rey and F. A. Tomás- Barberán, *J. Agric. Food Chem.*, 2007, **55**, 10476.
57. M. Larrosa, M. T. García-Conesa, J. C. Espín and F. A. Tomás-Barberán, *Mol. Aspects Med.*, 2010, **31**, 513.
58. F. A. Tomás-Barberán, R. García-Villalba, A. González-Sarrías, M. V. Selma and J. C. Espín, *J. Agric. Food Chem.*, 2014, **62**, 6535.
59. A. K. Holtekjølen, C. Kinitz and S. H. Knutsen, *J. Agric. Food Chem.*, 2006, **54**, 2253.
60. L. Li, P. R. Shewry and J. L. Ward, *J. Agric. Food Chem.*, 2008, **56**, 9732.
61. C. Manach, A. Scalbert, C. Morand, C. Rémésy and L. Jiménez, *Am. J. Clin. Nutr.*, 2004, **79**, 727.
62. N. M. Anson, E. Selinheimo, R. Havenaar, A. M. Aura, I. Mattila, P. Lehtinen, A. Bast, K. Poutanen and G. R. M. M. Haenen, *J. Agric. Food Chem.*, 2009, **57**, 6148.
63. I. A. Ludwig, M. P. De Peña, C. Cid and A. Crozier, *BioFactors*, 2013, **39**, 326.
64. L. Calani, F. Ounnas, P. Salen, C. Demeilliers, L. Bresciani, F. Scazzina, F. Brighenti, C. Melegari, A. Crozier, M. De Lorgeril and D. Del Rio, *Food Funct.*, 2014, **5**, 1738.
65. A. Adam, V. Crespy, M. A. Levrat-Verny, F. Leenhardt, M. Leuillet, C. Demiqné and C. Rémésy, *J. Nutr.*, 2002, **132**, 1962.
66. S. M. Kern, R. N. Bennett, F. A. Mellon, P. A. Kroon and M. T. Garcia-Conesa, *J. Agric. Food Chem.*, 2003, **51**, 6050.
67. L. Bresciani, F. Scazzina, R. Leonardi, E. Dall'Aglio, M. Newell, M. Dall'Asta, C. Melegari, S. Ray, F. Brighenti and D. Del Rio, *Mol. Nutr. Food Res.*, 2016, **60**, 2343.
68. S. Bazzocco, I. Mattila, S. Guyot, C. M. G. C. Renard and A. M. Aura, *Eur. J. Nutr.*, 2008, **47**, 442.

69. L. Condezo-Hoyos, I. P. Mohanty and G. D. Noratto, *Food Chem.*, 2014, **161**, 208.
70. J. I. Mosele, A. Macià, M. P. Romero, M. J. Motilva and L. Rubió, *J. Funct. Foods*, 2015, **14**, 529.
71. A. Koutsos, M. Lima, L. Conterno, M. Gasperotti, M. Bianchi, F. Fava, U. Vrhovsek, J. A. Lovegrove and K. M. Tuohy, *Nutrients*, 2017, **9**, 533.
72. A. Serra, A. Macià, M. P. Romero, N. Anglés, J. R. Morelló and M. J. Motilva, *Food Chem.*, 2011, **126**, 1127.
73. I. Juániz, I. A. Ludwig, L. Bresciani, M. Dall'Asta, P. Mena, D. Del Rio, C. Cid and M. P. de Peña, *J. Funct. Foods*, 2017, **32**, 195.
74. L. Rubió, A. Macià, A. Castell-Auví, M. Pinent, M. T. Blay, A. Ardévol, M. P. Romero and M. J. Motilva, *Food Chem.*, 2014, **149**, 277–284.
75. J. I. Mosele, A. Macià, M. P. Romero and M. J. Motilva, *Food Chem.*, 2016, **201**, 120.
76. J. van Duynhoven, E. E. Vaughan, D. M. Jacobs, R. A. Kemperman, E. J. van Velzen, G. Gross, L. C. Roger, S. Possemiers, A. K. Smilde, J. Dore, J. A. Westerhuis and T. Van de Wiele, *Proc. Natl. Acad. Sci. U. S. A.*, 2011, **108**, 4531.
77. E. Barroso, F. Sánchez-Patán, P. J. Martín-Alvarez, B. Bartolomé, M. V. Moreno-Arribas, C. Peláez, T. Requena, T. Van De Wiele and M. C. Martínez-Cuesta, *J. Agric. Food Chem.*, 2013, **61**, 10163.
78. F. Sanchez-Patán, E. Barroso, T. Van De Wiele, A. Jiménez-Girón, P. J. Martín-Alvarez, M. C. Martínez-Cuesta, C. Peláez, T. Requena and B. Bartolomé, *Food Chem.*, 2015, **183**, 273.
79. D. N. Olennikov, N. I. Kashchenko and N. K. Chirikova, *Nutrients*, 2015, **7**, 8456.
80. D. Y. Low, M. P. Hodson, B. A. Williams, B. R. D'Arcy and M. J. Gidley, *Food Chem.*, 2016, **207**, 214.
81. P. Gaya, M. Medina, A. Sánchez-Jiménez and J. M. Landete, *Molecules*, 2016, **21**, 1034.
82. R. González-Barrio, C. A. Edwards and A. Crozier, *Drug Metab. Dispos.*, 2011, **39**, 1680.
83. J. Oua and Z. Sun, *J. Funct. Foods*, 2014, 7, 90.
84. S. M. Henning, P. H. Summanen, R. P. Lee, J. Yang, S. M. Finegold, D. Heber and Z. Li, *Anaerobe*, 2017, **43**, 56.

CHAPTER 12

Non-extractable Polyphenols and the Gut Microbiome

J. HAVLIK[a] AND C. A. EDWARDS*[b]

[a]Czech University of Life Sciences Prague, Department of Quality of Agricultural Products, Kamycka 129, 16500 Prague, Czech Republic;
[b]University of Glasgow, Human Nutrition, New Lister Building, Alexandra Parade, G312ER Glasgow, United Kingdom
*E-mail: Christine.Edwards@glasgow.ac.uk

12.1 Introduction

The potential role of the gut microbiome in promoting health has become increasingly evident, but these bacteria may also play a key role in the development of disease. Since the advent of next-generation sequencing techniques, non-targeted metabolomics and associated complex bioinformatics, the microbiome has been associated with obesity, metabolic and cardiovascular disease, autoimmune diseases, brain function and many other gut and systemic diseases.[1–5] This has resulted in an explosion of research investigating the gut microbiome and its impact on the human metabolome under different conditions and disease burdens. Similarly, polyphenol-rich foods have been tested for their effects on cardiovascular risk factors, including inflammation, and for their impact on cognitive function and decline. There are many different polyphenols in the human diet, but only few are well absorbed in the upper gastrointestinal tract. The rest reach the colon directly

Food Chemistry, Function and Analysis No. 5
Non-extractable Polyphenols and Carotenoids: Importance in Human Nutrition and Health
Edited by Fulgencio Saura-Calixto and Jara Pérez-Jiménez

Published by the Royal Society of Chemistry, www.rsc.org

or *via* excretion in bile, where they are subject to catabolism by the microbiota. The products – phenolic acids or other stable metabolic intermediates such as γ-valerolactones or urolithins – may be the main effectors of polyphenol actions in the body. For example, the phenolic acid 3,4-dihydroxyphenylacetic acid (3,4DHPAA) has shown greater effects *in vitro* in terms of antioxidant, anti-inflammatory and antiplatelet aggregation than its parent compounds rutin or quercetin.[6,7]

One major mechanism by which the microbiome can influence health is by modulation of the release of bioactive molecules such as short-chain fatty acids (SCFA) and phenolic compounds from foods rich in dietary fibre and polyphenols. In turn, the gut microbiota likely plays a crucial role in the potential health benefits of the polyphenols.[8] There is also a growing evidence base indicating that polyphenols shape the diversity and abundance of the bacteria in the human digestive tract.[9–13]

The exact mechanisms of these relationships are not understood and are not fully explored despite much recent investigation. The combination of plant-derived foods and drinks in our diet make up complex mixtures of polyphenols that have multiple additive or synergistic activities among their components. There is great variability in the response of individuals to polyphenols,[6,14,15] with differences in the metabolism of each polyphenol amongst responders and non-responders in each study group, which may be attributed to differences in the gut microbiota composition, perhaps related to different enterotypes and metabotypes.[10,15,16]

One issue with the association of polyphenols with disease is that many have low bioavailability as a parent compound, and this is especially true for the non-extractable polyphenols (NEPP), which are more closely associated with non-digestible plant cellular components and may not be released (if at all) until after the actions of bacteria in the large intestine. It is feasible to hypothesise that some of the impact of the gut microbiome on the human body could be enacted *via* the impact of polyphenols on the gut bacteria (pre- and pro-biotic). In addition, the metabolism of polyphenols by gut bacteria, along with the matrix in which the polyphenols are situated, may release more bioactive metabolites such as phenolic acids.

Recently, it has been shown that polyphenol content in foods has been largely underestimated, as common solvent extraction methods leave a large fraction of polyphenols bound in the extraction residue.[17] In addition, large-molecular-weight polymers such as tannins are difficult to analyse with common sample preparation and analytical techniques. This fraction that commonly escapes analysis has been termed macromolecular antioxidants (MACAN) and comprise non-extractable proanthocyanidins and hydrolysable polyphenols (HPP); both have been shown to release low-molecular-weight bioactive compounds in the human digestive tract.[17] After liberation from the matrix or depolymerisation in the human gut, they are usually hydrolysed through similar pathways to polyphenols extracted by standard methods (extractable polyphenols [EPP]), leading to similar metabolites.

This chapter will explore the interrelationships between the polyphenols (extractable and non-extractable) and the gut microbiota and how these could affect the production of bioactives and ultimately the function of the human body.

12.2 Colonic Microbial Catabolism of Polyphenols

12.2.1 Early Evidence of Microbial Catabolism of Polyphenols

The metabolism of polyphenols in the human digestive tract to metabolites such as phenolic acids was discovered 60 years ago, when 3,4DHPAA and 3-hydroxyphenylacetic acid (3HPAA) were found as the main quercetin metabolites in animals.[18] Elucidation of the main (+)-catechin and (−)-epicatechin colonic metabolites followed 5 years later.[19,20] This triggered interest in the microbial pathways and microbiota responsible for the catabolism of polyphenols,[21] and the role of individual bacterial strains in the degradation of aromatics has now been well established.

Early research on ruminant fodder resulted in an understanding of the metabolism of NEPP. Polyphenols were of special interest as they significantly affect forage digestibility, microbial protein synthesis and the production of SCFA, key components of a ruminant's nutritional needs. These studies provided early evidence of NEPP, but the phenolic acids in the urine of ruminants is much higher than in other animals and the concentrations drop when the animals shift their diet from roughage to concentrated feeds and was more evident for late pastures. This effect seemed to be independent of cellulose, crude fibre and protein intake and was explained by the content of ferulic and *p*-coumaric acid, as well as that of 'pro-lignins' or fermentable lignins from young grasses.[22]

12.2.2 Bacterial Catabolism of Known Extractable Polyphenols

Polyphenols present in plants are mostly conjugated with sugars and organic acids such as quinic acid, hydrolysable tannins or as a structural part of the cell wall. They may also occur as non-conjugated oligomers as in the case of condensed tannins or proanthocyanidins. Intestinal absorption of conjugates or polymers is low. After escaping small intestinal absorption, they reach the colon and are transformed by bacteria to lower-molecular-weight compounds. Colonic microbial transformations include the following catabolic processes: *hydrolysis* (*O*-deglycosylation and ester hydrolysis), *cleavage* (C-ring cleavage; delactonisation, demethylation), and *further reactions* (Figure 12.1).[10,23]

Hydrolysis is often the first step, in which aglycones of the phenolic acid esters, flavanol monomers or aglycone are released. Much knowledge has been accumulated on how these liberated monomers or oligomers are fermented in the colon, and *in vitro* fermentations provide a clear picture of the metabolites that may dominate in human or animal caeca.[6,24–26]

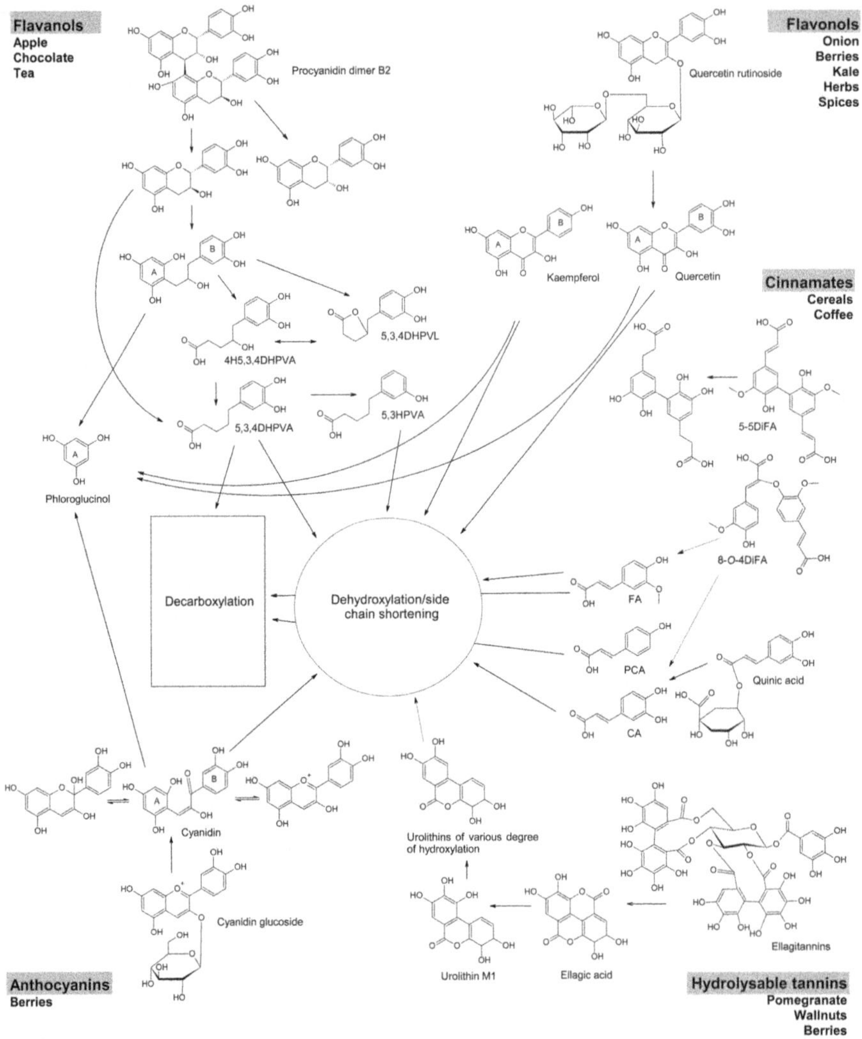

Figure 12.1 Colon catabolic pathways of common dietary polyphenols and aromatic amino acids – parent compounds and precursors. Solid lines represent common pathways, dotted lines represent hypothesised or minor pathways. 4H5,3,4DHPVA, 4-hydroxy- 5-(3,4-dihydroxyphenyl)-valeric acid; 5,3,4DHPVA: 5-(3,4-dihydroxyphenyl)-valeric acid; 5,3,4DHPVL: 5-(3,4-dihydroxyphenyl)-γ-valerolactone; 5,3HPVA: 5-(3-hydroxyphenyl)-valeric acid; 5-5DiFA: 5-5-ferulic acid dimer; 8-*O*-4DiFA: 8-*O*-4-ferulic acid dimer; CA: cinnamic acid; FA: ferulic acid; GA: gallic acid; PCA: *p*-coumaric acid.

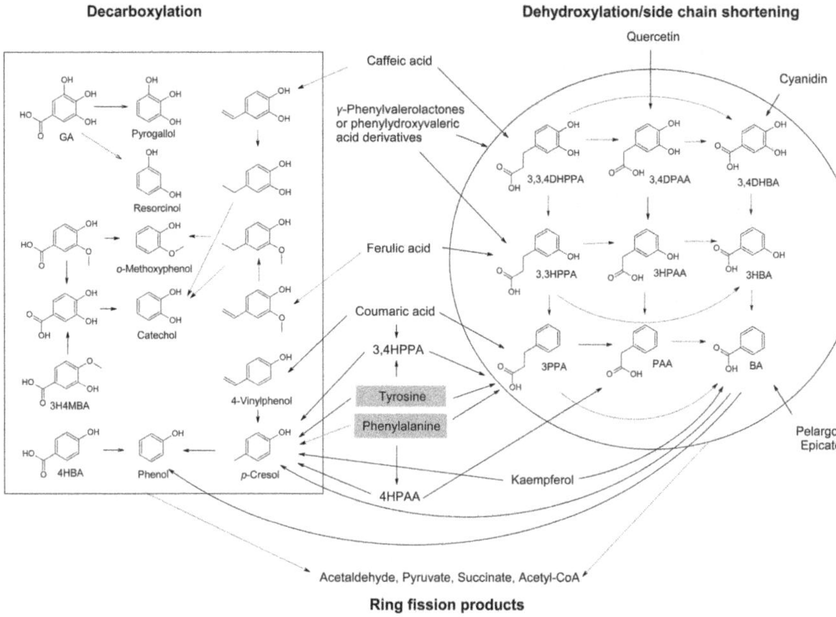

Figure 12.2 Details on the dehydroxylation, decarboxylation and side chain shortening reactions mentioned in Figure 12.1, leading to the most abundant low-molecular-weight metabolites in the colon. Solid lines represent common pathways, dotted lines represent hypothesized or minor pathways. 3,3,4DHPPA: 3-(3,4-dihydroxy)-phenylpropionic acid; 3,3HPPA: 3-(3-hydroxy)-phenylpropionic acid; 3,4DHBA: 3,4-dihydroxybenzoic acid; 3,4DHPAA: 3,4-dihydroxyphenylacetic acid; 3,4HPPA: 3-(4-hydroxy)-phenylpropionic acid; 3HBA: 3-hydroxybenzoic acid; 3HMBA: 3-hydroxy-4-methoxybenzoic acid; 3HPAA: 3-hydroxyphenylacetic acid; 3PPA: 3-phenylpropionic acid; 4HBA: 4-hydroxybenzoic acid; 4HPAA: 4-hydroxyphenylacetic acid; BA: benzoic acid; PAA: phenylacetic acid.

Further metabolism consists of various reactions or cleavages of functional groups (dehydroxylation, demethylation, decarboxylation and lactone formation)[27,28] and reduction of double bonds,[6,29] leading to aromatic products with various degrees of substitution (Figure 12.2).

Type of functional groups, degree of substitution of the aromatic ring and positions of the substituents further determine reaction the stability of the metabolite.[30,31] Some of the metabolites are stable, persisting for long periods of time in faecal fluid, and thus contributing to a relatively conserved pool of aromatic compounds in the lumen, subject to absorption.[32] Further decarboxylation may lead to products such as 4-ethylphenol or 4-vinylphenol, resorcinol or pyrogallol, and subsequent ring fission gives pyruvate, acetaldehyde, succinate and acetyl coenzyme A.[33,34] The final product of colonic polyphenol catabolism is thus CO_2, as shown in ^{14}C isotope studies. Details of the catabolic degradation pathways of the major polyphenols have been published in several reviews[10,27,35] and are shown in Figures 12.1 and 12.2. An overview of the major metabolites originating from catabolism of dietary polyphenols found in batch culture fermentation using human or animal stool is shown in Table 12.1.

12.2.3 Bacterial Catabolism of Non-extractable Polyphenols

12.2.3.1 Hydrolysable Cinnamates in the Cell Wall

Ferulic and *p*-coumaric acids are highly ubiquitous polyphenols in all vegetal matter. They are major non-core lignin monomers, usually esterified with arabinose. In plant cell walls, these components are found forming peroxidase-catalysed dimers (diferulates) or even higher-degree polymers and cross-linked cell wall hemicellulose with core lignin. Cross-linking of cell wall components with esters of cinnamic acid derivatives may have marked influence on numerous wall properties such as accessibility, extensibility, plasticity and, in turn, digestibility.[36]

A diverse range of diferulates have been detected in various plant materials, the most abundant characterised as 5-5, 8-*O*-4, 8-5, 8-8 and 4-*O*-5 coupled dehydrodimers.[37] To a much lesser extent, derivatives such as sinapic, cinnamic and *p*-hydroxybenzoic acids may also be found in structures of higher degrees of polymerisation.[38]

Cinnamate content has been largely underestimated, although it significantly impacts forage utilisation or may significantly contribute to polyphenol intake. Ferulic or *p*-coumaric acids and their derivatives have very low extractability with conventional methods (*e.g.* from wheat bran, 2–10 $\mu g\ g^{-1}$ was extractable with a conventional method, whereas up to 5786 $\mu g\ g^{-1}$ was extracted after acidic or alkaline hydrolysis).[39,40] Under the new classification of macromolecular polyphenols proposed recently,[17,41] ferulates make up a large part of the HPP fraction of MACAN.

Ferulic acid as a monomer can be cleaved from ester bonds to saccharides in incubations with human stool and is quickly catabolised. Approximately 200 μM of phenolic acids was formed from 50 mg of wheat bran, 115 μM of which was 3HPPA, catabolised to 167 μM 3HPAA at a later fermentation time point.[40] Even for dimers, ester bonds between ferulates and polysaccharides can be partially cleaved by the intestinal esterases of both humans and rats, passing from the small intestine, and already here, free diferulates enter the circulation.[42] Moreover, they can also be released from cell walls by bacterial hydrolytic activity in the caecum.[42] Incubations of 8-*O*-4-DiFA and 5-5-DiFA by the human intestinal microbiota showed that 8-*O*-4-DiFA was catabolised to 3-(3,4-dihydroxyphenyl)propionic acid (3,3,4DHPPA), homovanillic acid, 3,4DHPAA and later to 3-(3-hydroxyphenyl)propionic acid (Table 12.1).[43] In contrast, 5-5-DiFA is more stable. The dimer was not cleaved by human intestinal bacteria, only the side-chain double bonds were reduced and the methoxy groups were partially *O*-demethylated, forming caffeic acid-5-5-caffeic acid and caffeic acid-5-5-dihydrocaffeic acid with transient intermediates[43] (Figure 12.1).

Table 12.1 Metabolites of common phenolic compounds mostly found in *ex vivo* fermentation studies.

(Poly)phenol	Metabolites	Model	Species/ faecal donor	Ref.
Flavonols and flavanones				
Quercetin	3,4-Dihydroxyphenylacetic acid, 3-hydroxyphenylacetic acid	*In vitro* incubation	Human faeces	26
	3,4-Dihydroxyphenylacetic acid, 3-hydroxyphenylacetic acid, phenylacetic acid, 3,4-dihydroxybenzoic acid	*In vitro* incubation	Rat faeces	125
	3,4-Dihydroxyphenylacetic acid, 3-hydroxyphenylacetic acid, 3-(3-hydroxyphenyl)propionic acid (traces) Some subjects preferably forming 3,4-dihydroxybenzoic acid as the main metabolite	*In vitro* incubation	Human faeces	6
Kaempferol	Phenylacetic acid, benzoic acid	*In vitro* incubation	Rat faeces	125
Myricetin	3,5-Dihydroxyphenylacetic acid, 3,4,5-trihydroxyphenyl-acetic acid, 3-hydroxyphenylacetic (trace)	*In vivo*	Rat faeces	126
Naringenin	3-(4-Hydroxyphenyl)propionic acid, 3-phenylpropionic acid	*In vitro* incubation	Human faeces	35
Flavanols				
(−)-Epicatechin	3-Phenylpropionic acid, 3-hydroxyphenylpropionic acid, 3-(3,4-dihydroxyphenyl)propionic acid 5-(3-hydroxyphenyl)valeric acid, 5-(3,4-fihydroxyphenyl) valeric acid; valerolactones not tested	*In vitro* incubation	Human faeces	64
(+)-Catechin	3-Phenylpropionic acid, 3-hydroxyphenylpropionic acid, 3-(3,4-dihydroxyphenyl)propionic acid, 5-(3,4-dihy-droxy phenyl)valeric acid; valerolactones not tested	*In vitro* incubation	Human faeces	64
(−)-Epicatechin and (+)-catechin	5-(3′,4′-Dihydroxyphenyl)-γ-valerolactone, 5-phenyl-γ-valerolactone, 3-phenylpropionic acid	pH-controlled, stirred human colon model	Human faeces	127
Hydrolysable tannins				
Gallic acid and gallotannins	Pyrogallol, resorcinol	*In vitro* ruminal fermentation	Cow (ruminal fluid)	76

(*continued*)

Table 12.1 (*continued*)

(Poly)phenol	Metabolites	Model	Species/ faecal donor	Ref.
	Pyrogallol, resorcinol	*In vivo*	Rats	77
Ellagic acid, ellagitannins	Urolithins	*In vitro* fermentation	Human faeces	70
Non-hydrolysable tannins				
Procyanidin dimers	3,4-Dihydroxyphenylacetic acid, 5-(3,4-dihydroxyphenyl)-γ-valerolactone, 3-hydroxyphenylacetic acid, 4-hydroxyphenylacetic acid, 3-hydroxyphenylpropionic acid, phenylvaleric acids, monohydroxylated phenylvalerolactone, 1-(3′,4′-dihydroxyphenyl)-3-(2″,4″,6″-trihydroxyphenyl)propan-2-ol	*In vitro* fermentation	Human faeces	25
Anthocyanins				
Cyanidin	3,4-Dihydroxybenzoic acid	*In vitro* fermentation	Human faeces	24,128
Peonidin	3-Methoxy-4-hydroxybenzoic acid			128
Pelargonidin	4-Hydroxybenzoic acid			128
Malvidin	3,4-Dimethoxybenzoic acid			128
Stilbenoids				
Trans-resveratrol	Dihydroresveratrol, 3,4- dihydroxy-*trans*-stilbene, 3,4-dihydroxybibenzyl (lunularin)	*In vitro* fermentation	Human faeces	29
Cinnamates				
Caffeic acid, chlorogenic acid, caffeoyltartaric acid	3-(3-Hydroxyphenyl)propionic acid, benzoic acid	*In vitro* fermentation	Human faeces	129

Caffeic acid	3-(3,4-Dihydroxyphenyl)acetic acid, 3-(3-hydroxyphenyl) propionic acid, 4-vinylphenol, 4-ethylcatechol	*In vitro* fermentation	Rat caeca	130
3-(3,4-Dihydroxyphenyl) propionic acid	3-(3-Hydroxyphenyl)propionic acid			
3-(2-Hydroxyphenyl)propionic acid	–			
3-Hydroxy-4-methoxycinnamic acid	3-Methoxy-4-hydroxyphenylacetic acid, 3-(3-hydroxyphenyl)propionic acid			
3-Methoxy-4-hydroxycinnamic acid	3-(3-Methoxy-4-hydroxy)phenylpropionic acid, 3-(3-hydroxyphenyl)propionic acid, 3-(3,4-dihydroxyphenyl)propionic acid, 4-ethylcatechol, 2-methoxy-4-vinylphenol			
3-(3-Methoxy-4-hydroxy)propionic acid	3-(3-Hydroxyphenyl)propionic acid			
3-(3-Hydroxy,4-methoxyphenyl) propionic acid	3-(3-Hydroxyphenyl)propionic acid			
3-Hydroxycinnamic acid	3-(3-Hydroxyphenyl)propionic acid			
4-Hydroxycinnamic acid	3-(4-Hydroxyphenyl)propionic acid, 4-ethenylphenol, 4-ethylphenol			
3,4-Dihydroxyphenylacetic acid	3-Hydroxyphenylacetic acid, 4-methylcatechol			
2,5-Dihydroxyphenylacetic acid	–			
3-Methoxy-4-hydroxyphenylacetic acid	3,4-Dihydroxyphenylacetic acid, 3-hydroxyphenylacetic acid, 4-methylcatechol			
2-Hydroxyphenylacetic acid	–			
3-Hydroxyphenylacetic acid	–			
4-Hydroxyphenylacetic acid	*p*-Cresol			
5-5-Dehydrodiferulic acid	Caffeic acid-5-5-caffeic acid	*In vitro* fermentation	Human faeces	43
8-*O*-4-Dehydrodiferulic acid	Homovanillic acid, 3-(3,4-dihydroxyphenyl)propionic acid, 3,4-dihydroxyphenylacetic acid	*In vitro* fermentation	Human faeces	43
Non-phenolic cyclohexanes				
Quinic acid, shikimic acid	Cyclohexanecarboxylic acid; converted to benzoic acid in liver and excreted as hippuric acid	*In vivo*	Rat faeces	131

Ferulic acid and its polymers are ubiquitous parts of the cell wall. The extraction residues of ferulates in fibres might be partially responsible for the published evidence that dietary fibre potentiates the effect of polyphenols when administered to animals alone or together with the original matrix.[44–46]

In our current research, we have also recently seen an indication that even commercially available semi-purified soluble fibres used by the food industry, but not cellulose, release a spectrum of phenolic metabolites, when incubated with human stool. These probably originate from cinnamates present in the fibre.[47]

The fibre component of foods may also affect the phenolic acid production in terms of the rate of production, persistence of intermediates and range of phenolic acids produced.[6,48]

12.2.3.2 Tannins

Tannins are widely present in plant-based foods, particularly fruits, legumes, cereals and beverages (wine, tea, cocoa and cider). These water-soluble phenolic compounds (molecular weight 500–3000) precipitate proteins, which produces an astringent taste when tannin-rich foods are consumed.[49] This appears to be a defence mechanism against pests and herbivores.[50] Unlike ferulates present in the plant cell wall, tannins are localised in the vacuole and are largely soluble in methanol and acetone.[50,51] However, precipitation during processing for analysis makes their quantification difficult and, as such, they are considered as MACAN.[17,52]

Tannins are divided into two groups: (1) hydrolysable tannins – esters of phenolic acids and carbohydrate, usually glucose and phenolic acids, either gallic acid in gallotannins or other phenolic acids derived from oxidation of galloyl residues in ellagitannins;[49] and (2) proanthocyanidins, which are far more common in human diets. Proanthocyanidins are polymers of elementary flavan-3-ol units.

12.2.3.3 Proanthocyanidins

Proanthocyanidins are composed of more than one flavanol monomeric unit linked through carbon–carbon and ether linkages. The degree of polymerisation for extractable proanthocyanidins varies between 3 and 11, but may reach as high as 17 as in apple cider extract.[49] In most foods consumed by humans, the monomeric units are primarily linked through a C4 → C8 or C4 → C6 bond (B-type), which can coexist with an additional C2–O–C7 bond (A-type), making them more rigid.[25,53] Flavanol units can also bear various acyl or glycosyl substituents, such as gallic acid.[49] Extraction with methanol or acetone leaves large, non-extractable residues.[54] Several methods have been used to determine this fraction, including acid-catalysed depolymerisation in the presence of nucleophilic reagents, and enzymatic treatments have also been used to degrade the cell wall; however, depolymerisation by butanolysis under strong acidic conditions seems to provide a more realistic picture.[55]

The non-extractable proanthocyanidin contents of foods were 1751, 591 and 2716, 149 and 416 mg/100 g of the fresh edible parts of bananas, cooked pinto beans, apples,[55] peaches and pears,[17] respectively.

The intestinal absorption of proanthocyanidin decreases with degree of polymerisation. Dimers are still believed to be absorbed intact, albeit at much lower efficiency than the monomeric units.[25] In ileostomists, 62% of total proanthocyanidins was recovered in ileal effluent from an apple smoothie[56] and 90.3% of procyanidins was recovered from apple juice, but no B1–B4 dimers appeared in the effluents,[57,58] suggesting that oligomers with polymerisation higher than 2 reach the colon. Originally, it was thought that all dimers were poorly absorbed.[59,60] Non-absorbed proanthocyanidin polymers in the colon are readily metabolised to monomeric units and catabolised to corresponding phenolic acids. Larger oligomers might, however, inactivate the microbiota, as they have pronounced antimicrobial properties.[61,62] The dimers are broken down to the monomeric units of (epi)catechins by the human microbiota, some showing low stability in the medium alone. However, phenolic acids may be formed directly from the various sites on the polymer and, not first being cleaved into monomeric units,[25] similar derivatives have been identified from fermentations of (epi)catechin: 5-(3,4-dihydroxyphenyl)-γ-valerolactone, 5-(3-hydroxyphenyl)-γ-valerolactone and a series of dihydroxylated or 3-hydroxyphenolic acids with varying saturated side-chain lengths, but 3,4DHPAA is the most abundant (Table 12.1).[25,63,64]

12.2.3.4 Hydrolysable Tannins

Hydrolysable tannins are usually multiple esters of gallic acid with glucose, but other carbohydrate monomers can be found. When treated with dilute acids, they undergo hydrolytic cleavage to the respective sugar and acid moiety. If the acid component is gallic acid, the compounds are termed gallotannins. Esters with hexahydroxydiphenic acid (forming ellagic acid when hydrolysed through elimination of water) are called ellagitannins. Most ellagitannins are mixed esters both with hexahydroxydiphenic acid and gallic acid.[65]

Gallotannins can be degraded by gut bacteria, while ellagitannins are more difficult to degrade because of their complex structures with the further C–C coupling. Some bacteria, fungi and yeasts can hydrolyse only the galloyl residues of galloyl esters in ellagitannins.[66] However, when hydrolysed, the hydroxydiphenoyl residue undergoes lactonisation to produce ellagic acid, which remains relatively stable and is only partially degraded after ring fission to form bioactive metabolites. Ellagitannins, in contrast, have been studied mainly for their positive effect on human health.[67–69] In the gut, ellagic acid released from ellagitannins is transformed by the microbiota into urolithins. These have been found in caecal contents, in *in vitro* models[28,70,71] and in plasma.[72,73]

High concentrations of tannins in fodder plants inhibit gastrointestinal bacteria; however, ruminants on high-tannin diets develop a different set of tannin-resistant bacteria.[74] Gallotannins such as tannic acid significantly reduced fermentation and protein utilisation in rats and ruminants.[75,76] Gallic acid, pyrogallol and resorcinol are the main metabolites formed by the hydrolysis of tannic acid in rats or ruminants.[76,77] In humans, the same pathway is suggested by the urinary output of the metabolites after gallic acid administration.[78] Common strains of human gut bacteria possess tannin-degrading tannases.[79–81]

12.3 Role of Polyphenols in Shaping the Gut Microbiota

Various mechanisms for the impact of polyphenols on the gut microbiota have been proposed and investigated. Some polyphenols serve as prebiotics,[23] may affect bacterial attachment,[82–84] have antimicrobial effects (which might be very local for the micro-niche of the food particle)[13,62] and may inhibit microbial enzymes such as hydrolases or sterically hinder bacterial hydrolytic activity. Free polyphenols may also affect the viscosity of mucus,[85–87] influencing the colonising properties of bacteria in the inner and outer mucus layers or the expression of various bacterial genes,[88,89] altering their metabolism. Various polyphenols or their metabolites have also been shown to disrupt quorum sensing and therefore interact with bacterial virulence.[90–92] Antimicrobial assays, batch fermentations, studies in mice, human trials but also research in ruminants exposed to high levels of NEPP in forage provide overviews of the interactions of microbiota and dietary polyphenols.

12.3.1 Antimicrobial Activity of Polyphenols

Diffusion and dilution methods have been established to screen for plant-derived compounds.[93] Microdilution methods offer comparison between antimicrobial potencies based on minimum inhibitory concentrations (MIC) or the nearest serial dilutions responsible for 90% growth inhibition. However, these methods were developed for antibiotics used at much lower concentrations, with better solubility and with no rapid catabolism. They use isolated microbial strains, ignoring the potentially complex mutualistic relationships in bacterial communities. Comparisons between studies can be limited due to differences in media, solubilisers, inoculum size and other factors.[94,95] Generalisation of results might be difficult, as the higher bacterial densities found in the colon are more resistant to antimicrobial compounds[94,96] than in these assays. Bacteria are more susceptible to antimicrobials during their growth phase and are several orders in magnitude more tolerant in the steady state found in nutrient-deprived environments.[97,98] In complex environments, bacteria cooperate to alleviate antimicrobial pressure.[99]

In broth dilution studies, phenolic compounds with marked activities against human intestinal bacteria show MIC typically between 0.25 and

1.00 mg mL^{-1},[100,101] with some exceptions. In a study by Duda-Chodak,[13] quercetin showed a strong inhibitory effect against *Bacteroides galacturonicus*, *Ruminococcus gauvreauii* and *Lactobacillus* spp. with a MIC of 20–50 μg mL^{-1}. Naringenin was also active against the same species, but only above 250 μg mL^{-1},[13] while other polyphenols were less active. Phenolic acids showed higher activity against *Escherichia coli* than catechin and epicatechin. Phenolics not substituted on the aromatic ring, such as benzoic acid, phenylacetic acid and phenylpropionic acid, as well as those hydroxy-substituted in the *para*-position, had a broader effect.[101] As *para*-hydroxylated phenolic acids are quickly dehydroxylated in the colon,[102] this effect is expected.

In the study of Lee *et al.*,[103] the phenolic metabolites 3,3,4DHPPA, 3-phenylpropionic acid and 4-hydroxyphenylacetic acid were active against *E. coli*, *Salmonella enteritidis* and *Streptococcus aureus*; while catechin and epicatechin showed no effect, these metabolites often almost completely inhibited growth at 1 mg mL^{-1}. However, lack of methodological uniformity in these studies often led to conflicting results.[95] The mean concentrations of benzoic, phenylacetic and 3-phenylpropionic acids in faecal water were 51, 479 and 165 μM, respectively,[32] or 1, 160 and 55 μg g^{-1} in human faeces.[101] The sum of benzoic, phenylacetic and 3-hydroxyphenylpropionic acids reached an average 700 μg mL^{-1} in the faecal water of healthy donors.[32] Moreover, when an individual donor was followed for four consecutive days, the sum of these acids reached up to 1517 μg mL^{-1}.[32] Assuming an additive effect, this is above MIC values found in *in vitro* studies and might be of physiological relevance as a factor suppressing more susceptible bacterial species. However, as tyrosine and phenylalanine also contribute to the phenolic acid pool, a high-protein diet may lead to higher abundances of these antimicrobial forms.[104]

12.3.2 Effect on Microbiota in Gastrointestinal Models

12.3.2.1 In Vitro Studies

In the study of Parkar *et al.*,[105] chlorogenic acid, caffeic acid, rutin and quercetin were fermented in batch fermentations with human stool at 10, 30 and 100 μg mL^{-1} in Wilkins–Chalgren broth and the bacteria were assessed by real-time polymerase chain reaction and SCFA profiles. Rutin and quercetin at the highest concentrations suppressed growth of *Bifidobacterium* spp. by 4 log units. Any polyphenol added increased both Firmicutes and Bacteroides groups equally, showing dose–response relationships in many cases. Whereas production of acetate remained similar, the production of butyrate and propionate together increased by approximately 20–40 μmol mL^{-1} compared to controls, suggesting an indirect impact on butyrate and propionate producers.

When (−)-epicatechin and (+)-catechin at 150 μg mL^{-1} and 1 mg mL^{-1} were incubated with faecal bacteria, (−)-epicatechin significantly increased *Eubacterium* spp. (0.2 log units) at the lower but not the higher concentration.

(+)-Catechin increased Eubacteria at both concentrations (0.4 log units), increased bifidobacteria (0.2 log units) and showed a trend to inhibit *Clostridium coccoides.*[106]

Ex vivo fermentations of red wine extract (600 μg mL^{-1} or 284 μg mL^{-1} gallic acid equivalent) were compared to raftilose fermentation and inoculated medium only. The microbiota were evaluated by fluorescence *in situ* hybridisation. There was no effect on the microbiota, except a non-significant decrease in Clostridia. Black tea extract and red wine grape extract were tested in the SHIME gut model for their effects on the complex microbiota[107] using a single and continuous dose three times daily of 1 g of polyphenols (equivalent to 2 mg mL^{-1}). In all sections of the gut model, bifidobacteria gene copies dropped by 2 log units 5 hours after administration of both extracts, with grape extract being more powerful. Other bacterial groups decreased in the caecal compartment, but not in the distal part of the model. Total bacterial counts and *Bacteroides* spp. gene copies decreased, but recovered after 15 hours. In the caecal part of the model, *Klebsiella* spp. increased with both polyphenols (*e.g.* from 1.2% to 69.0% for wine extract and from 8.8% to 41.8% for tea extract) and remained abundant even after washout. *Akkermansia muciniphila* was also stimulated with both polyphenol extracts and continuous and single-dose stimulation. However, the concentration used was very high and not achievable with normal diet. Thus, these polyphenols appear to increase the potentially beneficial strains and reduce Firmicutes/Bacteroidetes.

12.3.2.2 Effects on the Microbiota in Human Intervention Studies

In a well-designed four-arm crossover trial,[108] subjects received red wine, the equivalent amount of de-alcoholised red wine or gin. Daily consumption of de-alcoholised red wine significantly increased faecal Fusobacteria ten-fold. The alcoholic red wine period increased Fusobacteria, Bacteroidetes, Firmicutes and Proteobacteria, whereas alcohol alone (gin) had no effect on the bacteria. With red wine, there was a major increase in *Enterococcus* spp. (1.7 log units), the *Blautia coccoides–Eubacterium rectale* group, bifidobacteria (2.8 log units) and *Eggerthella lenta* (2.1 log units), while Clostridia tended to decrease. Thus, wine polyphenols increased the abundance of species with possible health implications and this effect may be potentiated by alcohol. This is in line with another randomised clinical trial[109] in which cocoa flavanols (494 mg, of which 89 mg was [−]-epicatechin) administered daily for 4 weeks significantly increased gene copies of bifidobacteria in stool samples (0.1 log units) and lactobacilli populations (0.8 log units), while decreasing Clostridia counts (0.5 log units). In a randomised clinical trial,[110] human subjects consumed a mixture of strawberries, raspberries and cloudberries (300 g) for 12 weeks. PCR denaturing gel electrophoresis revealed that the intervention induced progression of Ruminococcaceae (clostridial cluster IV; *Clostridium leptum* group) or Lachnospiraceae taxa (clostridial cluster XIV; *Eubacterium rectale* group), with both taxa being butyrate producers. Administration of green tea for 10 days increased bifidobacteria counts up to six-fold, which often remained elevated 7 days after green tea withdrawal.[111]

12.3.2.3 Polyphenols and Microbiota in Animal Studies

Similar effects have been seen in animal studies. In a 13 week murine feeding trial, Concord grape juice (26.2 mg gallic acid equivalent per day per animal; mostly cyanidin polymers) increased growth of *Akkermansia muciniphila* and decreased the proportion of Firmicutes/Bacteroidetes.[112] Resveratrol (200 mg kg^{-1} day^{-1}) administered to mice for 12 weeks[113] reversed the adverse effect of high-fat intake on mice microbiota, decreased the Firmicutes/Bacteroides ratio, decreased *Enterococcus faecalis* abundance, increased the abundance of bifidobacteria and shifted the microbial balance to that seen with the control chow. In another trial,[114] mice were given diets containing apple powder (native and one with increased polyphenols, both pulp or peel, up to 1 mg polyphenols g^{-1}, mostly cyanidin galactoside and epicatechin) for 7 and 21 weeks. Mice fed the experimental diets had higher total microbial abundances as determined by RT-PCR of up to 0.6 log units, with the best effect seen in the polyphenol-enriched apples compared with the control. Both the pulp and peel from native apples and not the enriched extracts increased bifidobacteria counts by five-fold. All apple polyphenol-rich diets decreased lactobacilli.

12.3.3 Relevance of the Doses Used in Studies of the Actions and Metabolism of Colonic Polyphenols

The majority of studies used concentrations exceeding that achievable with normal diet and there is no agreement on what constitutes a physiological dose. This could be estimated from current knowledge of the bioavailability and physiological conditions in the gut. The intestinal absorption of polyphenols varies, but studies suggest that they mostly escape intestinal absorption depending on their molecular properties as well as the food matrix in which they are ingested. Studies with ileostomy subjects provide good estimates of what reaches the colon. After consumption of 1 L of cloudy apple juice, the polyphenol recovery in ileostomy subjects was low and only 14% (or 91.8 μmol) of the original 669.1 μmol (249.9 mg) of polyphenols was excreted.[57] When analysing each phenolic constituent of the juice separately, proportional rutin excretion was 10%. In another study, 62% of 1140 mg of polyphenol in an apple smoothie was excreted in ileostomy effluent, with 60% recovery of rutin and 100% of quercetin.[115] A study of tomato juice, fortified with rutin to 176 μM, suggested 86% reached the colon.[116] The main polyphenols of apple juice are chlorogenic acid, contributing to two-thirds of the juice's polyphenols, and their excretion was 10–33%.[57] In the same study, ileostomy subjects consumed 300 g of blueberries, corresponding to an anthocyanin content of 7834 mg. Between 28.3% and 85.1% of anthocyanins were recovered (median 45%).[57] This was similar to the 40% recovery of anthocyanins after raspberry consumption.[117] Recovery of ellagitannins from raspberries was 23% and a significant proportion was hydrolysed in the upper gut to ellagic acid. In a quercetin study, recoveries after consumption of 89–100 mg quercetin equivalent from either onions, pure quercetin or rutin ranged from 48% to 83%; the highest absorption was seen for onions and the lowest for rutin.[118] Additionally, recoveries of 2.8 mmol of chlorogenic or caffeic acid administered as pure compounds in another study were 33% and 95%, respectively.[119] Recovery of flavan-3-ol administered as a green tea extract (200 g) in ileal fluid was ~40%.[120]

In normal subjects, chyme enters the caecum in small batches, with larger amounts released after meals.[121] An approximation of the physiological concentrations of the polyphenols that might be found in the caecum and ascending colon can be demonstrated with quercetin. Daily intake of quercetin has been reported to be 20 mg day^{-1} in a French cohort.[122] After consumption of quercetin-rich food in various forms containing 20 mg of quercetin aglycone (the amount normally consumed in a day), 10 mg (50%) may escape absorption and 75% of this reaches the caecum within 4 hours of digestion and is diluted in 200 mL of effluent,[123] giving a concentration of quercetin in the caecum and proximal colon of 37.5 µg mL^{-1} (or 124 µM), far below the concentrations used in studies (*e.g.* assays for antimicrobial activity discussed above). Moreover, these polyphenols are quickly metabolised as they gradually reach the caecum where further dilution may occur, so achievable physiological concentrations will probably be much lower. The need for physiological doses in flavonoid experiments of between 1 and 10 µM has been previously stressed.[124] Thus, an intake of 500 mg of total polyphenols in polyphenol-rich food would lead to colonic concentrations of approximately 1 mg mL^{-1}. More research is needed at these much lower, dietary-relevant doses in terms of both antibacterial and pro-health effects.

12.4 Conclusions and Future Directions

It is clear that the gut microbiome plays a crucial role in the bioactivities of polyphenols and their metabolites and alterations of microbiome composition may explain the inter-individual variation observed in nearly all studies. While there is considerable knowledge of the degradation pathways of the major polyphenols *in vitro*, the effect of the polyphenols on the gut microbiota remains largely unexplored and all interactions between bacteria and polyphenols need to be considered at physiological doses. It is also crucial that the health benefits of the phenolic acids produced by bacteria from both EPP and NEPP are more comprehensively explored. A true and detailed understanding of the interactions between gut bacteria and polyphenol-rich food sources is required in order to fully elucidate their impacts on health and to allow the production of better dietary guidelines and improved dietary and food formulations.

References

1. P. J. Turnbaugh, M. Hamady, T. Yatsunenko, B. L. Cantarel, A. Duncan, R. E. Ley, M. L. Sogin, W. J. Jones, B. A. Roe and J. P. Affourtit, *Nature*, 2009, **457**, 480.
2. P. J. Turnbaugh, F. Bäckhed, L. Fulton and J. I. Gordon, *Cell Host Microbe*, 2008, **3**, 213–223.
3. P. J. Turnbaugh, R. E. Ley, M. A. Mahowald, V. Magrini, E. R. Mardis and J. I. Gordon, *Nature*, 2006, **444**, 1027–1131.
4. H. Tilg and A. Kaser, *J. Clin. Invest.*, 2011, **121**, 2126.
5. A. L. Kau, P. P. Ahern, N. W. Griffin, A. L. Goodman and J. I. Gordon, *Nature*, 2011, **474**, 327.
6. I. B. Jaganath, W. Mullen, M. E. J. Lean, C. A. Edwards and A. Crozier, *Free Radical Biol. Med.*, 2009, **47**, 1180–1189.
7. M. Monagas, N. Khan, C. Andrés-Lacueva, M. Urpí-Sardá, M. Vázquez-Agell, R. M. Lamuela-Raventós and R. Estruch, *Br. J. Nutr.*, 2009, **102**, 201–206.
8. A. Crozier, I. B. Jaganath and M. N. Clifford, *Nat. Prod. Rep.*, 2009, **26**, 1001–1043.
9. F. Cardona, C. Andrés-Lacueva, S. Tulipani, F. J. Tinahones and M. I. Queipo-Ortuño, *J. Nutr. Biochem.*, 2013, **24**, 1415–1422.
10. M. V. Selma, J. C. Espín and F. A. Tomás-Barberán, *J. Agric. Food Chem.*, 2009, **57**, 6485–6501.
11. U. Etxeberria, A. Fernández-Quintela, F. I. Milagro, L. Aguirre, J. A. Martínez and M. P. Portillo, *J. Agric. Food Chem.*, 2013, **61**, 9517–9533.
12. D. Hervert-Hernandez and I. Goñi, *Food Rev. Int.*, 2011, **27**, 154–169.
13. A. Duda-Chodak, *J. Physiol. Pharmacol.*, 2012, **63**, 497–503.
14. G. Gross, D. M. Jacobs, S. Peters, S. Possemiers, J. van Duynhoven, E. E. Vaughan and T. Van de Wiele, *J. Agric. Food Chem.*, 2010, **58**, 10236–10246.
15. A. González-Sarrías, R. García-Villalba, M. Romo-Vaquero, C. Alasalvar, A. Örem, P. Zafrilla, F. A. Tomás-Barberán, M. V. Selma and J. C. Espín, *Mol. Nutr. Food Res.*, 2017, **61**, 1600830.
16. R. Vázquez-Fresno, R. Llorach, A. Perera, R. Mandal, M. Feliz, F. J. Tinahones, D. S. Wishart and C. Andres-Lacueva, *J. Nutr. Biochem.*, 2016, **28**, 114–120.
17. J. Pérez-Jiménez and F. Saura-Calixto, *Food Res. Int.*, 2015, **74**, 315–323.
18. A. N. Booth, C. W. Murray, F. T. Jones and F. DeEds, *J. Biol. Chem.*, 1956, **223**, 251–257.
19. H. Watanabe, *Bull. Agric. Chem. Soc. Jpn.*, 1959, **23**, 260–262.
20. Y. Ohshima, H. Watanabe and S. Kuwazuka, *Bull. Agric. Chem. Soc. Jpn.*, 1960, **24**, 497–500.
21. H. G. Krishnamurty, K. J. Cheng, G. A. Jones, F. J. Simpson and J. E. Watkin, *Can. J. Microbiol.*, 1970, **16**, 759–767.
22. A. K. Martin, *Br. J. Nutr.*, 2007, **23**, 389–399.
23. J. C. Espín, A. González-Sarrías and F. A. Tomás-Barberán, *Biochem. Pharmacol.*, 2017, **139**, 82–93.

24. A. M. Aura, P. Martin-Lopez, K. A. O'Leary, G. Williamson, K. M. Oksman-Caldentey, K. Poutanen and C. Santos-Buelga, *Eur. J. Nutr.*, 2005, **44**, 133–142.
25. M. M. Appeldoorn, J.-P. Vincken, A.-M. Aura, P. C. H. Hollman and H. Gruppen, *J. Agric. Food Chem.*, 2009, **57**, 1084–1092.
26. A. M. Aura, K. A. O'Leary, G. Williamson, M. Ojala, M. Bailey, R. Puupponen-Pimiä, A. M. Nuutila, K. M. Oksman-Caldentey and K. Poutanen, *J. Agric. Food Chem.*, 2002, **50**, 1725–1730.
27. A.-M. Aura, *Phytochem. Rev.*, 2008, **7**, 407–429.
28. R. García-Villalba, D. Beltrán, J. C. Espín, M. V. Selma and F. A. Tomás-Barberán, *J. Agric. Food Chem.*, 2013, **61**, 8797–8806.
29. L. M. Bode, D. Bunzel, M. Huch, G.-S. Cho, D. Ruhland, M. Bunzel, A. Bub, C. M. A. P. Franz and S. E. Kulling, *Am. J. Clin. Nutr.*, 2013, **97**, 295–309.
30. G. A. Olah, *Acc. Chem. Res.*, 1971, **4**, 240–248.
31. J. Heider, A. M. Spormann, H. R. Beller and F. Widdel, *FEMS Microbiol. Rev.*, 1998, **22**, 459–473.
32. A. M. Jenner, J. Rafter and B. Halliwell, *Free Radical Biol. Med.*, 2005, **38**, 763–772.
33. E. Díaz, A. Ferrández, M. A. Prieto and J. L. García, *Microbiol. Mol. Biol. Rev.*, 2001, **65**, 523–569.
34. S. Dagley and D. T. Gibson, *Biochem. J.*, 1965, **95**, 466–474.
35. A. R. Rechner, M. A. Smith, G. Kuhnle, G. R. Gibson, E. S. Debnam, S. K. S. Srai, K. P. Moore and C. A. Rice-Evans, *Free Radical Biol. Med.*, 2004, **36**, 212–225.
36. J. Ralph, R. D. Hatfield, J. H. Grabber, H.-J. G. Jung, S. Quideau and R. F. Helm, *Lignin and Lignan Biosynthesis*, ed. N. G. Lewis and S. Sarkanen, American Chemical Society, USA, 1998, vol. 697, ch. 16, pp. 209–236.
37. M. Bunzel, J. Ralph, J. M. Marita, R. D. Hatfield and H. Steinhart, *J. Sci. Food Agric.*, 2001, **81**, 653–660.
38. J. Bidlack, M. Malone and R. Benson, *Proc. Okla. Acad. Sci.*, 1992, **72**, 51–56.
39. K.-H. Kim, R. Tsao, R. Yang and S. W. Cui, *Food Chem.*, 2006, **95**, 466–473.
40. E. Nordlund, A.-M. Aura, I. Mattila, T. Kössö, X. Rouau and K. Poutanen, *J. Agric. Food Chem.*, 2012, **60**, 8134–8145.
41. F. Saura-Calixto, *J. Agric. Food Chem.*, 2012, **60**, 11195–11200.
42. M. F. Andreasen, P. A. Kroon, G. Williamson and M.-T. Garcia-Conesa, *Free Radical Biol. Med.*, 2001, **31**, 304–314.
43. A. Braune, M. Bunzel, R. Yonekura and M. Blaut, *J. Agric. Food Chem.*, 2009, **57**, 3356–3362.
44. S. Auclair, M. Silberberg, E. Gueux, C. Morand, A. Mazur, D. Milenkovic and A. Scalbert, *J. Agric. Food Chem.*, 2008, **56**, 5558–5563.
45. B. Metzger, D. Barnes and J. Reed, *J. Med. Food*, 2009, **12**, 854–860.
46. M. Tamura, H. Nakagawa, T. Tsushida, K. Hirayama and K. Itoh, *J. Food Sci.*, 2007, **72**, S648–S651.

47. C. A. Edwards, J. Havlik, W. Cong, W. Mullen, T. Preston, D. J. Morrison and E. Combet, *BNF Nutr. Bull.*, 2017, **42**, 356–360.
48. B. Mansoorian, A. L. Garcia, E. Combet and C. A. Edwards, *Proc. Nutr. Soc.*, 2015, **74**, E47.
49. C. Santos-Buelga and A. Scalbert, *J. Sci. Food Agric.*, 2000, **80**, 1094–1117.
50. H. A. Stafford, *Phytochemistry*, 1988, **27**, 1–6.
51. J. Montes-Ávila, G. López-Angulo and F. Delgado-Vargas, *Fruit and Vegetable Phytochemicals: Chemistry and Human Health* ed. E. M. Yahia, Wiley, USA, 2017, vol. 2, ch. 13, pp. 221–268.
52. A. González-Sarrías, J. C. Espín and F. A. Tomás-Barberán, *Trends Food Sci. Technol.*, 2017, **69**, 281–288.
53. K. Ou and L. Gu, *J. Funct. Foods*, 2014, **7**, 43–53.
54. M. L. Mateos-Martín, J. Pérez-Jiménez, E. Fuguet and J. L. Torres, *Br. J. Nutr.*, 2011, **108**, 290–297.
55. J. Zurita, M. E. Díaz-Rubio and F. Saura-Calixto, *Int. J. Food Sci. Nutr.*, 2012, **63**, 936–939.
56. S. Hagl, H. Deusser, B. Soyalan, C. Janzowski, F. Will, H. Dietrich, F. W. Albert, S. Rohner and E. Richling, *Mol. Nutr. Food Res.*, 2011, **55**, 368–377.
57. K. Kahle, M. Kraus, W. Scheppach, M. Ackermann, F. Ridder and E. Richling, *Mol. Nutr. Food Res.*, 2006, **50**, 418–423.
58. K. Kahle, M. Kraus, W. Scheppach and E. Richling, *Mol. Nutr. Food Res.*, 2005, **49**, 1143–1150.
59. L. Y. Rios, R. N. Bennett, S. A. Lazarus and C. Rémésy, *Am. J. Clin. Nutr.*, 2002, **76**, 1106–1110.
60. J. L. Donovan, A. Lee, C. Manach, L. Rios, C. Morand, A. Scalbert and C. Rémésy, *Br. J. Nutr.*, 2002, **87**, 299–306.
61. A. Scalbert, *Phytochemistry*, 1991, **30**, 3875–3883.
62. M. Daglia, *Curr. Opin. Biotechnol.*, 2012, **23**, 174–181.
63. S. Déprez, C. Brezillon, S. Rabot, C. Philippe, I. Mila, C. Lapierre and A. Scalbert, *J. Nutr.*, 2000, **130**, 2733–2738.
64. A.-M. Aura, I. Mattila, T. Seppänen-Laakso, J. Miettinen, K.-M. Oksman-Caldentey and M. Orešič, *Phytochem. Lett.*, 2008, **1**, 18–22.
65. J. Serrano, R. Puupponen-Pimiä, A. Dauer, A.-M. Aura and F. Saura-Calixto, *Mol. Nutr. Food Res.*, 2009, **53**, S310–S329.
66. L. Mingshu, Y. Kai, H. Qiang and J. Dongying, *J. Basic Microbiol.*, 2006, **46**, 68–84.
67. A. Aguilera-Carbo, C. Augur, L. A. Prado-Barragan, E. Favela-Torres and C. N. Aguilar, *Appl. Microbiol. Biotechnol.*, 2008, **78**, 189–199.
68. J. C. Espín, M. Larrosa, M. T. García-Conesa and F. Tomás-Barberán, *Evidence-Based Complementary Altern. Med.*, 2013, 27041869.
69. M. Larrosa, A. González-Sarrías, M. T. García-Conesa, F. A. Tomás-Barberán and J. C. Espín, *J. Agric. Food Chem.*, 2006, **54**, 1611–1620.
70. B. Cerda, P. Periago, J. C. Espin and F. A. Tomas-Barberan, *J. Agric. Food Chem.*, 2005, **53**, 5571–5576.

71. P. Mena, M. Dall'Asta, L. Calani, F. Brighenti and D. Del Rio, *Eur. J. Nutr.*, 2017, **56**, 99–106.
72. N. P. Seeram, S. M. Henning, Y. Zhang, M. Suchard, Z. Li and D. Heber, *J. Nutr.*, 2006, **136**, 2481–2485.
73. M. A. Nuñez-Sánchez, R. García-Villalba, T. Monedero-Saiz, N. V. García-Talavera, M. B. Gómez-Sánchez, C. Sánchez-Álvarez, A. M. García-Albert, F. J. Rodríguez-Gil, M. Ruiz-Marín and F. A. Pastor-Quirante, *Mol. Nutr. Food Res.*, 2014, **58**, 1199–1211.
74. A. H. Smith, E. Zoetendal and R. I. Mackie, *Microb. Ecol.*, 2005, **50**, 197–205.
75. L. Bravo, R. Abia, M. A. Eastwood and F. Saura-Calixto, *Br. J. Nutr.*, 2007, **71**, 933–946.
76. B. Singh, T. K. Bhat and O. P. Sharma, *Livest. Prod. Sci.*, 2001, **68**, 259–262.
77. Y. Nakamura, S. Tsuji and Y. Tonogai, *J. Agric. Food Chem.*, 2003, **51**, 331–339.
78. S. Shahrzad and I. Bitsch, *J. Chromatogr. B: Biomed. Sci. Appl.*, 1998, **705**, 87–95.
79. H. Rodríguez, B. d. l. Rivas, C. Gómez-Cordovés and R. Muñoz, *Food Chem.*, 2008, **107**, 664–670.
80. R. Osawa, K. Kuroiso, S. Goto and A. Shimizu, *Appl. Environ. Microbiol.*, 2000, **66**, 3093–3097.
81. N. Noguchi, T. Ohashi, T. Shiratori, K. Narui, T. Hagiwara, M. Ko, K. Watanabe, T. Miyahara, S. Taira, F. Moriyasu and M. Sasatsu, *J. Gastroenterol.*, 2007, **42**, 346–351.
82. I. Bustos, T. Garcia-Cayuela, B. Hernández-Ledesma, C. Peláez, T. Requena and M. C. Martínez-Cuesta, *J. Agric. Food Chem.*, 2012, **60**, 9082–9088.
83. T. Volstatova, P. Marsik, V. Rada, M. Geigerova and J. Havlik, *J. Funct. Foods*, 2017, **35**, 391–397.
84. A. Yamanaka, T. Kouchi, K. Kasai, T. Kato, K. Ishihara and K. Okuda, *J. Periodontal Res.*, 2007, **42**, 589–592.
85. G. B. Gonzales, G. Smagghe, A. Mackie, C. Grootaert, B. Bajka, N. Rigby, K. Raes and J. Van Camp, *J. Funct. Foods*, 2015, **16**, 403–413.
86. P. Georgiades, P. D. A. Pudney, S. Rogers, D. J. Thornton and T. A. Waigh, *PLoS One*, 2014, **9**, e105302.
87. G. B. Gonzales, J. Van Camp, G. Smagghe, K. Raes and A. Mackie, *Food Res. Int.*, 2016, **88**, 342–347.
88. Y.-M. Zhang and C. O. Rock, *J. Biol. Chem.*, 2004, **279**, 30994–31001.
89. S. K. Sharma, P. Parasuraman, G. Kumar, N. Surolia and A. Surolia, *J. Med. Chem.*, 2007, **50**, 765–775.
90. H. Yin, Y. Deng, H. Wang, W. Liu, X. Zhuang and W. Chu, *Sci. Rep.*, 2015, **5**, 16158.
91. B. Huber, L. Eberl, W. Feucht and J. Polster, *Z. Naturforsch., C*, 2003, **58**, 879–884.
92. F. Nazzaro, F. Fratianni and R. Coppola, *Int. J. Mol. Sci.*, 2013, **14**, 12607–12619.

93. P. Cos, A. J. Vlietinck, D. V. Berghe and L. Maes, *J. Ethnopharmacol.*, 2006, **106**, 290–302.
94. I. Wiegand, K. Hilpert and R. E. W. Hancock, *Nat. Protoc.*, 2008, **3**, 163.
95. T. P. T. Cushnie and A. J. Lamb, *Int. J. Antimicrob. Agents*, 2005, **26**, 343–356.
96. D. L. Stevens, S. Van and A. E. Bryant, *J. Infect. Dis.*, 1993, **167**, 1401–1405.
97. R. H. Eng, F. T. Padberg, S. M. Smith, E. N. Tan and C. E. Cherubin, *Antimicrob. Agents Chemother.*, 1991, **35**, 1824–1828.
98. D. Nguyen, A. Joshi-Datar, F. Lepine, E. Bauerle, O. Olakanmi, K. Beer, G. McKay, R. Siehnel, J. Schafhauser and Y. Wang, *Science*, 2011, **334**, 982–986.
99. E. Giaouris, N. Chorianopoulos, A. Doulgeraki and G.-J. Nychas, *PLoS One*, 2013, **8**, e77276.
100. S. G. Parkar, D. E. Stevenson and M. A. Skinner, *Int. J. Food Microbiol.*, 2008, **124**, 295–298.
101. F. Sánchez-Patán, C. Cueva, M. Monagas, G. E. Walton, G. R. Gibson, P. J. Martín-Álvarez, M. V. Moreno-Arribas and B. Bartolomé, *Food Chem.*, 2012, **131**, 337–347.
102. Y. Nakagawa, M. R. Shetlar and S. H. Wender, *Biochim. Biophys. Acta, Gen. Subj.*, 1965, **97**, 233–241.
103. H. C. Lee, A. M. Jenner, C. S. Low and Y. K. Lee, *Res. Microbiol.*, 2006, **157**, 876–884.
104. W. R. Russell, S. H. Duncan, L. Scobbie, G. Duncan, L. Cantlay, A. G. Calder, S. E. Anderson and H. J. Flint, *Mol. Nutr. Food Res.*, 2013, **57**, 523–535.
105. S. G. Parkar, T. M. Trower and D. E. Stevenson, *Anaerobe*, 2013, **23**, 12–19.
106. X. Tzounis, J. Vulevic, G. G. C. Kuhnle, T. George, J. Leonczak, G. R. Gibson, C. Kwik-Uribe and J. P. E. Spencer, *Br. J. Nutr.*, 2008, **99**, 782–792.
107. R. A. Kemperman, G. Gross, S. Mondot, S. Possemiers, M. Marzorati, T. Van de Wiele, J. Doré and E. E. Vaughan, *Food Res. Int.*, 2013, **53**, 659–669.
108. M. I. Queipo-Ortuño, M. Boto-Ordóñez, M. Murri, J. M. Gomez-Zumaquero, M. Clemente-Postigo, R. Estruch, F. C. Diaz, C. Andrés-Lacueva and F. J. Tinahones, *Am. J. Clin. Nutr.*, 2012, **95**, 1323–1334.
109. X. Tzounis, A. Rodriguez-Mateos, J. Vulevic, G. R. Gibson, C. Kwik-Uribe and J. P. E. Spencer, *Am. J. Clin. Nutr.*, 2011, **93**, 62–72.
110. R. Puupponen-Pimiä, T. Seppänen-Laakso, M. Kankainen, J. Maukonen, R. Törrönen, M. Kolehmainen, T. Leppänen, E. Moilanen, L. Nohynek and A. M. Aura, *Mol. Nutr. Food Res.*, 2013, **57**, 2258–2263.
111. J. S. Jin, M. Touyama, T. Hisada and Y. Benno, *Microbiol. Immunol.*, 2012, **56**, 729–739.
112. D. E. Roopchand, R. N. Carmody, P. Kuhn, K. Moskal, P. Rojas-Silva, P. J. Turnbaugh and I. Raskin, *Diabetes*, 2015, **64**, 2847–2858.
113. Y. Qiao, J. Sun, S. Xia, X. Tang, Y. Shi and G. Le, *Food Funct.*, 2014, **5**, 1241–1249.
114. R. V. Espley, C. A. Butts, W. A. Laing, S. Martell, H. Smith, T. K. McGhie, J. Zhang, G. Paturi, D. Hedderley and A. Bovy, *J. Nutr.*, 2014, **144**, 146–154.

115. S. Hagl, H. Deusser, B. Soyalan, C. Janzowski, F. Will, H. Dietrich, F. W. Albert, S. Rohner and E. Richling, *Mol. Nutr. Food Res.*, 2011, **55**, 368–377.
116. I. B. Jaganath, W. Mullen, C. A. Edwards and A. Crozier, *Free Radical Res.*, 2006, **40**, 1035–1046.
117. R. González-Barrio, G. Borges, W. Mullen and A. Crozier, *J. Agric. Food Chem.*, 2010, **58**, 3933–3939.
118. P. C. Hollman, J. H. de Vries, S. D. van Leeuwen, M. J. Mengelers and M. B. Katan, *Am. J. Clin. Nutr.*, 1995, **62**, 1276–1282.
119. M. R. Olthof, P. C. H. Hollman and M. B. Katan, *J. Nutr.*, 2001, **131**, 66–71.
120. A. Stalmach, W. Mullen, H. Steiling, G. Williamson, M. E. J. Lean and A. Crozier, *Mol. Nutr. Food Res.*, 2010, **54**, 323–334.
121. E. M. M. Quigley, T. J. Borody, S. F. Phillips, M. Wienbeck, R. L. Tucker and A. Haddad, *Gastroenterology*, 1984, **87**, 857–866.
122. J. Pérez-Jiménez, L. Fezeu, M. Touvier, N. Arnault, C. Manach, S. Hercberg, P. Galan and A. Scalbert, *Am. J. Clin. Nutr.*, 2011, **93**, 1220–1228.
123. S. D. Ladas, P. E. Isaacs, G. M. Murphy and G. E. Sladen, *Gut*, 1986, **27**, 906–912.
124. P. A. Kroon, M. N. Clifford, A. Crozier, A. J. Day, J. L. Donovan, C. Manach and G. Williamson, *Am. J. Clin. Nutr.*, 2004, **80**, 15–21.
125. A. Serra, A. Macià, M.-P. Romero, J. Reguant, N. Ortega and M.-J. Motilva, *Food Chem.*, 2012, **130**, 383–393.
126. L. A. Griffiths and G. E. Smith, *Biochem. J.*, 1972, **130**, 141.
127. X. Tzounis, J. Vulevic, G. G. C. Kuhnle, T. George, J. Leonczak, G. R. Gibson, C. Kwik-Uribe and J. P. E. Spencer, *Br. J. Nutr.*, 2007, **99**, 782–792.
128. J. Fleschhut, F. Kratzer, G. Rechkemmer and S. E. Kulling, *Eur. J. Nutr.*, 2006, **45**, 7–18.
129. M. P. Gonthier, C. Remesy, A. Scalbert, V. Cheynier, J. M. Souquet, K. Poutanenc and A. M. Aura, *Biomed. Pharmacother.*, 2006, **60**, 536–540.
130. R. R. Scheline, *Acta Pharmacol. Toxicol.*, 1968, **26**, 189–205.
131. D. Brewster, R. S. Jones and D. V. Parke, *Biochem. J.*, 1978, **170**, 257–264.

CHAPTER 13

Antioxidant Capacity of Non-extractable Polyphenols as a Key Dietary Parameter

J. PÉREZ-JIMÉNEZ

Dpt. Metabolism and Nutrition, Institute of Food Science, Technology and Nutrition (ICTAN-CSIC), c/José Antonio Novais, 10, 28040 Madrid, Spain
*E-mail: jara.perez@ictan.csic.es

13.1 Introduction: Polyphenols as Dietary Antioxidants

The role of increased oxidation in the development of non-communicable chronic diseases, such as cardiovascular diseases, type 2 diabetes, cancer or neurodegenerative disorders, is widely known.[1] It is derived from an imbalance between, on the one hand, free radicals produced as a result of different physiological processes and also from exogenous sources, and, on the other hand, the capacity of the organism to neutralise those free radicals with both endogenous and exogenous antioxidant compounds. Exogenous antioxidants are provided by the diet and are composed of different compounds, such as vitamin C, vitamin E, certain carotenoids, selenium and, finally, polyphenols: the most abundant class.

Polyphenols are a wide class of secondary plant metabolites grouped into different families (phenolic acids, flavonoids, stilbenes, lignans and others), with structures that vary from very simple phenolic acids to

Food Chemistry, Function and Analysis No. 5
Non-extractable Polyphenols and Carotenoids: Importance in Human Nutrition and Health
Edited by Fulgencio Saura-Calixto and Jara Pérez-Jiménez

Published by the Royal Society of Chemistry, www.rsc.org

high-molecular-weight proanthocyanidins or ellagitannins. They are known to have several health-related properties (*e.g.*, hypolipidaemic, anti-inflammatory and hypotensive effects, among others). These effects are derived from different physiological processes, including the regulation of gene expression, interaction with gut microbiota and, as is widely known, antioxidant action.[2] The antioxidant effects of polyphenols have mostly been studied, as have other aspects of these phytochemicals, in extractable polyphenols (EPP). It has only been during the last few years that the contribution of non-extractable polyphenols (NEPP) to this aspect of the dietary intake of polyphenols has started to be considered.

There are several approaches to determining the modifications in oxidative status that result from dietary antioxidant intake: (a) measuring oxidised biomolecules (malondialdehyde for lipids, carbonyl groups for proteins and 8-hydroxydeoxyguanosine for DNA); (b) determining levels of endogenous antioxidants (the glutathione system and enzymes such as catalase); and (c) studying the circulating levels of antioxidant compounds. The last method may involve measuring individual compounds or antioxidant capacity (AC), which is considered to be the combined activity of all the antioxidants present in a biological sample and that may involve synergic, additive or antagonistic effects.[3]

AC may also be determined in foods and beverages by *in vitro* techniques. During recent decades, these techniques have often been used rather indiscriminately, and some dubious direct associations between high AC values in individual foods and *in vivo* beneficial effects have been drawn. The relationship is not so simple, but at the same time, individual *in vitro* AC determinations may certainly be used as a first approach to establishing the potential biological effects of certain diets. In this context, the concept of dietary AC is particularly relevant: it corresponds to the AC provided by all the food and beverages consumed in a diet while taking synergism into account.[4,5] Interestingly, polyphenols are the most consumed dietary antioxidants (accounting for some 90% of dietary antioxidant intake)[5] and therefore they are the main contributors to dietary AC. Nevertheless, most clinical and epidemiological studies still focus on just a few milligrams of single dietary antioxidants (usually vitamins C and E, selenium or beta-carotene), which represent a minor part of all the antioxidants present in human biological fluids, tissues and cells.

Up to now, an important number of studies have found significant inverse associations between dietary AC and several adverse health outcomes. The first study to show the direct relevance of dietary AC as a biological marker reported an inverse association between this parameter and plasma C reactive protein: an inflammation marker.[4] This result was further confirmed in a study in postmenopausal women; indeed, when the subjects followed a 9 month diet with high AC, it caused an additional decrease in plasma C reactive protein.[6] In the same way, a study of more than 4000 US adults showed that the levels of plasma homocysteine, a validated marker for cardiovascular disease risk, decreased across quartiles of AC consumption, either from diet or from supplements.[7] Similarly, in a group of 153 young, healthy subjects, it was found that higher dietary AC was associated with lower systolic blood pressure, blood glucose and serum free fatty acid levels.[8]

The beneficial effects of diets high in AC have been proven, not only indirectly in terms of decreases in risk factors, but also directly as a reduced risk of developing certain chronic diseases. Thus, an inverse association was reported between high-AC diets and the incidence of ischaemic stroke[9] and age-related cataracts.[10] Similarly, a large case-control study (2000 cancer patients and 4000 control subjects) found that the risk of suffering colorectal cancer decreases when one's diet exhibits high AC.[11] Moreover, it has recently been reported[12] that a diet with high AC is correlated with a decrease in all-cause mortality, cancer mortality and cardiovascular mortality in a study performed in the well-known nutritional cohort, EPIC (European Prospective Investigation into Cancer and Nutrition study). Finally, higher adherence to the Mediterranean diet, a dietary pattern that has been validated for its beneficial health effects, is characterized by both high dietary AC and high plasma AC.[13]

Nevertheless, it should be noted that all the studies cited considered only the contribution to dietary AC of EPP and other low-molecular-weight antioxidants (chiefly vitamins C and E), thereby neglecting the role of NEPP. We may expect, however, that these latter compounds would make a relevant contribution to dietary AC from both a quantitative and a qualitative point of view. Firstly, this is because they are important contributors to the total polyphenol content in common foods. Secondly, this is because they may give rise to particular physiological effects due to their specific metabolism, which completely takes place in the large intestine and is delayed in relation to that of EPP,[14] as discussed in other sections of the book. This chapter reports the procedure adopted to determine AC associated with NEPP and the current evidence concerning the contribution of NEPP to the total AC of foods and diets, and also to *in vivo* oxidative status.

13.2 Method for Determining the AC of NEPP in Food and Beverages

In order to evaluate the AC associated with NEPP, the following steps must be followed: release NEPP from the food matrix; measure the AC of the resulting solutions; and (if dietary AC is to be evaluated) match the AC data for each food item with its consumption in a specific population. The details of each step are discussed below.

13.2.1 Preparation of NEPP Solutions

Figure 13.1 shows the method to be followed in order to determine the AC derived from NEPP in foods, based on a specific study.[15]

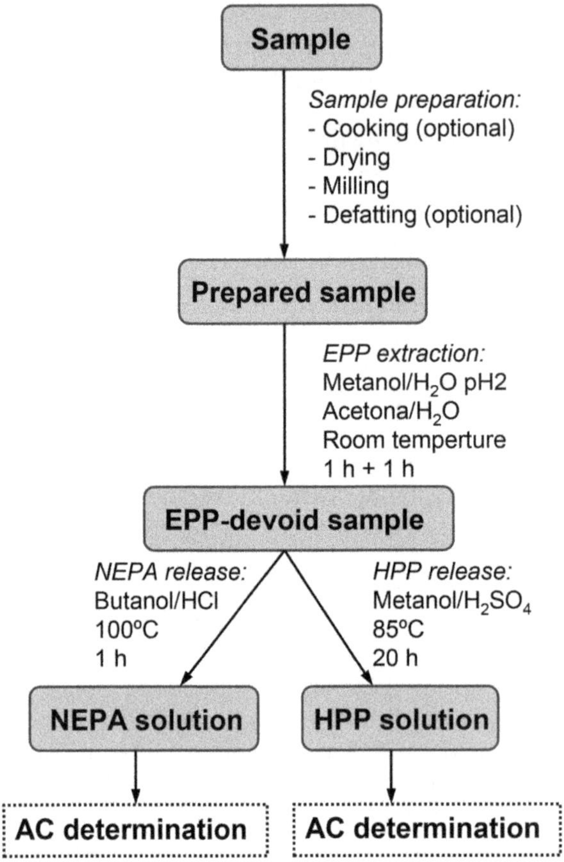

Figure 13.1 Scheme of the procedure for determining the AC associated with NEPP in food samples. AC: antioxidant capacity; EPP: extractable polyphenols; HPP: hydrolysable polyphenols; NEPA: non-extractable proanthocyanidins; NEPP: non-extractable polyphenols.

After several steps to prepare the sample (cooking, if the sample is consumed cooked; freeze-drying, if possible; milling with a system that avoids heating; and defatting, if the fat content is over 20%), the NEPP must be released from the food matrix. General considerations of the advantages and drawbacks of each strategy have been presented previously in this book (Chapter 4), so the one adopted by our team is shown here. Briefly, after removing the EPP *via* aqueous–organic extraction, the hydrolysates containing the two major fractions of NEPP (hydrolysable polyphenols [HPP] and non-extractable proanthocyanidins [NEPA]) must be obtained from the corresponding extraction residues.

Regarding HPP, the residues from the extraction of EPP are hydrolysed with H_2SO_4 in a methanol/water medium. It should be borne in mind that when carrying out this hydrolysis, the compounds originally present in the sample are transformed. Therefore, when analysing the AC of these hydrolysed NEPP, the values obtained may partially correspond not to the actual AC of the original NEPP, but rather to the AC of the phenolic compounds derived from their hydrolysis. For NEPA, the residues from polyphenol extractions are subjected to treatment with HCl/butanol, releasing the corresponding anthocyanins.

Finally, beverages constitute a specific case for the determination of the AC associated with NEPP (Figure 13.2). They are first subjected to different preparatory procedures (alcohol removal and/or degasification if needed, as well as concentration). Then, isolation of the dietary fibre through treatments with digestive enzymes under physiological conditions followed by dialysis is needed. Finally, the dietary fibre must be hydrolysed with cellulase and the AC may be measured in the corresponding hydrolysate.[16]

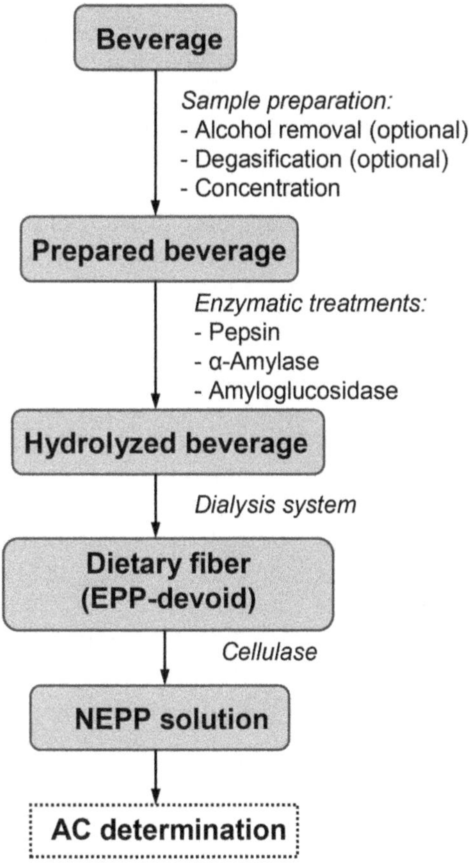

Figure 13.2 Scheme of the procedure for determining the AC associated with NEPP in beverages. AC: antioxidant capacity; EPP: extractable polyphenols; NEPP: non-extractable polyphenols.

13.2.2 Measurement of the AC of NEPP Solutions

Over the last few decades, many analytical methods have been developed to determine the AC of foods based on the different mechanisms by which food antioxidants (mostly polyphenols) may act. The most common methods are the ferric reducing/antioxidant power (FRAP), 2,2-azino-bis(3-ethylbenzothiazoline-6-sulphonate) (ABTS), 2,2-diphenyl-1-picrylhydrazyl radical (DPPH) and oxygen radical absorbance capacity (ORAC) assays. While FRAP evaluates the reducing capacity of polyphenols using an iron complex as a probe,[17] the other three methods are based on the free radical-scavenging capacity of polyphenols. In particular, DPPH and ABTS use synthetic radicals,[18,19] while ORAC uses peroxyl radicals.[20] In all these methods, Trolox, a water-soluble analogue of vitamin E, is used as the reference standard, and results are expressed as Trolox equivalents; that is, as the μmol of Trolox necessary to provide the same AC as 1 g of the sample.

All of these methods were initially developed to be applied in aqueous–organic extractions from foods (*i.e.*, to determine the AC associated with EPP). However, when the relevance of NEPP emerged, the AC associated with NEPP was also determined in order to obtain the overall data on food AC.[21] DPPH is the method that is applied least for the evaluation of the AC derived from NEPP, because it cannot be applied to biological fluids, which limits potential extrapolation of the results. Of the other methods, although ORAC has been applied, for instance, to evaluate the AC associated with NEPP in the fruit and vegetables most commonly consumed in four European countries,[22] it is known that this method is more susceptible to interference than the others.[23] Therefore, the most common methods used to evaluate AC from NEPP are FRAP and ABTS.

When measuring the AC associated with NEPP, it must be evaluated in the two classes of NEPP described above (*i.e.*, NEPA and HPP). Nevertheless, the butanol medium needed to release NEPA constitutes a problem for the application of most AC methods. Thus, ORAC cannot be applied to this fraction and FRAP has only been applied in some specific cases,[24] making ABTS the only AC method suitable for routine analysis of both the HPP and NEPA fractions. Therefore, if a food does not contain NEPA, the AC derived from the NEPP could be estimated using either the FRAP or ABTS assays. However, if it does contain NEPA, then the ABTS assay should be chosen in order to consider the total AC derived from the NEPP.

13.2.3 Evaluation of Dietary AC

Once data for the AC from NEPP (and also EPP in order to estimate total dietary AC) have been obtained, they must be matched with food consumption data in order to obtain dietary AC. When dietary AC is estimated in a specific cohort, consumption data will come from food records (24 hours recalls or Food Frequency Questionnaires) provided by the participants.[4,9] Meanwhile, to estimate the dietary AC for whole countries, other kinds of data are used: food questionnaires in a representative selection of subjects, official agriculture statistics, household expenditure surveys, *etc.*[22]

Some specific aspects should be considered when matching the AC of individual food items with food consumption data. First, the determination of AC is performed in dried foods, while consumption data correspond to fresh foods. Also, data from agriculture statistics or expenditure surveys correspond to whole products (*i.e.*, including non-edible parts), while the AC determinations are for the edible parts: the only ones that should be considered when performing the calculations. Also, the data may correspond to raw foods, while some foods are consumed cooked; thus, yield factors between the amount of raw food and the amount of cooked food should be applied.[25] Finally, in all of the systems for collecting food consumption data, it is common to find an 'others' heading for each food class, grouping all of the less common fruits, vegetables, *etc.* It is therefore recommended to calculate a mean AC value for this group, obtained from the other items detailed for the food class.

13.3 AC of NEPP in Common Foods

13.3.1 AC of NEPP in Fruit and Vegetables

Fruit and vegetables are rich sources of polyphenols, considered as the sum of EPP and NEPP, both of which make an important contribution to overall AC. Table 13.1 shows a selection of AC values derived from NEPP in common fruit and vegetables, obtained using the ABTS assay.[26]

All the analysed samples yielded important AC values derived from HPP. In particular, the highest values were determined in kiwi and banana in the group of fruits and in artichoke and chard in the group of vegetables. AC derived from NEPA was detected in about half of the fruits analysed, with the highest values in kiwi and strawberry. No AC derived from NEPA was found in vegetables, since common vegetables do not contain this class of NEPP.[27] The most remarkable fact revealed by these data, however, is that in these samples, AC derived from NEPP ranged from 20% to 81% (with a mean value of 64%) of the total AC, including EPP.[26]

Additionally, AC derived from NEPP was determined by the ORAC assay in fruit and vegetables commonly consumed in four European countries.[22] The lowest AC value for fruits was obtained in watermelon, while the highest corresponded to orange (11.1 and 109.1 μmol Trolox equivalents $(g\ dw)^{-1}$, respectively). In vegetables, the lowest AC value was found in cucumber and onion, while the highest was determined in Brussels sprouts (9.5 and 93.5 μmol Trolox equivalents $(g\ dw)^{-1}$, respectively).

Interestingly, some tropical fruits, such as acerola (*Malpighia emarginata*), cashew apple (*Anacardium occidentale*) or tamarind (*Tamarindus indica* L.), have very high AC values associated with NEPP.[28,29] The relevance of tropical fruits as dietary sources of NEPP is discussed in detail in Chapter 6.

Table 13.1 Selected AC values derived from HPP and NEPA in fruits and vegetables by ABTS assay (μmol Trolox equivalents (g dw)$^{-1}$).[a,b,c]

Food group	Sample	AC derived from HPP	AC derived from NEPA
Fruits	Apple	8.7 ± 0.9	n.d.
	Banana	39.9 ± 2.2	8.4 ± 1.0
	Grape	6.1 ± 0.3	2.6 ± 1.1
	Kiwi	21.4 ± 0.8	60.1 ± 5.6
	Mandarin	3.0 ± 0.1	n.d.
	Melon	4.6 ± 0.3	n.d.
	Orange	3.5 ± 0.08	n.d.
	Peach	5.3 ± 0.3	n.d.
	Pear	8.8 ± 0.9	16.7 ± 1.6
	Strawberry	4.5 ± 0.3	42.5 ± 3.6
	Watermelon	5.8 ± 0.3	n.d.
Vegetables	Artichoke	41.1 ± 2.7	n.d.
	Asparagus	5.3 ± 1.0	n.d.
	Cabbage	9.7 ± 2.3	n.d.
	Carrot	3.1 ± 0.3	n.d.
	Chard	15.2 ± 1.2	n.d.
	Cucumber	3.3 ± 0.8	n.d.
	Green bean	2.9 ± 0.6	n.d.
	Lettuce	6.0 ± 0.4	n.d.
	Onion	3.1 ± 0.4	n.d.
	Pepper, red	0.8 ± 0.2	n.d.
	Tomato	2.1 ± 0.3	n.d.

[a]Data from Pérez-Jiménez and Saura-Calixto.[26]
[b]All values correspond to AC released from the residues obtained after performing aqueous–organic extractions in the samples.
[c]ABTS: 2,2′-Azino-bis(3-ethylbenz-thiazoline-6-sulfonic acid); AC: antioxidant capacity; HPP: hydrolysable polyphenols; n.d.: not detected; NEPA: non-extractable proanthocyanidins.

Besides comparing the AC of different food items, the AC associated with NEPP has also been evaluated in specific studies in fruit and vegetables. For instance, the influence of the ripening stage on the AC derived from NEPP has been studied in plums, showing only minor differences in this parameter during the fruit ripening period.[30] Also, the AC derived from NEPP (using the FRAP assay) was determined in different varieties of peppers.[31] It has also been reported that high hydrostatic pressures may be used as a strategy to increase the AC derived from NEPP by up to three-fold (measured by the FRAP or ABTS assays) in plantain pulp.[32]

All of these results show that NEPP in fruit and vegetables are important contributors to total AC, and in many cases they contribute even more than EPP. Indeed, NEPP may contribute to the health effects associated with the consumption of these food groups. However, they are generally not taken into account in chemical and nutritional studies.[27] Indeed, even studies devoted to AC from NEPP in fruit and vegetables usually focus on different derived by-products[33–35] and not on the original foods. Although these by-products are undoubtedly NEPP concentrates that deserve specific studies, it should not be overlooked that the original fruit and vegetables are also important sources of AC derived from NEPP.

13.3.2 AC of NEPP in Cereals, Pulses and Nuts

Cereals are the food group for which a major fraction of the AC derived from NEPP was first reported and for which most data are available. This is due to the high content of HPP in these food products, present as phenolic acids – mostly ferulic acid – strongly associated with the food matrix. Traditionally, HPP in cereals have been called insoluble or bound phenolic acids. Meanwhile, NEPA have not been detected in the most commonly consumed cereals,[21,27] although extractable proanthocyanidins have been found in some specific cereal products.[36]

A selection of AC values derived from NEPP in cereal products is shown in Table 13.2. By the FRAP assay, the highest AC values were found for wheat flour, followed by millet and oat flour, for which similar values were reported; by the ABTS assay, the strongest activity was reported for red rice and the weakest corresponded to barley and black rice. Significant correlations have been reported between the AC derived from NEPP and the NEPP content in several cereals, determined either by the spectrophotometric Folin assay or by high-performance liquid chromatography–diode array detector (HPLC-DAD) analysis.[47,48]

Table 13.2 Selected AC values derived from HPP in cereals, nuts and pulses (μmol Trolox equivalents (g dw)$^{-1}$).[a,b]

Food group	Sample	Antioxidant capacity	Method	Reference
Cereals	Barley	7.44–9.88	ABTS	37
	Black rice	4.11 ± 0.02	ABTS	38
	Red rice	11.8–21.7	ABTS	39
	Millet	6.77–86.13	FRAP	40
	Oat bran	30.59 ± 1.71	FRAP	21
	Spelta	2.05	ABTS	41
	Wheat flour	108.30 ± 0.46	FRAP	21
Pulses	Black bean	4.97 ± 0.04	FRAP	42
	Broad bean	5.56	ABTS	41
	Chickpea	101.25 ± 6.19	FRAP	43
	Lentils	40.32 ± 4.65	FRAP	43
	Soybean	1.39 ± 0.06	FRAP	43
Nuts	Almond	4.13	ABTS	41
	Brazil nut	6.90 ± 0.60	ABTS	44
	Chestnut	66.49 ± 3.57	FRAP	43
	Defatted walnut	154.88 ± 15.26	FRAP	45
	Peanut skin	13.4 ± 0.8	ABTS	46

[a]All values correspond to AC released from the residues obtained after performing aqueous–organic extractions in the samples.
[b]ABTS: 2,2′-Azino-bis(3-ethylbenz-thiazoline-6-sulfonic acid); AC: antioxidant capacity; HPP: hydrolysable polyphenols; FRAP: ferric/reducing antioxidant power; HPP: hydrolysable polyphenols.

The AC from NEPP in cereal products may be affected by processing, either at home or industrially. In this way, rice lost 71% of its AC derived from NEPP due to boiling, as determined by the FRAP assay.[21] In contrast, baking may increase the AC derived from NEPP in the case of cereals containing proanthocyanidins, such as barley. This is due to the fact that this class of phenolic compounds becomes associated with the food matrix during this process through, among other mechanisms, condensation with sugars.[36] Also, as with the AC derived from EPP, that from NEPP may be highly influenced by several agronomical factors. Thus, when the AC from NEPP was measured in six varieties of wheat bran, differences of up to 50% were observed,[49] and they were as much as 200% in 20 varieties of barley.[50] Also, milling causes a decrease in the AC derived from NEPP, as reported for black and red rice.[38] Finally, due to their strong association with dietary fibre, the fractions of food with the highest dietary fibre content will also exhibit the highest AC derived from NEPP; in cereals, this applies to bran tissues.[51]

Interestingly, when a chemical extraction for the release of NEPP and further determination of their AC was compared with an enzymatic extraction (with digestive enzymes emulating physiological conditions), it was found that the enzymes were able to release a higher fraction of AC from cereal samples.[21] Thus, under physiological conditions, the fraction of AC that is bioaccessible in the small intestine may be higher than that obtained under laboratory conditions. For instance, boiled rice had an AC (using the FRAP assay) derived from NEPP of 1.07 μmol Trolox equivalents $(g\ dw)^{-1}$ when an aqueous–organic procedure was applied, while this value was 4.30 when an enzymatic procedure was carried out, indicating a higher release of phenolic compounds from the food matrix. This aspect should be considered when analysing data on the AC derived from NEPP in cereals (and other food products), since they are usually based on chemical extractions. Nonetheless, under physiological conditions, a fraction of the AC derived from NEPP is not released in the corresponding supernatants and reaches the colon intact *in vivo*. Most of this AC would ultimately become available, since the colonic microbiota is able to release the fraction of phenolic acids associated with dietary fibre from the food matrix,[52] thus causing local increases of AC.[53] This process is described in detail in other chapters of this book.

Some studies have revealed the structural features of NEPP in cereals that contribute to their AC. Thus, after isolating different complexes of ferulic acid associated with arabinoxylans in rice, it was found that the AC value was not only determined by the ferulic acid content, but also by the content of uronic acids and the molecular weight of the complex.[54] Similar conclusions regarding the role of uronic acids were found in another study in millet after isolation of complexes of hydroxycinnamic acids associated with arabinoxylans.[55]

Recently, a strategy to increase the AC derived from NEPP in cereal products was reported.[56] Briefly, a combination of free polyphenols and wheat bran under alkaline conditions caused the association of the phenolic compounds with the dietary fibre, since they were transformed to quinones capable of linking to the free amino acids present on the surface of the wheat bran.

In pulses, the AC derived from NEPP is due to both HPP and NEPA, since both classes of NEPP are commonly present in this food group.[27] However, most available data on the AC derived from NEPP in pulses correspond to HPP; some of them are shown in Table 13.2. Chickpea is the pulse for which the highest AC derived from NEPP has been reported of about 100 μmol Trolox equivalents (g dw)$^{-1}$, while lentils have about 40 and black beans or soybeans are below 5 μmol Trolox equivalents (g dw)$^{-1}$. Regarding the chemical nature of these HPP, it has been described that in dicotyledons (including pulses) ester and ether linkages are formed between feruloyl residues and lignin, other cell wall components and proteins.[57] Again, agronomical and post-harvest processes affect the AC derived from NEPP in pulses. Thus, it was observed that the AC derived from NEPP in lentils increased significantly during the germination process due to a condensation of previously soluble phenolic compounds, since it was concomitant with a significant decrease in AC from EPP.[58]

It has been suggested to partially replace wheat flour with lentils or carob flours during bread production, as when doing so, the AC derived from NEPP in the final products was between 46% and 84% by the FRAP assay. This shows once again the important role of this fraction of polyphenols in the total AC.[59]

Several studies have addressed the AC derived from NEPP in nut samples, where previous defatting is needed, as explained above. In this food group, the AC is derived from both HPP and NEPA. In the particular case of walnuts and chestnuts, the HPP consist of phenolic acids associated with the food matrix and ellagitannins – polyphenols of high molecular weight known to reach the colon intact, where they are extensively transformed by the microbiota.[60] Indeed, these two are the nuts that exhibit the highest AC values, while the lowest have been reported for peanut skin and Brazil nuts (Table 13.2).

As is the case with cereals, the nut fractions with the highest dietary fibre contents also show the highest AC derived from NEPP; thus, the values obtained for the brown skin of Brazil nuts were ten-fold higher than those in the kernel.[44] Regarding the influence of processing, when peanut skin was subjected to γ-irradiation as a way to reduce microbiological modifications, the AC derived from NEPP was significantly increased – up to 48% – due to the formation of new cross-links with the cell wall.[46]

In summary, as happens with other food groups, the AC derived from NEPP in cereals, pulses and nuts is at least equivalent to, and often greater than, that derived from EPP,[21,40,61] highlighting the important contribution of this fraction to the pool of bioactive compounds in plant foods.

13.3.3 Antioxidant Capacity of NEPP in Beverages

For decades, it was considered that beverages did not contain dietary fibre. However, about 10 years ago, it was shown that soluble dietary fibre is present in beverages such as wine.[62] The next step was to determine the presence of NEPP associated with this dietary fibre and their derived AC. Thus, it was reported that AC derived from NEPP was about 100 μmol Trolox equivalents L^{-1} in white wine and 200–600 μmol Trolox equivalents L^{-1} in red wine by the FRAP assay.[16] Indeed, significant correlations between these values and the polyphenol content of wine were found, showing that these AC values were mostly due to these compounds and not other molecules present in the food matrix. Besides small phenolic structures associated with the cell wall, wine also contains NEPP as high-molecular-weight polyphenols; indeed, it has been reported that their AC is similar to that of free low-molecular-weight phenolic compounds,[63] usually considered the only antioxidant compounds in wine.

In coffee, the AC derived from NEPP is much higher than in wine, with values from 4500 to 6100 μmol Trolox equivalents L^{-1} by the FRAP assay. It was shown that these AC values were mostly due to polyphenols and that the contribution of melanoidins, formed during coffee roasting, was low. Also, the mode of preparation of coffee affects total AC values, with the highest ones detected in soluble coffee, followed by espresso preparations and filter coffee. Interestingly, when the NEPP complex in coffee was treated with cellulase, a release of AC was observed, indicating that a fraction of these phenolic compounds was previously associated with cellulose.[64] The AC derived from NEPP in red tea is much lower: about 1200 μmol Trolox equivalents L^{-1} by the FRAP assay.[65] Nevertheless, this parameter remains to be determined in black tea, where a higher value would be expected, due to the condensation processes that take place in that beverage. Finally, the presence of the AC derived from NEPP was also reported in a less common herbal tea infusion of the roselle flower (*Hibiscus sabdariffa*).[66]

In beer, the AC derived from NEPP is about 320 μmol Trolox equivalents L^{-1} by the FRAP assay in blonde beer and 1190 in dark beer. This variation in AC accords with the fact that the dietary fibre content of dark beer is three-fold higher than that of blonde beer.[67] In cider, the AC derived from NEPP was reported as 500 μmol Trolox equivalents L^{-1} by the FRAP assay.[65]

In juices, there is high variability in the AC values derived from NEPP: from 50 to 700 μmol Trolox equivalents L^{-1} by the FRAP assay. The highest values were obtained for peach or grape juice, intermediate values were reported for tomato or orange juice and the lowest were obtained for pineapple juice.[65]

Two recent studies[68,69] focused on the development of a water-extracted cacao–coffee beverage and new edible bean beverages. In both cases, it was found that when the processing temperature was increased, the AC of the beverages also increased, indicating a partial release of the AC derived from NEPP. This suggests a route to explore in the development of new beverages (*i.e.*, particular proportions between the AC derived from EPP and NEPP could be achieved using different processing conditions).

An *in vitro* study[70] emulated the *in vivo* effects of NEPP from beverages. In particular, it is known that when the more exposed fractions of the non-digestible parts of food are in the presence of free radicals during the digestion process, the AC derived from NEPP is decreased, since they react with those molecules. Nevertheless, the study found that if this non-digestible fraction is in the presence of soluble antioxidant compounds (as happens in the digestive tube), they can help to regenerate the AC derived from NEPP. Thus, the following order of regeneration capacity, as measured by the FRAP and DPPH assays, was established: green tea > espresso coffee > black tea > instant coffee > orange juice > red wine. Also, it should not be forgotten that, as discussed above for other food groups, the AC derived from the beverage matrix when using enzymatic treatments is higher than those obtained from chemical extractions.[63] Nonetheless, a significant fraction of the AC associated with NEPP still reaches the colon intact.

13.4 Contribution of NEPP to Total Dietary AC

The individual determination of the AC of foods is a necessary step in order to determine dietary AC – a parameter associated with positive modifications to several markers of disease risk or health outcomes, as discussed above. Although studies of dietary AC have mostly focused on the AC derived from EPP, since many more data have been available for these compounds until recently, during the last couple of years some studies have also dealt with the dietary AC derived from NEPP.

An Italian study[41] collected samples of plant food consumed in Italy, in accordance with the EPIC cohort from Varese province, and calculated the AC for all food items, including NEPP in the case of cereals, pulses and nuts. Thus, they provided a database that could be used to estimate AC intake in different populations, although they did not apply those data to evaluate it in their cohort. Further studies did use those data to calculate AC intake in several populations, but the specific contributions of NEPP were not mentioned.[8,71]

In contrast, another recent study explored the contribution of EPP and NEPP to the foods consumed in Spain, calculating the total dietary AC in the country.[26] In particular, the AC from EPP and NEPP were determined in the 41 most commonly consumed food items, grouped into several categories: fruit, vegetables, cereals, nuts, legumes, cocoa products and oils. AC data for the 13 most consumed beverages were taken from previous studies.[16,62,64,65] All of these data, combined with official food consumption surveys, were used to calculate the total AC intake in the Spanish diet, which was estimated at 8000 μmol Trolox equivalents day^{-1}. Figure 13.3 shows the contribution of the different food groups to this AC intake, with separate data for EPP and NEPP. It can be seen that the groups with the highest contributions to AC intake were beverages and fruit for the AC derived from EPP, while in the case of the AC derived from NEPP, the highest contributors were cereals, followed by similar contributions from fruit, beverages and cocoa products. Again, the most remarkable aspect is that NEPP contribute 61% to the total AC intake. Therefore, studies of AC intake that focus exclusively on EPP are ignoring at least half of the total AC intake. Interestingly, in this study, the authors also estimated AC intake based on food consumption data from the 1960s, when the dietary pattern was closer to that of the Mediterranean diet. They found

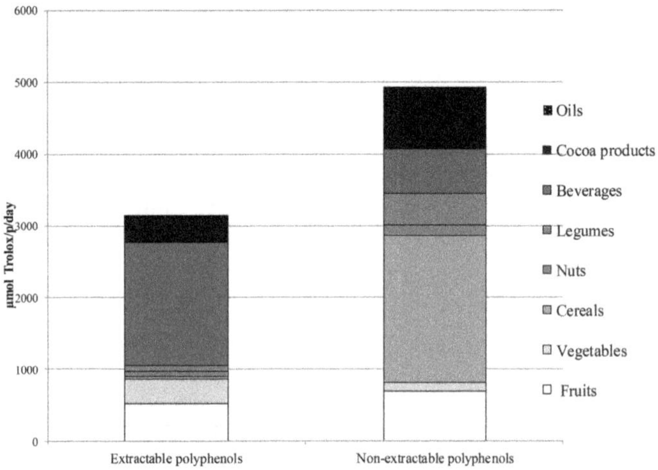

Figure 13.3 Daily dietary AC (ABTS assay) derived from EPP and NEPP in the Spanish diet (adapted from Pérez-Jiménez and Saura-Calixto[26]). ABTS: 2,2-Azino-bis(3-ethylbenzothiazoline-6-sulphonate); AC: antioxidant capacity; EPP: extractable polyphenols; NEPP: non-extractable polyphenols.

that modifications in the Spanish diet had caused a decrease of about 30% in the total AC intake, but the proportional contributions of EPP and NEPP had remained similar.

Other studies have compared dietary AC intake in different diets (*i.e.*, in European regions with high [Copenhagen, Denmark] and low [Murcia, Spain] incidence rates of colorectal cancer). These diets show significant differences, among other parameters, in the AC derived from NEPP, which was higher in Murcia than in Copenhagen.[72] Some differences in the partial contributions of the different food groups to AC intake were also observed; thus, the contribution of cereals was higher in Copenhagen than in Murcia due to the higher consumption of wholegrain cereals in the former region.[73]

The AC intake derived from NEPP was also determined in a rural population in Mexico.[74] The authors focused on fruit and vegetables, determining the AC from EPP and NEPP in these food groups. For both EPP and NEPP, the AC intake from fruit was higher than that from vegetables by both the FRAP and ABTS assays. For instance, the AC intake derived from NEPP in fruit was about 750 µmol Trolox equivalents day^{-1}, while in vegetables the value was 450 µmol Trolox equivalents day^{-1} (both by the FRAP assay). The specific contributions of individual foods could not be ascertained since the determinations were carried out in two composites representing either fruit or vegetable intake in the study population.

The common procedure used to evaluate the AC of foods in all these studies is to analyse them as they are consumed (*e.g.*, pulses or cereals are boiled before carrying out the procedures for the release of NEPP and the evaluation of their AC). Nevertheless, these food items are not consumed individually but in dishes, which means that interactions (additive, synergistic or antagonistic) may appear between the different food ingredients.[75] In this way, a recent study[76] evaluated the AC derived from EPP and NEPP in common Italian dishes analysed as a whole. They found that the contribution of NEPP to the total AC (by FRAP assay) ranged from 42% in *carciofi alla romana* (stuffed artichoke with garlic, mint and parsley) to 95% in *pan di Spagna* (sponge cake). Interestingly, a significant correlation ($R^2 = 0.8947$, $p = 0.015$, Pearson assay, own calculation based on data included in the paper) was found between the AC derived from NEPP and the dietary fibre content of these dishes, highlighting the strong association between these two dietary constituents.

Although all of these data consistently show the important (and neglected) contribution of NEPP to total dietary AC, further studies are needed in more populations. In particular, possible associations between the dietary AC derived from NEPP and disease biomarkers or health outcomes should be explored, as has previously been done for the dietary AC based exclusively on EPP.

13.5 Contribution of NEPP to *in vivo* Oxidative Status

Most data on the role polyphenols play in *in vivo* oxidative status are based, as discussed above, on studies of EPP. Studies on this topic that focus on NEPP, although scarce, show promising results.

The potential effects of NEPP on *in vivo* oxidative status are based on their metabolic fate. First, they may exert local effects inside the upper part of the gastrointestinal tract, which is described as a major site of action of dietary polyphenols.[77] Thus, they would be able to scavenge free radicals in the stomach and the small intestine, especially when NEPP are associated with dietary fibre.[78] However, the AC derived from NEPP is mostly released in the colon through the action of the microbiota, which is capable of either breaking the linkage between phenolic compounds and the food matrix[79] or transforming them into other metabolites.[80] This gives rise to two different processes: (1) local improvement of oxidative status in the colon; and (2) absorption of released metabolites, which may also exhibit AC that will be available in the bloodstream, as well as in target organs and tissues.

Regarding the local effects, it has been reported that NEPP are capable of increasing the caecum AC in rats. In particular, supplementation with a NEPP-rich grape pomace derivative for 5 weeks caused a five-fold increase in the caecum AC derived from NEPP.[53] At the same time, this kind of supplementation improved oxidative status by other mechanisms: enhancing the endogenous antioxidant system, with significant increases in the activity of mitochondrial superoxide dismutase, as well as in the ratio between the reduced and oxidised glutathione forms.[81,82] Finally, a significant decrease in lipid oxidation in colonic mucosa, as measured by the sum of malondialdehyde and 4-hydroxynonenal, was observed.[82] Overall, these effects on oxidative status in the colon may be quite important, since it is known that along the gastrointestinal tract there is a tendency towards a decrease in AC and the activity of antioxidants enzymes, with the lowest values in the colon.[83] This capacity of NEPP may be related with the potential prevention of certain diseases, such as colorectal cancer; indeed, it has been reported that patients with that disease present with a heightened oxidant status in the colon.[84]

Although measurements of colonic AC are difficult to carry out in humans, the AC of faeces may be used as an indirect marker of oxidative status in the colon, since increases in caecum AC are concomitant with increases in faecal AC.[53] In this way, when healthy subjects followed two diets with high or low AC (being similar in both macro- and micro-nutrient characteristics, as well as in dietary fibre content), a significant increase in the faecal AC of those subjects following the high-AC diet was observed.[85] The design of these diets was exclusively based on EPP (with a significant fraction of them also reaching the colon and therefore locally modifying oxidative status), but these promising results show the need to develop similar studies that include the AC derived from NEPP when designing diets with low or high AC.

Regarding the second process that takes place after NEPP transformation in the colon by the microbiota (*i.e.*, the absorption of the released metabolites), it is notable that 8 hours after supplementation of healthy humans with a NEPP-rich product, significant increases in plasma AC were observed, while these variations are usually observed 2–4 hours after EPP intake.[86] These results suggest that consumption of NEPP gives rise to prolonged (compared to EPP intake) circulation in the body of metabolites exhibiting AC. Along the same lines, a recent study[87] in which rats were supplemented for 2 weeks either with tea catechin (an EPP) or NEPP from dried persimmon showed that the plasma AC values were similar in both groups and significantly higher than those in the control group, indicating the transformation of NEPP into absorbable antioxidant compounds.

Since these colonic metabolites would ultimately be responsible for the modification of *in vivo* oxidative status, their AC has been explored. Thus, metabolites such as 3,4-dihydroxybenzoic acid, 3,4-dihydroxyphenylacetic acid, coumaric acid or dihydroferulic acid exhibited AC similar to those of their parent phenolic compounds, while other colonic metabolites (3-hydroxyphenylacetic acid or 4-hydroxybenzoic acid) had no measurable AC.[88] Some key structural aspects for the AC of these compounds have been studied; for instance, it was shown that the AC exhibited by the microbial metabolite 5-(3′,4′-dihydroxyphenyl)-γ-valerolactone was present in an isomer, but not in γ-valerolactone, which indicated the important contribution of the catechol moiety to the AC of the molecule.[89] Eventually, these metabolites may be related to other physiological processes; for instance, consumption of wholegrain wheat, as compared to refined wheat – therefore providing a higher dose of NEPP – led to a two-fold increase in faecal and urinary ferulic acid and a four-fold increase in microbial-derived dihydroferulic acid. All of this was concomitant with a decrease in plasma levels of TNF-α, an inflammation marker.[90]

Overall, the consideration of NEPP as dietary antioxidants, taking into account their particular metabolic fate, might modify previous results concerning the *in vivo* effects of dietary antioxidants. Moreover, synergism is another aspect that is rarely considered in clinical trials dealing with dietary antioxidants and has not been explored in the specific case of NEPP interacting with other dietary antioxidants. Therefore, new clinical studies should be designed, specifically aiming to elucidate the potential effects of NEPP – alone and in combination with other dietary antioxidants – on *in vivo* oxidative status and diseases associated with it.

13.6 Final Remarks

- Dietary AC has been shown to have significant beneficial associations with several disease biomarkers and health outcomes in many studies. However, they are mostly based exclusively on EPP.
- Studies of AC derived from NEPP in individual foods (as a previous step in order to determine total dietary AC) have shown that, in all groups of food and beverages, NEPP contribute greatly to total AC.
- Studies in different populations have shown that NEPP provide about half of the dietary AC. The association of the dietary AC derived from NEPP with health outcomes remains to be elucidated.
- Preliminary animal and human studies have shown that the AC provided by NEPP gives rise to significant improvements in *in vivo* oxidative status.
- A new generation of clinical trials should be designed that takes into account the role of NEPP as dietary antioxidants. Then the actual complete network of dietary antioxidants consumed daily would be considered. This might resolve some problems with previous studies of dietary antioxidants.

List of Abbreviations

AAPH	2,2′-Azobis(2-amidinopropane) dihydrochloride
ABTS	2,2-Azinobis (3-ethylbenzothiazoline-6-sulfonate)
AC	Antioxidant capacity
DPPH	2,2-Diphenyl-1-picrylhydrazyl
EPIC	European Prospective Investigation into Cancer and Nutrition study
EPP	Extractable polyphenols
FRAP	Ferric reducing/antioxidant power
HPLC-DAD	High-performance liquid chromatography–diode array detector
HPP	Hydrolysable polyphenols
NEPA	Non-extractable proanthocyanidins
NEPP	Non-extractable polyphenols
ORAC	Oxygen radical absorbance capacity

Acknowledgements

Financial support from the Spanish Ministry of Economy and Competitiveness (*via* project AGL2014-55102-JIN) is acknowledged.

References

1. D. Giustarini, I. Dalle-Donne, D. Tsikas and R. Rossi, *Crit. Rev. Clin. Lab. Sci.*, 2009, **46**, 241–281.
2. A. M. Rodríguez-Mateos, D. Vauzour, C. G. Krueger, D. Shanmuganayagam, J. Reed, L. Calani, P. Mena, D. del Rio and A. Crozier, *Arch. Toxicol.*, 2014, **88**, 1803–1853.
3. M. Serafini, C. Miglio, I. Peluso and T. Petrosino, *Curr. Top. Med. Chem.*, 2011, **11**, 1821–1846.
4. F. Brighenti, S. Valtuena, N. Pellegrini, D. Ardigo, D. Del Rio, S. Salvatore, P. Piatti, M. Serafini and I. Zavaroni, *Br. J. Nutr.*, 2005, **93**, 619–625.
5. F. Saura-Calixto and I. Goñi, *Food Chem.*, 2006, **94**, 442–447.
6. Y. Wang, M. Yang, S.-G. Lee, C. G. Davis, S. I. Koo, M. L. Fernandez, J. S. Volek and O. K. Chun, *Eur. J. Nutr.*, 2014, **53**, 1363–1369.
7. M. Yang, S.-J. Chung, A. Floegel, W. O. Song, S. I. Koo and O. K. Chun, *Eur. J. Nutr.*, 2013, **52**, 1901–1911.
8. B. Puchau, M. A. Zulet, A. G. de Echávarri, H. H. M. Hermsdorff and J. A. Martínez, *Nutrition*, 2010, **26**, 534–541.
9. D. Del Rio, C. Agnoli, N. Pellegrini, V. Krogh, F. Brighenti, T. Mazzeo, G. Masala, B. Bendinelli, F. Berrino and S. Sieri, *J. Nutr.*, 2011, **141**, 118–123.
10. J. Mares, *JAMA, J. Am. Med. Assoc.*, 2015, **313**, 1048–1049.
11. M. Malvezzi, P. Bertuccio, F. Levi, C. La Vecchia and E. Negri, *Ann. Oncol.*, 2013, **24**, 792–800.
12. N. Bastide, L. Dartois, V. Dyevre, L. Dossus, G. Fagherazzi, M. Serafini and M.-C. Boutron-Ruault, *Eur. J. Nutr.*, 2017, **56**, 1233–1243.
13. A. I. Kolomvotsou, L. S. Rallidis, K. C. Mountzouris, J. Lekakis, A. Koutelidakis, S. Efstathiou, M. Nana-Anastasiou and A. Zampelas, *Eur. J. Nutr.*, 2013, **52**, 37–48.
14. J. Pérez-Jiménez, M. E. Díaz-Rubio and F. Saura-Calixto, *Nutr. Res. Rev.*, 2013, **26**, 118–129.
15. J. Pérez-Jiménez, S. Arranz, M. Tabernero, M. E. Díaz-Rubio, J. Serrano, I. Goñi and F. Saura-Calixto, *Food Res. Int.*, 2008, **41**, 274–285.
16. F. Saura-Calixto and M. E. Díaz-Rubio, *Food Res. Int.*, 2007, **40**, 613–619.
17. I. F. Benzie and J. Strain, *Anal. Biochem.*, 1996, **239**, 70–76.
18. R. Re, N. Pellegrini, A. Proteggente, A. Pannala, M. Yang and C. Rice-Evans, *Free Radical Biol. Med.*, 1999, **26**, 1231–1237.
19. W. Brand-Williams, M. Cuvelier and C. Berset, *LWT–Food Sci. Technol.*, 1995, **28**, 25–30.
20. B. Ou, M. Hampsch-Woodill and R. L. Prior, *J. Agric. Food Chem.*, 2001, **49**, 4619–4626.
21. J. Pérez-Jiménez and F. Saura-Calixto, *J. Agric. Food Chem.*, 2005, **53**, 5036–5040.

22. J. Pérez-Jiménez and F. Saura-Calixto, *Food Res. Int.*, 2015, **74**, 315–323.
23. J. Pérez-Jiménez and F. Saura-Calixto, *Food Res. Int.*, 2016, **39**, 791–800.
24. M. Tabernero, J. Serrano and F. Saura-Calixto, *Int. J. Food Sci. Technol.*, 2006, **41**, 28–32.
25. A. Bognar, *Ber. Bundesforschungsanst. Ernaehr.*, 2002, 1–98.
26. J. Pérez-Jiménez, M. E. Díaz-Rubio and F. Saura-Calixto, *Plant Foods Hum. Nutr.*, 2015, **70**, 365–370.
27. S. Arranz, J. M. Silván and F. Saura-Calixto, *Mol. Nutr. Food Res.*, 2010, **54**, 1646–1658.
28. M. d. S. M. Rufino, J. Pérez-Jiménez, M. Tabernero, R. E. Alves, E. S. De Brito and F. Saura-Calixto, *Int. J. Food Sci. Technol.*, 2010, **45**, 2227–2233.
29. M. E. Páez-Peñuñuri, G. Mercado-Mercado, F. J. Blancas-Benitez, R. B. Villegas-González and S. G. Sáyago-Ayerdi, *BIOtecnia*, 2016, **18**, 10–21.
30. J. Kristl, M. Slekovec, S. Tojnko and T. Unuk, *Food Chem.*, 2011, **125**, 29–34.
31. D. Hervert-Hernández, S. G. Sáyago-Ayerdi and I. Goñi, *J. Agric. Food Chem.*, 2010, **58**, 3399–3406.
32. M. C. Jiménez-Martínez, E. Montalvo-González, S. G. Sáyago-Ayerdi, G. Mercado-Mercado, J. A. Ramírez de León, E. Paz-Gamboa and M. A. Vivar-Vera, *J. Sci. Food Agric.*, 2017, **97**, 2508–2515.
33. I. Navarro-González, V. García-Valverde, J. García-Alonso and M. J. Periago, *Food Res. Int.*, 2011, **44**, 1528–1535.
34. W. W. Tow, R. Premier, H. Jing and S. Ajlouni, *J. Food Sci.*, 2011, **76**, 163–172.
35. F. J. Esparza-Martínez, R. Miranda-López and S. H. Guzman-Maldonado, *Ind. Crops Prod.*, 2016, **84**, 1–6.
36. A. K. Holtekjølen, A. Bævre, M. Rødbotten, H. Berg and S. H. Knutsen, *Food Chem.*, 2008, **110**, 414–421.
37. T. Madhujith and F. Shahidi, *Food Chem.*, 2009, **117**, 615–620.
38. F. F. Paiva, N. L. Vanier, J. D. J. Berrios, J. Pan, F. de Almeida Villanova, G. Takeoka and M. C. Elias, *J. Food Compos. Anal.*, 2014, **35**, 10–17.
39. D. Sumczynski, E. Kotásková, H. Družbíková and J. Mlček, *Food Chem.*, 2016, **211**, 339–346.
40. A. Chandrasekara and F. Shahidi, *J. Agric. Food Chem.*, 2010, **58**, 6706–6714.
41. N. Pellegrini, M. Serafini, S. Salvatore, D. Del Rio, M. Bianchi and F. Brighenti, *Mol. Nutr. Food Res.*, 2006, **50**, 1030–1038.
42. Y.-K. Wang, X. Zhang, G.-L. Chen, J. Yu, L.-Q. Yang and Y.-Q. Gao, *J. Funct. Foods*, 2016, **24**, 359–372.
43. A. Durazzo, V. Turfani, E. Azzini, G. Maiani and M. Carcea, *Food Chem.*, 2013, **140**, 666–671.
44. J. A. John and F. Shahidi, *J. Funct. Foods*, 2010, **2**, 196–209.
45. S. Arranz, J. Pérez-Jiménez and F. Saura-Calixto, *Eur. Food Res. Technol.*, 2008, **227**, 425–431.
46. A. C. de Camargo, M. A. B. Regitano-d'Arce, C. R. Gallo and F. Shahidi, *J. Funct. Foods*, 2015, **12**, 129–143.

47. B. Min, L. Gu, A. M. McClung, C. J. Bergman and M.-H. Chen, *Food Chem.*, 2012, **133**, 715–722.
48. A. Begum, A. Goswami and P. Chowdhury, *Int. J. Food Sci. Technol.*, 2015, **50**, 2529–2536.
49. B. Verma, P. Hucl and R. Chibbar, *Food Chem.*, 2009, **116**, 947–954.
50. A. Yoshida, K. Sonoda, Y. Nogata, T. Nagamine, M. Sato, T. Oki, S. Hashimoto and H. Ohta, *Food Sci. Technol. Res.*, 2010, **16**, 215–224.
51. C. M. Liyana-Pathirana and F. Shahidi, *J. Agric. Food Chem.*, 2006, **54**, 1256–1264.
52. L. Poquet, M. N. Clifford and G. Williamson, *Drug Metab. Dispos.*, 2008, **36**, 190–197.
53. I. Goñi and J. Serrano, *J. Sci. Food Agric.*, 2005, **85**, 1877–1881.
54. R. S. P. Rao and G. Muralikrishna, *Phytochemistry*, 2006, **67**, 91–99.
55. V. Bijalwan, U. Ali, A. K. Kesarwani, K. Yadav and K. Mazumder, *Int. J. Biol. Macromol.*, 2016, **88**, 296–305.
56. E. Doğan and V. Gökmen, *Food Res. Int.*, 2015, **69**, 189–193.
57. V. V. Lozovaya, T. A. Gorshkova, E. V. Yablokova, N. I. Rumyantseva, A. Valieva, A. Ulanov and J. M. Widholm, *Phytochemistry*, 1999, **50**, 395–400.
58. J. Yeo and F. Shahidi, *J. Agric. Food Chem.*, 2015, **63**, 379–381.
59. V. Turfani, V. Narducci, A. Durazzo, V. Galli and M. Carcea, *LWT–Food Sci. Technol.*, 2017, **78**, 361–366.
60. B. Cerdá, F. A. Tomás-Barberán and J. C. Espín, *J. Agric. Food Chem.*, 2005, **53**, 227–235.
61. A. Serpen, E. Capuano, V. Fogliano and V. Gökmen, *J. Agric. Food Chem.*, 2007, **55**, 7676–7681.
62. M. E. Díaz-Rubio and F. Saura-Calixto, *Am. J. Enol. Vitic.*, 2006, **57**, 69–72.
63. M. Noguer, A. B. Cerezo, M. Rentzsch, P. Winterhalter, A. M. Troncoso and M. C. García-Parrilla, *J. Agric. Food Chem.*, 2008, **56**, 8879–8884.
64. M. E. Díaz-Rubio and F. Saura-Calixto, *J. Agric. Food Chem.*, 2007, **55**, 1999–2003.
65. M. E. Diaz-Rubio and F. Saura-Calixto, *Int. J. Food Sci. Nutr.*, 2011, **62**, 715–718.
66. S. G. Sáyago-Ayerdi, S. Arranz, J. Serrano and I. Goñi, *J. Agric. Food Chem.*, 2007, **55**, 7886–7890.
67. M. E. Díaz-Rubio and F. Saura-Calixto, *J. Am. Soc. Brew. Chem.*, 2009, **67**, 38–43.
68. R. Y. Gan, M. F. Wang, W. Y. Lui, K. Wu and H. Corke, *Int. J. Food Sci. Technol.*, 2016, **51**, 2090–2098.
69. H. S. Kwak, M. Kim and Y. Jeong, *CyTA–J. Food*, 2016, **14**, 248–254.
70. E. E. Çelik, V. Gökmen and V. Fogliano, *J. Agric. Food Chem.*, 2013, **61**, 10329–10334.
71. K. Kim, T. M. Vance and O. K. Chun, *Nutrients*, 2016, **8**, 15.
72. M. Tabernero-Urbieta, PhD thesis, Complutense University of Madrid, 2008.
73. F. Saura-Calixto, J. Pérez-Jiménez and I. Goñi, *J. Cereal Sci.*, 2009, **50**, 291–294.

74. D. Hervert-Hernández, O. P. García, J. L. Rosado and I. Goñi, *Food Res. Int.*, 2011, **44**, 1182–1189.
75. M. Nicoli, M. Anese and M. Parpinel, *Trends Food Sci. Technol.*, 1999, **10**, 94–100.
76. A. Durazzo, S. Lisciani, E. Camilli, P. Gabrielli, S. Marconi, L. Gambelli, A. Aguzzi, M. Lucarini, G. Maiani and G. Casale, *Food Chem.*, 2017, **218**, 70–77.
77. B. Halliwell, K. Zhao and M. Whiteman, *Free Radical Res.*, 2000, **33**, 819–830.
78. F. Saura-Calixto, *J. Agric. Food Chem.*, 2011, **59**, 43–49.
79. M. F. Andreasen, P. A. Kroon, G. Williamson and M.-T. Garcia-Conesa, *J. Agric. Food Chem.*, 2001, **49**, 5679–5684.
80. M. L. Mateos-Martín, J. Pérez-Jiménez, E. Fuguet and J. L. Torres, *Br. J. Nutr.*, 2012, **108**, 290–297.
81. M. E. López-Oliva, A. Agis-Torres, I. Goni and E. Muñoz-Martínez, *Br. J. Nutr.*, 2010, **103**, 1110–1117.
82. M. E. López-Oliva, M. J. Pozuelo, R. Rotger, E. Muñoz-Martínez and I. Goni, *Br. J. Nutr.*, 2013, **109**, 4–16.
83. S. Blau, A. Rubinstein, P. Bass, C. Singaram and R. Kohen, *Mol. Cell. Biochem.*, 1999, **194**, 185–191.
84. D. Chang, F. Wang, Y.-S. Zhao and H.-Z. Pan, *Biomed. Environ. Sci.*, 2008, **21**, 286–289.
85. M. A. Bianchi, F. Scazzina, D. Del Rio, S. Valtuena, N. Pellegrini, L. Franzini, M. L. Callegari, C. Pellacani, A. Buschini and I. Zavaroni, *Br. J. Nutr.*, 2010, **104**, 1500–1507.
86. J. Pérez-Jiménez, J. Serrano, M. Tabernero, S. Arranz, M. E. Díaz-Rubio, L. García-Diz, I. Goñi and F. Saura-Calixto, *Plant Foods Hum. Nutr.*, 2009, **64**, 102–107.
87. Y. Matsumara, T. Ito, H. Yano, E. Kita, K. Mikasa, M. Okada, A. Furutani, Y. Murono, M. Shibata, Y. Nishii and S. I. Kayano, *Food Chem.*, 2016, **202**, 99–103.
88. I. B. Jaganath, W. Mullen, M. E. Lean, C. A. Edwards and A. Crozier, *Free Radical Biol. Med.*, 2009, **47**, 1180–1189.
89. F. Sánchez-Patán, M. Chioua, I. Garrido, C. Cueva, A. Samadi, J. Marco-Contelles, M. V. Moreno-Arribas, B. Bartolomé and M. Monagas, *J. Agric. Food Chem.*, 2011, **59**, 7083–7091.
90. P. Vitaglione, I. Mennella, R. Ferracane, A. A. Rivellese, R. Giacco, D. Ercolini, S. M. Gibbons, A. La Storia, J. A. Gilbert and S. Jonnalagadda, *Am. J. Clin. Nutr.*, 2015, **101**, 251–261.

CHAPTER 14

Non-extractable Polyphenols in Fruit: Distribution, Changes, and Potential Health Effects

Y. HAMAUZU

Shinshu University, Department of Agricultural and Life Sciences, 8304 Minamiminowa, Kamiina, 399-4598, Japan
*E-mail: hamauzu@shinshu-u.ac.jp

14.1 Introduction

The benefits of fruit consumption for human health are well known and supported by various kinds of research, including epidemiological and intervention studies. For example, consumption of adequate amounts of fruit in the daily diet is associated with lower risks of coronary heart disease and cancer.[1–10] Intervention studies have shown that the consumption of fresh apples increases the overall antioxidant potential of plasma, increases the levels of antioxidant enzymes such as SOD and glutathione peroxidase in red blood cells,[11] and decreases serum low-density lipoprotein (LDL)-cholesterol concentrations.[12]

Different fruits contain various types of bioactive compounds (phytochemicals), including polyphenols, carotenoids, ascorbic acid, pectin and lignin. These compounds are at least partly associated with the health-promoting effects of fruit consumption. Phytochemicals derived from cell wall components, which make up dietary fibre, seem to be especially beneficial

Food Chemistry, Function and Analysis No. 5
Non-extractable Polyphenols and Carotenoids: Importance in Human Nutrition and Health
Edited by Fulgencio Saura-Calixto and Jara Pérez-Jiménez

Published by the Royal Society of Chemistry, www.rsc.org

for gastrointestinal health and the prevention of cardiovascular disease (CVD).[13–16] Dietary fibre intake is inversely associated with the risks of coronary heart disease[17] and large intestine cancer.[18] Dietary fibre is associated, to varying degrees, with non-extractable polyphenols (NEPP) and therefore the beneficial effects of dietary fibre may actually be due to the combined effects of fibre and the associated NEPP.

NEPP are defined as polyphenols that are not significantly released from the food matrix by mastication, the acid pH of the stomach, or the action of digestive enzymes.[19] Thus, NEPP reach the colon nearly intact.[19] While the extractable polyphenols (EPP) in fruits have been well studied, this is not the case for NEPP. NEPP are strongly associated with the extraction residues in EPP extraction procedures and have often been ignored in polyphenol research.[20–22]

In this chapter, I review studies of the NEPP content of different fruits, their changes during post-harvest handling, and the health-related properties of NEPP, especially those associated with gastrointestinal health and CVD risk reduction.

14.2 Distribution and Contents of NEPP in Fruit

NEPP contents and their percentages relative to total polyphenols (EPP + NEPP) differ from one fruit species to another. In some cases, NEPP make up more than 50% of total polyphenols. Sun *et al.*[23] studied the EPP and NEPP contents in the edible portions of 11 commercial types of fruit: apple, pear, peach, strawberry, cranberry, red grape, banana, orange, grapefruit, lemon, and pineapple. Pineapples had the highest NEPP content, with 53.9 mg/100 g fresh weight (FW), and these NEPP showed the highest proportion of the total polyphenol content (57.2%) (Table 14.1). Bananas and oranges also contained high levels of NEPP (34.3 and 24.4 mg/100 g FW, respectively) with high proportions of the total polyphenols (37.9% and 30.0%, respectively). In that report, the NEPP contents (and their proportions of the total polyphenols) of cranberries and apples were relatively low: 20.1 mg/100 g FW (3.8% of total polyphenols) and 24.1 mg/100 g FW (8.1% of the total), respectively; however, the total polyphenol contents of these fruits were quite high (527 mg/100 g FW and 296 mg/100 g FW, respectively).

Meanwhile, other researchers have found different percentages of NEPP relative to total polyphenols in apples (65.2%), pears (83.1%), peaches (42.7%), grapes (24.7%), bananas (91.7%), and oranges (58.9%),[24] and these percentages are far higher than those reported by Sun *et al.*[23] (Table 14.2). These large differences may be due to differences in the conditions of the hydrolysing (solubilising) treatments applied to the extraction residues and the determination methods after these treatments. Characteristics of the different fruits, such as the cultivar, harvested stage, and post-harvest conditions, might also affect the proportions of NEPP.

Polyphenols are structurally diverse compounds. They can be classified as phenolic acids (hydroxybenzoic and hydroxycinnamic acids), flavonoids

Table 14.1 Extractable (soluble) and non-extractable (bound) polyphenol contents of fruits and their proportions to total polyphenols.[a,b,c,d,e]

Fruits	Total PP (soluble + bound) (mg/100 g FW)	EPP (soluble) (mg/100 g FW)	NEPP (bound) (mg/100 g FW)	%EPP	%NEPP
Apple	296	272	24.1	91.9	8.1
Pear	70.6	53.6	17.0	75.9	24.1
Peach	84.6	65.3	19.3	77.2	22.8
Strawberry	160	148	12.2	92.4	7.6
Cranberry	527	507	20.1	96.2	3.8
Grape (red)	201	182	19.0	90.5	9.5
Banana	90.4	56.1	34.3	62.1	37.9
Orange	81.2	56.8	24.4	70.0	30.0
Grapefruit	49.6	30.7	18.9	61.9	38.1
Lemon	82.0	66.3	15.7	80.9	19.1
Pineapple	94.3	40.4	53.9	42.8	57.2

[a]Adapted with permission from J. Sun, Y.-F. Chu, X. Wu and R. H. Liu, *J. Agric. Food Chem.*, 2002, **50**, 7449. Copyright 2002 American Chemical Society.
[b]The extractable polyphenols (EPP) and non-extractable polyphenols (NEPP) were separated by 80% acetone extraction. NEPP were solubilised by base digestion using 4 M NaOH.
[c]The Folin–Ciocalteu colorimetric assay was used (mg gallic acid equivalent).
[d]The NEPP quantities shown in the table are the totals of 'Bound-E' and 'Bound-W' in the literature.
[e]FW: fresh weight; PP: polyphenol.

Table 14.2 Extractable polyphenol (EPP) and non-extractable polyphenol (NEPP) contents (mg/100 g dry weight) of fruits and their proportions to total polyphenols.[a,b,c]

Sample	TPP (mg) (EPP + NEPP)	EPP (mg)	NEPP (mg)		%HPNE	%NEPA	%NEPP
			HPNE	NEPA			
Apple	1747	609.1	1063	76.0	60.8	4.4	65.2
Pear	2219	375.8	1427	415.8	64.3	18.7	83.1
Peach	1293	741.3	403.4	148.5	31.2	11.5	42.7
Grape	2484	1870	319.7	294.1	12.9	11.8	24.7
Banana	3581	297.3	1415	1868	39.5	52.2	91.7
Orange	1911	786.4	1125	ND	58.9	ND	58.9

[a]Adapted from *Food Research International*, 74, J. Pérez-Jiménez and F. Saura-Calixto, Macromolecular antioxidants or non-extractable polyphenols in fruit and vegetables: Intake in four European countries, Copyright 2015, 315–323, with permission from Elsevier.
[b]Extractable polyphenols (EPP) were extracted successively with MeOH/water/HCl (50 : 50, v/v, pH 2) and acetone/water (70 : 30, v/v). HPNE were extracted from the residue by hydrolysis with MeOH/H_2SO_4 (90 : 10, v/v) before analysis by Folin–Ciocalteu assay (mg gallic acid equivalent). NEPA were analysed by direct treatment with BuOH/HCl reagent (carob pod tannin was used as the standard).
[c]NEPP: non-extractable polyphenols; TPP: total polyphenols; HPNE: hydrolysed polyphenols in non-extractable form; NEPA: non-extractable proanthocyanidins.

(flavonols, flavones, isoflavones, flavanones, anthocyanidins, and flavan-3-ols), stilbenoids, or lignans. Lignins and tannins, including hydrolysable tannins and proanthocyanidins (PA; polymeric flavan-3-ols), are polyphenols derived from phenolic acids or flavan-3-ol units. The NEPP in fruit are mainly PA polymers and hydrolysable polyphenols, including hydrolysable tannins with high molecular weights and covalently bound phenolic acids.[19,25]

PA are a major group of polyphenols that are oligomers and polymers of flavan-3-ol units (epicatechin, catechin, epigallocatechin, gallocatechin, and their gallic acid esters). The larger polymers can strongly bind to cell wall components, mainly pectic polysaccharides and proteins, and thus are difficult to extract. These non-extractable PA (NEPA) mostly bind to cell wall components *via* hydrogen bonding and hydrophobic interactions,[19,26] although covalent bonding might also occur.[19,27] Fruit is the major source of PA in the diet.[28]

Hellström *et al.*[29] reported on the contents of NEPA and extractable PA (EPA) in various plant foods and the products. They extracted the EPA with acetone/methanol/water (2:2:1, v/v/v) and analysed them using normal-phase high-performance liquid chromatography (HPLC). The NEPA in the extraction residues were determined by thioacidolytic degradation followed by reversed-phase HPLC. The only vegetable sample with detectable levels of either EPA or NEPA was rhubarb (*Rheum rhaponticum*). However, most of the different fruits and fruit products contained PA, although none were detected in watermelon (*Citrullus lanatus*) or the citrus samples (grapefruit, mandarin, and orange). In that study, the highest NEPA contents (mg/100 g FW) and proportions (of total PA) were found in blackcurrants (*Ribes nigrum*; 202 mg, 50.2%), sea buckthorn (*Hippophae rhamnoides*; 199 mg, 81.5%), cranberries (*Vaccinium oxycoccos*; 181 mg, 44.6%), and redcurrants (*Ribes rubrum*; 86.8 mg, 70.8%) (Figure 14.1). Bananas (*Musa sapientum*) contained only NEPA (64.3 mg/100 g FW, 100%). Strawberries (*Fragaria ananassa*), apples (*Malus domestica*), cherries (*Prunus avium*), plums (*Prunus domestica*), and raspberries (*Rubus idaeus*) contained relatively small amounts of NEPA (<6 mg/100 g FW), representing low proportions of total PA (<10%). However, other researchers have indicated considerably higher proportions of NEPA in total PA using a direct degradation treatment of the extraction residues with a butanol/HCl reagent.[30] According to one report, the percentages of NEPA in the total PA in apples (var. Golden Delicious, with peel), peaches (var. Royal, with peel), and nectarines (var. Royal, with peel) were approximately 93%, 98%, and 85%, respectively. Because the analytical method adopted can greatly affect the results,[31] methodology is still an important issue for NEPP research.

We have analysed the PA contents of Chinese quince (*Pseudocydonia sinensis*) and quince (*Cydonia oblonga*) and showed that these fruits contain high amounts of EPA and NEPA. Analysis using thioacidolysis revealed that the PA in these fruits were extractable procyanidins (EPC) and non-extractable procyanidins (NEPC). We found that quince contained 190 mg/100 g FW of NEPC[32] and 248 mg/100 g FW of EPC,[33] and thus the NEPC proportion was

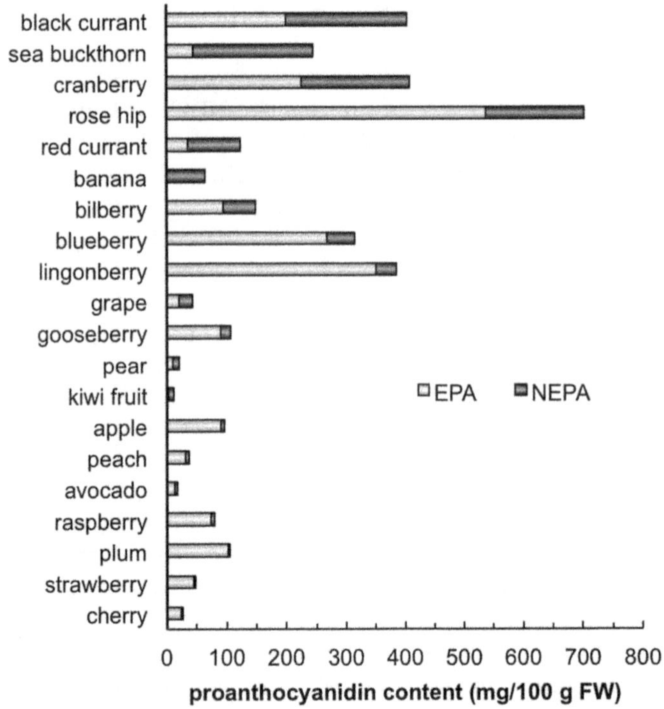

Figure 14.1 Contents of extractable proanthocyanidin (EPA) and non-extractable proanthocyanidin (NEPA) in various fruits. EPA were extracted with acetone/methanol/water (2 : 2 : 1) and analysed by normal-phase high-performance liquid chromatography (HPLC). NEPA remaining in the extraction residue were analysed by thiolysis (with benzyl mercaptan) followed by reversed-phase HPLC. FW: fresh weight. Adapted with permission from J. K. Hellström, A. R. Törrönen and P. H. Mattila, *J. Agric. Food Chem.*, 2009, **57**, 7899. Copyright 2009 American Chemical Society.

approximately 43.4%. In Chinese quince, the NEPC content (267 mg/100 g FW) was relatively high, but the proportion of NEPC in total procyanidins was relatively low (16.1%) because the EPC content was 1390 mg/100 g FW. The mean degree of polymerisation (mDP) of the procyanidins was also high in these fruits. The EPC in quince and Chinese quince had mDP values of 35 and 25, respectively,[33] and the NEPC had values of 93 and 40, respectively.[32] The higher levels of polymerisation of the NEPC resulted in a higher capacity to bind strongly with cell wall polysaccharides, and this is consistent with the non-extractable nature of the NEPC. For comparison, in the same study, we found that apples (var. Fuji) contained 27.8 mg/100 g FW of EPC (mDP = 3) and 1.8 mg/100 g FW of NEPC (mDP = 8).

Hydrolysable polyphenols are another important group that is composed of different classes of phenolic compounds (mainly phenolic acids), and some of these are non-extractable. The hydrolysable polyphenols in non-extractable form (HPNE) are covalently bound to structural cell wall components such as cellulose, hemicellulose, pectin, lignin, and rod-shaped structural proteins.[25] For example, phenolic acids such as hydroxycinnamic acid and hydroxybenzoic acid form ether linkages with lignin *via* hydroxyl groups in their aromatic rings and they form ester linkages with structural carbohydrates and proteins through their carboxylic groups. Hydrolysable tannins are also likely to form covalent and hydrogen bonds with cell wall components in the food matrix.[19]

The HPNE in fruit have been studied less extensively than the NEPA. This might be due to the lack of a convenient method for the analysis of HPNE. Chemical hydrolysis is usually concomitant with decomposition of NEPA. The resultant compounds interfere with the colorimetric assay of phenolic acids derived from HPNE. Therefore, the hydrolysate of the HPNE must be separated from NEPA-derived decomposing compounds if a conventional colorimetric assay for phenolic compounds (such as the Folin–Ciocalteu method) is to be used. Arranz *et al.*[20] analysed fruit by performing successive extractions with MeOH/water (50 : 50, v/v, pH 2) and acetone/water (70 : 30, v/v) then hydrolysis of the residue with MeOH/H_2SO_4 (90 : 10, v/v). The hydrolysate was then examined by HPLC with a diode-array detector and mass spectrometry. They found that apples (var. Golden Delicious, with peel), peaches (var. Royal, with peel), and nectarines (var. Royal, with peel) contained approximately 73, 50, and 60 mg/100 g FW of HPNE-derived phenolic acids, respectively. In all three types of fruit, the predominant component was ferulic acid (68%), followed by *p*-hydroxybenzoic acid (24–27%). More recently, Pérez-Jiménez and Saura-Calixto[24] reported analysis data for EPP, HPNE, and NEPA in various types of fruit and vegetables. Those contents in fruit are shown in Table 14.2. The ratio of HPNE to total polyphenols is quite large.

The distribution of NEPP in different fruit parts also varies from one fruit species to another. In apples, the NEPA content of the peel seems to be higher than that of the flesh.[34,35] Hellström and Mattila[34] compared the EPA and NEPA contents of cored apples (flesh with peel) with those of the peel in four cultivars (Lobo, Valkeakuulas, Red Delicious, and Granny Smith) and found that in all four cultivars the peel contained higher amounts of EPA and NEPA and had a higher proportion of NEPA in the total PA (6.5–18.5%) than the cored apples (4.3–8.9%). The Granny Smith peel seemed to be the richest source of NEPA (37.1 mg/100 g FW) among these four apple cultivars. Studies have indicated that, in contrast to apples, European pear and peach pulps contain higher levels of NEPA than their peels.[35] In a study of two Japanese plum cultivars, the proportion of NEPP (bound phenolics) in total polyphenols (free and bound phenolics) was higher in the flesh than in the peel: approximately 20% and 30% of the total polyphenols were NEPP in the flesh of Queen Garnet and Black Diamond, respectively.[36]

14.3 Changes in the NEPP Contents of Fruit

Chemical components in fruit generally change with their maturation, ripening, and post-harvest handling, which includes storage, processing, and cooking. However, scientific papers dealing with the changes in NEPP during fruit ripening and post-harvest handling remain scarce.

14.3.1 Biological/Physiological Changes

Changes in the EPP and NEPP contents and proportions seem to vary in different types of fruit. In plantain bananas (*Musa paradisiaca*), EPP and NEPP contents of unripe fruit (full green, stage 2) were 5.74 mg gallic acid equivalents (GAE)/g DW and 3.45 mg as cyanidin/g DW, respectively.[37] In fully ripe fruit (yellow, stage 6), the contents were 2.84 mg GAE/g DW for EPP and 1.95 mg g^{-1} DW for NEPP, indicating that both the EPP and NEPP contents decreased during ripening. In contrast, the NEPP (hydrolysable tannins and NEPA) of plums (*Prunus domestica*) seem to increase during ripening. Kristl *et al.*[38] studied changes in the total antioxidant activity (TAA; by 2,2-azinobis (3-ethylbenzothiazoline-6-sulfonic acid) [ABTS] assay) of four plum cultivars (Stanley, Valor, Hanita, and Tophit). They reported that in the last week of on-tree ripening (to eating quality), the TAA (μmol Trolox equivalent [TE]/g FW) increased by up to 38% and showed a similar pattern for all the cultivars studied. In that study, the extractable antioxidants were minor contributors to the TAA in all the cultivars at all ripening stages, representing less than 18% of the TAA, while both hydrolysable tannins and NEPA contributed considerably more. In the case of grape (*Vitis vinifera* var. Shiraz), the EPA and NEPA contents of seeds, pulp, and skin increased with berry development and reached maximum levels at around the onset of ripening (veraison).[39,40]

14.3.2 Changes During Post-harvest Handling and Processing

There has been very little research into post-harvest changes in the NEPP content of fruit. However, it appears that the drying process can increase NEPP contents, at least in some cases. For example, particular varieties of astringent (sour) persimmons (*Diospyros kaki*) contain about 1–2% soluble tannins (PA polymers).[41,42] During dehydration, these soluble tannins are converted to insoluble tannins, and the dried persimmons lose their astringency and present a strong, sweet taste. Although the mechanism of insoluble tannin formation has not yet been fully determined, the process may involve the formation of complexes of soluble tannins and pectin.[43] Matsumura *et al.*[44] reported that a large amount of NEPP (4%, w/w as GAE) was determined in dried persimmons after hydrolysing the non-extractable (90% ethanol-insoluble) fraction. The antioxidant activity derived from these NEPP contributed 81.3% (based on the oxygen radical absorbance capacity [ORAC] assay) or 96.1% (based on the 1,1-diphenyl-2-picrylhydrazyl radical [DPPH] assay) to the total activity. The EPP and NEPP of dried persimmon (var. Ichidagaki) have also been separated using 70% (v/v) acetone, and considerable amounts of NEPA were detected after direct BuOH–HCl degradation of the non-extractable fraction. In this experiment, the NEPA content was 90-times higher than that of EPA (unpublished data). Thus, it was also found that dried persimmon is an example of a fruit with a high NEPP content in which the polyphenols are mostly in the form of NEPP.

An increase in NEPP content has also been reported in the traditional sun-drying of Portuguese pears (*Pyrus communis* L. var. S. Bartolomeu).[45] According to the report, sun-drying caused a decrease of 64% (on a dry pulp basis) in the total quantity of native polyphenols. However, the proportion of NEPC in the total procyanidins increased from 6% in the fresh pear pulp to 86% in the sun-dried pulp. The NEPC in the dried pears could not be extracted, even with an acetone/8 M urea (6 : 4, v/v) solvent.

Studies of citrus fruit peels revealed that the drying temperature affected the EPP and NEPP contents.[46–48] Esparza-Martínez *et al.*[48] measured the contents of NEPP (as GAE), non-extractable flavonoids (as rutin equivalents [RE]), and non-extractable condensed tannins (as catechin equivalents [CE]) in mandarin waste (*Citrus reticulata* var. Blanco). They found that the content in samples dried at 120 °C was higher than in fresh mandarin waste or in samples dried at lower temperatures (60 and 90 °C). The data for samples dried at 120 °C were 74.56 mg GAE/g DW (77.5% of total phenolics), 21.56 mg RE/g DW (62.6% of the total), and 4.76 mg CE/g DW (79.5% of the total). The mandarin waste contained 45.15 mg GAE/g (71.2%), 12.01 mg RE/g DW (45.5%), and 2.59 mg CE/g DW (85.5%).

Because NEPP can bind strongly to cell wall polysaccharides, the breakdown of these cell wall components during processing might cause the release of NEPP from the food matrix in either solubilised or insoluble suspended forms. Pectin is the most important polysaccharide that binds to PA.[27,49,50] Ruiz-Garcia *et al.*[50] reported that in grape skins, more than 54% of cell wall-bound PA were localised within the chelator-soluble pectic fraction. They used purified skin PA in a model adsorption system and confirmed that the removal of pectic polysaccharides by a chelator significantly reduced the adsorption of PA to cell walls. They also found a high residual binding capacity of hemicellulose, whereas cellulose and residual lignin, which represent approximately 30% of the cell wall by weight, had low PA adsorption capacities.

Enzymatic degradation of pectin may release PA from cell walls. Oszmiański *et al.*[51] examined the treatment of apple pomace with pectinases such as Pectinex XXL and Pectinex Ultra SPL. The treatment had positive effects on the puree-enriched cloudy apple juices produced by increasing their polyphenolic contents (especially procyanidins in extractable and non-extractable forms) and juice yields.

Cooking and heat processing can also affect the NEPP and EPP contents of different fruits in various ways, including the release of polyphenols from the food matrix, thermal degradation of polyphenols, and the formation of bonds between polyphenols and macromolecules.[27,52–55] Grilling for 30 minutes caused increases in the NEPP (HPNE + NEPA) contents and decreases in the EPP contents of both unripe and ripe plantain bananas.[37] In contrast, boiling the fruit (10 minutes in twice the volume of water) caused increases in the EPP contents, but had no effects on the NEPA contents. However, other studies showed that after prolonged boiling of pears (*Pyrus communis*), PA were strongly adsorbed to cell wall materials, especially pectin,[52,54] and formed strong bonds (possibly covalent linkages) with cell wall polysaccharides.[54] Additionally, in the case of dried citrus peel (*Citrus reticulata*), heat treatment (oven drying, 100 °C for 2 hours) seems to cause the release of NEPP that are beneficial for health, especially polymethoxy flavones.[56] The treatment at this relatively low temperature (100 °C) may be important because a higher temperature (150 °C) has a destructive impact on hesperidin, which is the most abundant flavanone.

14.4 *In vitro* and Animal Studies of Health-related Properties of NEPP Together with Dietary Fibre

When we consume fruit, NEPP pass through the gastrointestinal tract along with fibre components from cell walls, and there are some health benefits of this passage. Since NEPP bind strongly to macromolecules such as pectic polysaccharides in dietary fibre and are consumed together with the polysaccharides, it is difficult to assess functions that can be attributed solely to the NEPP. However, it is likely that NEPP enhance some functions of the fibre components. Important effects of NEPP that should be discussed include: imparting antioxidant (free radical-scavenging) activity to fibre matrices; enhancement of the bile acid-binding capacity of the fibre; and the gradual release of both polyphenolic molecules that become bioaccessible and microbial metabolites that can be absorbed and act in systemic circulation. Here, we will focus on: (1) the contribution of NEPP to bile acid binding; and (2) microbial fermentation and increases in bioaccessibility, leading to the production of absorbable metabolites. The antioxidant and free radical-scavenging activities are covered in Chapter 13.

14.4.1 Bile Acid-binding Capacity

Binding bile acids and increased excretion of them are thought to be important functions of dietary fibre that contribute to reductions in cholesterol in the body. Many studies have shown positive relationships between diets rich in soluble dietary fibre (such as β-glucan from oats and barley, pectin from fruit, guar gum, and psyllium) and reductions in total cholesterol and LDL-cholesterol in serum.[57] The cholesterol-lowering effects of dietary fibre (and potentially the accompanying NEPP) are believed to results from: (1) reductions in intestinal cholesterol absorption; and (2) reductions in bile acid uptake by interference with the enterohepatic circulation of bile acids.[58] Bile acids are acidic steroids synthesised in the liver from cholesterol.

After conjugation with glycine or taurine, they are secreted into the duodenum. Bile acids are actively reabsorbed by the terminal ileum and undergo enterohepatic circulation.[59–61] Dietary fibre can bind bile acids, leading to an increase in their faecal excretion, reduces the enterohepatic bile acid pool. This may stimulate the conversion of cholesterol to additional bile acids in the liver. Thus, bile acid-binding and increasing their faecal excretion has been hypothesised as a possible mechanism by which dietary fibre lowers cholesterol.[60,62,63] The effect of fruit fibre intake on bile acid excretion has been confirmed in animal studies involving the supplementation of diets with apple fibre.[64,65] Sembries *et al.*[64] fed rats diets supplemented with either juice colloids or alcohol-insoluble solids (AIS) derived from apples. They found that up to 30% and 88% more primary bile acids were excreted in the faeces, respectively, compared to the control group. The primary bile acid contents in the caecum also increased by 15% after eating juice colloids and 37% after eating AIS.

Soluble fibres, such as pectin, and lignin are well known as bile acid-binding agents,[66–68] but the associated polyphenols, including NEPP, seem to contribute significantly to the bile acid-binding capacity of dietary fibre. Even the EPP weakly associated with cell wall materials, such as soluble tannins or EPC, make important contributions to this large binding capacity. For example, the removal of tannin from carob fibre with a dimethylformamide treatment resulted in a sharp decrease in the bile acid-binding capacity of the fibre.[69] Similar results were observed in a study of AIS (cell wall materials) from Chinese quince and quince. Removal of EPC from the AIS with 70% (v/v) acetone caused a significant reduction in their bile acid-binding capacity.[70] Moreover, administration of a diet containing 2–5% dried young persimmons (rich in kaki-tannin) to mice enhanced their bile acid excretion and lowered their concentrations of hepatic lipids and plasma cholesterol.[71] In that study, the tannin was thought to be an important contributor to the cholesterol-lowering effect, because tannin extracted from the fruit exhibited a strong *in vitro* bile acid-binding capacity, comparable to that of cholestyramine. These studies involved EPP rather than NEPP, but the importance of NEPP in bile acid binding could be at least partly predicted based on these results with EPP. Unfortunately, at this time, the actual *in vivo* contributions of NEPP to bile acid binding and cholesterol lowering are unknown, because studies with NEPP have not yet been performed.

We performed an *in vitro* analysis of the bile acid-binding activity and the DPPH radical-scavenging activity of AIS from eight fruit samples (apple, unripe pear, ripe pear, quince, Chinese quince, hawthorn, unripe blueberry, and ripe blueberry) after removal of the EPC with 70% (v/v) acetone.[32] We analysed the NEPC contents, mDP of the NEPC, carbohydrate contents, and lignin contents of the samples. The mDP × NEPC value was found to be a good index of the radical-scavenging activity ($R^2 = 0.9194$). However, only moderate relationships were found between the bile acid-binding activity and the lignin content ($R^2 = 0.3407$) or the NEPC

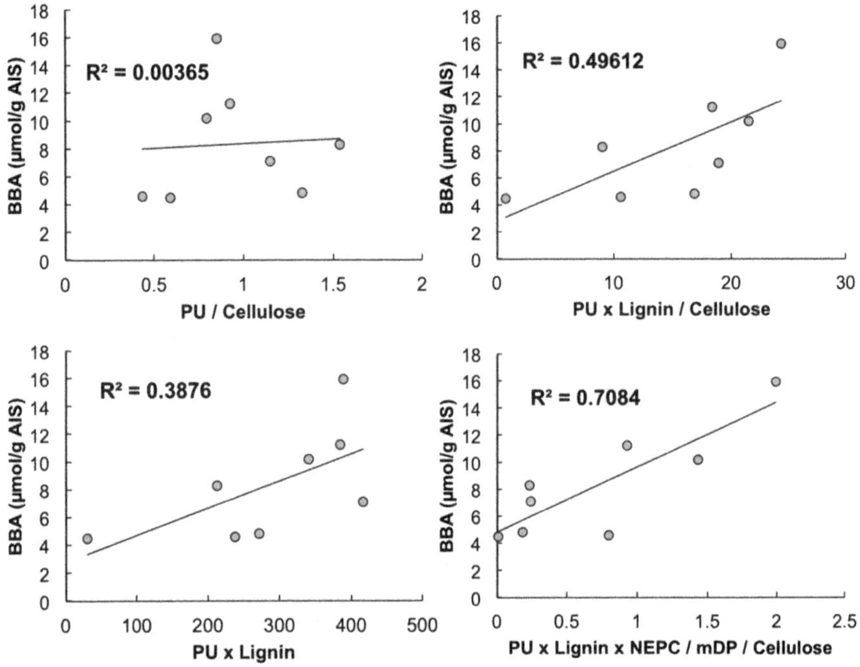

Figure 14.2 Relationships between components of alcohol-insoluble solids (AIS) and the bile acid binding activities of AIS prepared from various fruits. Each data point represents a different fruit sample. Apples, immature pears, ripe pears, quinces, Chinese quinces, hawthorns, immature blueberries, and ripe blueberries are included. BBA: bile acid binding activity; PU: polyuronides; NEPC: non-extractable procyanidins; mDP: mean degree of polymerisation of the NEPC.

content ($R^2 = 0.2074$). After that, we reanalysed the data and found that the bile acid-binding activity had a strong positive relationship ($R^2 = 0.7084$) with an index of cell wall components (polyuronides × lignin × NEPC × mDP^{-1} × $cellulose^{-1}$) (Figure 14.2). This indicates that the AIS (cell wall materials) composed of large amounts of pectin (polyuronides), lignin, and NEPC with low mDP values and small amounts of cellulose tended to have large bile acid-binding capacities. The existence of NEPC in the AIS had an impact on the bile acid-binding activity, because the relationship was relatively weak if only the pectin, lignin, and cellulose contents were taken into account ($R^2 = 0.49612$).

Another important aspect of the bile acid-binding effect (resulting in excretion) is the elimination of secondary bile acids, such as deoxycholic acid and lithocholic acid, which are risk factors for colon cancer.[72] AIS from quince and Chinese quince bind secondary bile acids (taurodeoxycholic acid and glycodeoxycholic acid), and this effect is enhanced if the EPP are not removed from the AIS.[70] This indicates that fruit AIS, containing EPP and NEPP, potentially reduce the risk of colon cancer. This effect should be further explored using animal and intervention studies.

14.4.2 Fermentation, Bioaccessibility, and Metabolic Fate

The fermentation, bioaccessibility, and metabolic fates of NEPP in the gastrointestinal tract, as well as their health effects, have been reviewed by Pérez-Jiménez *et al.*[19] NEPP are at least partly solubilised (becoming bioaccessible) during their passage through the gastrointestinal tract and are subjected to absorption and metabolism with the aid of intestinal microbes and/or enzymes in the body. Therefore, certain amounts of NEPP in the diet meet similar metabolic fates to those of EPP. However, because NEPP are tightly bound to dietary fibre components, they have certain characteristics that differ from those of EPP.

14.4.2.1 Fermentation and Bioaccessibility

Bioaccessibility is defined as the amount of a food constituent that is solubilised in the gastrointestinal fluids and available for absorption by the body. Some NEPP become bioaccessible as a consequence of physicochemical conditions and the action of either digestive enzymes or the intestinal microbiota. The level of bioaccessibility varies depending on the type of NEPP. In an *in vitro* study, Saura-Calixto *et al.*[73] found that a high proportion of hydrolysable polyphenols (58%) were released by treatment with digestive enzymes, whereas NEPA were not. This indicates that NEPA arrive nearly intact in the colon, where those NEPP that are not solubilised in the small intestine can become bioaccessible by microbial degradation (fermentation) of the molecules they are associated with (such as carbohydrates or proteins) or the action of intestinal enzymes that are capable of breaking covalent bonds, such as esterases.[19,74] About 95% of NEPA arriving in the colon can be released from the food matrix by these two mechanisms.[73] Of various groups of plant food (cereals, vegetables, legumes, fruit, nuts, beverages, and oils), fruit has relatively high amounts of polyphenols that become bioaccessible in both the small intestine and the large intestine.[73]

14.4.2.2 Metabolic Transformation and Bioavailability of NEPP

The NEPP that are solubilised but not absorbed in the small intestine and most NEPA will reach the colon, then undergo colonic fermentation. Colonic fermentation by intestinal microbes results in the metabolic transformation of NEPP and the production of potentially absorbable metabolites.[19] These include similar phenolic acids and other metabolites that differ depending on the type of NEPP.

(1) Metabolites derived from NEPA include the monomeric constituents of NEPA polymers (mainly epicatechin) and a wide variety of compounds derived from the breakdown of the basic flavanol skeleton, followed by further degradation. These include valerolactones, hydroxyphenylpropionic acids, hydroxyphenylacetic acids, hydroxybenzoic acids, and hippuric acid.

(2) Metabolites derived from hydrolysable phenolics include free phenolic acids released from the food matrix and various other phenolic acids produced by the colonic microbiota (coinciding with the degradation of flavanols).
(3) Metabolites derived from hydrolysable tannins include urolithins, transformed by the microbiota from ellagic acid, which can be released from ellagitannins in the gastrointestinal tract.

These metabolites are generally the same compounds that are produced in the metabolism of EPP. Since EPP usually exist together with NEPP in fruit tissues, it is difficult to distinguish which metabolites originated from the NEPP.

Research focusing specifically on NEPP metabolism is scarce. Mateos-Martín *et al.*[75] reported that rats fed a diet supplemented with grape NEPA excreted polyphenol-derived metabolites in urine (phase II conjugates of epicatechin, valerolactones, and various phenolic acids, including hydroxyphenylpropionic, hydroxyphenylacetic, hydroxybenzoic and coumaric acids). They also found monomeric and dimeric flavan-3-ols and hydroxyphenylpropionic acids in the faeces. These results indicate that NEPA undergo depolymerisation into smaller PA dimers and monomeric catechin units during their transit along the intestinal tract and are later metabolised by the intestinal microbiota into smaller units.

Various research publications report the biological functions of absorbed polyphenols and their metabolites.[76–80] For instance, monomeric catechins and dimeric procyanidins can modulate pro-inflammatory and oncogenic signals by being transported inside cells, where they directly interact and modulate the activity of signalling proteins and/or prevent oxidation.[79] Moreover, phenolic acid metabolites such as protocatechuic acid and isovanillic acid have anti-inflammatory effects that reduce the risk of CVD.[80] The precursor flavonoids had no such function when assessed by reductions in the secretion of soluble vascular adhesion molecule-1 protein. These effects of absorbed metabolites derived from NEPP can be considered to be the same as those of metabolites derived from EPP.

14.4.2.3 Characteristics of NEPP Metabolism

Although phenolic metabolites originating from NEPP may be the same as those from EPP, their availability is different.[19] Because of their high molecular weight or association with other macromolecules such as pectic polysaccharides, NEPP have two specific features that distinguish them from EPP during their colonic transformation. The first is the time and duration of metabolite availability: fermentation of the macromolecules associated with NEPP takes time, and thus the release, metabolism, and absorption of NEPP takes more time than for EPP. Therefore, the metabolites derived from NEPP circulate in the body for longer periods than those derived from EPP.[81]

The second feature is the extent of their fermentation. Because NEPP are bound to other fermentable dietary components such as polysaccharides and proteins, the activity of the colonic microbiota can be affected, and this can result in higher levels of fermentation. Saura-Calixto *et al.*[81] reported that the *in vitro* fermentability (assessed by the production of short-chain fatty acids [SCFA]) of polymeric NEPA isolated from carob pods was 23%, whereas that of NEPA from grape seeds and peel associated with dietary fibre was 50%. Juśkiewicz *et al.*[82] also reported that polyphenols and fructo-oligosaccharides, when consumed together, are important dietary factors that affect the susceptibility of quercetin glycosides to microbial metabolism in the caecum. Thus, enhanced bioavailability due to the accompanying prebiotic cell wall fibre seems to be a characteristic of NEPP.

14.5 Potential Relationship of NEPP with the Beneficial Health Effects of Fruit Consumption

The consumption of fruit has various beneficial health effects derived from the functional compounds that it contain. These include vitamins (mainly ascorbic acid and provitamin A carotenoids), minerals (such as potassium), antioxidants (such as ascorbic acid, tocopherols, carotenoids, EPP, and NEPP), and dietary fibre (insoluble and soluble dietary fibre, including pectin). Furthermore, fruit enhances the eating experience with bright colours, rich aromas, and pleasant textures and flavours; these can all be related to improved mental and physical human health. Since these beneficial health effects result from interactions between different components and properties, it is difficult to attribute an overall health effect to a specific functional component. Nevertheless, the contribution of some functional compounds has been partly elucidated by scientific research.

The contributions of NEPP to the health effects of fruit intake might be mainly in relation to gastrointestinal health and CVD risk reduction. In these effects, NEPP may act together with dietary fibre in a synergistic manner. The fate of the NEPP–dietary fibre complex after intake and the potential effects on gastrointestinal health and CVD risk reduction are summarised in Figure 14.3.

14.5.1 Gastrointestinal Health

The gastrointestinal tract may be the principal site where NEPP exert their health-promoting effects, together with dietary fibre and EPP. Most NEPP (and EPP) reach the colon after fruit intake. There, the NEPP–fibre complex may exhibit various functions relating to gastrointestinal health, such as antioxidant activity,[83] anti-tumour activities,[84,85] stimulation of colonic peristalsis (reducing the period of contact between toxic compounds and the colon epithelium),[19] and stimulation of *Lactobacillus* bacterial growth.[86] Other beneficial bacterial species in the colon may also be stimulated by

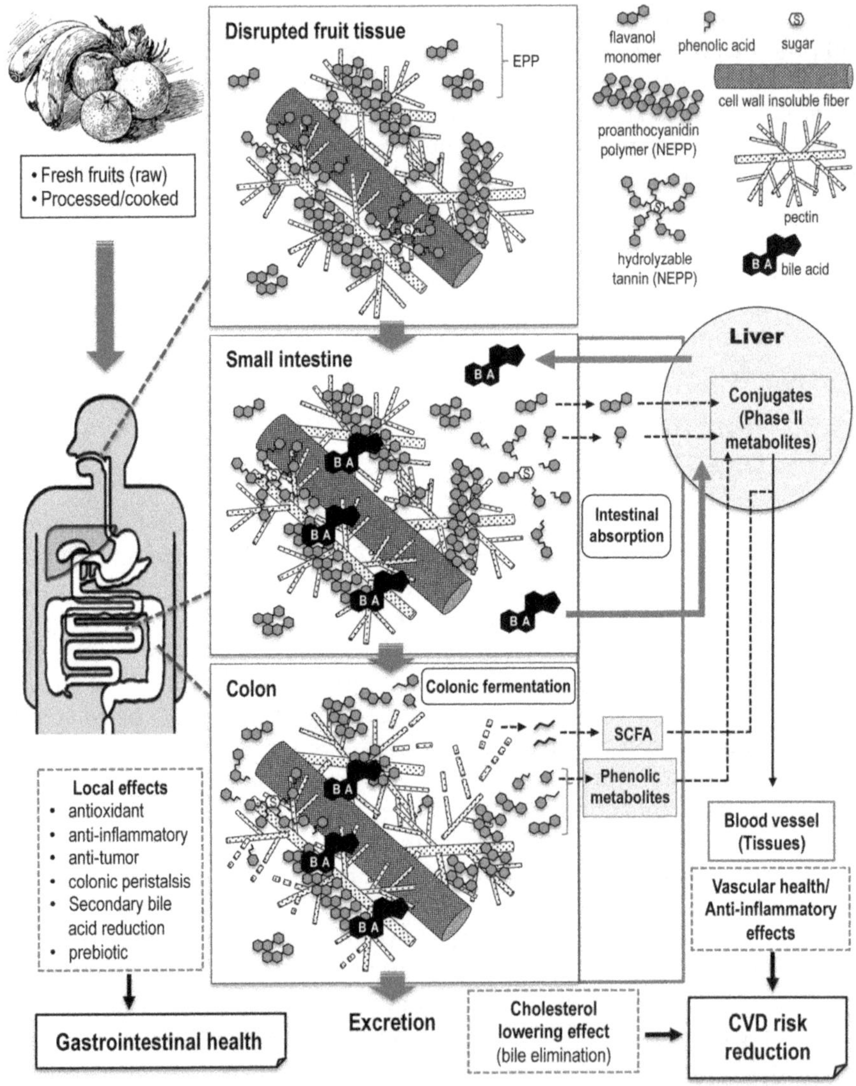

Figure 14.3 The fates of cell wall fibres, their associated non-extractable polyphenols (NEPP) after fruit ingestion, and their potential effects on cardiovascular disease (CVD) and gastrointestinal health. The NEPP–fibre complexes in fresh or processed fruit tissues are transferred to the small intestine. There, hydrolysable polyphenols in non-extractable form become partly bioaccessible. The released phenolic acids and extractable polyphenols (EPP) can be absorbed. Bile acids (BA) in the intestine bind to the remaining complex. In the colon, pectin and polyphenols are fermented by the intestinal microbiota into metabolites that can be absorbed. Short-chain fatty acids (SCFA) and phenolic metabolites may have local and systemic effects.

the NEPP–fibre complex. *In vitro* studies showed that two *Bifidobacterium* strains (*B. adolescentis* and *B. bifidum*) were stimulated by rutin, sinapic acid, caffeic acid, and vanillic acid,[87] while *Clostridium coccoides* was stimulated by flavanol monomers.[88] Some of these compounds can be released from NEPP. Dietary fibre (and the associated polyphenols) can bind to secondary bile acids (deoxycholic acid and lithocholic acid), which are risk factors for colorectal cancer.[72] Since NEPP (and EPP) can increase the bile acid-binding capacity of dietary fibre, the intake of EPP- and NEPP-rich fibre may be effective in the elimination of secondary bile acids. In an epidemiological study, Rossi *et al.*[89] found inverse associations between PA intake and the risk of colorectal cancer. The associations appeared stronger for PA with higher degrees of polymerisation (odds ratio = 0.69 for ≥10 mers). Those authors did not use databases with specific information for NEPA, but their results point to the importance of NEPA, because they have higher degrees of polymerisation than EPA. Furthermore, the colonic microbial metabolism of dietary fibre releases SCFA, including butyrate, which has a remarkable array of colonic health-promoting and antineoplastic properties that contribute to the prevention of colonic cancer.[61,90,91] Butyrate production may be increased by synergy between polyphenols and dietary fibre.[82,92] For example, Aprikian *et al.*[92] reported that in rats fed a diet containing apple polyphenols and/or pectin, the caecal butyrate pool was highest in the rats fed the pectin plus polyphenol diet. In another study, the addition of apple polyphenols (from pomace) to a rat diet containing fructo-oligosaccharides also resulted in an increase in the caecal butyrate pool.[82] Barth *et al.*[93] used a rat model of colon carcinogenesis to examine the effects of a diet supplemented with cloudy apple juice formulations. They found that administration of a diet containing both polyphenols and a cloudy juice fraction composed of lipids, proteins, pectin, and bound polyphenols was more effective in terms of cancer reduction than a diet containing the polyphenols or the cloudy juice fraction alone. Synergistic effects have also been observed between EPP and NEPP in terms of their antioxidant activities,[94–96] with NEPP exerting longer periods of activity than soluble antioxidants (EPP).

Because of these combined effects, the intake of polyphenol (EPP and NEPP)-rich dietary fibre may greatly reduce the risk of colorectal cancer. Among plant foods, fruit contains relatively high amounts of NEPP (especially NEPA), which can become bioaccessible in the intestinal tract.[73] Fruit also contains large amounts of pectic polysaccharides, which have beneficial prebiotic and adsorbing effects. These characteristics might explain the epidemiological observations that dietary fibre from fruit is more effective in the prevention of colorectal cancer than fibre from vegetables or legumes.[5,18,97–99]

14.5.2 CVD Risk Reduction

Some epidemiological studies have highlighted the superiority of fruit or fruit-derived fibre over other groups of plant foods in the prevention of CVD and coronary heart disease.[6,7,9,17,100,101]

Specific characteristics, including the components of fruit, contribute to CVD prevention. As described above, fruits contain relatively high amounts of NEPP (mainly NEPA), EPP (mainly EPA), and pectic polysaccharides. These mixtures have enhanced capacities to bind bile acids, thereby leading to their excretion and to lower levels of cholesterol. In addition, the absorbable polyphenols and their metabolites that originate from EPP and NEPP can protect the cardiovascular system through various biological actions. These include anti-inflammatory action,[78,80] protecting or stimulating endothelial nitric oxide metabolism,[76,77] and inhibition of platelet aggregation.[102] Pectic polysaccharides also undergo colonic fermentation, resulting in the production of SCFA[103] that can benefit vascular health through blood pressure regulation[104] and anti-inflammatory action.[105] Fruit also has a low sodium : potassium ratio,[106,107] which helps prevent increases in blood pressure.[108] Thus, the various components in fruit can promote vascular health in a comprehensive manner.

Apples and apple products have been used to study the combined effects of polyphenols and pectin on CVD risk reduction.[12,65,92,103] In a rat experiment, Aprikian *et al.*[92] observed that apple pectin and the polyphenol-rich fraction of an apple extract were more beneficial for blood lipid metabolism when fed together than when fed separately. Similar results were obtained in a human crossover study: LDL-cholesterol was lowered after the intake of whole apples (by 6.7%), apple pomace (by 7.9%), and cloudy apple juice (by 2.2%), whereas it increased after the intake of clear apple juice.[12] Clear apple juice has no pectin or other cell wall components including NEPP, so the results indicate that these components are necessary for the cholesterol-lowering effect. Therefore, processing treatments that remove pectin/NEPP mixtures from fruit may decrease their beneficial effects on CVD risk reduction.

14.6 Conclusions

Different fruits are important sources of NEPP, especially NEPA. Dried fruit may have higher proportions of NEPP than fresh. As our knowledge of the health-related functions of polyphenols and dietary fibre increases, studies on NEPP might provide a key to understanding the overall health benefits of fruit consumption.

The health-related functions of NEPP are mainly exerted in combination with dietary fibre, such as pectin, which strongly binds to NEPP. Although some NEPP, such as HPNE, will become bioaccessible in the small intestine, most NEPP (and most NEPA) will reach the colon. The NEPP–fibre complex can stimulate intestinal microbial activity. Fermentation by the colonic microbiota releases various polyphenols that can be absorbed with or without further metabolism by the microbes. The released polyphenols can stimulate the growth of beneficial bacterial species (the prebiotic effect). The absorbed phenolic metabolites may have systemic effects on health, including vascular health. These effects are similar to those of EPP, but the metabolism of NEPP is slower, resulting in the circulation of metabolites over a longer period of time. Pectin also undergoes colonic fermentation and this results in increases in SCFA, including butyrate. The NEPP–fibre complex can bind bile acids, including secondary bile acids, and accelerate their excretion. Reductions in the enterohepatic bile acid pool lead to a lowering of cholesterol. The combined activities of the NEPP–fibre complex, including antioxidant, anti-tumour, and prebiotic activities, along with binding secondary bile acids, may contribute to colon cancer prevention.

Thus, NEPP appear to play important roles in risk reduction for both CVD and colon cancer, but their effects cannot be separated from those of the associated dietary fibre. Studies indicate that polyphenols and dietary fibre consumed together (or as a complex) have a superior effect on CVD and colon cancer risk reduction than the consumption of polyphenols or dietary fibre alone. It seems that our bodies use this natural phytochemical complex effectively through a symbiotic relationship with the gut microbiota. In addition to the NEPP–dietary fibre complex, fruit contains other beneficial components, such as EPP, ascorbic acid (vitamin C), and carotenoids, including provitamin A. Fruit also has beneficial properties, including low calorie contents, low sodium : potassium ratios, and pleasant sensory qualities that encourage us to access its phytochemicals. Along with the relatively high contents of NEPP and pectic polysaccharides, all of these properties contribute to the beneficial health effects of fruit consumption.

Although the importance of NEPP for human health has started to become apparent, more studies are needed to clarify how the NEPP content of fruit changes during ripening, post-harvest storage, and processing. Ripening and/or processing may affect the proportions and nature of the EPP, NEPP, and pectin in fruit. For example, thermal processing or drying of fruit can lead to strong interactions between polyphenols and cell wall materials (including pectin) that might affect their colonic fermentation and bioaccessibility. Additional research is also needed to further elucidate the importance of NEPP for, and their impact on, the overall health effects of fruit consumption.

References

1. E. Riboli and T. Norat, *Am. J. Clin. Nutr.*, 2003, **78**, 559S.
2. I. Soerjomataram, D. Oomen, V. Lemmens, A. Oenema, V. Benetou, A. Trichopoulou, J. W. Coebergh, J. Barendregt and E. de Vries, *Eur. J. Cancer*, 2010, **46**, 2563.
3. J. V. Woodside, I. S. Young and M. C. McKinley, *Proc. Nutr. Soc.*, 2013, **72**, 399.
4. O. Oyebode, V. Gordon-Dseagu, A. Walker and J. S. Mindell, *J. Epidemiol. Community Health*, 2014, **68**, 856.
5. Q. Ben, J. Zhong, J. Liu, L. Wang, Y. Sun, L. Yv and Y. Yuan, *Medicine*, 2015, **94**, e1599.

6. H. T. Lai, D. E. Threapleton, A. J. Day, G. Williamson, J. E. Cade and V. J. Burley, *Eur. J. Epidemiol.*, 2015, **30**, 1035.
7. H. Du, L. Li, D. Bennett, Y. Guo, T. J. Key, Z. Bian, P. Sherliker, H. Gao, Y. Chen, L. Yang, J. Chen, S. Wang, R. Du, H. Su, R. Collins, R. Peto and Z. Chen, *N. Engl. J. Med.*, 2016, **374**, 1332.
8. B. Nguyen, A. Bauman, J. Gale, E. Banks, L. Kritharides and D. Ding, *Int. J. Behav. Nutr. Phys. Act.*, 2016, **13**, 9.
9. P. Buil-Cosiales, M. A. Martinez-Gonzalez, M. Ruiz-Canela, J. Díez-Espino, A. García-Arellano and E. Toledo, *Nutrients*, 2017, **9**, 295.
10. D. Aune, E. Giovannucci, P. Boffetta, L. T. Fadnes, N. Keum, T. Norat, D. C. Greenwood, E. Riboli, L. J. Vatten and S. Tonstad, *Int. J. Epidemiol.*, 2017, **46**, 1029.
11. A. Avci, T. Atli, I. Eruder, M. Varli, E. Devrim, S. Turgay and I. Durak, *Exp. Aging Res.*, 2007, **33**, 429.
12. G. Ravn-Haren, L. O. Dragsted, T. Buch-Andersen, E. N. Jensen, R. I. Jensen, M. Németh-Balogh, B. Paulovicsová, A. Bergström, A. Wilcks, T. R. Licht, J. Markowski and S. Bügel, *Eur. J. Nutr.*, 2013, **52**, 1875.
13. W. J. Dahl and M. L. Stewart, *J. Acad. Nutr. Diet.*, 2015, **115**, 1861.
14. K. Maćkowiak, N. Torlińska-Walkowiak and B. Torlińska, *Postepy Hig. Med. Dosw.*, 2016, **70**, 104.
15. V. H. Ozyurt and S. Ötles, *Acta Sci. Pol., Technol. Aliment.*, 2016, **15**, 233.
16. A. Padayachee, L. Day, K. Howell and M. J. Gidley, *Crit. Rev. Food Sci. Nutr.*, 2017, **57**, 59.
17. M. A. Pereira, E. O'Reilly, K. Augustsson, G. E. Fraser, U. Goldbourt, B. L. Heitmann, G. Hallmans, P. Knekt, S. Liu, P. Pietinen, D. Spiegelman, J. Stevens, J. Virtamo, W. C. Willett and A. Ascherio, *Arch. Intern. Med.*, 2004, **164**, 370.
18. S. A. Bingham, N. E. Day, R. Luben, P. Ferrari, N. Slimani, T. Norat, F. Clavel-Chapelon, E. Kesse, A. Nieters, H. Boeing, A. Tjønneland, K. Overvad, C. Martinez, M. Dorronsoro, C. A. Gonzalez, T. J. Key, A. Trichopoulou, A. Naska, P. Vineis, R. Tumino, V. Krogh, H. B. Bueno-de-Mesquita, P. H. Peeters, G. Berglund, G. Hallmans, E. Lund, G. Skeie, R. Kaaks and E. Riboli, *Lancet*, 2003, **361**, 1496.
19. J. Pérez-Jiménez, M. E. Díaz-Rubio and F. Saura-Calixto, *Nutr. Res. Rev.*, 2013, **26**, 118.
20. S. Arranz, F. Saura-Calixto, S. Shaha and P. A. Kroon, *J. Agric. Food Chem.*, 2009, **57**, 7298.
21. F. Saura-Calixto, *J. Agric. Food Chem.*, 2012, **60**, 11195.
22. G. Domínguez-Rodríguez, M. L. Marina and M. Plaza, *J. Chromatogr. A*, 2017, **1514**, 1.
23. J. Sun, Y.-F. Chu, X. Wu and R. H. Liu, *J. Agric. Food Chem.*, 2002, **50**, 7449.
24. J. Pérez-Jiménez and F. Saura-Calixto, *Food Res. Int.*, 2015, **74**, 315.
25. B. A. Acosta-Estrada, J. A. Gutiérrez-Uribe and S. O. Serna-Saldívar, *Food Chem.*, 2014, **152**, 46.

26. C. Le Bourvellec, S. Guyot and C. M. Renard, *Biochim. Biophys. Acta*, 2004, **1672**, 192.
27. C. M. G. C. Renard, A. A. Watrelot and C. Le Bourvellec, *Trends Food Sci. Technol.*, 2017, **60**, 43.
28. R. L. Prior and L. Gu, *Phytochemistry*, 2005, **66**, 2264.
29. J. K. Hellström, A. R. Törrönen and P. H. Mattila, *J. Agric. Food Chem.*, 2009, **57**, 7899.
30. J. Pérez-Jiménez, S. Arranz and F. Saura-Calixto, *Food Res. Int.*, 2009, **42**, 1381.
31. J. Pérez-Jiménez and J. L. Torres, *J. Agric. Food Chem.*, 2011, **59**, 12713.
32. Y. Hamauzu and Y. Mizuno, *Plant Foods Hum. Nutr.*, 2011, **66**, 70.
33. Y. Hamauzu, H. Yasui, T. Inno, C. Kume and M. Omanyuda, *J. Agric. Food Chem.*, 2005, **53**, 928.
34. J. K. Hellström and P. H. Mattila, *J. Agric. Food Chem.*, 2008, **56**, 7617.
35. A. Delpino-Rius, J. Eras, F. Vilaró, M. Á. Cubero, M. Balcells and R. Canela-Garayoa, *Food Chem.*, 2015, **172**, 575.
36. A. Bobrich, K. J. Fanning, M. Rychlik, D. Russell, B. Topp and M. Netzel, *Food Res. Int.*, 2014, **65**, 20.
37. O. Ebun and K. Santosh, *World J. Dairy Food Sci.*, 2011, **6**, 189.
38. J. Kristl, M. Slekovec, S. Tojnko and T. Unuk, *Food Chem.*, 2011, **125**, 29.
39. M. O. Downey, J. S. Harvey and S. P. Robinson, *Aust. J. Grape Wine Res.*, 2003, **9**, 15.
40. C. Verries, J. L. Guiraud, J. M. Souquet, S. Vialet, N. Terrier and D. Ollé, *J. Agric. Food Chem.*, 2008, **56**, 5896.
41. S. Taira, I. Satoh and S. Watanabe, *J. Jpn. Soc. Hortic. Sci.*, 1992, **60**, 1003.
42. M. S. Butt, M. T. Sultan, M. Aziz, A. Naz, W. Ahmed, N. Kumar and M. Imran, *EXCLI J.*, 2015, **14**, 542.
43. S. Taira, M. Ono and N. Matsumoto, *Postharvest Biol. Technol.*, 1997, **12**, 265.
44. Y. Matsumura, T. Ito, H. Yano, E. Kita, K. Mikasa, M. Okada, A. Furutani, Y. Murono, M. Shibata, Y. Nishii and S. Kayano, *Food Chem.*, 2016, **202**, 99.
45. D. Ferreira, S. Guyot, N. Marnet, I. Delgadillo, C. M. Renard and M. A. Coimbra, *J. Agric. Food Chem.*, 2002, **50**, 4537.
46. G. Xu, X. Ye, J. Chen and D. Liu, *J. Agric. Food Chem.*, 2007, **55**, 330.
47. M.-L. Chen, D.-L. Yang and S.-C. Liu, *Int. J. Food Sci. Technol.*, 2011, **46**, 1179.
48. F. J. Esparza-Martínez, R. Miranda-López, S. M. Mata-Sánchez and S. H. Guzmán-Maldonado, *Plant Foods Hum. Nutr.*, 2016, **71**, 294.
49. C. Le Bourvellec, B. Bouchet and C. M. Renard, *Biochim. Biophys. Acta*, 2005, **1725**, 10.
50. Y. Ruiz-Garcia, P. A. Smith and K. A. Bindon, *Carbohydr. Polym.*, 2014, **114**, 102.
51. J. Oszmiański, A. Wojdyło and J. Kolniak, *Food Chem.*, 2011, **127**, 623.
52. C. M. Renard, *J. Sci. Food Agric.*, 2005, **85**, 310.

53. M. J. González, J. L. Torres and I. Medina, *J. Agric. Food Chem.*, 2010, **58**, 4274.
54. C. Le Bourvellec, B. Gouble, S. Bureau, M. Loonis, Y. Plé and C. M. Renard, *J. Agric. Food Chem.*, 2013, **61**, 6679.
55. M. Palermo, N. Pellegrini and V. Fogliano, *J. Sci. Food Agric.*, 2014, **94**, 1057.
56. S. C. Ho and C. C. Lin, *J. Agric. Food Chem.*, 2008, **56**, 7976.
57. P. Gunness and M. J. Gidley, *Food Funct.*, 2010, **1**, 149.
58. M. L. Fernandez, *Curr. Opin. Lipidol.*, 2001, **12**, 35.
59. A. F. Hofmann, *Front. Biosci., Landmark Ed.*, 2009, **14**, 2584.
60. T. S. Kahlon and G. E. Smith, *Food Chem.*, 2007, **100**, 1182.
61. I. Rowland, G. Gibson, A. Heinken, K. Scott, J. Swann, I. Thiele and K. Tuohy, *Eur. J. Nutr.*, 2017, DOI: 10.1007/s00394-017-1445-8.
62. A. M. van Bennekum, D. V. Nguyen, G. Schulthess, H. Hauser and M. C. Phillips, *Br. J. Nutr.*, 2005, **94**, 331.
63. T. C. Rideout, S. V. Harding, P. J. Jones and M. Z. Fan, *Vasc. Health Risk Manage.*, 2008, **4**, 1023.
64. S. Sembries, G. Dongowski, K. Mehrländer, F. Will and H. Dietrich, *J. Nutr. Biochem.*, 2004, **15**, 296.
65. E. N. Jensen, T. Buch-Andersen, G. Ravn-Haren and L. O. Dragsted, *J. Hortic. Sci. Biotechnol.*, 2009, 34.
66. F. Garcia-Diez, V. Garcia-Mediavilla, J. E. Bayon and J. Gonzalez-Gallego, *J. Nutr.*, 1996, **126**, 1766.
67. C. Liu, X. L. Lin, Z. Wan, Y. Zou, F. F. Cheng and X. Q. Yang, *Food Funct.*, 2016, 7, 4830.
68. S. Sayar, J. L. Jannink and P. J. White, *J. Agric. Food Chem.*, 2006, **54**, 5142.
69. P. Würsch, *J. Nutr.*, 1979, **109**, 685.
70. Y. Hamauzu and Y. Mizuno, *Acta Hortic.*, 2014, **1040**, 125.
71. K. Matsumoto, S. Yokoyama and N. Gato, *Phytother. Res.*, 2010, **24**, 205.
72. H. Ajouz, D. Mukherji and A. Shamseddine, *World J. Surg. Oncol.*, 2014, **12**, 164.
73. F. Saura-Calixto, J. Serrano and I. Goñi, *Food Chem.*, 2007, **101**, 492.
74. P. A. Kroon, C. B. Faulds, P. Ryden, J. A. Robertson and G. Williamson, *J. Agric. Food Chem.*, 1997, **45**, 661.
75. M. L. Mateos-Martín, J. Pérez-Jiménez, E. Fuguet and J. L. Torres, *Br. J. Nutr.*, 2012, **108**, 290.
76. T. Wallerath, H. Li, U. Gödtel-Ambrust, P. M. Schwarz and U. Förstermann, *Nitric Oxide*, 2005, **12**, 97.
77. T. Schewe, Y. Steffen and H. Sies, *Arch. Biochem. Biophys.*, 2008, **476**, 102.
78. Y. Kawai, T. Nishikawa, Y. Shiba, S. Saito, K. Murota, N. Shibata, M. Kobayashi, M. Kanayama, K. Uchida and J. Terao, *J. Biol. Chem.*, 2008, **283**, 9424.
79. C. G. Fraga and P. I. Oteiza, *Free Radical Biol. Med.*, 2011, **51**, 813.
80. E. F. Warner, Q. Zhang, K. S. Raheem, D. O'Hagan, M. A. O'Connell and C. D. Kay, *J. Nutr.*, 2016, **146**, 465.

81. F. Saura-Calixto, J. Pérez-Jiménez, S. Touriño, J. Serrano, E. Fuguet, J. L. Torres and I. Goñi, *Mol. Nutr. Food Res.*, 2010, **54**, 939.
82. J. Juśkiewicz, J. Milala, A. Jurgoński, B. Król and Z. Zduńczyk, *Nutrition*, 2011, **27**, 351.
83. F. Saura-Calixto, *J. Agric. Food Chem.*, 2011, **59**, 43.
84. D. Lizarraga, M. P. Vinardell, V. Noé, J. H. van Delft, G. Alcarraz-Vizán, S. G. van Breda, Y. Staal, U. L. Günther, J. B. Carrigan, M. A. Reed, C. J. Ciudad, J. L. Torres and M. Cascante, *J. Nutr.*, 2011, **141**, 1597.
85. S. Sánchez-Tena, D. Lizárraga, A. Miranda, M. P. Vinardell, F. García-García, J. Dopazo, J. L. Torres, F. Saura-Calixto, G. Capellà and M. Cascante, *Carcinogenesis*, 2013, **34**, 1881.
86. M. J. Pozuelo, A. Agis-Torres, D. Hervert-Hernández, M. E. López-Oliva, E. Muñoz-Martínez, R. Rotger and I. Goñi, *J. Food Sci.*, 2012, **77**, H59.
87. D. Gwiazdowska, K. Juś, J. Jasnowska-Małecka and K. Kluczyńska, *Acta Biochim. Pol.*, 2015, **62**, 895.
88. X. Tzounis, J. Vulevic, G. G. Kuhnle, T. George, J. Leonczak, G. R. Gibson, C. Kwik-Uribe and J. P. Spencer, *Br. J. Nutr.*, 2008, **99**, 782.
89. M. Rossi, C. Bosetti, E. Negri, P. Lagiou and C. La Vecchia, *Nutr. Cancer*, 2010, **62**, 871.
90. J. M. Wong, R. de Souza, C. W. Kendall, A. Emam and D. J. Jenkins, *J. Clin. Gastroenterol.*, 2006, **40**, 235.
91. S. J. O'Keefe, *Nat. Rev. Gastroenterol. Hepatol.*, 2016, **13**, 691.
92. O. Aprikian, V. Duclos, S. Guyot, C. Besson, C. Manach, A. Bernalier, C. Morand, C. Rémésy and C. Demigné, *J. Nutr.*, 2003, **133**, 1860.
93. S. W. Barth, C. Faehndrich, A. Bub, B. Watzl, F. Will, H. Dietrich, G. Rechkemmer and K. Briviba, *J. Agric. Food Chem.*, 2007, **55**, 1181.
94. E. E. Çelik, V. Gökmen and V. Fogliano, *J. Agric. Food Chem.*, 2013, **61**, 10329.
95. E. E. Çelik, V. Gökmen and L. H. Skibsted, *J. Agric. Food Chem.*, 2015, **63**, 2338.
96. E. D. Cömert and V. Gökmen, *Compr. Rev. Food Sci. Food Saf.*, 2017, **16**, 382.
97. Q. Ben, Y. Sun, R. Chai, A. Qian, B. Xu and Y. Yuan, *Gastroenterology*, 2014, **146**, 689.
98. A. T. Kunzmann, H. G. Coleman, W.-Y. Huang, C. M. Kitahara, M. M. Cantwell and S. I. Berndt, *Am. J. Clin. Nutr.*, 2015, **102**, 881.
99. A. Vulcan, J. Brändstedt, J. Manjer, K. Jirström, B. Ohlsson and U. Ericson, *Br. J. Nutr.*, 2015, **114**, 959.
100. D. E. Threapleton, D. C. Greenwood, C. E. Evans, C. L. Cleghorn, C. Nykjaer, C. Woodhead, J. E. Cade, C. P. Gale and V. J. Burley, *BMJ*, 2013, **347**, f6879.
101. Y. Wu, Y. Qian, Y. Pan, P. Li, J. Yang, X. Ye and G. Xu, *Clin. Nutr.*, 2015, **34**, 603.
102. T. Yasuda, A. Takasawa, T. Nakazawa, J. Ueda and K. Ohsawa, *J. Pharm. Pharmacol.*, 2003, **55**, 239.

103. A. Koutsos, K. M. Tuohy and J. A. Lovegrove, *Nutrients*, 2015, **7**, 3959.
104. X. Cui and J. Cai, *Austin Hypertens*, 2016, **1**, 1005.
105. H. Ohira, W. Tsutsui and Y. Fujioka, *J. Atheroscler. Thromb.*, 2017, **24**, 660.
106. P. Meneton, L. Lafay, A. Tard, A. Dufour, J. Ireland, J. Ménard and J. L. Volatier, *Eur. J. Clin. Nutr.*, 2009, **63**, 1169.
107. S. A. O'Halloran, C. A. Grimes, K. E. Lacy, K. J. Campbell and C. A. Nowson, *Nutrients*, 2016, **8**, pii: E496.
108. A. A. McDonough, L. C. Veiras, C. A. Guevara and D. L. Ralph, *Am. J. Physiol. Endocrinol. Metabol.*, 2017, **312**, E348.

CHAPTER 15

In vivo Studies of Non-extractable Polyphenols: a Case Study with Grape Products

J. C. E. SERRANO*, M. MARTÍN-GARI AND
A. B. GRANADO-SERRANO

Lleida University, Department of Experimental Medicine, Lleida, 25198, Spain
*E-mail: jceserrano@mex.udl.cat

15.1 Introduction

Since 1992, after Renaud and de Lorgeril[1] published their work on wine consumption by the French population and the low incidence of coronary heart disease – the so-called French paradox – the study of dietary antioxidants and especially the consumption of grape products has aroused considerable interest among the scientific community. Initially, it was unclear whether the observed beneficial effects were due to the antioxidant or the alcohol content of wine. The beneficial effects of alcohol consumption have been reported to follow a J-shaped curve: a modest intake is beneficial and either no intake or an increased intake is harmful.[2,3] However, the protective effects reported in the French population seem to be specific to wine, rather than other alcoholic beverages, including beer with a lower alcoholic content. In a recent meta-analysis, the consumption of wine and beer, which is also rich in antioxidants, but not spirits, was found to confer protection against all-cause

Food Chemistry, Function and Analysis No. 5
Non-extractable Polyphenols and Carotenoids: Importance in Human Nutrition and Health
Edited by Fulgencio Saura-Calixto and Jara Pérez-Jiménez

Published by the Royal Society of Chemistry, www.rsc.org

mortality.[4] Later, some evidence suggested that red wine is more protective than white wine. For example, in healthy volunteers, drinking red wine, but not white wine or vodka, increased coronary flow velocity reserve,[5] high-density lipoprotein (HDL) cholesterol and plasma antioxidant status;[6] while *in vitro*, red wine inhibited the platelet-derived growth factor β-receptor,[7] which is crucial in the development of atherosclerosis. These effects were attributed to antioxidants contained in red wine, supporting the beneficial effects of the red wine antioxidant fraction in cardiovascular diseases.

Grapes contain a wide variety of antioxidants, the principal ones being phenolic compounds, flavonoids and stilbenes. Chromatographic studies of grape have confirmed the presence of different polyphenol families, most notably flavanols (catechin, epicatechin and gallocatechin), flavonols (quercetin and kaempferol), flavones (luteolin) and proanthocyanidins.[8,9] The last of these, also known as condensed tannins, are particularly abundant in grape seeds. Proanthocyanidins are polyhydroxyflavan oligomers and polymers usually linked by carbon–carbon bonds in the 4→6 or the 4→8 position (B-type proanthocyanidins). B-type proanthocyanidins can be classified according to the hydroxylation pattern(s) of the chain extender unit(s) as procyanidins (3,5,7,3′,4′-pentahydroxylation), prodelphinidins (3,5,7,3′,4′,5′-hexahydroxylation), propelargonidins (3,5,7,4′-tetrahydroxylation), profisetinidins (3,7,3′,4′-tetrahydroxylation), prorobinetinidins (3,7,3′,4′,5′-pentahydroxylation), proteracacinidins (3,7,8,4′-tetrahydroxylation), promelacacinidins (3,7,8,3′,4′-pentahydroxylation), procassinidins (7,4′-dihydroxylation) and probutinidins (7,3′,4′-trihydroxylation).[10] This class of polyphenols may be present in grapes as extractable proanthocyanidins (EPA) or non-extractable proanthocyanidins (NEPA), according to their solubility in acetone/water, the most common solvent used for their extraction. The NEPA content of grapes is in the same range as that of EPA, and it may be higher in grape-derived products such as grape pomace. However, as happens with the analysis of non-extractable polyphenols (NEPP) in other food matrices, the NEPA content usually has not been taken into account when characterising grape or grape-derived products.[11]

In general, proanthocyanidins are hardly absorbed *via* the small intestine, mainly due to their large molecular size and the lack of gut enzymes to hydrolyse them. Furthermore, high-molecular-weight proanthocyanidins can form complexes with protein, starch and digestive enzymes, including pectinase, amylase, lipase, protease and β-galactosidase, resulting in the formation of less digestible complexes.[12] However, it has been observed that small-molecule proanthocyanidins such as dimeric and trimeric proanthocyanidins can be more readily absorbed. For example, procyanidin B1 (epicatechin-[4β→8]-catechin) with a concentration of 11 nmol L^{-1} is detected in human blood 2 hours after the intake of 2.0 g of proanthocyanidin-rich grape seed extract.[13] Later, during colonic digestion, proanthocyanidins are extensively degraded by gut microflora into simpler phenolic compounds including phenolic acids. The main proanthocyanidin colonic metabolites identified in plasma are 3,4-dihydroxyphenyl acetic acid (also detected in its methylated and sulphated forms), followed by hydroxyphenyl valeric and hydroxyphenyl proponic acids.[14]

Epidemiological studies of the biological effects of grape proanthocyanidins suggest that tannin intake may prevent the onset of chronic diseases.[4,15–17] However, clinical data on humans is limited and it is not clear how, or even whether, these complex polyphenols biologically affect several chronic diseases.

The aim of this chapter is to summarise the available evidence of the biological effects of grape NEPP on human health. In particular, most studies have focused on high-molecular-weight grape seed proanthocyanidins. So, a literature search was performed as illustrated in Figure 15.1. The search term was 'grape seed', for which a total of 2130 papers are indexed (PubMed, Scopus and Science Direct). To understand the growing interest in this topic over time, it is worth noting that in 1996, the number of published studies involving grape seed was just six, while in 2016 the number had risen to 216. Notwithstanding this, the number of clinical trial studies is low: 71, of which 22 were excluded from the analysis for this chapter, mostly because the compound studied was not sufficiently characterised.

Figure 15.2 summarises the evidence and the effectiveness of treatment with grape seed proanthocyanidins *via* several disease parameters. Most of the compounds studied correspond to high-molecular-weight proanthocyanidins, thus NEPP. The x-axis of Figure 15.2 represents the evidence of the effect through the number of publications describing a significant change, while the y-axis represents the effectiveness of supplementation, indicating the percentage change in the physiological parameter. Additionally, the average dose used to observe the change in each physiological parameter is included. The doses employed in the studies reviewed ranged from the equivalent of 50 to 1000 mg of proanthocyanidins, while the treatment period

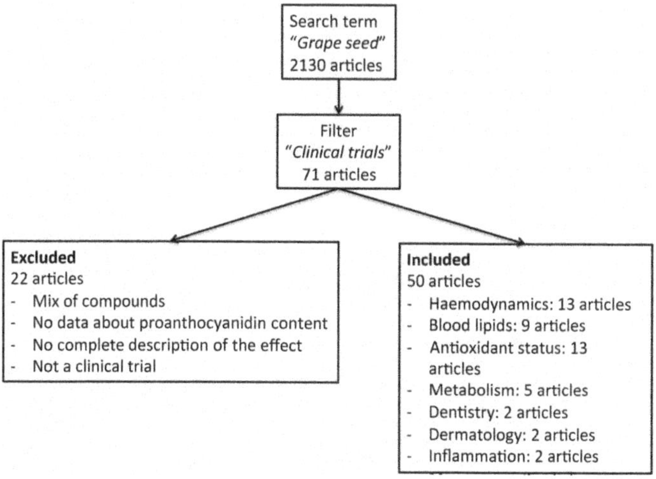

Figure 15.1 Flow diagram of study selection.

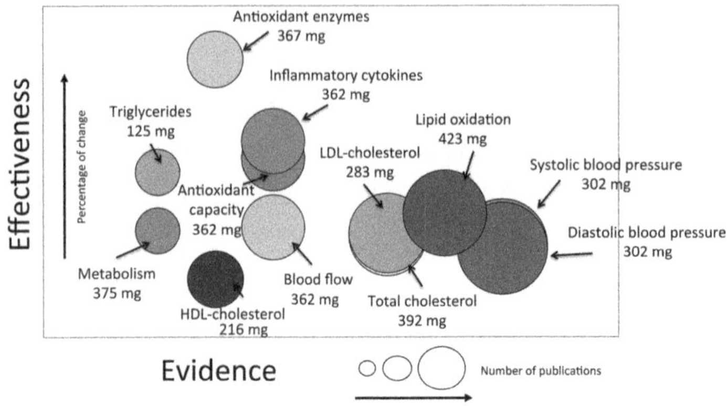

Figure 15.2 Evidence and effectiveness of the modification of physiological parameters by the supplementation of grape seed proanthocyanidins.

covered was from 3 weeks to 1 year. We found that at least 12 physiological parameters could be significantly modified by grape seed proanthocyanidin supplementation, with reductions in blood pressure, blood lipids (mainly total cholesterol and low-density lipoprotein [LDL] cholesterol) and lipid oxidative markers being those for which the most studies showed beneficial effects. The effectiveness of the interventions was around 6–10%. Modification of the activity of antioxidant enzymes, mainly glutathione peroxidase, was the physiological parameter for which the greatest effect was observed, with an increase in activity of 22–62%. This was followed by the reduction in the levels of inflammatory cytokines: 19–37% of C-reactive protein, TNF-α or plasminogen activator inhibitor-1.

The main difficulty in the analysis and comparison of the effects of grape seed products on several biological parameters is the lack of exhaustive descriptions of the product tested. Few studies describe the distribution of proanthocyanidin classes or preparation of the test extract properly, which may limit the standardisation of the effect in relation to the dose. For example, some of the compounds tested were used as a liquid formula,[18,19] which may limit the amount of NEPA. Moreover, whereas most of the studies employed powder supplements with a higher concentration of NEPA, they only described the composition of EPA,[20–22] or the composition was not reported at all.[23–25]

15.2 Haemodynamics

Oxidative stress has come to be seen as one of the fundamental conditions leading to the development of hypertension.[26,27] Several studies demonstrated that essential hypertensive patients produce excessive amounts of radical oxygen species.[28,29] Later, a strong association between blood pressure and some oxidative stress-related parameters such as 8-isoprostane and malondialdehyde was demonstrated.[30] Similarly, antihypertensive agents such as β-adrenergic blockers, angiotensin-converting enzyme inhibitors, angiotensin receptor antagonists and calcium channel blockers have been reported to induce an increase in the antioxidant capacity of plasma and a reduction in lipid hydroperoxides and malondialdehyde.[31] Thus, grape seed proanthocyanidins, as an antioxidant agent, could be a therapeutic option for the management and treatment of hypertension.

Table 15.1 summarises the clinical trials reviewed in this chapter that showed significant effects on blood pressure. Only the studies by Ras *et al.*[20] and Park *et al.*[19] were chiefly designed to evaluate the effect of grape seed proanthocyanidins on hypertensive subjects; the others focused on subjects exhibiting at least one metabolic syndrome component[21,32,33] or healthy subjects.[34] The doses of grape seed products employed varied from the equivalent of 50 mg to 1 g of proanthocyanidins, and the treatment time was from 3 to 16 weeks. Despite the differences in subjects, doses and treatment times, the reductions in systolic and diastolic blood pressure were similar in the studies reviewed: from −3.3% to −8.2% and from −3.0% to −8.9%, respectively. Surprisingly, higher doses/effects were observed in studies with shorter treatment durations, as shown in Figure 15.3. This may reveal homeostatic regulation after prolonged grape seed proanthocyanidin treatment, whereby further improvement is diminished. Unfortunately, none of the studies performed time course analysis to follow the changes in blood pressure during the treatment period, while only one evaluated the effect of grape seed extract on hypertensive subjects (between 120 and 159 mmHg of systolic blood pressure), and the reduction in blood pressure was lower[20] than in healthy or metabolic syndrome volunteers.

The mechanism of action of proanthocyanidins on blood pressure is not well understood, but may be related to modifications that could influence haemodynamics, such as: stimulation of nitric oxide release, which has vasorelaxing and anti-aggregating effects on the vascular endothelium;[35] inhibition of lipoprotein oxidation, macrophage cholesterol accumulation or foam cell formation; and enhancing cholesterol efflux from vessel walls. Several authors have reported that extracts from both grape and wine are capable of inducing endothelium-dependent relaxation *via* enhanced generation or increased biological activity of nitric oxide.

Two meta-analyses of the impact of grape seed treatment on blood pressure are available. The first, which included nine studies from 2000 to 2010, was performed to elucidate the effect of grape seed extract on cardiovascular risk markers. It found a statistically significant reduction in systolic blood pressure of −1.54 mmHg (95% confidence interval: −2.85 to −0.22 mmHg), but no effect on diastolic blood pressure.[36] The same authors compared the reduction observed in grape seed extract supplementation with the prescription of antihypertensive drugs and considered it to be quite modest. The second, published in 2016, included six more studies and concluded that grape seed extract has a beneficial impact on blood pressure and that this impact is more obvious in younger (age <50 years) or obese subjects,

Table 15.1 Characteristics of clinical trials of grape seed proanthocyanidins that showed significant effects in blood pressure.

Author	Type of population	Dose	Period of treatment	Percentage of change
Sivaprakasapillai *et al.*, 2009[21]	Subjects with metabolic syndrome	150 and 300 mg day^{-1} of grape seed extract	4 weeks	150 mg day^{-1} dose •−8.2% systolic •−7.2% diastolic 300 mg day^{-1} dose •−8.0% systolic •−8.9% diastolic
Pérez-Jiménez *et al.*, 2008[32]	Normo- and hyper-cholesterolaemic subjects	7.5 g of grape dietary fibre, with around 1 g of proanthocyanidins	16 weeks	•−6.3% systolic •−4.8% diastolic
Urquiaga *et al.*, 2015[33]	Male subjects with metabolic syndrome	20 g of grape pomace fluor	16 weeks	•−3.3% systolic •−6.6% diastolic
Queipo-Ortuño *et al.*, 2012[34]	Healthy adults	Around 50 mg day^{-1} of proanthocyanidins	3 weeks	•−7.0% systolic •−6.5% diastolic
Ras *et al.*, 2013[20]	Pre- and stage I hypertension	Grape seed extract 300 mg day^{-1}	8 weeks	•−3.8% systolic •−3.0% diastolic
Park *et al.*, 2016[19]	Pre-hypertension	Grape seed extract 300 mg day^{-1}	6 weeks	•−5.6% systolic •−4.7% diastolic

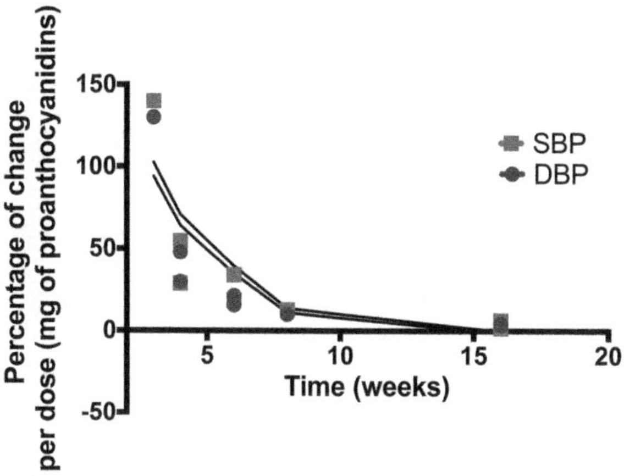

Figure 15.3 Relationship between the percentage of change in systolic blood pressure (SBP) and diastolic blood pressure (DBP) related to the dose of supplement and the duration of the treatment.

as well as in patients with metabolic syndrome.[37] The overall analysis reported reductions in systolic and diastolic blood pressure of 6.077 and 2.803 mmHg, respectively, after grape seed extract treatment relative to placebo. The reduction in blood pressure was more obvious in longer trials (>8 weeks) and with lower doses (<800 mg day^{-1}). Unfortunately, whereas this marginal reduction in blood pressure was observed in patients with pre- and stage 1 hypertension, no significant change was observed in hypertensive patients. However, the reduction in blood pressure is of clinical relevance since it has been estimated that for every 3 mmHg of systolic blood pressure that is reduced, the risk of all-cause mortality could diminish by 4%, after stroke by 8% and after coronary artery disease by 5%.[38]

Besides blood pressure, a growing body of research shows that grape seed proanthocyanidins may have beneficial effects on other factors related to haemodynamics. It is not clear whether the effect of grape seed proanthocyanidins on lowering blood pressure is dependent on a reduction of heart rate or the mechanisms described above. None of the studies reviewed in this chapter reported a significant effect on heart rate, while in two of them, significant effects on blood flow parameters were observed. Vaisman and Niv[18] observed that daily intake of 400 mg of red grape cell powder containing 33.6 mg g^{-1} of dry weight of proanthocyanidins for 12 weeks increased forearm flow-mediated dilation in all the study subjects, while this parameters was increased in only 22% of the subjects in the placebo group. Similarly, other authors showed an increase in flow-mediated dilation of the brachial artery after 30 days of grape powder intake, which represented a supplementation of 267 mg of proanthocyanidins per day.[39] In this study, it was suggested that the increase in blood flow was due to an increase in nitric oxide availability, which was favoured by the antioxidant properties of grape proanthocyanidin metabolites. Interestingly, since both studies used a powder obtained from whole grapes, they were providing not only EPA, but also NEPA.

15.3 Blood Lipid Profile

The lipid-lowering properties of proanthocyanidin have been widely reported in both *in vitro* and animal models, not solely after chronic exposure, but also acute postprandial effects have been observed. For example, reductions in postprandial triglyceride levels have been observed as a result of the administration of 25 mg kg^{-1} of grape seed procyanidin extract to normolipidaemic animals[40,41] and chronically in animals that received a high-fat diet.[42] *In vitro*, the synthesis of cholesterol esters, free cholesterol and triglycerides was reduced in HepG2 cells treated with grape seed proanthocyanidin serum metabolites.[43] Interestingly, in this study, it was observed that grape seed proanthocyanidins induce a greater reduction in the described parameters than pycnogenol and cocoa proanthocyanidin extracts. Different mechanisms have been proposed for the lipid-lowering and anti-hyperlipidaemic activities of grape seed proanthocyanidins, including: the inhibition of lipase, of cellular cholesterol uptake or of pancreatic lipase and cholesterol esterase; the formation of cholesterol micelles and binding to bile acids; and the depletion of transition metals.[44]

A total of nine clinical trials reporting significant changes in blood lipid profiles are available in the literature (Table 15.2). Changes in the levels of total cholesterol and LDL-cholesterol have most commonly been observed, ranging from −5.2% to −13.9% for total cholesterol and from −6.8% to −14.4% for LDL-cholesterol. The results in relation to HDL-cholesterol levels are contradictory, since both increases[24] and decreases[34] in this parameter have been observed. Additionally, effects have been observed on other parameters related to blood lipids, such as an increase in serum paraoxonase,[25] which is mainly associated with ApoAI that is located in HDL-cholesterol particles and may inhibit the oxidation of LDL. In relation to particle size analysis, Zunino *et al.*[45] observed a reduction in the concentration of large LDL-cholesterol particles after 3 weeks of intervention and a reduction in very-low-density lipoprotein size after 2 weeks of intervention.

A common feature in all the studies reviewed is that the lipid-lowering effect of grape seed proanthocyanidin supplementation is particularly observed in hypercholesterolaemic subjects.[32,46,47] Average decreases of 14.2% for total cholesterol, 11.6% for LDL-cholesterol and 18.6% for triglycerides were observed in hypercholesterolaemic subjects, but not in normocholesterolaemic ones.[32] Similar results were reported for normocholesterolaemic subjects after 12 weeks of supplementation at different doses of proanthocyanidins (200 and 400 mg day^{-1}).[24] In addition, a study evaluated the adjuvant effect of grape seed extract supplementation in hypercholesterolaemic patients receiving statins.[48] In this case, only a slight reduction in LDL-cholesterol levels (−3.3 mg dL^{-1}) was observed, with no effect on total

Table 15.2 Characteristics of clinical trials of grape seed proanthocyanidins that showed significant effects in blood lipid profile.[a]

Author	Type of population	Dose	Period of treatment	Percentage of change
Pérez-Jiménez *et al.*, 2008[32]	Normo- and hyper-cholesterolaemic subjects	7.5 g of grape dietary fibre, with around 1 g of proanthocyanidins	16 weeks	Total cholesterol: −8.4%
Sano *et al.*, 2007[24]	Healthy subjects	200 or 400 mg day^{-1} of proanthocyanidins	12 weeks	HDL: +8.1% for 200 mg day^{-1} HDL: +9.2% for 400 mg day^{-1}
Argani *et al.*, 2016[25]	Mild to moderate hyperlipidaemia	Grape seed extract 200 mg day^{-1}	8 weeks	Total cholesterol: −6.5% LDL: −9.2% Triglycerides: −9.7% Paraoxonase: +3.9% ApoA1: +8.1%
Queipo-Ortuño *et al.*, 2012[34]	Healthy adults	Around 50 mg day^{-1} of proanthocyanidins	3 weeks	Total cholesterol: −6.7% HDL: −16% Triglycerides: −30%
Zunino *et al.*, 2014[45]	Obese adults	46 g day^{-1} of dried grapes	9 weeks	Large LDL particle: −9.0%
Weseler *et al.*, 2011[46]	Male non-obese smokers	200 mg day^{-1} monomeric and polymeric flavanols	8 weeks	Total cholesterol: −9% LDL: −11%
Yubero *et al.*, 2013[56]	Healthy volunteers	700 mg day^{-1} grape seed extract	8 weeks	Total cholesterol: −13.9% LDL: −14.4%
Razavi *et al.*, 2013[47]	Mild hyperlipidaemia	200 mg day^{-1} red grape seed extract	8 weeks	Total cholesterol: −5.2% LDL: −6.8%
Tomé-Carneiro *et al.*, 2012[48]	Patients with primary prevention of cardiovascular disease	350 mg day^{-1} grape extract	24 weeks	LDL: −2.9%

[a]LDL: low-density lipoprotein; HDL: high-density lipoprotein.

cholesterol, triglyceride or HDL-cholesterol levels. However, it is necessary to emphasise that the initial levels of LDL-cholesterol of the subjects treated with statin were low, and supplementation with grape seed extract was capable of increasing the cholesterol-lowering effect of the statins.

In contrast, a meta-analysis of cardiovascular risk markers showed no statistically significant effect on the levels of any lipid, including total cholesterol, triglycerides, LDL-cholesterol and HDL-cholesterol.[36] One explanation for the lack of effect in some human intervention studies when there is one in some *in vitro* and animal models is that many of the latter studies are performed with grape seed proanthocyanidin doses that are much higher than those to which humans are exposed through intervention. For example, the inhibition of cholesterol esterase and pancreatic lipase, the formation of cholesterol micelles and the reduction in serum cholesterol and triglyceride levels were observed in male Wistar rats after doses higher than 250 mg kg^{-1} of body weight,[49] which could represent doses of over 15 g of proanthocyanidins in a person of 60 kg.

15.4 Antioxidant Status

The antioxidant capacities of proanthocyanidins are limited by their accessibility with regard to target tissues. Some studies have demonstrate that dimeric, trimeric and tetrameric proanthocyanidins, also referred to as EPA, are bioavailable.[50,51] These low-molecular-weight proanthocyanidins are also known as sustained-release antioxidants: they can remain in plasma and tissues for as long as 7–10 days and exert antioxidant properties. This is mechanistically different from the effects of other water-soluble antioxidants. In contrast, high-molecular-weight polymeric NEPA are very potent scavengers of peroxyl radicals, but they are not absorbable or bioavailable at all. So, in order to explain how these high-molecular-weight, non-absorbable (*per se*) antioxidants induce positive biological effects, it was suggested that they could exert their antioxidant activity in the digestive tract and protect lipids, proteins and carbohydrates from oxidative damage during digestion, thus sparing soluble antioxidants. It has also been reported that metabolites of colonic fermentation of proanthocyanidins exhibit antioxidant capacities, in a similar way to quercetin.[52] For example, the main metabolite of colonic fermentation of proanthocyanidins, 3,4-dihydroxyphenyl acetic acid, has reported values from ferric reducing/antioxidant power (FRAP) and 1,1-diphenyl-2-picryl-hydrazil (DPPH) assays of between 2.11–4.03 and 1.76 μM Trolox equivalents.[53,54]

Evidence from clinical trials suggests that grape seed proanthocyanidin supplementation modifies several parameters related to oxidative status, as shown in Table 15.3. Five studies demonstrate a reduction in lipid oxidation parameters such as oxidised cholesterol or plasma malondialdehyde (a by-product of lipid peroxidation) of between 8.8% and 14.7%, with treatment times of 4–12 weeks.[24,47,55–57] The dose-dependent reduction in LDL-malondialdehyde and its time course have been reported. For example, low doses of proanthocyanidins are not effective until the twelfth week of supplementation, whereas higher doses are effective at the sixth.[24] Interestingly, the antioxidant-protective

Table 15.3 Characteristics of clinical trials of grape seed proanthocyanidins that showed significant effects in the inhibition of lipid oxidation biomarkers.

Author	Type of population	Dose	Period of treatment	Percentage of change
Vigna *et al.*, 2003 [55]	Heavy smokers	150 mg of grape seed extract	4 weeks	TBARS-LDL:[a] −14.7%
Sano *et al.*, 2007 [24]	Healthy subjects	200 or 400 mg day^{-1} of proanthocyanidins	12 weeks	MDA-LDL: [b] −12.2% for 200 mg day^{-1} MDA-LDL: [b] −13.9% for 400 mg day^{-1}
Yubero *et al.*, 2013 [56]	Healthy volunteers	700 mg day^{-1} grape seed extract	8 weeks	LDLox: [c] −13.6%
Mellen *et al.*, 2010 [57]	Adults with coronary disease	1300 mg day^{-1} muscadine grape seed	4 weeks	Plasma MDA: −18.5%
Razavi *et al.*, 2013 [47]	Mild hyperlipidaemia	200 mg day^{-1} red grape seed extract	8 weeks	LDLox: [c] −8.8%

[a]TBARS-LDL: thiobarbituric acid reactive substances-low density lipoprotein.
[b]MDA-LDL: malondialdhyde-low density lipoprotein.
[c]LDLox: oxidazed-low density lipoprotein.

effect of grape seed proanthocyanidin supplementation was also observed in high-oxidative stress conditions such as haemodialysis patients[58] and heavy smokers.[55] Additionally, a reduction in the number of carbonyls (oxidation by-products) was also found in patients exhibiting at least one component of metabolic syndrome[33] and in total lipid peroxides present in plasma.[18]

However, in spite of all this evidence, the mechanism of action is not clearly understood (Figure 15.4). None of the studies that have been reviewed for this chapter determined the levels of absorbable proanthocyanidin compounds, either as intact low-molecular-weight compounds or as colonic fermentation metabolites, so that the observed effects in these studies cannot be attributed to them. A total 28% increase in the plasma antioxidant capacity has been observed after supplementation with 700 mg of grape seed extract for 8 weeks.[56] This increase seems to be triggered postprandially, 30 minutes after a single dose of 300 mg of grape seed extract,[59] and is maintained for the following 3 hours, while no modifications were observed in the placebo group. Antioxidant-sparing properties are also suggested as a potential mechanism in some studies, where the levels of dietary antioxidants such as γ-tocopherol, δ-tocopherol and vitamin C,[33] as well as endogenous antioxidants such as glutathione,[46] were increased after grape seed supplementation. Evidence of indirect mechanisms of action has also been observed in clinical trials where grape seed extract supplementation induced higher levels of endogenous antioxidant enzymes, mainly glutathione peroxidase.[22,58] Additionally, since inflammatory processes induce an increase in oxidative stress,

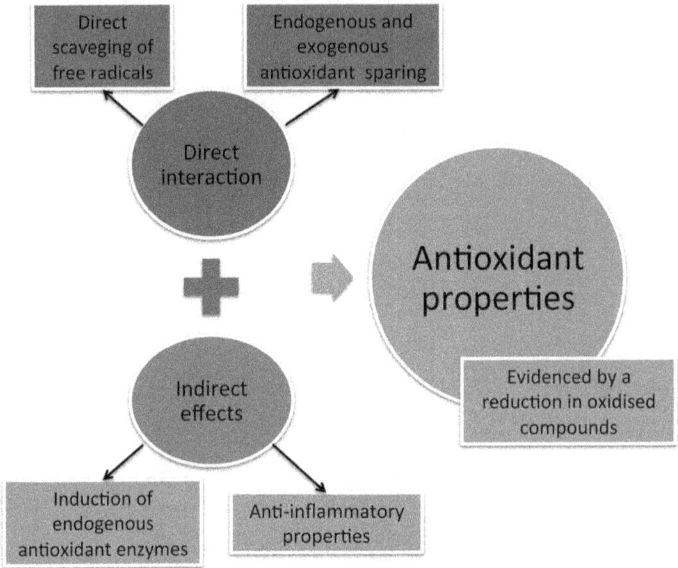

Figure 15.4 Scheme of the potential mechanism of action of proanthocyanidins in terms of antioxidant status.

the reported anti-inflammatory properties of grape seed proanthocyanidins could also explain the reduction in oxidative stress observed after supplementation. For example, reductions in the levels of C-reactive protein (−26% to −37%),[48,60] TNF-α (−20%)[48] and plasminogen activator inhibitor-1 (−19%)[60] have been observed, as has an increase in the anti-inflammatory cytokine IL-10 (+20%).[48]

15.5 Metabolism

Proanthocyanidins from grape seed products have the widely demonstrated capacity to modulate the increase in body weight in animal models fed a high-fat diet. With respect to the mechanism of action of grape seed extract in animal models, the proanthocyanidins present in grape seeds have been shown to reduce food intake. Additionally, *in vitro* studies have shown that grape proanthocyanidins are associated with significant inhibition of the activity of α-amylase and α-glucosidase, with the potential to prevent the breakdown of starch into its glucose components, thus blocking a required step for the absorption of carbohydrates available from the diet.[61,62] Furthermore, during colonic fermentation, due to the antimicrobial activity of proanthocyanidins, microbial catabolic activity could be modified and consequently affect the bioavailability of fermentable products in the colon. However, evidence of a positive effect on body weight in humans is scarce.

No modification in body weight was reported in any clinical trials reviewed in this chapter, although the evaluation period (4–16 weeks) is relatively short for the observation of a statistical difference in body weight. Only one study evaluated – in normal to overweight men and women – the potential effect of grape seeds on energy intake and the macronutrient composition of food ingested, taking attitudes towards eating, appetite ratings, hedonics, sensory-specific satiety, mood and tolerance into account. As a result, grape seed extract (300 mg containing >90% procyanidins) taken 30 minutes before meals appeared to be effective at reducing energy intake in normal to overweight subjects with an energy requirement above the median,[63] while satiety rakings were sustained. The decrease in energy intake was on average 4%, which could have long-term consequences for body weight, notwithstanding, the fact that the postprandial effects on appetite are not clear in humans, even with doses of proanthocyanidins as high as 500–1500 mg.[64]

In relation to metabolic hormones, two studies observed an 8–10% increase in circulating levels of adiponectin.[24,60] Adiponectin is a protein hormone that modulates a number of metabolic processes, including glucose regulation and fatty acid oxidation, and is exclusively secreted from adipose tissue (and also from the placenta during pregnancy). Its secretion is stimulated during situations of caloric restriction,[65] which is a proposed mechanism of action of proanthocyanidin supplementation. Additionally, blood adiponectin levels have been reported to correlate negatively with several oxidative biomarkers,[66] which may represent a link between the antioxidant properties of proanthocyanidins and the observed effects on the metabolism.

15.6 Dermatology

Three clinical trials demonstrated beneficial effects of the administration of grape seed proanthocyanidin extract on skin health. One of them assessed the effect of its topical administration on wound healing, while the other two evaluated the effects of oral and topical administration of grape seed proanthocyanidins on skin pigmentation.

Regarding wound healing, which has been described as a complex and dynamic process involving several local and systemic factors such as age, the topical administration of a 2% grape seed cream to surgery wounds induced on average a full recovery of the surgical lesion on day 8 (8.06 ± 1.45) compared to day 14 (14.40 ± 3.94) for the placebo group – clearly significant in terms of statistical analysis.[67] It has been suggested that these wound-healing properties may be related to the capacity of grape seed proanthocyanidins to release endothelial growth factors, thereby increasing cell density and the deposition of connective tissue at the wound site. In general, this improves the cellular structure of the wound.[68] Furthermore, topical use may cause wound contraction, which is an important part of the natural healing process.

In relation to skin pigmentation, the effect of a formulation containing 2% of a soluble Muscat Hamburg seed extract (a blue–black table grape cultivar grown under subtropical, Mediterranean and continental climate conditions) using occlusive patches on the cheek skin of healthy male adults has been assessed. It was observed that the skin melanin content decreased over the period of study (8 weeks).[69] Although the molecular mechanism was not reported in the study, it was suggested that the effect may be due to the inhibition of tyrosinase, a critical enzyme involved in melanin synthesis and skin pigmentation. In the same context, another study observed that oral administration of a proanthocyanidin-rich extract from grape seeds (162 mg of proanthocyanidins for 12 months) improved chloasma.[70] Chloasma (melasma) is abnormal facial pigmentation characterised by irregular light to dark brown macules and patches on sun-exposed areas of the skin. Sun exposure is the most important factor, since UV radiation can act on melanocytes through the release of keratinocyte-derived factors that are known to increase melanisation and even induce its proliferation. The maximum beneficial effects of the grape seed extract were achieved after 6 months and it was suggested that the effects were due to the inhibition of melanogenesis or even melanocyte proliferation. Additionally, it was observed that the topical application of a cream containing grape seed extract increased skin elasticity by over 10% compared to the baseline measure.[69] In the same manner as explained for the melanin content, it was suggested that the observed effect might be due to the inhibition of collagenases and elastases, and also a decrease in sebum content, mediated by the inhibition of 5α-reductase. Finally, grape seed extract may also work as a sunscreen in topical formulations; extracts from Burgund Mare grapes have been reported to protect against multiple doses of UV-B irradiation in mice.[71]

15.7 Dentistry

There is some evidence that proanthocyanidins from grape seed extract affect the stability of teeth dentin. Dentin is a calcified tissue, which has an organic matrix containing fibrillar type-I collagen and non-collagenous proteins such as phosphoproteins and proteoglycans. Ultrastructural studies of root caries suggest that during the caries process, the mineral part of dentine is dissolved by acids produced by bacteria, thereby exposing the organic matrix to breakdown by bacterially derived enzymes, as well as by host-derived enzymes such as matrix metalloproteinases present in the dentin and in the gingival crevicular fluid.

Grape seed extract has been reported to be a low-toxicity compound that can induce exogenous cross-links in the type-I collagen in tooth dentin.[72] Moreover, it has been reported that glutaraldehyde and cross-linking induced by grape seed proanthocyanidins improved the mechanical properties of dentin.[73] In freshly extracted human molars with non-cavitated occlusal caries, the etched dentin surface was treated with 6.5% w/v grape seed extract for 1 hour and followed by the induction of collagen degradation. Compared to a control group and teeth treated with glutaraldehyde, an increase in bond strength and lower collagen digestibility were observed.[23] In this last study, it was suggested that the increased resistance to degradation might be attributed to the masking of the recognition site by the cross-links or to the retention of the cleaved peptide fragments by the newly formed cross-links. In relation to caries prevention, teeth treated with a pH-cycling artificial caries protocol for 18 days with a 30 minutes application of grape seed proanthocyanidin extract (6.5% solution) showed decreased rates of root demineralisation and possibly increased remineralisation.[74] The hydrogen bonding between the amine, carboxyl and phenolic hydroxyl groups may be the primary mechanism of proanthocyanidin and collagen interactions. Notwithstanding this, another study has reported an anti-caries effect using a bacterial model, and therefore antibacterial properties cannot be ruled out.[75]

15.8 Conclusions

The evidence of the biological effects of grape seed proanthocyanidins from clinical trials is still scarce. The main drawbacks to the analysis of the available information are the diversity of the formulations employed in the studies and the lack of satisfactory descriptions of the proanthocyanidin content in the products used, particularly NEPA, as significant contributors to total grape proanthocyanidin content. However, it seems evident that grape seed proanthocyanidin supplementation exerts antioxidant effects *via* several mechanisms. Almost all of the studies reviewed that reported physiological effects also reported modifications in antioxidant status or biomarkers. This last effect could explain the results observed in clinical trials regarding the reduction of blood pressure, lipid homeostasis and metabolism. Additionally, the topical uses of grape seed proanthocyanidins in dermatology and dentistry also deserve further attention. Finally, it is worth mentioning that the doses employed in all of the clinical studies (from 50 mg to 1.5 g of proanthocyanidins) were reported to be tolerable. However, it seems that these doses are lower than those employed in animal studies.

References

1. S. Renaud and M. de Lorgeril, *Lancet*, 1992, **339**, 1523–1526.
2. J. R. Emberson, A. G. Shaper, S. G. Wannamethee, R. W. Morris and P. H. Whincup, *Am. J. Epidemiol.*, 2005, **161**, 856–863.
3. A. Di Castelnuovo, S. Costanzo, V. Bagnardi, M. B. Donati, L. Iacoviello and G. de Gaetano, *Arch. Intern. Med.*, 2006, **166**, 2437–2445.
4. S. Costanzo, A. Di Castelnuovo, M. B. Donati, L. Iacoviello and G. de Gaetano, *Eur. J. Epidemiol.*, 2011, **26**, 833–850.
5. K. Shimada, H. Watanabe, K. Hosoda, K. Takeuchi and J. Yoshikawa, *Lancet*, 1999, **354**, 1002.

6. D. P. van Velden, E. P. G. Mansvelt, E. Fourie, M. Rossouw and A. D. Marais, *Ann. N. Y. Acad. Sci.*, 2002, **957**, 337–340.
7. S. Rosenkranz, D. Knirel, H. Dietrich, M. Flesch, E. Erdmann and M. Böhm, *FASEB J.*, 2002, **16**, 1958–1960.
8. M. J. Cho, L. R. Howard, R. L. Prior and J. R. Clark, *J. Sci. Food Agric.*, 2004, **84**, 1771–1782.
9. E. Cantos, J. C. Espín and F. A. Tomás-Barberán, *J. Agric. Food Chem.*, 2002, **50**, 5691–5696.
10. J. Serrano, R. Puupponen-Pimiä, A. Dauer, A.-M. Aura and F. Saura-Calixto, *Mol. Nutr. Food Res.*, 2009, **53**(suppl. 2), S310–S329.
11. J. Perez-Jiménez, S. Arranz and F. Saura-Calixto, *Food Res. Int.*, 2009, **42**, 1381–1388.
12. K.-T. Chung, C.-I. Wei and M. G. Johnson, *Trends Food Sci. Technol.*, 1998, **9**, 168–175.
13. A. Sano, J. Yamakoshi, S. Tokutake, K. Tobe, Y. Kubota and M. Kikuchi, *Biosci., Biotechnol., Biochem.*, 2003, **67**, 1140–1143.
14. F. Saura-Calixto, J. Pérez-Jiménez, S. Touriño, J. Serrano, E. Fuguet, J. L. Torres and I. Goñi, *Mol. Nutr. Food Res.*, 2010, **54**, 939–946.
15. M. Biagi and A. A. E. Bertelli, *Life Sci.*, 2015, **131**, 19–22.
16. A. L. Klatsky, G. D. Friedman, M. A. Armstrong and H. Kipp, *Am. J. Epidemiol.*, 2003, **158**, 585–595.
17. M. Grønbaek, U. Becker, D. Johansen, A. Gottschau, P. Schnohr, H. O. Hein, G. Jensen and T. I. Sørensen, *Ann. Intern. Med.*, 2000, **133**, 411–419.
18. N. Vaisman and E. Niv, *Int. J. Food Sci. Nutr.*, 2015, **66**, 342–349.
19. E. Park, I. Edirisinghe, Y. Y. Choy, A. Waterhouse and B. Burton-Freeman, *Br. J. Nutr.*, 2016, **115**, 226–238.
20. R. T. Ras, P. L. Zock, Y. E. M. P. Zebregs, N. R. Johnston, D. J. Webb and R. Draijer, *Br. J. Nutr.*, 2013, **110**, 2234–2241.
21. B. Sivaprakasapillai, I. Edirisinghe, J. Randolph, F. Steinberg and T. Kappagoda, *Metabolism*, 2009, **58**, 1743–1746.
22. D. De Groote, K. Van Belleghem, J. Devière, W. Van Brussel, A. Mukaneza and L. Amininejad, *Ann. Nutr. Metab.*, 2012, **61**, 15–24.
23. G. V. Macedo, M. Yamauchi and A. K. Bedran-Russo, *J. Dent. Res.*, 2009, **88**, 1096–1100.
24. A. Sano, R. Uchida, M. Saito, N. Shioya, Y. Komori, Y. Tho and N. Hashizume, *J. Nutr. Sci. Vitaminol.*, 2007, **53**, 174–182.
25. H. Argani, A. Ghorbanihaghjo, H. Vatankhahan, N. Rashtchizadeh, S. Raeisi and H. Ilghami, *Sao Paulo Med. J.*, 2016, **134**, 234–239.
26. J. González, N. Valls, R. Brito and R. Rodrigo, *World J. Cardiol.*, 2014, **6**, 353.
27. D. G. Harrison and M. C. Gongora, *Med. Clin. North Am.*, 2009, **93**, 621–635.
28. F. Lacy, M. T. Kailasam, D. T. O'Connor, G. W. Schmid-Schönbein and R. J. Parmer, *Hypertension*, 2000, **36**, 878–884.
29. J. Redon, M. R. Oliva, C. Tormos, V. Giner, J. Chaves, A. Iradi and G. T. Saez, *Hypertension*, 2003, **41**, 1096–1101.

30. R. Rodrigo, H. Prat, W. Passalacqua, J. Araya, C. Guichard and J. P. Bächler, *Hypertens. Res.*, 2007, **30**, 1159–1167.
31. L. Ghiadoni, A. Magagna, D. Versari, I. Kardasz, Y. Huang, S. Taddei and A. Salvetti, *Hypertension*, 2003, **41**, 1281–1286.
32. J. P. Jiménez, J. Serrano, M. Tabernero, S. Arranz, M. E. Díaz-Rubio, L. García-Diz, I. Goñi and F. Saura-Calixto, *Nutrition*, 2008, **24**, 646–653.
33. I. Urquiaga, S. D'Acuña, D. Pérez, S. Dicenta, G. Echeverría, A. Rigotti and F. Leighton, *Biol. Res.*, 2015, **48**, 49.
34. M. I. Queipo-Ortuno, M. Boto-Ordonez, M. Murri, J. M. Gomez-Zumaquero, M. Clemente-Postigo, R. Estruch, F. Cardona Diaz, C. Andres-Lacueva and F. J. Tinahones, *Am. J. Clin. Nutr.*, 2012, **95**, 1323–1334.
35. E. J. Chou, J. G. Keevil, S. Aeschlimann, D. A. Wiebe, J. D. Folts and J. H. Stein, *Am. J. Cardiol.*, 2001, **88**, 553–555.
36. H. H. H. Feringa, D. A. Laskey, J. E. Dickson and C. I. Coleman, *J. Am. Diet. Assoc.*, 2011, **111**, 1173–1181.
37. H. Zhang, S. Liu, L. Li, S. Liu, S. Liu, J. Mi and G. Tian, *Medicine*, 2016, **95**, e4247.
38. P. K. Whelton, J. He, L. J. Appel, J. A. Cutler, S. Havas, T. A. Kotchen, E. J. Roccella, R. Stout, C. Vallbona, M. C. Winston and J. Karimbakas, National High Blood Pressure Education Program Coordinating Committee, *JAMA*, 2002, **288**, 1882–1888.
39. J. Barona, J. C. Aristizabal, C. N. Blesso, J. S. Volek and M. L. Fernandez, *J. Nutr.*, 2012, **142**, 1626–1632.
40. J. M. Del Bas, M.-L. Ricketts, M. Vaqué, E. Sala, H. Quesada, A. Ardevol, M. J. Salvadó, M. Blay, L. Arola, D. D. Moore, G. Pujadas, J. Fernandez-Larrea and C. Bladé, *Mol. Nutr. Food Res.*, 2009, **53**, 805–814.
41. J. M. Del Bas, M. L. Ricketts, I. Baiges, H. Quesada, A. Ardevol, M. J. Salvadó, G. Pujadas, M. Blay, L. Arola, C. Bladé, D. D. Moore and J. Fernandez-Larrea, *Mol. Nutr. Food Res.*, 2008, **52**, 1172–1181.
42. H. Quesada, J. M. del Bas, D. Pajuelo, S. Díaz, J. Fernandez-Larrea, M. Pinent, L. Arola, M. J. Salvadó and C. Bladé, *Int. J. Obes.*, 2009, **33**, 1007–1012.
43. L. Guerrero, M. Margalef, Z. Pons, M. Quiñones, L. Arola, A. Arola-Arnal and B. Muguerza, *J. Nutr. Biochem.*, 2013, **24**, 2092–2099.
44. M. Akaberi and H. Hosseinzadeh, *Phytother. Res.*, 2016, **30**, 540–556.
45. S. J. Zunino, J. M. Peerson, T. L. Freytag, A. P. Breksa, E. L. Bonnel, L. R. Woodhouse and D. H. Storms, *Br. J. Nutr.*, 2014, **112**, 369–380.
46. A. R. Weseler, E. J. B. Ruijters, M.-J. Drittij-Reijnders, K. D. Reesink, G. R. M. M. Haenen and A. Bast, *PLoS One*, 2011, **6**, e28460.
47. S.-M. Razavi, S. Gholamin, A. Eskandari, N. Mohsenian, A. Ghorbanihaghjo, A. Delazar, N. Rashtchizadeh, M. Keshtkar-Jahromi and H. Argani, *J. Med. Food*, 2013, **16**, 255–258.
48. J. Tomé-Carneiro, M. Gonzálvez, M. Larrosa, F. J. García-Almagro, F. Avilés-Plaza, S. Parra, M. J. Yáñez-Gascón, J. A. Ruiz-Ros, M. T. García-Conesa, F. A. Tomás-Barberán and J. C. Espín, *Mol. Nutr. Food Res.*, 2012, **56**, 810–821.

49. S. Adisakwattana, J. Moonrat, S. Srichairat, C. Chanasit, H. Tirapongporn, B. Chanathong, S. Ngamukote, K. Mäkynen and S. Sapwarobol, *J. Med. Plants Res., Acad. J.*, 2010, **4**, 2113–2120.
50. R. R. Holt, S. A. Lazarus, M. C. Sullards, Q. Y. Zhu, D. D. Schramm, J. F. Hammerstone, C. G. Fraga, H. H. Schmitz and C. L. Keen, *Am. J. Clin. Nutr.*, 2002, **76**, 798–804.
51. S. Deprez, I. Mila, J.-F. Huneau, D. Tome and A. Scalbert, *Antioxid. Redox Signaling*, 2001, **3**, 957–967.
52. Y. Tang, S. Nakashima, S. Saiki, Y. Myoi, N. Abe, S. Kuwazuru, B. Zhu, H. Ashida, Y. Murata and Y. Nakamura, *Food Res. Int.*, 2016, **89**, 716–723.
53. J. A. Giménez-Bastida, H. Zielinski, M. Piskula, D. Zielinska and D. Szawara-Nowak, *Mol. Nutr. Food Res.*, 2017, **61**, 1600475.
54. M. Dueñas, F. Surco-Laos, S. González-Manzano, A. M. González-Paramás and C. Santos-Buelga, *Eur. Food Res. Technol.*, 2011, **232**, 103–111.
55. G. B. Vigna, F. Costantini, G. Aldini, M. Carini, A. Catapano, F. Schena, A. Tangerini, R. Zanca, E. Bombardelli, P. Morazzoni, A. Mezzetti, R. Fellin and R. Maffei Facino, *Metabolism*, 2003, **52**, 1250–1257.
56. N. Yubero, M. Sanz-Buenhombre, A. Guadarrama, S. Villanueva, J. M. Carrión, E. Larrarte and C. Moro, *Int. J. Food Sci. Nutr.*, 2013, **64**, 400–406.
57. P. B. Mellen, K. R. Daniel, K. B. Brosnihan, K. J. Hansen and D. M. Herrington, *J. Am. Coll. Nutr.*, 2010, **29**, 469–475.
58. A. G. de P. R. Janiques, V. de O. Leal, M. B. Stockler-Pinto, N. X. Moreira and D. Mafra, *J. Bras. Nefrol.*, 2014, **36**, 496–501.
59. S. L. Nuttall, M. J. Kendall, E. Bombardelli and P. Morazzoni, *J. Clin. Pharm. Ther.*, 1998, **23**, 385–389.
60. J. Tomé-Carneiro, M. Gonzálvez, M. Larrosa, M. J. Yáñez-Gascón, F. J. García-Almagro, J. A. Ruiz-Ros, F. A. Tomás-Barberán, M. T. García-Conesa and J. C. Espín, *Cardiovasc. Drugs Ther.*, 2013, **27**, 37–48.
61. S. Adisakwattana, P. Jiphimai, P. Prutanopajai, B. Chanathong, S. Sapwarobol and T. Ariyapitipan, *Int. J. Food Sci. Nutr.*, 2010, **61**, 295–305.
62. M. Yilmazer-Musa, A. M. Griffith, A. J. Michels, E. Schneider and B. Frei, *J. Agric. Food Chem.*, 2012, **60**, 8924–8929.
63. N. Vogels, I. M. T. Nijs and M. S. Westerterp-Plantenga, *Eur. J. Clin. Nutr.*, 2004, **58**, 667–673.
64. H.-S. Shin, S. Kindleysides, W. Yip, S. C. Budgett, J. R. Ingram and S. D. Poppitt, *Nutr. J.*, 2015, **14**, 96.
65. Q. Ding, C. Ash, T. Mracek, B. Merry and C. Bing, *J. Nutr. Biochem.*, 2012, **23**, 867–874.
66. S. Gustafsson, L. Lind, S. Söderberg, M. Zilmer, J. Hulthe and E. Ingelsson, *Obesity*, 2013, **21**, 1467–1473.
67. A. A. Hemmati, M. Foroozan, G. Houshmand, Z. B. Moosavi, M. Bahadoram and N. S. Maram, *Global J. Health Sci.*, 2014, **7**, 52–58.
68. S. Khanna, M. Venojarvi, S. Roy, N. Sharma, P. Trikha, D. Bagchi, M. Bagchi and C. K. Sen, *Free Radical Biol. Med.*, 2002, **33**, 1089–1096.
69. A. Sharif, N. Akhtar, M. S. Khan, A. Menaa, B. Menaa, B. A. Khan and F. Menaa, *Int. J. Cosmet. Sci.*, 2015, **37**, 253–258.

70. J. Yamakoshi, A. Sano, S. Tokutake, M. Saito, M. Kikuchi, Y. Kubota, Y. Kawachi and F. Otsuka, *Phytother. Res.*, 2004, **18**, 895–899.
71. A. Filip, S. Clichici, D. Daicoviciu, C. Catoi, P. Bolfa, I. D. Postescu, A. Gal, I. Baldea, C. Gherman and A. Muresan, *J. Physiol. Pharmacol.*, 2011, **62**, 385–392.
72. B. Han, J. Jaurequi, B. W. Tang and M. E. Nimni, *J. Biomed. Mater. Res.*, 2003, **65A**, 118–124.
73. A. K. B. Bedran-Russo, D. H. Pashley, K. Agee, J. L. Drummond and K. J. Miescke, *J. Biomed. Mater. Res., Part B*, 2008, **86**, 330–334.
74. S. Pavan, Q. Xie, A. T. Hara and A. K. Bedran-Russo, *Caries Res.*, 2011, **45**, 443–447.
75. R. Walter, P. A. Miguez, R. R. Arnold, P. N. R. Pereira, W. R. Duarte and M. Yamauchi, *Caries Res.*, 2008, **42**, 263–268.

CHAPTER 16

Sorghum Non-extractable Polyphenols: Chemistry, Extraction and Bioactivity

GANGCHENG WU[a,b], STUART K. JOHNSON[c] AND ZHONGXIANG FANG*[d]

[a]State Key Laboratory of Food Science and Technology, Jiangnan University, Wuxi 214122, China; [b]School of Food Science and Technology, Jiangnan University, Wuxi 214122, China; [c]Food Science and Technology Program, School of Public Health, Curtin University, Perth, WA 6845, Australia; [d]Faculty of Veterinary and Agricultural Sciences, The University of Melbourne, Parkville, Victoria 3010, Australia
*E-mail: zhongxiang.fang@unimelb.edu.au

16.1 Introduction

Sorghum (*Sorghum bicolor* [L.] Moench) is a member of the grass family and ranks as the fifth most consumed cereal crop in the world after rice, wheat, maize and barley. The global production of sorghum was 66.71 million metric tons in 2016, around 2.82 million metric tons higher than that in 2015.[1] In many African (Sudan, Nigeria and Ethiopia), Asian (China and India) and Central American countries (Honduras and Nicaragua), sorghum is widely planted and also contributes to the staple diets of local people. In Western societies such as Australia and the USA, the majority of sorghum grain has been used to feed animals or produce biofuel.[2]

Food Chemistry, Function and Analysis No. 5
Non-extractable Polyphenols and Carotenoids: Importance in Human Nutrition and Health
Edited by Fulgencio Saura-Calixto and Jara Pérez-Jiménez

Published by the Royal Society of Chemistry, www.rsc.org

Recently, sorghum has been considered as a potentially sustainable crop due to its agronomic advantages, such as drought and high-temperature tolerance.[3]

Compared to other cereal grains, the nutritional quality of sorghum foods can be negatively affected by polyphenols, especially condensed tannins, through inhibiting digestive enzyme activity (*e.g.* amylase in starch digestion) and binding to dietary proteins and minerals reducing their digestibility and bioavailability.[4,5] Sorghum is a gluten-free cereal grain with a reputation as a source of bioactive phytochemicals including polyphenols.[6] Not surprisingly, sorghum's appeal as global human food is slowly but steadily increasing due to its polyphenolics with potential health-beneficial properties, such as antioxidant activity, which in turn can help reduce the risks of some chronic diseases, including cardiovascular diseases, cancer and type 2 diabetes.[7–9]

In nature, polyphenols generally exist in extractable and non-extractable forms. Phenolic acids mainly occur in non-extractable forms, while flavonoids are generally present in the C-glycoside or O-glycoside forms. The linkage between flavonoids and multiple or single sugar moieties can be through an OH group to form O-glycosides or through carbon–carbon bonds to form C-glycosides.[10] Numerous assays have been developed to release NEPP, including alkaline, acid,[11] enzymatic,[12] microwave-assisted[13] and ultrasound-assisted hydrolysis,[14] with alkaline and acidic hydrolysis being the most common. In addition, an accurate procedure for the determination of NEPP was developed widely in cereal grains.[15,16]

The understanding of the profile and content of polyphenols in sorghum grain is essential to fully appreciate the nutritional and health properties of this grain. However, in many previous studies, only extractable polyphenols have been the focus since it was believed that only this fraction of the polyphenols was bioavailable in humans.[17] Therefore, the NEPP have often been ignored.[18–20] The NEPP still remain in the residue after extraction by standard methods using aqueous–organic solvents; however, the percentage of NEPP can reach 50% in sorghum grain[21–24] and hence its evaluation is of great importance. More recently, it has become understood that the NEPP, when entering the human colon, can be released from the food matrix though the action of the diverse colonic microbiota, and then microbiologically degraded to simple phenolic acids or lactone structures, which are transformed to microbial metabolites that exhibit antioxidant activity.[25,26] In light of this, the high levels of NEPP in sorghum grain may contribute to gastrointestinal health, such as the prevention of colon cancer and other digestive cancers.[22,27] In addition, some *in vitro* studies indicate that the antioxidant activities of NEPP is higher than those of extractable ones.[28,29]

Therefore, this chapter will highlight research on NEPP of sorghum grain and indicate how different physical, chemical and food processes influence their bioactivities. The extraction methods to release NEPP from the cell wall matrix will also be discussed.

16.2 Non-extractable Polyphenols in Sorghum Grain

The concept of NEPP emerged to explain a paradoxical result of the dietary fibre analysis in carob pods.[30] The dietary fibre of this leguminous plant was 40.4% (dry weight), which must correspond to non-starch polysaccharides and lignin, based on the traditional definition of dietary fibre. However, the actual content of polysaccharides and lignin accounted for only 18.9% (dry weight) after specific analysis of these compounds. Therefore, a gap of 21.5% was identified as polymeric proanthocyanidins, a major type of NEPP.

All polyphenols that cannot be directly extracted with the solvents of choice, such as acetone–water and methanol–water mixtures, are defined as NEPP.[31] The NEPP concept can also be applied to other polyphenols that are not chemically complex or oligomeric in nature, such as the orange hesperidin and ellagic acid. In addition, this concept can also be extended to other high-molecular-weight polyphenols, such as ellagitannins. The ellagitannin composition has structural complexity that, in addition to the limitations of poor resolution of chromatographic methods, makes accurate quantification of it difficult. These polymeric molecules can be better determined after acid hydrolysis treatment,[32] which will be discussed in detail in the following section. NEPP have been found in many natural plants, such as fruits, vegetables and whole cereal grains.

16.2.1 Total Non-extractable Polyphenols in Sorghum Grain

Polyphenols are a type of chemical compound that contain one or more aromatic rings substituted by one or more hydroxyl groups. In sorghum grain, polyphenols are mainly located in the pericarp and influence the appearance, flavour and antioxidant activity of the grain. Sorghum is genetically very diverse with many different grain colours (Figure 16.1), which affects their polyphenol content. The types of NEPP occurring in sorghum grain and their levels are shown in Table 16.1. NEPP comprise 18.18–63.92% of the total polyphenols present in sorghum grains. White- and red-coloured sorghum have lower contents of NEPP than brown or black; however, the percentage contributions of the non-extractable fractions of the white and red sorghums to total polyphenols were higher than for brown or black grains. All of these results highlight that the levels of NEPP are significantly influenced by sorghum genotype.

Polyphenols of other cereal grains are also mainly found in non-extractable forms. NEPP as a percentage of the total in barley varied from 47.88% to 89.46%, depending on the genotype.[35] In total, 86.36%, 76.35%, 72.89% and 62.23% of the polyphenols present in corn, wheat, oats and rice are in non-extractable form, respectively.[22] Up to 90% of NEPP of the total was reported in finger millet.[36] In addition, the determination of the content and profile of NEPP in fruit and vegetables from four European countries showed

Figure 16.1 Sorghum genotypes with different grain colours. Left: white colour sorghum 'QL12', contains relatively low levels of polyphenols; right: black colour sorghum 'Shawaya short black 1', contains high levels of polyphenols.

Table 16.1 Non-extractable and total polyphenol contents of different-coloured sorghum grain (mg gallic acid equivalents g^{-1}, dry base).

Sorghum colour	Non-extractable polyphenols	Total polyphenols	Percentage of non-extractable fraction	References
White	1.2	2.0	60.0	33
Red	1.0	3.0	33.3	33
Brown	2.1	9.6	21.9	33
Black	2.7	15.1	17.9	34

that the total values of polyphenols and antioxidant activity were mainly contributed to by NEPP, with an average value 57%.[37]

16.2.2 Individual Non-extractable Polyphenols in Sorghum Grain

Polyphenols of sorghum grain mainly consist of phenolic acids, flavonoids and condensed tannins.[22] Phenolic acids consist of hydroxybenzoic and hydroxycinnamic acids and their derivatives. Hydroxybenzoic acids, with a C6–C1 structure, in sorghum include gallic, protocatechuic, vanillic, *p*-hydroxybenzoic and syringic acids. Hydroxycinnamic acids have the common C6–C3 structure, and these in sorghum include sinapic, ferulic, *p*-coumaric and caffeic acids. Both non-extractable and extractable individual phenolic acids have been identified in sorghum grains, the details of which are shown in Table 16.2. Most of these individual non-extractable phenolic

Table 16.2 Non-extractable individual and total polyphenol contents of sorghum grains (μg g^{-1}, dry base).[a]

Individual polyphenol	Non-extractable	Total	Percentage in non-extractable fraction	Reference
Hydroxybenzoic acids				
p-Hydroxy-benzoic	23.7	33.8	70.1	50
Vanillic	27.4	50.7	54.0	50
Protocatechuic	11.5	34.0	33.8	50
Gallic	13.2	13.2	100.0	50
Syringic	TINQ	TINQ	TINQ	51
Gentistic	TINQ	TINQ	TINQ	52
Salicylic	TINQ	TINQ	TINQ	52
Hydroxycinnamic acids				
Caffeic	20.4	30.9	66.0	50
Ferulic	213.0	287.0	74.2	50

[a]TINQ= Temporarily identified but not quantified

acids were quantified, while the others were tentatively identified but not quantified, due to a lack of authentic standards. The majority of individual phenolic acids, such as ferulic and *p*-hydroxybenzoic acids, mainly occur as non-extractable forms, whereas gallic and cinnamic acids have only been reported in the non-extractable forms.

The non-extractable phenolic acids of sorghum can influence the grain hardness, due in large part to the linkages between these phytochemicals and the plant cell walls, with ferulic acid and its oligomers being the most prevalent polyphenolic in this role.[38] It has been shown that ferulic acid could simultaneously form ester–ether linkages between the lignin and the non-starch polysaccharide arabinoxylan.[39] During early maturation, the ferulic acid ester linkages are formed,[40] and then the benzyl ether linkages in lignified cell walls are formed at maturity after reacting with lignin quinonemethide intermediates.[41] In terms of content, *p*-coumaric acid has been considered as the second most significant cross-linking phenolic acid after ferulic acid (Table 16.2). *p*-Coumaric acid can be strongly esterified to lignin or arabinoxylan cell walls at seed maturity.[42,43] Therefore, a higher content of non-extractable ferulic and *p*-coumaric acids in sorghum is associated with higher grain hardness.[38]

16.2.3 Individual Non-extractable Flavonoids in Sorghum Grain

The majority of polyphenols are composed of flavonoids, with the common three-ring structure of a C6–C3–C6 skeleton with two aromatic rings (A and B) linked with three carbons.[44] In nature, over 8000 flavonoids have been reported, and flavonoids are composed of different classes of compounds, including flavones, flavonols, flavonones, flavanols and anthocyanins.[45] The anthocyanidins, 3-deoxyanthocyanidins (loss of the hydroxyl group at the C3 position), have been identified as unique to sorghum grain. Compared with anthocyanidins, the 3-deoxyanthocyanidins are more stable under acidic conditions.[46]

Several individual flavonoids, both non-extractable and extractable, have been identified in sorghum grain (Table 16.2), but not all of them have been quantified, due to the absence of analytical standards. The content of non-extractable flavonoids of coloured sorghum grain is higher than in white grain.[34] Two major individual non-extractable 3-deoxyanthocyanidins, luteolinidin and apigeninidin, were identified and quantified in sorghum grain,[47] and black sorghum grain was a good source of these compounds, which was in agreement with other studies.[46,48] In addition, two major individual flavanones, named eriodictyol and naringenin, were identified and quantified in sorghum grain, and lemon yellow-coloured sorghums had higher levels of total flavanones than grains of other colours.[49] However, the levels of non-extractable flavanones were not reported.

16.3 Extraction Techniques for Non-extractable Polyphenols in Sorghum

16.3.1 Phenolic Covalent Bonds

The NEPP are covalently bound to cell wall substances such as structural proteins, arabinoxylans, hemicellulose, cellulose, pectin and lignin.[53,54] Phenolic acids, such as cinnamic and benzoic acids, can form ester bonds with hydroxyl groups of cell wall structural components, such as structural carbohydrates and proteins, through their carboxylic group. In addition, the hydroxyl groups of polyphenols can also form ether linkages with hydroxyl groups of cell wall structural components, such as lignin.[27,55] Once formed, these covalent bonds play an important role in binding cell wall structural components, and the structure of the cell wall matrix is significantly enhanced. Additionally, these NEPP can provide both chemical and physical barriers, including protection against pathogens and penetration of fungi, as well as antifungal, antibacterial and antioxidant activities.[56,57]

16.3.2 Release of Non-extractable Polyphenols

The most common methods that have been used to release NEPP from the cell wall matrix of sorghum are alkaline and acid hydrolysis. Recently, numerous assays have been developed to improve the release of NEPP, including enzymatic, microwave-assisted and ultrasound-assisted hydrolysis.[10] The released polyphenols are then usually extracted with organic solvents, including ethanol, methanol, ethyl acetate, diethyl ether and acetone.[58–60]

Additionally, the NEPP are also frequently extracted with acidified solvents.[61,62] As the pure organic solvents cannot effectively and completely extract some extremely polar phenolic acids, including cinnamic and benzoic acids, the mixtures of acetone–water and alcohol–water have been applied.[11]

16.3.2.1 Acid Hydrolysis

Acid hydrolysis has been widely carried out to release NEPP, and the concentration of hydrochloric acid has ranged from 1% to 8% in water–methanol.[50,61] Compared to other methods, it is convenient and simple to extract with acid hydrolysis, and the released NEPP can be directly analysed. However, the glycosidic bonds and solubilised sugars could be easily broken, while ester bonds generally exist after acid treatment.[11] In addition, the lower pH can decrease the stability of polyphenols, or even degrade some types during the acid treatment. For example, the degradation of flavonols has been reported in cactus pear after acidic hydrolysis.[63] Some non-extractable phenolic acids, including protocatechuic, syringic, hydroxycinnamic and caffeic acids, have not been detected in acid hydrolysates, but have been after alkaline hydrolysis.[64] Acid hydrolysis is normally conducted at elevated temperatures, which can cause the loss of some polyphenols, such as *p*-coumaric, *trans*-ferulic and *trans*-sinapic acids.[15] However, different results were obtained by other authors,[65] and results showed that acid hydrolysis released more rosmarinic acid than did alkaline hydrolysis in *Perilla frutescens* seed flour. It is reported that both syringic and caffeic acids were found in the acid-labile fraction, but not in the alkali fraction.[66] These contradictory results suggest that the release of NEPP might depend on the types of individual polyphenols. For sorghum grain, both acidic and alkaline hydrolysis have been used to release NEPP,[67,68] whereas no study has been reported directly comparing the effects of acid and alkali hydrolysis on the release of polyphenols from sorghum grain.

16.3.2.2 Alkaline Hydrolysis

The NEPP are commonly extracted with alkaline treatments. It has been clearly demonstrated that alkaline hydrolysis can better release the NEPP from the cell wall matrix than acid hydrolysis, due to the cleavage of ether and ester bonds occurring under alkaline hydrolysis conditions, but not under acidic hydrolysis conditions.[69] Additionally, alkaline hydrolysis is usually conducted at the room temperature, which can decrease the degradative loss of polyphenols during the extraction when comparing to acid hydrolysis. For example, the total NEPP content of wheat bran was 1.9 mg g^{-1} by acid hydrolysis, while it could reach 4.0 mg g^{-1} after alkaline hydrolysis.[66] In brief, 2 M sodium hydroxide treatment for 1 hours at room temperature has been used to release the NEPP of cereal grain, followed by neutralising with an appropriate amount of hydrochloric acid.[15] Therefore, the NEPP of cereal grain can be virtually totally released under this alkaline treatment.[22] However, the procedure for extraction with alkaline treatment is more complex, since further extraction steps are required before sample analysis when compared to acid treatment.[70]

16.3.2.3 *Enzymatic Hydrolysis*

Some NEPP are retained in the cell cytoplasm or in the polysaccharide–lignin network by hydrophobic or hydrogen bonding after conventional acid or alkaline extraction. Enzymatic hydrolysis is a method that can more efficiently release these polyphenols.[71] The addition of carbohydrate-hydrolysing enzymes, including glucanases, amylases, cellulases, pectinases and hemicellulases, is commonly used during enzymatic hydrolysis extraction of polyphenols, since they can effectively degrade the cell wall structure.[12] For example, both pectinolytic and cellulolytic enzymes have been used to release NEPP from grape pomace, and higher contents of these compounds were obtained compared to a non-enzymatic control method.[72] The successful release of polyphenols from the cell walls of cereal grains, such as barley, rice and maize, has been reported when using enzymatic hydrolysis.[73–75] There are many factors that influence the effectiveness of enzymatic hydrolysis, such as enzyme types, concentration, solid to water ratio, hydrolysis time and particle size of the plant material. For example, different types of commercial enzymes were used to release the polyphenols from grape waste, and higher contents of polyphenols were released from grape waste with Novoferm®, followed by Pectinex® Ultra and Celluclast® 1.5 L.[76] The advantages of enzymatic hydrolysis extraction include reduced loss of polyphenols and use water as a solvent instead of organic chemicals.[77] The measured level of total polyphenols in apples was increased after hydrolysing with carbohydrases such as cellulases, β-glucanase and pectinase, while the contents of several individual polyphenols, such as phloridzin and chlorogenic acid, were reduced when using enzymatic treatment.[78]

16.3.2.4 *Microwave-assisted Extraction*

Microwave-assisted extraction is recognised as a green technology due to it reducing the amount of organic solvent required for effective extraction. It is considered a novel method for rupturing plant tissue and releasing the constituents into the solvent. There are three proposed sequential steps of microwave-assisted extraction. Firstly, solutes are separated from active sites of the sample matrix under high pressure and temperature. Then, solvent is diffused across the sample matrix, followed by release of solutes from the sample matrix into the solvent.[78] Microwave-assisted extraction has various advantages, including increased polyphenol yield, reduced equipment size and faster heating for extraction.[79] The effectiveness of microwave-assisted extraction for NEPP is influenced by several key variables, such as solvent type and ratio (solid/liquid), temperature, irradiation power and extraction time. For example, higher polyphenol extraction occurred when using microwave-assisted extraction with methanol compared to when acetone or hexane were used.[80] They also found that compared to different extraction temperatures, when the extraction temperature was higher than 80 °C, the level of total phenolic compounds extracted from wheat bran was significantly increased with microwave-assisted methanol extraction.

There are many studies focusing on the combination of microwaves with alkaline hydrolysis to increase the level and speed of release of NEPP. For example, the same level of release of NEPP in flaxseed was accomplished in 15 minutes with microwave-assisted extraction in 0.1 M NaOH instead of 6 hours without the microwave assistance.[81] Similarly, when extracting non-extractable phenolic acids of sorghum grain with microwave-assisted extraction (1400 W and 45 seconds) using 2 M NaOH, it was found that ether bonds were broken, and ferulic and coumaric acids were completely released at 190 °C.[82]

16.3.2.5 Ultrasound-assisted Extraction

Ultrasound is a type of high-frequency sound beyond human hearing detection. Ultrasound power levels in the frequency range of 20–100 kHz have been widely used to physically or chemically alter food properties.[71] Ultrasound can pass through a medium by the production of compressive and expansive forces to produce a phenomenon named 'cavitation', which is defined as the process of bubble production, growth and collapse. The temperature, pressure and heating and cooling rate of these bubbles are about 5000 K, 1000 atm and 10^{10} K $second^{-1}$, respectively. Due to these extremely rapid changes in temperature and pressure, cellular membranes are disrupted, and then the desired components can leach out of the cell wall. According to this principle, ultrasound-assisted extraction has been developed to release NEPP. The mechanism of ultrasound-assisted extraction involves two principal kinds of physical phenomenon: (a) diffusion across the cell wall; and (b) increasing the cell contents after breaking walls.[83] The effectiveness of ultrasound-assisted extraction is governed by the action of ultrasound frequency and time of sonication, as well as the resulting sample temperature and pressure. It is also influenced by the solvent, particle size and moisture content of the material. The extraction time, level of solvent use and energy use can be reduced with ultrasound-assisted extraction.

In addition, other advantages of ultrasound energy for extraction include reduced thermal gradients, reduced extraction temperature, smaller equipment size, faster energy transfer, rapid response to extraction process control, increased production and quicker start-up. In grape pomace, more NEPP were released and extracted with ultrasound-assisted extraction when the temperature was increased from 20 °C to 50 °C.[84] Numerous individual polyphenols of strawberries, such as ellagic acid, quercetin and kaempferol, naringenin, rutin and naringin, have also been effectively extracted by ultrasound.[14]

16.3.2.6 Pulsed Electric Field Extraction

Pulsed electric fields (PEF), as a non-thermal processing technology, have been widely used to extract bioactive compounds due to their effectiveness at destroying the cell membrane structure. For PEF extraction, short-duration electric field pulses of high intensity are applied to materials located between the two electrodes and a transmembrane potential difference across the cell membrane is induced by PEF. Electrical breakdown can lead to the electroporation of the cell membrane at a critical value of potential difference.[85] The permeability of cytoplasmic membranes is increased after PEF treatment, assisting in the release of the intracellular contents.[84] The effectiveness of PEF treatment is mainly influenced by pulse number, specific energy input, field strength, treatment temperature and properties of the materials to be treated.[86]

The NEPP of sorghum grain were largely released by mild-intensity PEF (2 kV cm^{-1} electric field intensity and 875 μs treatment time), and the total phenolic content and antioxidant activity were higher than the control sorghum under this condition.[87] Tomato juices were subjected to high-intensity PEF and thermal treatment (90 °C for 60 seconds), and the PEF influenced the profile and content of polyphenols in these juices.[88] The results showed that the extraction efficiency was significantly improved under an electric field strength of 2 kV cm^{-1} compared to thermal treatment. However, a higher electric field strength of 5 kV cm^{-1} and a pulse number of 100 were found to degrade the polyphenols of apple tissue.[89]

16.4 Bioactivities of Non-extractable Polyphenols

The physical and chemical properties of polyphenols within food are altered upon ingestion by the chemical environment (*e.g.* acidic pH of stomach), the multi-enzyme reactions in the digestive tract and the action of microflora in the colon. The extractable polyphenols are released from the food matrix and absorbed into the circulation in the small intestine.[32] However, it has been estimated that only up to ~10% of polyphenols can be absorbed in the small intestine and the remainder move on to the colon.[90] Conversely, as NEPP are commonly bound to indigestible macromolecules such as cellulose, hemicellulose and pectin, they cannot be released and absorbed in the small intestine.[22] Thus, these compounds are released and can be partially metabolised in the colon through the fermentation action of the colon microbiota on the non-starch polysaccharides. In the colon, there are large numbers of microorganisms of about 1000 different bacterial species at approximately

10^{14} colony-forming units. The microorganisms, including those such as *Bifidobacterium* spp. and *Lactobacillus* spp., secrete various extracellular enzymes, including carbohydrases and proteases, which disrupt the cell wall matrix of the undigested materials reaching the colon and can hydrolyse the covalent bonds between the NEPP and the food cellular matrix.[91] In addition, the released polyphenols can modulate the fermentation environment of the colon by decreasing the pH and inhibiting the growth of cancer-inducing microorganisms, which will have health-beneficial effects.[70]

During fermentation in the colon, NEPP can be catabolised by gut microbes to release small molecules that can exert additional health benefits in the gut. For example, proanthocyanidins, flavonoids and phenolic microbial metabolites were studied in humans and rats and the authors found that these phenolics can be degraded to simple phenolic acids, such as protocatechuic acid, vanillic acid, 3-hydroxyphenylacetic acid, *m*-coumaric acid and 3,4-dihydroxyphenylpropionic acid.[92] Some reports also provide strong evidence for the important role of the gut microbe in the aetiology of some chronic diseases, including obesity, cardiovascular disease and Parkinson's disease. The function and composition of the gut microbiota can be modulated by NEPP in the colon; therefore, the risk of some chronic diseases may be indirectly impacted by these phytochemicals.[93–95] The efficiency of absorption of NEPP of sorghum grain in the colon is not currently well understood; therefore, the bioactivities of NEPP from other plant foods will be discussed in this chapter. This information is valuable to inform future studies on the absorption and bioactivities of NEPP of sorghum grain.

16.4.1 Bioavailability and Bioactivity of Non-extractable Phenolic Acids

Recently, various bioactivities of phenolic acids have been reported. The growth of human epidermoid carcinoma (A431) skin cancer cells was reduced by gallic acid and methyl gallate extracted from *Givotia rottleriformis*, a tree species belonging to Euphorbiaceae family.[96] In a range of cell line systems, ferulic acid also exhibited inhibitory activity against cancer, diabetes, neurodegenerative and cardiovascular diseases.[97–99] The arabinoxylan-bound ferulic acid can be released by human colonic microbiota and intestinal esterases, and thus both the microbiota and intestinal tissues play key roles in the release and uptake of phenolic acids esterified to the food matrix. Another phenolic acid, caffeic acid, can be transformed to 3-(3-hydroxyphenyl) propionic acid or 4-ethylcatechol by human faecal microbiota.[100] As all of these non-extractable phenolic acids have been identified in sorghum grain, it is hypothesised that the bioactivities of non-extractable phenolic acids might occur in sorghum grain. Therefore, more research will be needed in this area. Moreover, a multitude of bioactivities of ellagic acid, such as antioxidant and cancer-protective activities, have been reported in cell lines, animal models and humans, which have been critically reviewed by Cardona *et al.*[101]

16.4.2 Bioavailability and Bioactivity of Non-extractable Flavonoids

The effective antioxidant activity of flavonoids is linked to their multiple hydroxyl groups.[44] Flavonoids have demonstrated bioactivities such as anti-cancer, anti-inflammation and antivirus activities, and their increased intake reduces the risk cardiovascular diseases, type 2 diabetes and cholesterol.[70] A previous study found that the colonic microflora *Clostridium* spp. and *Eubacterium* spp. could degrade non-extractable flavonols to simpler polyphenols that may then have bioactivity in the colonic tissues.[102] The bioactivity of (+)-epigallocatechin 3-*O*-gallate extracted from the roots of *Limonium sinense*, belonging to the Plumbaginaceae family, was studied,[103] and the results indicated that this compound exhibited potent inhibitory activity against virus type-1 infection. It was proposed that non-extractable wholegrain polyphenols released and partially degraded in the colon may partly explain the protective effects of increased wholegrain consumption against colon cancer and other digestive cancers.[22]

The negative effects of NF-κB activation, inflammation and healthy cell apoptosis caused by cisplatin (an anticancer drug) were abated by some types of flavonoids, such as hesperidin, rutin, quercetin and naringin.[70] A derivative of rutin, troxerutin, is able to bind DNA by interacting at the minor groove of DNA and enhance cancer cell killing in response to radiation-induced sensitising, thereby acting as a chemopreventive agent.[104] The influence of apigenin on the radiosensitivity of SQ-5 cells, derived from a human lung carcinoma, was studied by Watanabe *et al.*,[105] and the results suggested that apigenin could be a promising radiosensitizing agent in treating human carcinomas. Apigenin can also ameliorate cisplatin-induced apoptosis through reduction of p53 activation and promotion of the PI3K/Akt pathway in epithelial HK-2 cells.[106]

Kaempferol, a common dietary flavonoid present in many wholegrains, fruits and vegetables, is well known for its bioactivities, such as antidiabetic, cardioprotective, analgesic, anti-inflammatory, osteoporotic and anti-allergic activities.[70] Additionally, kaempferol could induce nuclear condensation and mitochondria dysfunction in order to suppress human cancer cell lines, including human lung carcinoma (A549), human stomach carcinoma (SGC-7901), human cervical carcinoma (HeLa) and human breast carcinoma (MCF-7). Additionally, other flavonoids such as quercetin have exhibited potential anticancer effects, including inhibition of induction of apoptosis and cell proliferation.[107]

16.5 Release of Non-extractable Polyphenols Through Food Processing

NEPP can be liberated from the cell wall matrix of foods through several forms of processing, including both non-thermal and thermal processing. Non-thermal processing includes fermentation and germination, while thermal processing includes extrusion cooking, boiling and baking.

16.5.1 Non-thermal Processing

Fermentation, as a traditional food processing method, is defined as the microorganism-mediated bioconversion of sugar (or monosaccharide) into ethanol, lactic acid and gases. Fermentation has been widely used to extend the shelf-life of foods, and has also been suggested to release NEPP from the cell wall matrix of foods.[70] Various extracellular enzymes such as carbohydrases, proteases and lipases are secreted by microorganisms during fermentation, and these enzymes also include cell wall matrix-disintegrating enzymes, including pectinase, amylase, glucanase, hemicellulase and cellulase, which can release the NEPP bound to the cell wall non-starch polysaccharides.

The total and free polyphenol contents and antioxidant activities have been reported to increase during food fermentation. The values of 2,2-diphenyl-1-picrylhydrazyl (DPPH) and 2,2′-azinobis (3-ethylbenzothiazoline-6-sulphonic acid) (ABTS) radical scavenging of wheat koji were enhanced after incubation with the fungi *Aspergillus oryzae* during 4 days of fermentation.[108] Tijana *et al.*[109] also found that the total phenolic content increased by 17.2–39.4% when buckwheat, wheat germ, barley and rye were fermented with the bacteria *Lactobacillus rhamnosus* and *Saccharomyces cerevisiae*. Similarly, the content of extractable polyphenols of fermented finger millet increased more than two-fold, while NEPP decreased after fermentation for 24 hours compared to unfermented finger millet.[110]

However, fermentation has also been reported to significantly reduce the values of total polyphenol and antioxidant activity (ABTS and DPPH) when compared to unprocessed whole sorghum grain.[111] This difference might be caused by changes occurring during the fermentation process that influenced the extractability of polyphenols. It was proposed that polyphenols might bind to proteins, carbohydrate and other components in the aqueous phase during the fermentation process, which makes them less extractable.[112]

Germination is the bioprocess that induces breaking of the dormancy of seeds by sprouting and growth and is another useful way to release NEPP from grains. For example, an increase in extractable phenolic and antioxidant capacity with a corresponding decrease in non-extractable polyphenols was reported after germination of barley,[113] whereas the levels of free forms of the individual polyphenols – ferulic and sinapic acids – increased by two-fold and nearly ten-fold, respectively, after fermentation. Similar results have also been observed in other cereal grains, such as brown rice.[114]

16.5.2 Thermal Processing

Sorghum grain can be prepared for human consumption by a variety of thermal processing methods. Several types of foods, such as sorghum bread, cakes, cookies and pasta, are made from sorghum by thermal processing.[115]

During thermal processing, it has been reported that the levels of NEPP and the antioxidant activity of sorghum grain can either increase or decrease. For example, the changes in the polyphenol content and antioxidant activity of sorghum grain by soaking, steaming and roasting were assessed by Wu *et al.*,[116] and the results indicated that both soaking and steaming caused significant losses of polyphenols and antioxidant activity, while roasting promoted increases of phenolic compound levels and antioxidant activity. The mechanisms of these opposing effects might include the release of NEPP from the food matrix, polymerisation and oxidation of phenolics, thermal degradation and depolymerisation of high-molecular-weight phenolics such as condensed tannins.[112]

The contents of NEPP of wheat flour were found to be significantly reduced after baking at 180 °C for 40 minutes, while the extractable polyphenol content increased.[117] For finger millet, the values of antioxidant activity and extractable polyphenols significantly increased after boiling, probably due to the release of NEPP.[110]

With regard to roasting wholegrain sorghum, increases in total polyphenols, flavonoids, procyanidins and all detected phenolic acids of sorghum grain were found.[116] The increases were probably due to the degradation of high-molecular-weight polyphenols such as condensed tannins to simple polyphenols or the release of NEPP from cell walls under roasting, which made more polyphenols extractable.

The measurable total polyphenol and ABTS radical-scavenging activity of whole sorghum grain was found to be reduced after extrusion cooking.[11] It is proposed that the specific effect of extrusion may be dependent on feed moisture during the extrusion process. When the feed moisture was higher than 18%, polyphenols could be polymerised, possibly through oxidative coupling, to decrease their extractability and antioxidant activity. Furthermore, extrusion cooking can change protein sequestration, which can promote the formation of insoluble protein–polyphenol complexes, thus increasing the levels of NEPP.[112]

16.6 Conclusion

Sorghum grains contain abundant levels of polyphenols, the majority of which are classified as non-extractable. Due to methodological limitations, only some of many individual NEPP in sorghum grain have been identified. In order to efficiently utilise these NEPP of sorghum grain in the food and pharmaceutical industries, it will be necessary to develop methods to extract all of these compounds from agro-industrial products such as sorghum grains without using organic solvents. Since little information is available on the influences of both non-thermal and thermal processing on the levels and profiles of non-extractable polyphenols in sorghum grain, more research will be needed in this area. In addition, despite a variety of studies on the levels of polyphenols in sorghum grain and the antioxidant activities of NEPP, their release from the food matrix in the human gastrointestinal tract, their catabolism by human colonic microflora and their bioavailability are not well understood and thus need in-depth investigation in order to better understand the mechanisms by which these polyphenols might impact on human health.

References

1. https://www.usda.gov/oce/commodity/wasde/latest.pdf, last accessed April 2016.
2. A. Stefoska-Needham, E. J. Beck, S. K. Johnson and L. C. Tapsell, *Food Rev. Int.*, 2015, **31**, 401.
3. T. J. Schober, M. Messerschmidt, S. R. Bean, S.-H. Park and E. K. Arendt, *Cereal Chem.*, 2005, **82**, 394.
4. G. Wu, J. F. Bornman, S. J. Bennett, M. W. Clarke, Z. Fang and S. K. Johnson, *J. Cereal Sci.*, 2017, **77**, 17.
5. N. M. Emmambux and J. Taylor, *J. Sci. Food Agric.*, 2003, **83**, 402–407.
6. A. Stefoska-Needham, E. J. Beck, S. K. Johnson, M. J. Batterham, R. Grant, J. Ashton and L. C. Tapsell, *J. Am. Coll. Cardiol.*, 2017, **36**, 184.
7. L. Dykes and L. Rooney, *Cereal Foods World*, 2007, **52**, 105.
8. L. Dykes and L. Rooney, *J. Cereal Sci.*, 2006, **44**, 236.
9. J. K. Vanamal, A. R. Massey, S. R. Pinnamaneni, L. Reddivari and K. F. Reardon, *Crit. Rev. Food Sci. Nutr.*, 2017, DOI: 10.1080/10408398.2017.1344186.
10. B. A. Acosta-Estrada, J. A. Gutiérrez-Uribe and S. O. Serna-Saldívar, *Food Chem.*, 2014, **152**, 46.
11. C. D. Stalikas, *J. Sep. Sci.*, 2007, **30**, 3268.
12. A.-K. Landbo and A. S. Meyer, *J. Agric. Food Chem.*, 2001, **49**, 3169.
13. S. Jokić, M. Cvjetko, Đ. Božić, S. Fabek, N. Toth, J. Vorkapić-Furač and I. R. Redovniković, *Int. J. Food Sci. Technol.*, 2012, **47**, 2613.
14. M. Herrera and M. L. De Castro, *Anal. Bioanal. Chem.*, 2004, **379**, 1106.
15. K. Krygier, F. Sosulski and L. Hogge, *J. Agric. Food Chem.*, 1982, **30**, 330.
16. F. Sosulski, K. Krygier and L. Hogge, *J. Agric. Food Chem.*, 1982, **30**, 337.
17. F. Saura-Calixto, *J. Agric. Food Chem.*, 2012, **60**, 11195.
18. A. Helal, D. Tagliazucchi, E. Verzelloni and A. Conte, *J. Funct. Foods*, 2014, 7, 506.
19. G. Hithamani and K. Srinivasan, *J. Agric. Food Chem.*, 2014, **62**, 11170.
20. B. Gullon, M. E. Pintado, X. Barber, J. Fernández-López, J. A. Pérez-Álvarez and M. Viuda-Martos, *Food Res. Int.*, 2015, **78**, 169.
21. M. Nardini, E. Cirillo, F. Natella, D. Mencarelli, A. Comisso and C. Scaccini, *Food Chem.*, 2002, **79**, 119.
22. K. K. Adom and R. H. Liu, *J. Agric. Food Chem.*, 2002, **50**, 6182.
23. G. Wu, S. K. Johnson, J. F. Bornman, S. J. Bennett and Z. Fang, *Food Chem.*, 2016, **214**, 199.
24. A. Yousif, D. Nhepera and S. K. Johnson, *Food Chem.*, 2012, **134**, 880.
25. C. Felgines, S. Talavéra, M.-P. Gonthier, O. Texier, A. Scalbert, J.-L. Lamaison and C. Rémésy, *J. Nutr.*, 2003, **133**, 1296.

26. M. Natsume, N. Osakabe, M. Oyama, M. Sasaki, S. Baba, Y. Nakamura, T. Osawa and J. Terao, *Free Radical Biol. Med.*, 2003, **34**, 840.
27. R. H. Liu, Whole grain phytochemicals and health, *J. Cereal Sci.*, 2007, **46**, 207.
28. A. Chandrasekara and F. Shahidi, *J. Agric. Food Chem.*, 2010, **58**, 6706.
29. C. M. Liyana-Pathirana and F. Shahidi, *J. Agric. Food Chem.*, 2006, **54**, 1256.
30. F. Saura-Calixto, *J. Food Sci.*, 1988, **53**, 1769.
31. S. Matthews, I. Mila, A. Scalbert and D. M. Donnelly, *Phytochemistry*, 1997, **45**, 405.
32. R. García-Villalba, J. C. Espín, K. Aaby, C. Alasalvar, M. Heinonen, G. Jacobs, S. Voorspoels, T. Koivumäki, P. A. Kroon and E. Pelvan, *J. Agric. Food Chem.*, 2015, **63**, 6555.
33. G. Wu, S. K. Johnson, J. F. Bornman, S. Bennett, V. Singh and Z. Fang, *Cereal Chem.*, 2016, **93**, 419.
34. G. Wu, S. K. Johnson, J. F. Bornman, S. J. Bennett and Z. Fang, *Food Chem.*, 2017, **214**, 199.
35. E.-S. M. Abdel-Aal, T.-M. Choo, S. Dhillon and I. Rabalski, *Cereal Chem.*, 2012, **89**, 199.
36. M. Gabaza, H. Shumoy, M. Muchuweti, P. Vandamme and K. Raes, *J. Agric. Food Chem.*, 2016, **64**, 7615.
37. J. Pérez-Jiménez and F. Saura-Calixto, *Food Res. Int.*, 2015, **74**, 315.
38. C. Chiremba, J. R. Taylor, L. W. Rooney and T. Beta, *Food Chem.*, 2012, **134**, 81.
39. T. B. T. Lam, K. Iiyama and B. A. Stone, *Phytochemistry*, 1992, **31**, 1179–1183.
40. T. Ishii, Isolation and characterization of a diferuloyl arabinoxylan hexasaccharide from bamboo shoot cell-walls, *Carbohydr. Res.*, 1991, **219**, 15–22.
41. T. B. T. Lam, K. Iiyama and B. A. Stone, *Phytochemistry*, 1992, **31**, 2655.
42. R. Sun, X. Sun, S. Wang, W. Zhu and X. Wang, *Ind. Crops Prod.*, 2002, **15**, 179.
43. M. Bunzel, J. Ralph, F. Lu, R. D. Hatfield and H. Steinhart, *J. Agric. Food Chem.*, 2004, **52**, 6496.
44. L. Bravo, *Nutr. Rev.*, 1998, **56**, 317.
45. D. Ghosh and T. Konishi, *Asia Pac. J. Clin. Nutr.*, 2007, **16**, 200.
46. J. M. Awika, L. W. Rooney and R. D. Waniska, *J. Agric. Food Chem.*, 2004, **52**, 4388.
47. G. Wu, S. J. Bennett, J. F. Bornman, M. W. Clarke, Z. Fang and S. K. Johnson, *Food Res. Int.*, 2017, **97**, 347.
48. J. M. Awika, L. W. Rooney and R. D. Waniska, *Food Chem.*, 2005, **90**, 293.
49. L. Dykes, G. C. Peterson, W. L. Rooney and L. W. Rooney, *Food Chem.*, 2011, **128**, 173.
50. D. Hahn, J. Faubion and L. Rooney, *Cereal Chem.*, 1983, **60**, 255.
51. A. E.-M. M. Afify, H. S. El-Beltagi, S. M. A. El-Salam and A. A. Omran, *Asian Pac. J. Trop. Biomed.*, 2012, **2**, 203.

52. R. Waniska, J. Poe and R. Bandyopadhyay, *J. Cereal Sci.*, 1989, **10**, 217.
53. G. Wu, S. K. Johnson, J. F. Bornman, S. J. Bennett, M. W. Clarke, V. Singh and Z. Fang, *Sci. Rep.*, 2016, **6**, 21835.
54. D. W. Wong, *Appl. Biochem. Biotechnol.*, 2006, **133**, 87.
55. S. R. McLusky, M. H. Bennett, M. H. Beale, M. J. Lewis, P. Gaskin and J. W. Mansfield, *Plant J.*, 1999, **17**, 523.
56. V. Atanasova-Penichon, C. Barreau and F. Richard-Forget, *Front. Microbiol.*, 2016, **7**, 566.
57. R. L. Nicholson and R. Hammerschmidt, *Annu. Rev. Phytopathol.*, 1992, **30**, 369.
58. J. Gutiérrez-Uribe, I. Romo-Lopez and S. Serna-Saldívar, *J. Funct. Foods*, 2011, **3**, 290.
59. J. M. Awika, *ACS Symp. Ser.*, 2011, **1089**, 171.
60. P. S. Belton and J. R. N. Taylor, *Trends Food Sci. Technol.*, 2004, **15**, 94.
61. L. Dykes, W. L. Rooney and L. W. Rooney, *J. Cereal Sci.*, 2013, **58**, 278.
62. J. S. Shelembe, D. Cromarty, M. Bester, A. Minnaar and K. G. Duodu, *J. Food Biochem.*, 2014, **38**, 110.
63. T. E. Moussa-Ayoub, S. K. El-Samahy, L. W. Kroh and S. Rohn, *Food Chem.*, 2011, **124**, 1177.
64. I. M. Sani, S. Iqbal, K. W. Chan and M. Ismail, *Molecules*, 2012, **17**, 7584.
65. L. Yan and G. Zheng, *Int. J. Food Sci. Technol.*, 2017, **52**, 2497.
66. B. Verma, P. Hucl and R. Chibbar, *Food Chem.*, 2009, **116**, 94.
67. L. Svensson, B. Sekwati-Monang, D. L. Lutz, A. Schieber and M. G. Gänzle, *J. Agric. Food Chem.*, 2010, **58**, 9214.
68. R. Licata, J. Chu, S. Wang, R. Coorey, A. James, Y. Zhao and S. Johnson, *Int. J. Food Sci. Technol.*, 2014, **49**, 1408.
69. K.-H. Kim, R. Tsao, R. Yang and S. W. Cui, *Food Chem.*, 2006, **95**, 466.
70. F. Shahidi and J. Yeo, *Molecules*, 2016, **21**, 1216.
71. J. Azmir, I. Zaidul, M. Rahman, K. Sharif, A. Mohamed, F. Sahena, M. Jahurul, K. Ghafoor, N. Norulaini and A. Omar, *J. Food Eng.*, 2013, **117**, 426.
72. T. Maier, A. Göppert, D. R. Kammerer, A. Schieber and R. Carle, *Eur. Food Res. Technol.*, 2008, **227**, 267.
73. A. Sancho, B. Bartolomé, C. Gómez-Cordovés, G. Williamson and C. Faulds, *J. Cereal Sci.*, 2001, **34**, 173.
74. L. Saulnier, C. Marot, M. Elgorriaga, E. Bonnin and J.-F. Thibault, *Carbohydr. Polym.*, 2001, **45**, 269.
75. Z. Zhou, K. Robards, S. Helliwell and C. Blanchard, *Food Chem.*, 2004, **87**, 401.
76. R. Gómez-García, G. C. Martínez-Ávila and C. N. Aguilar, *Biotech*, 2012, **2**, 297.
77. M. Puri, D. Sharma and C. J. Barrow, *Trends Biotechnol.*, 2012, **30**, 37.
78. H.-z. Zheng, I.-W. Hwang and S.-K. Chung, *J. Zhejiang Univ., Sci., B*, 2009, **10**, 912.
79. G. Cravotto, L. Boffa, S. Mantegna, P. Perego, M. Avogadro and P. Cintas, *Ultrason. Sonochem.*, 2008, **15**, 899.

80. D. S. Oufnac, Z. Xu, T. Sun, C. Sabliov, W. Prinyawiwatkul and J. S. Godber, *Cereal Chem.*, 2007, **84**, 125.
81. V. Beejmohun, O. Fliniaux, É. Grand, F. Lamblin, L. Bensaddek, P. Christen, J. Kovensky, M. A. Fliniaux and F. Mesnard, *Phytochem. Anal.*, 2007, **18**, 275.
82. C. Chiremba, L. W. Rooney and T. Beta, *J. Agric. Food Chem.*, 2012, **60**, 4735.
83. T. Mason, L. Paniwnyk and J. Lorimer, The uses of ultrasound in food technology, *Ultrason. Sonochem.*, 1996, **3**, S253.
84. M. González-Centeno, F. Comas-Serra, A. Femenia, C. Rosselló and S. Simal, *Ultrason. Sonochem.*, 2015, **22**, 506.
85. D. Knorr and A. Angersbach, *Trends Food Sci. Technol.*, 1998, **9**, 185.
86. V. Heinz, S. Toepfl and D. Knorr, *Innovative Food Sci. Emerging Technol.*, 2003, **4**, 167.
87. U. C. Lohani and K. Muthukumarappan, *Innovative Food Sci. Emerging Technol.*, 2016, **35**, 29.
88. A. Vallverdú-Queralt, I. Odriozola-Serrano, G. Oms-Oliu, R. M. Lamuela-Raventós, P. Elez-Martínez and O. Martín-Belloso, *J. Agric. Food Chem.*, 2012, **60**, 9667.
89. A. Wiktor, M. Schulz, E. Voigt, D. Witrowa-Rajchert and D. Knorr, *J. Food Eng.*, 2015, **146**, 8.
90. M. Carbonaro, G. Grant and A. Pusztai, *Eur. J. Nutr.*, 2001, **40**, 84.
91. M. F. Andreasen, P. A. Kroon, G. Williamson and M.-T. Garcia-Conesa, *Free Radical Biol. Med.*, 2001, **31**, 304.
92. M. Urpi-Sarda, M. Monagas, N. Khan, R. M. Lamuela-Raventos, C. Santos-Buelga, E. Sacanella, M. Castell, J. Permanyer and C. Andres-Lacueva, *Anal. Bioanal. Chem.*, 2009, **394**, 1545.
93. V. K. Ridaura, J. J. Faith, F. E. Rey, J. Cheng, A. E. Duncan, A. L. Kau, N. W. Griffin, V. Lombard, B. Henrissat and J. R. Bain, *Science*, 2013, **341**, 1241214.
94. F. Koeth, H. G. Marques and T. Delbruck, *Biol. Inspired Cognit. Archit.*, 2013, **6**, 8.
95. T. R. Sampson, J. W. Debelius, T. Thron, S. Janssen, G. G. Shastri, Z. E. Ilhan, C. Challis, C. E. Schretter, S. Rocha and V. Gradinaru, *Cell*, 2016, **167**, 1469.
96. S. Kamatham, N. Kumar and P. Gudipalli, *Toxicol. Rep.*, 2015, **2**, 520.
97. A. Narasimhan, M. Chinnaiyan and B. Karundevi, *Eur. J. Pharmacol.*, 2015, **761**, 391.
98. Y. E. Choi and E. Park, *Biochem. Biophys. Res. Commun.*, 2015, **458**, 520.
99. T. Eitsuka, N. Tatewaki, H. Nishida, T. Kurata, K. Nakagawa and T. Miyazawa, *Biochem. Biophys. Res. Commun.*, 2014, **453**, 606.
100. L. Rondini, M. N. Peyrat-Maillard, A. Marsset-Baglieri, G. Fromentin, P. Durand, D. Tomé, M. Prost and C. Berset, *J. Agric. Food Chem.*, 2004, **52**, 4338.
101. F. Cardona, C. Andrés-Lacueva, S. Tulipani, F. J. Tinahones and M. I. Queipo-Ortuño, *J. Nutr. Biochem.*, 2013, **24**, 1415.

102. M. V. Selma, J. C. Espín and F. A. Tomásbarberán, *J. Agric. Food Chem.*, 2009, **57**, 6485.
103. L. Lin, Y. Kuo and C. Chou, *Planta Med.*, 2000, **66**, 333.
104. N. A. Panat, B. G. Singh, D. K. Maurya, S. K. Sandur and S. S. Ghaskadbi, *Chem.-Biol. Interact.*, 2016, **251**, 34.
105. N. Watanabe, R. Hirayama and N. Kubota, *J. Radiat. Res.*, 2007, **48**, 45.
106. S. M. Ju, J. G. Kang, J. S. Bae, H. O. Pae, Y. S. Lyu and B. H. Jeon, *J. Evidence-Based Complementary Altern. Med.*, 2015, **2015**, 186436.
107. W. Liao, L. Chen, X. Ma, R. Jiao, X. Li and Y. Wang, *Eur. J. Med. Chem.*, 2016, **114**, 24.
108. T. Bhanja, A. Kumari and R. Banerjee, *Bioresour. Technol.*, 2009, **100**, 2861.
109. Đ. Tijana, Š. Slavica and D. Suzanai, *Food Chem.*, 2010, **119**, 957.
110. M. Gabaza, H. Shumoy, M. Muchuweti, P. Vandamme and K. Raes, *J. Agric. Food Chem.*, 2016, **64**, 7615.
111. N. R. Dlamini, T. Jrn and L. W. Rooney, *Food Chem.*, 2007, **105**, 1412.
112. J. R. Taylor and K. G. Duodu, *J. Sci. Food Agric.*, 2015, **95**, 225.
113. M. Dvorakova, L. F. Guido, P. Dostalek, Z. Skulilova, M. M. Moreira and A. A. Barros, *J. Inst. Brew.*, 2008, **114**, 27.
114. S. Tian, K. Nakamura and H. Kayahara, *J. Agric. Food Chem.*, 2004, **52**, 4808.
115. J. R. Taylor and K. G. Duodu, *J. Sci. Food Agric.*, 2015, **95**, 225.
116. L. Wu, Z. Huang, P. Qin and G. Ren, *Food Res. Int.*, 2013, **53**, 678.
117. N. Nikolić, J. Stojanović, J. Mitrović, M. Lazić, I. Karabegović and G. Stojanović, *Int. J. Food Sci. Technol.*, 2016, **51**, 2019.

CHAPTER 17

Non-extractable Phenolic Compounds as Food Ingredients

S. SILVA[a], A. L. S. OLIVEIRA[a], T. RIBEIRO[a,b], A. R. MADUREIRA[a] AND M. M. PINTADO*[a]

[a]Centro de Biotecnologia e Química Fina – Laboratório Associado, Escola Superior de Biotecnologia, Universidade Católica Portuguesa/Porto, Rua Arquiteto Lobão Vital, 4202-401 Porto, Portugal; [b]BLC3 Association – Technology and Innovation Campus, Centre R&D, Agriculture and Food Technologies Department, Rua Nossa Senhora da Conceição, 2, Lagares, 3045-155 Oliveira do Hospital, Portugal
*E-mail: mpintado@porto.ucp.pt

17.1 Introduction

The rise in life expectancy has led to an increase in concerns regarding health and wellbeing. Fruits and vegetables have been systematically associated with reductions in the risks of degenerative and/or chronic diseases (*e.g.* cardiovascular and neurodegenerative conditions), namely through the improvement of blood flow and protection against oxidative stress and exacerbated inflammatory responses.[1–5] Several of the secondary metabolites produced by plants are phenolics that are derivatives of several metabolic pathways, such as shikimate (chorismate, an end product, can serve as a precursor for the biosynthesis of phenylalanine, tyrosine and tryptophan), phenylpropanoid (leads to the synthesis of flavonoids, monolignols,

Food Chemistry, Function and Analysis No. 5
Non-extractable Polyphenols and Carotenoids: Importance in Human Nutrition and Health
Edited by Fulgencio Saura-Calixto and Jara Pérez-Jiménez

Published by the Royal Society of Chemistry, www.rsc.org

hydroxycinnamic acids, stilbenes, coumarins and sinapoyls esters) and pentose phosphate (provides sugar intermediates to produce aromatic amino acids).[6,7] These compounds, usually found as esters or glycosides, are essential to the plant's physiology due to the myriad functions that they can perform, from structural support to involvement in the plant's reproductive and defense mechanisms (protection against radiation, plagues and other environmental stresses). Considering both the ubiquitous presence of phenolic compounds and their physiological roles in plant homeostasis, it stands to reason that they play, at least in part, a role in the health-promoting benefits of fruits and vegetables.[8–10]

From a chemical standpoint, phenolics are a group of compounds that possess one or more hydroxyl groups (–OH) attached to an aromatic hydrocarbon group, with phenol (the simplest member of this class of compounds) being the basic chemical structure upon which this group is based. Though there are several different systems used to classify these compounds, one of the most recognized has been proposed by Harborne and Simmonds and uses the different amounts of carbons in the molecule to establish a classification.[11,12]

Overall, there has been much evidence reporting the beneficial effects of polyphenolics (*e.g.* flavonoids, hydroxycinnamic acids, *etc.*), such as their antioxidant capacity or anticarcinogenic activity. However, when considering their dietary impact, it is important to consider that most phenolics are capable of interacting with other compounds in the environment, particularly proteins, which may alter not only the properties of the foodstuffs, but also their nutritional value (*e.g.* by altering protein or starch digestibility) and even their bioaccessibility, causing a loss of some of their beneficial properties.

Most of the research performed on the biological potential of polyphenolic compounds has focused on extractable polyphenols (EPP; *i.e.* compounds that are readily removed from the matrix using water and organic solvents). However, nowadays, it is known that this approach disregards a significant fraction of dietary polyphenolics: those that are not extracted using the conventional approach, the non-extractable polyphenols (NEPP).[13] In fact, when considering the total phenolic intake in diets, mainly through fruit and vegetable intake, the overall amount of NEPP consumed may be as high as the extractable function.[13–15] In fact, Arranz *et al.*[14] and Saura-Calixto *et al.*[15] reported that, considering an example of the Mediterranean diet, NEPP contributed ~50% of the total phenolic compounds consumed by the considered individuals. This highlights the importance of better understanding the potential of this fraction and its role when incorporated in food products. One of the most relevant classes of NEPP are those that are associated with dietary fiber (DF). This particular combination represents an interesting source for the food industry as it may combine a favorable perception by consumers (who perceive both DF and antioxidants as health-promoting ingredients) with European Food Safety Authority (EFSA)-approved nutritional claims (such as "high fiber") and technological advantages (such as increase of storage stability, improved water- and fat-holding capabilities or functioning as a non-caloric bulking agent), leading to new products with added value.[13,16–18] Moreover, the use of DF rich in NEPP allows for the avoidance of frequently expensive extraction protocols that may compromise the industrial application of phenolic-rich extracts.[13,19,20]

17.2 Non-extractable Polyphenolic Compounds: Health Promotion and Consumer Perspectives

Polyphenolics are ubiquitous plant metabolites; therefore, they are an integral part of human diet. In fact, several epidemiological studies, from as early as the 1990s, have established a link between the consumption of dietary polyphenols and a reduction of the risk of developing some chronic diseases.[21,22] Cardiovascular diseases are one example of this association. Mursu *et al.*[23] reported that high intakes of flavonoids were associated with a lower risk of suffering an ischemic stroke and cardiovascular disease-associated mortality in Finnish men aged from 42 to 60 years. Tresserra-Rimbau *et al.*[24] and McCullough *et al.*[25] reported a similar observation (a risk reduction for individuals with higher polyphenolic intakes) for men and women from Mediterranean countries and the USA. However, considering the different range of dietary polyphenolic consumption and the different types of polyphenolics, the results may not always be in agreement. In fact, Sesso *et al.*[26] reported that the consumption of flavonoids was not associated with the risk of cardiovascular disease in women. However, the group in question only ingested an estimate of 24.6 ± 18.5 mg day^{-1} of flavonoids, which is significantly lower than those reported for other groups of individuals, such as the inhabitants of Krakow (897 mg day^{-1}) or of South Korea (318.0 mg day^{-1}).[26–28] This possibility is further supported by the work of Mink *et al.*[29] that reported a significant relation in women between flavonoid consumption and reduction of cardiovascular disease risk when considering higher dietary flavonoid intakes. Cancer is another example of a pathology whose risk can be reduced through the consumption of polyphenolic compounds. A study on the impact of the consumption of dietary flavonoids upon the risk of developing lung cancer found that, in post-menopausal women, the polyphenolic intake leads to a reduction in the risk of developing this pathology, particularly among current and past smokers.[30] Similarly, Frankenfeld *et al.*[31] reported a reduction in the risk of developing non-Hodgkin lymphoma, and Zamora-Ros *et al.*[32,33] described a reduction in the risk of contracting gastric and colorectal cancer.

In spite of the need for further human studies in order to provide clear evidence of the beneficial health effects of dietary polyphenols, the information regarding the possible benefits has trickled down into the general public.[34] One of the most common positive perceptions of consumers stems from the antioxidant effects of polyphenolic compounds. Overall, consumers associate the aging process, and consequent health-related issues, with oxidative stress and, therefore, the antioxidant capacity of phenolic compounds appears to be a straightforward approach to mitigating age-related health problems. In turn, this means that consumers are more willing to buy polyphenolic/antioxidant-enriched products. As this knowledge was paired with the importance of a balanced diet for overall wellbeing, it led consumers to search for better, healthier, more natural and, if possible, health-promoting foodstuffs.[35] This has been demonstrated by Davidov-Pardo *et al.*,[36] who reported that consumers were willing to purchase cookies enriched with a polyphenolic extract, even at a higher cost, and by Markosyan *et al.*,[37] who reported that consumers were, on average, willing to pay a small premium price to purchase antioxidant-enriched apples.

NEPP, although not widely known amongst consumers, may capitalize on parallels drawn with their EPP counterparts, particularly if consumers perceive them as a good source of antioxidants. From this standpoint, one of the most promising sources of food ingredients is antioxidant DF (DF that is associated with antioxidant compounds like polyphenols, carotenoids or vitamins), as it can enjoy not only the health perceptions already described for antioxidants and polyphenols, but also those that stem from DF.[38–41] On its own, DF consumption has been associated with an array of gastrointestinal-positive aspects, like the amelioration of irritable bowel syndrome, Crohn's disease and constipation and the reduction of the risk of obesity and cardiovascular heart disease development, while some types of DF (like fructooligosaccharides and inulin) have also been reported to have a modulatory effect upon the gut microbiota.[42–52] Moreover, some of these benefits are already being perceived by consumers. In fact, a study regarding the presence on the label of a DF-related nutritional claim (such as "source of fiber" or "high fiber") showed that the consumption of such products (by the participants) increased by ~67.5%.[53]

17.3 Non-extractable Polyphenols in Foods

Current research into food polyphenols focuses on the study of polyphenols that can be extracted with common aqueous or organic solvents (particularly ethanol, ethanol:water or methanol). The phenolic compounds detected in such food extracts constitute the EPP and are frequently regarded by authors as the sole source from which the phenolic fraction of a foodstuff is characterized.[1] However, as has been reported by several research groups, these traditional, extraction-based methods do not allow for the complete extraction of the polyphenols of a given matrix; in fact, there is a significant amount of phenolic compounds that may be disregarded as they remain within the extraction residue, associated with different elements of the food matrix (mainly polysaccharides/DF and proteins). These compounds are referred to as NEPP.[2] Moreover, this means that the polyphenol content of plants and plant-based foods (such as cereals, fruits, vegetables, nuts and legumes), as well as total phenolic consumption, have been systematically underestimated by several studies.[3] Furthermore, in fruits, vegetables and cereals (as well as foodstuffs based on these products), the NEPP fraction is either similar or exceeds that of their EPP counterparts. In fact, in many plant-derived foods, the NEPP fraction may constitute approximately 50% of total dietary antioxidants.[5,6] From a chemical standpoint, NEPP are either high-molecular-weight molecules (proanthocyanidins) or phenolic compounds (like hydrolysable tannins and phenolic acids) associated with the different constituents of the cell wall, such as neutral polysaccharides, acidic pectic substances, lignin and/or structural proteins.[2,7] Given their association with these cellular structures, after being ingested, they are not released at significant levels from the food matrix by any of the different steps of the digestive process (*i.e.* mastication, the acidic pH values of the stomach, the alkaline pH values of the small intestine, digestive enzymes or bile salts). Instead, they pass relatively intact through the digestive system and are not absorbed in the small intestine, until they reach the colon, where they can be fermented by the gut microbiota[8] and contribute to a healthy colonic environment rich in antioxidant compounds that contribute to the maintenance of the oxidative homeostatic balance of the colon by scavenging free radicals and counteracting the effects of dietary pro-oxidants.[9]

Nowadays, the concept of DF has been extended to include all indigestible food constituents; it encompasses not only non-starchy polysaccharides and lignin,[10] but also different groups of compounds like resistant starch, oligosaccharides, resistant protein, polyphenols and others.[11] NEPP compounds and DF are intimately associated; in fact, phenolic compounds are frequently associated with DF up to a point where they may also be considered DF constituents. Moreover, NEPP and DF exert a mutual influence: (i) NEPP modify the structure of the DF, affecting its physiochemical properties;[2] (ii) DF acts as a carrier of NEPP, stabilizing them and reducing the impact that the digestive process may have on the phenolic compounds;[6] and (iii) NEPP play a significant role in the health properties reported for DF, characteristics that are frequently, and possibly erroneously, attributed only to polysaccharides, the major constituents of DF.[12] In fact, when combined, DF and NEPP should be considered as a particular type of DF – the antioxidant DF – which has specific health-promoting properties (derived from their phenolic fraction) as well as the technological properties of DF. Antioxidant DF is defined as "a DF concentrate containing significant amounts of natural antioxidants associated with non-digestible compounds".[13–15] This first definition was proposed by Saura-Calixto and defined the criteria that a DF should meet to be considered as an antioxidant DF and therefore able to be considered a potentially beneficial food ingredient.[149] The requirements for a DF to be considered as an antioxidant DF are: (i) the DF content corresponds to more than 50% (w/w) of the total dry matter of the material; (ii) 1 g of antioxidant DF should have a 2,2-diphenyl-1-picrylhydrazyl (DPPH) free radical-scavenging capacity that is equivalent to at least 50 mg vitamin E; and (iii) the antioxidant capacity of the DF must be an intrinsic property that results from the natural foodstuff/ingredient's constituents and not the result of any added antioxidants or by-products that are released due to previous chemical or enzymatic treatments.[13]

17.3.1 Sources of Non-extractable Polyphenols Associated with Dietary Fiber

There is growing interest in the beneficial properties of DF associated with NEPP in terms of both their biological potential and technological properties. Several authors have investigated plant foods as sources of DF with associated NEPP.[1] However, from an industrial standpoint, there is little incentive to transform foodstuffs that can be sold directly into food ingredients, as this implies the transformation of products that already have an interesting commercial value. Considering this, in recent years, the search for novel sources of DF with antioxidant properties has focused not necessarily on plant foods, but on the by-products that result from their processing, namely carambola or orange pomace, grape seeds and pomace, buckwheat hulls, tomato peels or wheat bran.[54–61] Moreover, the by-products that result from the processing of fruits and vegetable have been reported to be good sources of DF and non-extractable phenolic compounds, possessing similar or even higher contents of NEPP than the parts not used by the food transforming industry. This means that not only may they be used to produce food additives that have added functional and technological value, but this may also be done using raw materials that have little perceived intrinsic value. Table 17.1 shows the DF composition of several whole plant foods and some by-products, providing information in regard to their total DF content as well as their antioxidant DF content as an indirect means of providing some insight into their NEPP composition.

17.4 Technological Aspects of Non-extractable Polyphenol Incorporation in Foods

There has been an emerging application of DF in food products. Considering that, as mentioned previously, non-extractable phenolic compounds and DF are closely associated and exert a mutual influence, at least some of these DF can be considered antioxidant DF. Moreover, antioxidant DF is, in fact, a novel, promising ingredient for the food processing and nutrition industry because it combines the bioactive and technological properties of both types of compounds in an unique ingredient.[14] The phenolic compounds, regardless of whether they are extractable or non-extractable, cannot be used to establish any nutritional claim for a product (in Europe) as the EFSA has not approved any claim on this subject. However, the EFSA has approved two health-related claims in regard to polyphenols: one considering olive oil polyphenols (standardized in regard to the content of hydroxytyrosol and its derivatives) and another considering cocoa flavonoids.[18,62] Considering the specificity needed to make such claims on a label, such as the source of phenolic compounds, the exploitation of this approach to an increase of value is mostly linked with consumers' knowledge of the advantages of ingesting phenolic and antioxidant compounds. However, from a technological standpoint, the use of an antioxidant additive may delay the formation

Table 17.1 Dietary fiber, extractable polyphenols and non-extractable polyphenols in plant foods and by-products as sources of antioxidant dietary fiber.[a]

	Dietary fiber	Extractable polyphenols	Non-extractable polyphenols	Ref.
Fruit and vegetables				
Onion	IDF: 18.83% DM SDF: 0.53% DM		Hydrolysable tannins: 0.41% DM	21
Green beans	IDF: 27.51% DM SDF: 4.79% DM		Hydrolysable tannins: 0.80% DM	
Lettuce	IDF: 19.76% DM SDF: 1.20% DM		Hydrolysable tannins: 0.67% DM	
Tomato	IDF: 36.68% DM SDF: 1.26% DM		Hydrolysable tannins: 0.66% DM	
Cherry	IDF: 13.75% DM SDF: 1.40% DM		Condensed tannins: 0.11% DM Hydrolysable tannins: 0.18% DM	
Plum	IDF: 14.60% DM SDF: 3.60% DM		Condensed tannins: 0.11% DM Hydrolysable tannins: 0.23% DM	
Strawberry	IDF: 21.51% DM SDF: 3.39% DM		Condensed tannins: 0.12% DM Hydrolysable tannins: 0.44% DM	
Tangerine	IDF: 23.60% DM SDF: 1.20% DM		Hydrolysable tannins: 0.50% DM	
Orange	IDF: 23.52% DM SDF: 2.98% DM		Hydrolysable tannins: 0.34% DM	
Apple with peel	IDF: 14.89% DM SDF: 1.73% DM		Condensed tannins: 0.20% DM Hydrolysable tannins: 0.16% DM	
Pear with peel	IDF: 16.63% DM SDF: 1.20% DM		Condensed tannins: 0.29% DM Hydrolysable tannins: 0.49% DM	
Cereals				
Wholegrain wheat flour			Hydrolysable phenolics: 32 mg/100 g FW	
Wholegrain barley flour			Hydrolysable phenolics: 60–135 mg/100 g FW	

(continued)

Table 17.1 *(continued)*

	Dietary fiber	Extractable polyphenols	Non-extractable polyphenols	Ref.
Nuts				
Walnut			Hydrolysable tannins: 543 mg/100 g FW	12
Brazil nut			Hydrolysable tannins: 210 mg/100 g FW	
Antioxidant dietary fiber				
Açaí fruit pulp (*Euterpe oleracea*)	TDF: 71.22 g/100 g DM	1.50 g/100 g DM	Hydrolysable tannins: 1.59 g/100 g DM Condensed tannins: 1.24 g/100 g DM	22
Cocoa (fiber-rich cocoa product)	TDF: 60.51% DM SDF: 10.09% DM IDF: 50.42% DM	Soluble total phenols: 1.32% DM Soluble tannins: 1.04% DM	Condensed tannins: 4.46% DM	23
Guava peel/pulp (*Psidium guajava* L., *Psidium acutangulum* L.)	TDF: 40.86–49.42% DM	26.2–77.9 g GAE/kg DM	Condensed tannins: 1.06–2.87% DM	24
Red grape pomace and stem (*Vitis vinifera*)	TDF: 74.5–77.2% DM	2.63–11.6 g GAE/100 g DM	Condensed tannins: 10.3–22.3% DM	25
Red wine grape pomace (*Vitis vinifera*)	TDF: 61.32% DM SDF: 1.44% DM IDF: 59.88% DM	67.74 mg GAE/g DM	Condensed tannins: 12.11% DM	15
White grape stem and pomace (*Vitis vinifera*)	TDF: 715.6–790.5 g kg^{-1} DM SDF: 39.4–103.0 g kg^{-1} DM IDF: 612.6–751.1 g kg^{-1} DM	34.9–87.3 g GAE/kg DM	Condensed tannins: 79.0–168.3 g kg^{-1} DM	26

[a]TDF: total dietary fiber; SDF: soluble dietary fiber; IDF: insoluble dietary fiber; DM: dry mass; FW: fresh weight; GAE: gallic acid equivalents.

Table 17.2 Benefits of the incorporation of dietary fiber with associated polyphenols into food as technological ingredients.[14,27]

Contribution to nutritional value	Non-caloric bulking agents Increase dietary fiber content Increase polyphenols content
Contribution to storage stability	Inhibit lipid oxidation Antimicrobial activity Antioxidant activity
Contribution to rheological properties	Improve water-holding capacity Improve swelling capacity Increase viscosity, texture or gel formation Enhance water and oil retention Improve emulsion stability
Contribution to sensorial properties	Coloring Flavoring

of toxic oxidation products, control rancidity development, extend shelf-life and maintain the nutritional quality of a product.[63] While this can usually be achieved with any synthetic antioxidant, due to the rising consumer preference for natural products and the safety concerns regarding synthetic antioxidants, alternative sources of antioxidants, like antioxidant DF and NEPP, are becoming more relevant.[63] On the other hand, DF does allow for some nutritional claims to be made on the label (the EFSA has approved some claims regarding DF); therefore, its addition to a product represents a straightforward means of increasing its perceived intrinsic value. Nevertheless, the addition of DF also has several technological benefits, as it allows for the improvement of some food characteristics like an increase of viscosity or improved water retention, functioning as bulking agents (with no caloric impact) and acting like fat mimetics.[64–67] Considering the intrinsic potential of both phenolic compounds and DF, it stands to reason that the combination of both presents itself as a particularly interesting approach to be exploited for the simultaneous functional and technological aspects of the combinations.[14,16] Table 17.2 provides a list of the potential benefits of the incorporation of DF associated with NEPP into foodstuffs as technological and/or functional ingredients.

In recent decades, several works have focused on the application of DF with associated phenolic compounds. The use of red grape antioxidant DF has been reported to inhibit lipid oxidation in foodstuffs. As an example, in frozen minced mackerel patties, supplementation with 2% grape antioxidant DF has been reported to cause a 57% lipid oxidation inhibition (measured by thiobarbituric acid reactive substances (TBARS)) while also registering a lower reduction in water during thawing and cooking.[68] The considerable antioxidant potential of red grape antioxidant DF can be explained based on the combined actions of the NEPP (polymeric proanthocyanidins and high-molecular-weight hydrolysable tannins) and some EPP (anthocyanins, flavonols, flavan-3-ols and phenolic acids) that may remain. Regardless of the presence of some extractable phenolic compounds, the NEPP were reported to be 15–30 times more effective at quenching peroxyl radicals than their extractable counterparts. Moreover, these authors also reported that the observed inhibition of lipid oxidation could also be a result of some chelation action by the DF over some pro-oxidant metallic ions.[68] Similar results were observed for raw and cooked chicken hamburgers supplemented with grape antioxidant DF and stored for 14 days at 4 °C. In this particular work, the authors reported not only an inhibition of lipid oxidation (evaluated using TBARS), but this also actually led to an increase in the 2,2′-azino-bis(3-ethylbenzothiazoline-6-sulphonic acid) (ABTS) radical-scavenging capacity of the fortified hamburgers. Moreover, these authors also reported that the inclusion of the grape antioxidant DF, while not compromising consumer acceptability, led to a significant reduction in the yellowness that normally appears during storage.[69] In addition to its application in meat products, grape antioxidant DF has been added to yogurt and salad dressings as a means to enhance their nutritional value and effectively improve their shelf life while not compromising consumer acceptability of either of these products.[70] Other sources of antioxidant DF have been incorporated into products. For example, DF attained from mango peel has been used to enhance the nutritional composition of dough biscuits and pasta (while also improving the pasta's firmness after cooking). Apple juice solid by-product has been used to increase the DF and phenolic content of cakes, and watermelon rind powder has been added to noodles.[71–74]

From a different perspective, food constituents have been reported to interact with EPP, with interactions with proteins being the best described. Protein binding by flavonoids is often associated with the generation of characteristic sensorial properties in foods and beverages, including color, flavor and taste attributes like bitterness or astringency.[75–77] EPP have a significant affinity for proteins that leads to the formation of soluble complexes and even form sediments. Most of the models suggest that protein–polyphenol complexes are formed by multiple weak interactions (mainly hydrophobic) between amino acid side chains and the polyphenol aromatic rings, indicating that the association of polyphenols with proteins is mainly a surface phenomenon. Sometimes these interactions could be complemented by hydrogen bonding, which would play an important role in reinforcing and stabilizing the complexes.[78] Flavonoid–protein interactions have been well documented, and both covalent and non-covalent interactions have been the subject of several reviews.[75,79–82] Covalent binding of flavonoids and proteins typically results from the reaction between functional groups, such as proteins' amine groups on the one hand and quinones formed by flavonoid oxidation (chemically or enzymatically mediated) on the other hand. Oxidation of flavonoids in foods, fruits and vegetables in particular can occur through exposure to high pH values, heat and oxygen reactive species or during post-harvest handling or processing, as these processes facilitate the disruption of plant cellular materials and can, in some cases, release degradative enzymes such as polyphenol oxidases.[83] Interactions between flavonoids and non-enzymatic proteins are generally non-specific and can be affected by the overall matrix composition (*e.g.* flavonoids may also interact with polysaccharides and therefore be less available to interact with the proteins). Moreover, these phenolic–protein interactions may result in conformational changes and partial alteration of the protein structures, as well as lead to a reduction of phenolic compound bioavailability.[84–87] Overall, this indicates that interactions that occur between EPP and a food matrix may lead to the formation of insoluble complexes that may not be degraded during the digestive process and therefore become a part of the NEPP of the matrix they were added to.[87,88] This possibility has been demonstrated by Butler *et al.*,[88] who reported that the interaction between a specific group of rat salivary proteins (with elevated proline contents and lacking sulfur and aromatic amino acids) and sorghum tannins led to a reduction of the ill effects of these phenolic compounds, particularly as these proteins were only produced after the mice underwent a high-tannin sorghum diet.[88] Moreover, Mateus and Freitas[77] reported that tannins may interact with the proteins resulting from the formation of insoluble complexes that may not be degraded during the digestive process.

17.5 Non-extractable Polyphenols, Digestion and Gut Microbiota

Food microstructure may affect the final uptake of nutrients in the gut and consequently their presence in the blood plasma. Typical microstructural elements in foods are cell walls, starch granules, proteins, water and oil droplets, fat crystals, gas bubbles and so on. The concept of a "food matrix" points to the fact that nutrients are contained in a larger, continuous medium that may be of cellular origin (in fruits and vegetables) or a microstructure produced by processing, where they may interact at different timescales with the components and structures of the medium.[89] There is evidence that the physical state of the matrix plays a key role in the release, mass transfer, accessibility and biochemical stability of many food components.[90] Food processing such as grinding, fermentation and/or mild heating may improve bioavailability, most likely as a result of disruption of the cell walls of plant tissues, dissociation of the nutrient–matrix complexes or transformation into more active molecular structures. The total amount of a nutrient may decrease in the food chain due to chemical degradation during storage and physical losses in processing, while at the same time the bioavailability may increase by the aforementioned processes.[89] While true for most of the nutrients ingested, this is not so for NEPP as they pass through our digestive system unaltered by the enzymatic and pH variation and reach the colon nearly intact.[13,91,92] In fact, it is mostly in the large intestine, mediated by the gut microbiota and some specific enzymes, that NEPP may be transformed into other metabolites and/or released and therefore become able to be absorbed into the bloodstream.[92–94] While the effect of the digestive process is still somewhat poorly described in the literature, it has been reported that while non-extractable proanthocyanidins reached the colon nearly intact (suffering only a partial depolymerization in the small intestine), when they are there they are released by the food matrix through the action of enzymes capable of disrupting covalent bonds (*e.g.* esterases) and/or the gut microbiota.[91,95,96] When fermented by the gut microbiota, the fermentation products are frequently the same as those observed for EPP, although their high molecular weight results in delayed absorption of the metabolites (when compared to that of EPP) and frequent association of NEPP with other molecules such as fibers and proteins may result in the alteration of organic acid production (particularly short-chain fatty acids like butyric and propionic acids) by the microbiota.[97–100] This indicates that while NEPP (particularly those associated with DF; *i.e.* antioxidant DF) have an important role to play in the maintenance of colonic health (through the increase of stool bulk, reducing transit time, stimulating the growth of beneficial bacteria and a reduction of tumorigenesis), they may also allow for some systemic effects; for instance, the consumption of a NEPP-rich product derived from grapes resulted in a reduction of blood cholesterol levels and blood pressure, both of which being risk factors for cardiovascular disease.[58,94,99,101–113]

17.6 Improving the Functional and Physicochemical Properties of Non-extractable Polyphenols

The potential benefits of DF, antioxidant or not, are relatively well known. Therefore, the development of food ingredients rich in these compounds has increased, and so has the market looking for these products.[114] However, for poorly soluble substances like NEPP, bioaccessibility may be an important issue in regard to their overall biological potential.[75] Micronization, also known as ultrafine grinding, is a mechanical process that alters the structure of the source materials that, in turn, results in an increase of NEPP exposure to digestive enzymes and the gut microbiota (consequently improving their biological potential) and an enhancement of the physicochemical characteristics of antioxidant DF (as well as DF in general).[114,115] This is supported by the work of Zhu *et al.*,[60] who reported that the micronization of buckwheat hull DF using ultrafine grinding technology allowed for an increase of its water holding capacity, swelling capacity, water retention capacity, oil binding capacity and nitrite absorption capacity, while also showing an improvement of its overall antioxidant capacity and total phenolic content (these last two properties are probable consequences of different NEPP exposures), with similar results being reported by Zhu *et al.*[61] and Rosa *et al.*[116] for wheat bran DF, by Zhu *et al.*[117] for wine grape pomace DF and by Du *et al.*[118] for hull-less barley flour DF. Moreover, Wu *et al.*[56] reported that the micronization of insoluble DF from orange and carambola pomace also allowed for an improved excretion of cholesterol and bile acids in feces, as well a reduction of serum cholesterol and triglyceride levels. Although the authors do not evaluate the specific impact of NEPP in this study, it stands to reason that, considering that the sources themselves are rich in phenolic compounds, at least some of those are non-extractable.[59,119]

17.6.1 Micronization

Micronization is a process that involves the reduction of particles to the micrometer range, although it is sometimes also used to refer to a reduction to the nanometer scale. From an industrial standpoint, the usage of micronization allows for an improvement of the bioavailability, bioaccessability and physiological properties of the micronized compounds.[120,121] NEPP are an example of this, as this process allows for an improvement of their properties as described above, therefore allowing for the attainment of an ingredient/additive with improved antioxidant, rheological and nutritive functions.[116] There are two major approaches to micronization. The classical approaches are mechanic, friction-based methods, which involve milling, bashing and grinding a substance to the desired size. Although it is one of the most widespread techniques, it does not allow for any control of the final particle properties (like size standardization, shape and morphology) and it leads to friction-derived heating of the sample (which in turn may lead to the need for cooling systems and/or lubrication fluids to avoid damage to the milling equipment).[120] Modern techniques (like rapid expansion of supercritical solutions, supercritical antisolvents or particles from gas-saturated solutions) require the use of supercritical fluids and allow for better control over the properties of the final particles. However, these methods are usually employed to build up particles (and not break down matter) from saturated solution (*i.e.* using soluble compounds) or a melted solid (*i.e.* requiring that the solid is capable of melting and that the fibers are not), and therefore their application to NEPP is, to the best of our knowledge, unreported.[121,122]

17.7 Lignin: A Promising Prospect for Non-extractable Polyphenolic Food Ingredients

Lignin, popularly known as a natural plastic, is the third most abundant natural polymer after cellulose and hemicellulose.[123] It is a branched, highly reticulate, racemic, complex and three-dimensional phenolic macromolecule molecule that, in most plants, grants them protection against infection while also providing them with structural support and mechanical strength.[7,11,124,125] It is estimated that, on planet Earth, there is around 300 billion tons of lignin and an annual biosynthetic production rate of around 20 billion tons. Nowadays, the industrial production of lignin is of around 50 million metric tons, most of which is used as a combustion fuel for energy production, a traditional approach to utilize reduced-value industrial wastes.[126,127] In fact, only a small amount (~1–2% of all lignin produced) is used to produce added-value products like plastics (*e.g.* Arboform, a mixture of lignin, hemp or flax, fine wood fibers and wax that results in a plastic that can be injection molded) and lignosulfonates that can be used as dispersant, binding, emulsifying and chelating agents as well as food additives. However, the current approach to lignin usage falls far from the concept of a modern biorefinery (*i.e.* the use of the resources containing lignin/cellulose complexes in order to extract the highest value possible from them), particularly when dealing with industrial sub-products/by-products.[128,129] As examples, sugar cane bagasse (the fibrous matter that remains from sugarcane after juice extraction) is composed of ~20% lignin, while straw has ~14% lignin, and these by-products can be used when considering the aromatic structure of lignin as a potential and renewable sources of aromatic chemical products.[130]

From a chemical standpoint, lignin can be dissolved in low-molecular-weight alcohols, acetone, pyridine and dimethyl sulfoxide (though most of the technical lignins are rarely recuperated in the pure, original form), with high-temperature treatments resulting in the chemical softening of the molecule and the increase of the reaction speed for acid or alkaline depolymerization reactions.[131–133] To extract lignin from an organic source, the first step is typically to remove most of the cellulose and hemicellulose, though fragments of both may still be present in raw lignin fractions. Moreover, the removal of waxes before extracting lignin is of importance as its prior reactivity is influenced by these compounds. Similarly, the presence of polysaccharides may also influence the final properties of the lignin fractions, with their removal being a core step that occurs due to the breakage of the ester bond between the hemicellulose and lignin, yielding lignin fractions with almost no polysaccharide. Therefore, when considering polymer biosynthesis, the largest disadvantages of technical lignins are their low solubility (in organic solvents), their functional/chemical group heterogeneity and the relatively low accessibility of its hydroxyl groups. Additionally, when seeking to produce lignin materials with controlled properties while ensuring extraction repeatability, a profound knowledge of the lignin's structure and reactivity is needed.[134–138]

Structurally, lignin is a complex heteropolymer composed of three main canonical units that are derived from the phenylpropanoid pathway; *p*-coumaryl alcohol, coniferyl alcohol and sinapyl alcohol. These monolignol monomers are methylated, to various degrees, in their aromatic ring at the *ortho*-, *meta*- and *para*-positions.[139] Phenylalanin, the main substrate in lignin's biosynthetic pathway, is produced in the shikimate pathway and is then converted into cinnamic acid by the enzyme phenylalanine ammonia lyase. From this step onward, a series of hydroxylases (*e.g.* *p*-cumarate-3-hydroxylase, cinamate-4-hydroxylase or ferrulate-5-hydroxylase), methyl transferases (*e.g.* caffeic-*O*-methyl transferase), reductases (*e.g.* cinnamoyl-CoA-redutase) and deshydrogenases (*e.g.* cinnamyl alcohol deshydrogenase or synapilic alcohol deshydrogenase) are responsible for the metabolic pathways that yield the different lignin components in the form of alcohols, acids and aldehydes.[7,140] Besides being used to synthetize lignin, these compounds can also be used in the biosynthesis of other aromatic compounds. More importantly, several biosynthetic pathways have been described that transform lignin into naringenin, several flavonoids and anthocyanins – compounds that have color and flavor properties with elevated economical relevance and that are of great interest to the food industry.[140]

In its purified form, lignin contains several low-weight phenolic monomers that have biological potential that may also be native to lignin itself, such as the effective stabilization of chemical reactions induced by the presence of reactive oxygen species. Most of the technical lignins, as well as their purified counterparts, have some antioxidant capabilities, though different forms exhibit different activities. Lignins isolated from sugar cane bagasse have been shown to possess a significant antioxidant capacity and may therefore represent an interesting, natural and relatively safe source of antioxidant compounds to be incorporated into food products. Different radical-neutralizing capacities are described for different types of lignins originating from an array of botanical sources like miscanthus, oak or even bamboo. Overall, the antioxidant capacities of lignins stem from their capacity to remove free radicals through the general method associated with monomeric phenolic compounds. The hydroxyl groups that are part of lignin's structure act as proton donners and are therefore capable of stabilizing the radical within the quinone resonance structure. Overall, the antioxidant capacity of lignins is negatively influenced by the presence of the high heterogeneity of the molecules, their dispersibility, their molecular weight and the presence of other, carbon-based substances. Moreover, lignins extracted using an aqueous sodium hydroxide solution (alkaline hydrolysis lignins) have been reported to possess a higher antioxidant capacity that other types of lignins, namely lignosulfonates, Kraft or lignins extracted through steam explosion. In fact, alkaline lignins have been reported to possess antioxidant activities similar to those observed for several epicatechins that, in turn, are considered reference antioxidants.[141–146] As alkaline lignins contain a fraction of free, monomeric phenolic compounds (mostly ferrulic acid and *p*-coumaric acid) that exhibit strong antioxidant capacities, it is likely that this activity is due, at least in part, to their presence. Besides the antioxidant properties, monomers obtained from the degradation of lignin and its derivatives have been linked with several other potentially beneficial properties whose inclusion into a food product may allow for an increase in the foodstuff's intrinsic value, such as anti-inflammatory or anti-hypertensive properties.[147,148]

17.8 Conclusion

Polyphenols have an array of potentially beneficial properties that could be interesting to exploit by the food industry, both in extractable and the non-extractable form. From a biological standpoint, the inclusion of NEPP, particularly when associated with DF, represents a means of adding some functionally to products, allowing for the formulation of foodstuffs that have health benefits like reducing the glycemic response and plasmatic cholesterol levels. From a technological standpoint, the use of NEPP may allow for an improvement of the shelf life of products (through the prevention of nutrient oxidation or the inhibition of the production of toxic oxidative products) while also permitting, when combined with DF, some technological (*e.g.* increases of viscosity) and nutritional (*e.g.* non-caloric bulking agent) improvements to food products. Moreover, these NEPP-rich DF can be further processed through micronization techniques to yield products with improved bioaccessibility of the entrapped polyphenols.

Acknowledgements

The authors would like to acknowledge the funding provided by Fundação para a Ciência e a Tecnologia (FCT, Lisboa, Portugal) through the projects UID/Multi/50016/2013 and Multibiorefinery (reference SAICTPAC/0040/2015) and the PhD grant to T. Ribeiro SFRH/BDE/108271/2015.

References

1. C. Manach, A. Mazur and A. Scalbert, *Curr. Opin. Lipidol.*, 2005, **16**, 77–84.
2. Y. Wang, C.-F. Chang, J. Chou, H.-L. Chen, X. Deng, B. K. Harvey, J. L. Cadet and P. C. Bickford, *Exp. Neurol.*, 2005, **193**, 75–84.
3. R. Krikorian, M. D. Shidler, T. A. Nash, W. Kalt, M. R. Vinqvist-Tymchuk, B. Shukitt-Hale and J. A. Joseph, *J. Agric. Food Chem.*, 2010, **58**, 3996–4000.
4. R. González, I. Ballester, R. López-Posadas, M. D. Suárez, A. Zarzuelo, O. Martínez-Augustin and F. S. D. Medina, *Crit. Rev. Food Sci. Nutr.*, 2011, **51**, 331–362.
5. J. A. Vita, *Am. J. Clin. Nutr.*, 2005, **81**, 292S–297S.
6. N. Balasundram, K. Sundram and S. Samman, *Food Chem.*, 2006, **99**, 191–203.
7. W. Vermerris and R. L. Nicholson, *Phenolic Compound Biochemistry*, Springer, 2007, pp. 35–62.
8. B. Shukitt-Hale, F. C. Lau, A. N. Carey, R. L. Galli, E. L. Spangler, D. K. Ingram and J. A. Joseph, *Nutr. Neurosci.*, 2008, **11**, 172–182.
9. C. Del Bo', D. Fracassetti, C. Lanti, M. Porrini and P. Riso, *Mutagenesis*, 2015, **30**, 29–35.
10. X. Wu, G. Cao and R. L. Prior, *J. Nutr.*, 2002, **132**, 1865–1871.
11. W. Vermerris and R. Nicholson, in *Phenolic Compound Biochemistry*, ed. W. Vermerris and R. Nicholson, Springer Science+Business Media, Netherlands, 1st edn, 2007, pp. 1–34.
12. I. B. Jaganath and A. Crozier, *Plant Phenolics and Human Health*, John Wiley & Sons, Inc., 2009, pp. 1–49.
13. J. Perez-Jimenez, M. E. Diaz-Rubio and F. Saura-Calixto, *Nutr. Res. Rev.*, 2013, **26**, 118–129.
14. S. Arranz, J. M. Silván and F. Saura-Calixto, *Mol. Nutr. Food Res.*, 2010, **54**, 1646–1658.
15. F. Saura-Calixto, J. Serrano and I. Goñi, *Food Chem.*, 2007, **101**, 492–501.
16. A. Kataria, B. Chauhan and D. Punia, *Food Chem.*, 1989, **32**, 9–17.
17. P. J. Mink, C. G. Scrafford, L. M. Barraj, L. Harnack, C.-P. Hong, J. A. Nettleton and D. R. Jacobs, *Am. J. Clin. Nutr.*, 2007, **85**, 895–909.

18. C. Agostoni, J. Bresson, S. Fairweather Tait, A. Flynn, I. Golly, H. Korhonen, P. Lagiou, M. Løvik, R. Marchelli and A. Martin, *EFSA J.*, 2011, **9**, 2033.
19. S. Silva, E. Costa, C. Calhau, R. Morais and M. Pintado, *Crit. Rev. Food Sci. Nutr.*, 2017, **57**, 3072–3083.
20. C. D. Goldsmith, Q. V. Vuong, C. E. Stathopoulos, P. D. Roach and C. J. Scarlett, *Antioxidants*, 2014, **3**, 700–712.
21. M. G. Hertog, D. Kromhout, C. Aravanis, H. Blackburn, R. Buzina, F. Fidanza, S. Giampaoli, A. Jansen, A. Menotti and S. Nedeljkovic, *et al.*, *Arch. Intern. Med.*, 1995, **155**, 381–386.
22. M. G. L. Hertog, E. J. M. Feskens and D. Kromhout, *Lancet*, 1997, **349**, 699.
23. J. Mursu, S. Voutilainen, T. Nurmi, T. P. Tuomainen, S. Kurl and J. T. Salonen, *Br. J. Nutr.*, 2008, **100**, 890–895.
24. A. Tresserra-Rimbau, E. B. Rimm, A. Medina-Remon, M. A. Martinez-Gonzalez, R. de la Torre, D. Corella, J. Salas-Salvado, E. Gomez-Gracia, J. Lapetra, F. Aros, M. Fiol, E. Ros, L. Serra-Majem, X. Pinto, G. T. Saez, J. Basora, J. V. Sorli, J. A. Martinez, E. Vinyoles, V. Ruiz-Gutierrez, R. Estruch and R. M. Lamuela-Raventos, *Nutr., Metab. Cardiovasc. Dis.*, 2014, **24**, 639–647.
25. M. L. McCullough, J. J. Peterson, R. Patel, P. F. Jacques, R. Shah and J. T. Dwyer, *Am. J. Clin. Nutr.*, 2012, **95**, 454–464.
26. H. D. Sesso, J. M. Gaziano, S. Liu and J. E. Buring, *Am. J. Clin. Nutr.*, 2003, **77**, 1400–1408.
27. S. Jun, S. Shin and H. Joung, *Br. J. Nutr.*, 2016, **115**, 480–489.
28. G. Grosso, U. Stepaniak, R. Topor-Mądry, K. Szafraniec and A. Pająk, *Nutrition*, 2014, **30**, 1398–1403.
29. P. J. Mink, C. G. Scrafford, L. M. Barraj, L. Harnack, C. P. Hong, J. A. Nettleton and D. R. Jacobs Jr, *Am. J. Clin. Nutr.*, 2007, **85**, 895–909.
30. G. J. Cutler, J. A. Nettleton, J. A. Ross, L. J. Harnack, D. R. Jacobs Jr, C. G. Scrafford, L. M. Barraj, P. J. Mink and K. Robien, *Int. J. Cancer*, 2008, **123**, 664–671.
31. C. L. Frankenfeld, J. R. Cerhan, W. Cozen, S. Davis, M. Schenk, L. M. Morton, P. Hartge and M. H. Ward, *Am. J. Clin. Nutr.*, 2008, **87**, 1439–1445.
32. R. Zamora-Ros, A. Agudo, L. Lujan-Barroso, I. Romieu, P. Ferrari, V. Knaze, H. B. Bueno-de-Mesquita, M. Leenders, R. C. Travis, C. Navarro, E. Sanchez-Cantalejo, N. Slimani, A. Scalbert, V. Fedirko, A. Hjartaker, D. Engeset, G. Skeie, H. Boeing, J. Forster, K. Li, B. Teucher, C. Agnoli, R. Tumino, A. Mattiello, C. Saieva, I. Johansson, R. Stenling, M. L. Redondo, P. Wallstrom, U. Ericson, K. T. Khaw, A. A. Mulligan, A. Trichopoulou, V. Dilis, M. Katsoulis, P. H. Peeters, L. Igali, A. Tjonneland, J. Halkjaer, M. Touillaud, F. Perquier, G. Fagherazzi, P. Amiano, E. Ardanaz, L. Bredsdorff, K. Overvad, F. Ricceri, E. Riboli and C. A. Gonzalez, *Am. J. Clin. Nutr.*, 2012, **96**, 1398–1408.
33. R. Zamora-Ros, C. Not, E. Guino, L. Lujan-Barroso, R. M. Garcia, S. Biondo, R. Salazar and V. Moreno, *Canc. Causes Contr.*, 2013, **24**, 549–557.

34. A. Scalbert, C. Manach, C. Morand, C. Rémésy and L. Jiménez, *Crit. Rev. Food Sci. Nutr.*, 2005, **45**, 287–306.
35. J. Di Pasquale, F. Adinolfi and F. Capitanio, *Int. J. Food Syst. Dynam.*, 2011, **2**, 181–193.
36. G. Davidov-Pardo, M. Moreno, I. Arozarena, M. R. Marín-Arroyo, R. N. Bleibaum and C. M. Bruhn, *J. Food Sci.*, 2012, **77**, S430–S438.
37. A. Markosyan, J. J. McCluskey and T. I. Wahl, *Can. J. Agr. Econ.*, 2009, **57**, 325–341.
38. F. Saura-Calixto, *J. Agric. Food Chem.*, 2011, **59**, 43–49.
39. C. Toner, *J. Nutr.*, 2004, **134**, 3192S–3193S.
40. S. Fullmer, C. J. Geiger and C. Parent, *J. Am. Diet. Assoc.*, 1991, **91**, 166–171.
41. J. L. Slavin, *J. Am. Diet. Assoc.*, 2008, **108**, 1716–1731.
42. M. A. Pereira, E. O'reilly, K. Augustsson, G. E. Fraser, U. Goldbourt, B. L. Heitmann, G. Hallmans, P. Knekt, S. Liu and P. Pietinen, *Arch. Intern. Med.*, 2004, **164**, 370–376.
43. D. S. Ludwig, M. A. Pereira and C. H. Kroenke, *et al.*, *JAMA*, 1999, **282**, 1539–1546.
44. N. C. Howarth, E. Saltzman and S. B. Roberts, *Nutr. Rev.*, 2001, **59**, 129–139.
45. T. B. Van Itallie, *Am. J. Clin. Nutr.*, 1978, **31**, S43–S52.
46. G. Castillejo, M. Bulló, A. Anguera, J. Escribano and J. Salas-Salvadó, *Pediatrics*, 2006, **118**, e641–e648.
47. M. Sabater-Molina, E. Larque, F. Torrella and S. Zamora, *J. Physiol. Biochem.*, 2009, **65**, 315–328.
48. G. R. Gibson, *J. Nutr.*, 1999, **129**, 1438S–1441S.
49. J. Slavin, *Nutrients*, 2013, **5**, 1417–1435.
50. I. J. Cook, E. J. Irvine, D. Campbell, S. Shannon, S. N. Reddy and S. M. Collins, *Gastroenterology*, 1990, **98**, 66–72.
51. J. A. Marlett, M. I. McBurney and J. L. Slavin, *J. Am. Diet. Assoc.*, 2002, **102**, 993–1000.
52. A. N. Ananthakrishnan, H. Khalili, G. G. Konijeti, L. M. Higuchi, P. de Silva, J. R. Korzenik, C. S. Fuchs, W. C. Willett, J. M. Richter and A. T. Chan, *Gastroenterology*, 2013, **145**, 970–977.
53. M. N. Vella, L. M. Stratton, J. Sheeshka and A. M. Duncan, *Nutr. J.*, 2014, **13**, 44.
54. I. Navarro-González, V. García-Valverde, J. García-Alonso and M. J. Periago, *Food Res. Int.*, 2011, **44**, 1528–1535.
55. P.-K. Hsu, P.-J. Chien and C.-F. Chau, *Eur. Food Res. Technol.*, 2007, **225**, 199–204.
56. S.-C. Wu, S.-H. Wu and C.-F. Chau, *J. Agric. Food Chem.*, 2009, **57**, 5610–5614.
57. I. Goñi and J. Serrano, *J. Sci. Food Agric.*, 2005, **85**, 1877.
58. J. P. Jiménez, J. Serrano, M. Tabernero, S. Arranz, M. E. Díaz-Rubio, L. García-Diz, I. Goñi and F. Saura-Calixto, *Nutrition*, 2008, **24**, 646–653.
59. K. Mahattanatawee, J. A. Manthey, G. Luzio, S. T. Talcott, K. Goodner and E. A. Baldwin, *J. Agric. Food Chem.*, 2006, **54**, 7355–7363.

60. F. Zhu, B. Du, R. Li and J. Li, *Biocatal. Agric. Biotechnol.*, 2014, **3**, 30–34.
61. K. Zhu, S. Huang, W. Peng, H. Qian and H. Zhou, *Food Res. Int.*, 2010, **43**, 943–948.
62. EFSA Panel on Dietetic Products, Nutrition and Allergies, *EFSA J.*, 2012, **10**, 2809.
63. F. Shahidi and P. Ambigaipalan, *J. Funct. Foods*, 2015, **18**, 820–897.
64. C. S. Brennan and E. Samyue, *Int. J. Food Prop.*, 2004, **7**, 647–657.
65. A. G. J. Voragen, *Trends Food Sci. Technol.*, 1998, **9**, 328–335.
66. R. J. Redgwell and M. Fischer, *Mol. Nutr. Food Res.*, 2005, **49**, 521–535.
67. A. E. Quiros-Sauceda, H. Palafox-Carlos, S. G. Sayago-Ayerdi, J. F. Ayala-Zavala, L. A. Bello-Perez, E. Alvarez-Parrilla, L. A. de la Rosa, A. F. Gonzalez-Cordova and G. A. Gonzalez-Aguilar, *Food Funct.*, 2014, **5**, 1063–1072.
68. I. Sánchez-Alonso and A. J. Borderías, *Int. J. Food Sci. Technol.*, 2008, **43**, 1009–1018.
69. S. G. Sáyago-Ayerdi, A. Brenes and I. Goñi, *LWT-Food Sci. Technol.*, 2009, **42**, 971–976.
70. A. Tseng and Y. Zhao, *Food Chem.*, 2013, **138**, 356–365.
71. C. M. Ajila, M. Aalami, K. Leelavathi and U. J. S. P. Rao, *Innovative Food Sci. Emerging Technol.*, 2010, **11**, 219–224.
72. C. Ajila, K. Leelavathi and U. P. Rao, *J. Cereal Sci.*, 2008, **48**, 319–326.
73. M. Sudha, V. Baskaran and K. Leelavathi, *Food Chem.*, 2007, **104**, 686–692.
74. L.-H. Ho and N. Che Dahri, *CyTA-J. Food*, 2016, **14**, 465–472.
75. A. Bennick, *Crit. Rev. Oral Biol. Med.*, 2002, **13**, 184–196.
76. E. Jöbstl, J. O'Connell, J. P. A. Fairclough and M. P. Williamson, *Biomacromolecules*, 2004, **5**, 942–949.
77. G. Luck, H. Liao, N. J. Murray, H. R. Grimmer, E. E. Warminski, M. P. Williamson, T. H. Lilley and E. Haslam, *Phytochemistry*, 1994, **37**, 357–371.
78. A. J. Charlton, N. J. Baxter, M. L. Khan, A. J. G. Moir, E. Haslam, A. P. Davies and M. P. Williamson, *J. Agric. Food Chem.*, 2002, **50**, 1593–1601.
79. V. de Freitas and N. Mateus, *Curr. Org. Chem.*, 2012, **16**, 724–746.
80. A. Papadopoulou and R. A. Frazier, *Trends Food Sci. Technol.*, 2004, **15**, 186–190.
81. K. J. Siebert, *J. Agric. Food Chem.*, 1999, **47**, 353–362.
82. J. Xiao and G. Kai, *Crit. Rev. Food Sci. Nutr.*, 2012, **52**, 85–101.
83. S. Bittner, *Amino Acids*, 2006, **30**, 205–224.
84. J. Xi and R. Guo, *Int. J. Biol. Macromol.*, 2007, **40**, 305–311.
85. P. Bourassa, C. Kanakis, P. Tarantilis, M. Pollissiou and H. Tajmir-Riahi, *J. Phys. Chem. B*, 2010, **114**, 3348–3354.
86. L. Liang, H. Tajmir-Riahi and M. Subirade, *Biomacromolecules*, 2007, **9**, 50–56.
87. T. Ozdal, E. Capanoglu and F. Altay, *Food Res. Int.*, 2013, **51**, 954–970.
88. L. G. Butler, D. J. Riedl, D. G. Lebryk and H. J. Blytt, *J. Am. Oil Chem. Soc.*, 1984, **61**, 916–920.
89. J. Parada and J. M. Aguilera, *J. Food Sci.*, 2007, 72, R21–R32.

90. J. M. Aguilera, *J. Food Eng.*, 2005, **67**, 3–11.
91. S. Touriño, J. Pérez-Jiménez, M. L. Mateos-Martín, E. Fuguet, M. P. Vinardell, M. Cascante and J. L. Torres, *J. Agric. Food Chem.*, 2011, **59**, 5955–5963.
92. M. L. Mateos-Martín, J. Pérez-Jiménez, E. Fuguet and J. L. Torres, *Br. J. Nutr.*, 2012, **108**, 290–297.
93. P. A. Kroon, C. B. Faulds, P. Ryden, J. A. Robertson and G. Williamson, *J. Agric. Food Chem.*, 1997, **45**, 661–667.
94. A. González-Sarrías, J. C. Espín and F. A. Tomás-Barberán, *Trends Food Sci. Technol.*, 2017, **69**, 281–288.
95. F. Saura-Calixto, J. Serrano and I. Goni, *Food Chem.*, 2007, **101**, 492.
96. A. Crozier, I. B. Jaganath and M. N. Clifford, *Nat. Prod. Rep.*, 2009, **26**, 1001–1043.
97. L. Rondini, M.-N. Peyrat-Maillard, A. Marsset-Baglieri, G. Fromentin, P. Durand, D. Tomé, M. Prost and C. Berset, *J. Agric. Food Chem.*, 2004, **52**, 4338–4343.
98. F. Saura-Calixto, J. Pérez-Jiménez, S. Touriño, J. Serrano, E. Fuguet, J. L. Torres and I. Goñi, *Mol. Nutr. Food Res.*, 2010, **54**, 1.
99. O. Aprikian, V. Duclos, S. Guyot, C. Besson, C. Manach, A. Bernalier, C. Morand, C. Rémésy and C. Demigné, *J. Nutr.*, 2003, **133**, 1860–1865.
100. J. Juśkiewicz, J. Milala, A. Jurgoński, B. Król and Z. Zduńczyk, *Nutrition*, 2011, **27**, 351–357.
101. L. Bravo, F. Saura-Calixto and I. Goni, *Br. J. Nutr.*, 1992, **67**, 463–473.
102. N. Martin-Carron, A. Garcia-Alonso, I. Goni and F. Saura-Calixto, *Am. J. Enol. Vitic.*, 1997, **48**, 328–332.
103. N. Martín-Carrón, F. Saura-Calixto and I. Goñi, *J. Sci. Food Agric.*, 2000, **80**, 1183–1188.
104. L. A. Bravo, E. Mañas and F. Saura-Calixto, *J. Sci. Food Agric.*, 1993, **63**, 63–68.
105. M. E. López-Oliva, M. J. Pozuelo, R. Rotger, E. Muñoz-Martínez and I. Goni, *Br. J. Nutr.*, 2013, **109**, 4–16.
106. D. Lizarraga, M. P. Vinardell, V. Noé, J. H. van Delft, G. Alcarraz-Vizán, S. G. van Breda, Y. Staal, U. L. Günther, M. A. Reed and J. L. Torres, *J. Nutr.*, 2011, **141**, 1597–1604.
107. S. Sánchez-Tena, D. Lizárraga, A. Miranda, M. P. Vinardell, F. García-García, J. Dopazo, J. L. Torres, F. Saura-Calixto, G. Capellà and M. Cascante, *Carcinogenesis*, 2013, **34**, 1881–1888.
108. M. E. López-Oliva, A. Agis-Torres, P. García-Palencia, I. Goñi and E. Muñoz-Martínez, *Nutr. Res.*, 2006, **26**, 651–658.
109. M. E. López-Oliva, A. Agis-Torres, I. Goni and E. Muñoz-Martínez, *Br. J. Nutr.*, 2010, **103**, 1110–1117.
110. D. Hervert-Hernández, C. Pintado, R. Rotger and I. Goñi, *Int. J. Food Microbiol.*, 2009, **136**, 119–122.
111. A. Gunenc, C. Alswiti and F. Hosseinian, *J. Food Res.*, 2017, **6**, 1–10.
112. D. A. Luna-Vital, A. K. Ramírez-Jiménez, M. Gaytan-Martinez, L. Mojica and G. Loarca-Piña, *Dietary Fiber Functionality in Food and Nutraceuticals: From Plant to Gut*, 2016, pp. 95–122.

113. A. Aleixandre and M. Miguel, *Food Funct.*, 2016, **7**, 1864–1871.
114. C.-F. Chau, Y.-T. Wang and Y.-L. Wen, *Food Chem.*, 2007, **100**, 1402–1408.
115. D. Angelino, M. Cossu, A. Marti, M. Zanoletti, L. Chiavaroli, F. Brighenti, D. Del Rio and D. Martini, *Food Funct.*, 2017, **8**, 2368–2393.
116. N. N. Rosa, C. Barron, C. Gaiani, C. Dufour and V. Micard, *J. Cereal Sci.*, 2013, **57**, 84–90.
117. F.-M. Zhu, B. Du and J. Li, *Food Sci. Technol. Int.*, 2014, **20**, 55–62.
118. B. Du, F. Zhu and B. Xu, *Bioact. Carbohydr. Diet. Fibre*, 2014, **4**, 170–175.
119. M. C. Garau, S. Simal, C. Rosselló and A. Femenia, *Food Chem.*, 2007, **104**, 1014–1024.
120. N. Rasenack and B. W. Müller, *Pharm. Dev. Technol.*, 2004, **9**, 1–13.
121. E. Reverchon and D. Porta Giovanna, *Pure Appl. Chem.*, 2001, **73**, 1293.
122. E. Reverchon, G. Della Porta, A. Di Trolio and S. Pace, *Ind. Eng. Chem. Res.*, 1998, **37**, 952–958.
123. J. Pérez, J. Munoz-Dorado, T. de la Rubia and J. Martinez, *Int. Microbiol.*, 2002, **5**, 53–63.
124. S. H. Ghaffar and M. Fan, *Biomass Bioenergy*, 2013, **57**, 264–279.
125. K. Freudenberg and A. C. Neish, *Constitution and Biosynthesis of Lignin*, Springer, 1968.
126. N. R. Council, D. E. L. Studies, C. L. Sciences and C. B. I. Products, *Biobased Industrial Products: Research and Commercialization Priorities*, National Academies Press, 2000.
127. Z. Anwar, M. Gulfraz and M. Irshad, *J. Radiat. Res. Appl. Sci.*, 2014, 7, 163–173.
128. J. E. Holladay, J. F. White, J. J. Bozell and D. Johnson, *Top Value Added Chemicals from Biomass-volume II, Results of Screening for Potential Candidates from Biorefinery Lignin*, Pacific Northwest National Lab. (PNNL), National Renewable Energy Laboratory (NREL), Richland, WA (United States); Golden, CO (United States), 2007.
129. M. FitzPatrick, P. Champagne, M. F. Cunningham and R. A. Whitney, *Bioresour. Technol.*, 2010, **101**, 8915–8922.
130. E. A. Capanema, M. Y. Balakshin and J. F. Kadla, *J. Agric. Food Chem.*, 2004, **52**, 1850–1860.
131. N. Brosse, R. El Hage, M. Chaouch, M. Pétrissans, S. Dumarçay and P. Gérardin, *Polym. Degrad. Stab.*, 2010, **95**, 1721–1726.
132. J.-Y. Kim, H. Hwang, S. Oh, Y.-S. Kim, U.-J. Kim and J. W. Choi, *Int. J. Biol. Macromol.*, 2014, **66**, 57–65.
133. J. Li, G. Henriksson and G. Gellerstedt, *Bioresour. Technol.*, 2007, **98**, 3061–3068.
134. D. Fu, G. Mazza and Y. Tamaki, *J. Agric. Food Chem.*, 2010, **58**, 2915–2922.
135. A. Björkman, *Nature*, 1954, **174**, 1057–1058.
136. E. Adler, K. Lundquist and G. E. Miksche, *The Structure and Reactivity of Lignin*, ACS Publications, 1966.
137. N.-E. El Mansouri and J. Salvadó, *Ind. Crops Prod.*, 2006, **24**, 8–16.
138. A. G. Vishtal and A. Kraslawski, *BioResources*, 2011, **6**, 3547–3568.

139. A. J. Ragauskas, G. T. Beckham, M. J. Biddy, R. Chandra, F. Chen, M. F. Davis, B. H. Davison, R. A. Dixon, P. Gilna and M. Keller, *Science*, 2014, **344**, 1246843.
140. W. Boerjan, J. Ralph and M. Baucher, *Annu. Rev. Plant Biol.*, 2003, **54**, 519–546.
141. Z. Li and Y. Ge, *Int. J. Biol. Macromol.*, 2012, **51**, 1116–1120.
142. R. Kaur and S. Uppal, *Colloid Polym. Sci.*, 2015, **293**, 2585–2592.
143. V. Ugartondo, M. Mitjans and M. P. Vinardell, *Bioresour. Technol.*, 2008, **99**, 6683–6687.
144. A. Toledano, L. Serrano, A. M. Balu, R. Luque, A. Pineda and J. Labidi, *ChemSusChem*, 2013, **6**, 529–536.
145. X. Pan, J. F. Kadla, K. Ehara, N. Gilkes and J. N. Saddler, *J. Agric. Food Chem.*, 2006, **54**, 5806–5813.
146. H. Faustino, N. Gil, C. Baptista and A. P. Duarte, *Molecules*, 2010, **15**, 9308–9322.
147. E. Karvela, D. P. Makris, P. Kefalas and M. Moutounet, *Food Chem.*, 2008, **110**, 263–272.
148. M. P. Vinardell, T. Benavides, M. Mitjans, M. R. Infante, P. Clapés and R. Clothier, *Food Chem. Toxicol.*, 2008, **46**, 3837–3841.
149. F. Saura-Calixto, *J. Agric. Food Chem.*, 1998, **46**, 4303–4306.

Concluding Remarks

FULGENCIO SAURA-CALIXTO AND JARA PÉREZ-JIMÉNEZ

fsaura@nutriantioxidants.com; jara.perez@ictan.csic.es

The preceding chapters provide an overview of the current state of the art in the field of non-extractable polyphenols (NEPP) and non-bioavailable carotenoids as dietary antioxidants. Although research into these compounds is still rare, this book shows that a body of evidence has already accumulated regarding their important contribution to the total antioxidant content of plant foods and beverages. To a lesser extent, some preclinical and clinical trials have indicated their metabolic fate once ingested and their potential health-related properties. In general, it is important that these major dietary antioxidants are systematically included in future research in this field, including many different aspects: contents, intake, physiological effects, *etc.* Nonetheless, there are some specific aspects that should be considered particularly closely in research on this topic in the immediate future. These aspects are presented in this section, which aims to highlight current gaps in our knowledge of NEPP and bioavailable carotenoids, emphasising perspectives for research on these topics in the coming years.

Metabolic Fate of NEPP

Over recent decades, abundant and valuable information on the metabolic fate of extractable polyphenols (EPP) and NEPP has been gathered. Although initial research was devoted to EPP absorbed in the small intestine (less than 10% of total EPP) and paid little attention to non-bioavailable EPP and NEPP,

Food Chemistry, Function and Analysis No. 5
Non-extractable Polyphenols and Carotenoids: Importance in Human Nutrition and Health
Edited by Fulgencio Saura-Calixto and Jara Pérez-Jiménez

Published by the Royal Society of Chemistry, www.rsc.org

in the last decade, the importance of the colon as an organ that is key for the metabolic fate of polyphenols (PP), and also the biological activities of microbial-derived PP metabolites, have become hot topics in this field of research. Interestingly, many of the results come from human studies, which overcome the well-known limitations of animal studies. However, there are still many aspects of the metabolic fate of NEPP that deserve further exploration.

First, the interaction of tannins with salivary enzymes producing astringency is widely known, but the contribution of NEPP to this effect has not been explored. This is relevant not only from a sensory perspective, but also from a physiological one, since if NEPP also form such complexes with salivary enzymes, their potential permanence along the digestive tract should be explored. The influence of NEPP as intact structures on the metabolic fate of other food constituents should be assessed, once they have passed through the mouth and are transported along the upper digestive tract without modification. Thus, it is known that dietary fibre (DF) is capable of retaining a fraction of dietary fat, but this kind of effect (considering all food constituents together) has not been evaluated for NEPP.

Once NEPP reach the colon, it has already been shown that they are subjected to colonic transformation, but since they do not arrive in the colon alone, the influence that they exert on the colonic fermentation of DF and *vice versa*, as well as on other substrates reaching the colon, should be further explored. Moreover, most data on the colonic transformation of NEPP are qualitative, so to fill a gap here, the rate of fermentation of the different classes of NEPP should be estimated, as well as the rate of absorption of the metabolites so derived. Interestingly, even NEPP that are not fermented or are fermented but not absorbed might have beneficial local effects, depending on their location within faeces, since it is known that the aqueous fraction of faeces enters into deeper contact with the colonic mucosa.

Moreover, there is one particular aspect of colonic transformation of NEPP that should be explored in depth: how NEPP affect the microbiota profile. Over the last few years, some of the species involved in the metabolic fate of PP have been identified, although information is still scarce. Indeed, a huge research effort remains to be made in considering the two types of PP–microbiota interaction: the microbiota allows the release of new metabolites from PP at the same time as these compounds are a substrate for the microbiota, fostering its growth. Therefore, we should explore whether NEPP, as compared with EPP, cause specific modifications of colonic microbiota. This kind of determination should be performed not only after supplementation with NEPP for a short period of time (weeks or months), but also after following an NEPP-rich diet for longer periods; for the latter, more data on NEPP intake are needed. Interestingly, close associations between microbiota profile and the risk of several chronic diseases have been established in recent years, and the microbiota profile itself has even been suggested as a health biomarker. Within this framework, we expect new clinical trials of NEPP and microbiota to provide important information on the complex interaction between diet and microbiota.

Finally, the destination of the microbial-derived PP metabolites must still be better defined. In particular, their interactions with the colonic epithelium should be explored in depth. It is also important to know: how long they remain in circulation in the bloodstream; at what concentrations; whether they are capable of accumulating in certain organs or tissues (an aspect that can only be evaluated using animal models); in which kinds of cells they are absorbed; and their sites of action. Also, recent data show that microbial-derived PP metabolites (including those from NEPP) might be used as biomarkers of different health outcomes, another aspect that should be further explored.

Physiological Effects of NEPP: the Particular Case of Metabolic Syndrome

Preclinical and clinical trials on the physiological effects of NEPP, although promising, are still few and far between. Many more studies are needed in this field, especially to foster clinical trials. One ultimate objective of these studies, besides determining the effects that NEPP exert on human health, would be to accumulate enough evidence to be in a position to apply for the recognition of health claims from the European Food Safety Authority (EFSA) in the European Union, the Food and Drug Administration (FDA) in the USA and other competent agencies. It is important to recall the EFSA requirements for clinical trials before it will approve health claims, which, in the context of NEPP, include: (a) the design of the study should be robust; (b) there should be proper characterisation of the compounds tested (which emphasises the importance of the development of analytical methods); (c) there should be a clear definition of the claim, with specific clinical effects that can be measured by recognised, accepted methods; and (d) there should be a clear cause–effect relationship.

One particular situation in which NEPP might be especially relevant, although this has not yet been explored, is metabolic syndrome (MetS). MetS is a cluster of cardiometabolic risk factors (central obesity, hypertension, hyperglycaemia, hypertriglyceridaemia and low high-density lipoprotein cholesterol) that is itself a risk factor for the development of cardiovascular disease and type 2 diabetes. The main physiological conditions underlying MetS are increased oxidative stress (closely linked to sustained mild inflammation) and insulin resistance (IR). Both these may be regulated by PP, as has been shown for EPP, mostly in preclinical models to date, and not yet explored for NEPP.

During their transit along the digestive tract, NEPP can behave as antioxidants, quenching free radicals that may be present in the digestive bolus. Moreover, once in the colon, they help to shift the oxidative status towards an antioxidant environment. Although this is especially relevant at the local level, the potential systemic effects of this shift in the microbiota profile should be explored. Also, a decrease of oxidative stress in the colon averts the appearance of local inflammation that, in the context of MetS, may easily be propagated to other organs and tissues. Finally, microbial-derived NEPP metabolites may be absorbed, and although they usually show less direct antioxidant capacity than their parent compounds, they still exert some such effect. Moreover, they could also modulate systemic oxidative status, enhancing endogenous antioxidant systems – a well-known effect of many PP.

Similarly, NEPP could have an effect on IR, the other physiological condition behind MetS. Once again, this effect would start during transit along the digestive tract, since proanthocyanidins (PA) are known to inhibit different intestinal enzymes, thereby decreasing the rate of absorption of simple carbohydrates. This effect remains to be tested specifically for NEPP in the non-extractable PA class (NEPA); low-molecular-weight PA have already been tested and shown a specific effect and NEPA should be expected to have such an effect since they differ from single PA in the fact that they present a longer chain of monomeric units. After microbial transformation in the colon, NEPP would contribute to the formation of small, absorbable phenolic acids, which have been shown to stimulate insulin secretion as well as the response to this hormone, allowing glucose uptake in its target tissues.

Therefore, based on current evidence, we may expect that NEPP play a beneficial role in the modulation of MetS through amelioration of oxidative stress, inflammation and IR. Robust clinical trials should be designed to test this hypothesis.

Non-bioaccessible Carotenoids

Most of this book focuses on NEPP, since they are the class of macromolecular antioxidants for which more information is available. However, we should not ignore the fact that other dietary antioxidants reach the colon together with them. This is the case of carotenoids.

Although we know much concerning the food contents, dietary intakes, metabolism and physiological effects of carotenoids, information on their metabolic fate is still limited. Indeed, it is known that an important fraction of them reaches the colon, although detailed data on the transformations that take place there are still scarce. Similarly, just as in the case of NEPP, other aspects related to this topic should be explored: how the presence or absence of DF affects the colonic transformation of non-bioaccessible carotenoids; which microbial species are responsible for this transformation and how these compounds alter the microbiota profile; whether they improve oxidative status in the colon; *etc.* All these questions open up a wide range of research waiting to be undertaken, which we hope will be performed in coming years.

NEPP as Dietary Antioxidants

As highlighted in some of the chapters, NEPP are major dietary antioxidants – an aspect that has commonly been neglected. Nevertheless, now that their contribution to the pool of dietary antioxidants ingested daily has been established, this fact should be systematically considered in new studies in the field.

Thus, it has been shown in *in vitro* studies and some acute clinical trials that NEPP can exert antioxidant capacity. However, their long-term effect on human oxidative status *via* different potential mechanisms (enhancement of antioxidant capacity, stimulation of endogenous antioxidant systems and decreasing the oxidation of biomolecules) should be evaluated in clinical trials. Those studies should be designed for different populations (smokers, the aged, subjects with risk factors or chronic diseases, *etc.*), since it is known that different physiological situations affect oxidative status homeostasis. Also, now that dietary antioxidant capacity has been established as a quality parameter of diet, it should be estimated in different populations considering the contribution of NEPP and their potential association with health outcomes.

However, it is also important to bear in mind that NEPP are not consumed alone, so new studies of the topic should consider all dietary antioxidants, emulating the proportions at which they are commonly ingested, so that the different interactions between them can be ascertained. Moreover, these interactions may take place at different levels (sections of the gastrointestinal tract, biological fluids or cells) that should be independently evaluated. Some studies, in order to ascertain mechanistic aspects, may be performed *in vitro*, but they should be combined with robust clinical trials in order to determine the real role of NEPP as dietary antioxidants.

Other Macromolecular Antioxidants

The bioactive compounds reaching the colon are not restricted to DF, polyphenols and carotenoids. Maillardised DF, a specific complex present in foodstuffs subjected to the Maillard reaction during thermal processing, has already been reported in coffee and bread. This complex is characterised by its high content of DF, polyphenols and Maillard reaction products (MRP), in a structure with many kinds of associations between its different constituents and where current analytical methods are not able to discern what corresponds to one constituent and what to another. MRP include many different compounds, with molecular weights ranging from low to very high; beneficial physiological effects have been attributed to the heaviest ones – melanoidins – in recent years. The formation of Maillardised DF in other common products in diets, such as other cereal products, fried foods or dark beer, among others, has yet to be studied. Therefore, special health-related properties of these complexes could be expected once they reach the colon, although they remain to be evaluated.

An overview of research on dietary antioxidants over the last century shows how this field has evolved, from initial interest exclusively in a few small compounds, such as vitamins; passing through a second stage, when PP (but only EPP) were also considered; and finally, the discovery that foods also contain NEPP. Nevertheless, as we have said, NEPP do not reach the colon alone, but accompanied by other bioactive compounds present in plant foods. Today, we have some data, mostly on non-bioaccessible compounds, but the picture may fill out over the next few years with the discovery of new chemical species present in plant foods that have antioxidant capacity. In particular, the use of techniques such as untargeted metabolomics, guided by a data-driven approach instead of a hypothesis-driven approach, may help to identify such substances.

Chemical and Structural Analysis

Although there have been some advances in the techniques used to analyse NEPP, several aspects still have to be improved. These affect the main steps in the characterisation of NEPP: separation from EPP, release from food matrix, analysis of individual compounds in the hydrolysates and structural elucidation.

First of all, the most common way to start the determination of NEPP is to remove the EPP from the food matrix using aqueous–organic solvents, followed by different hydrolytic procedures performed on the residue; this is the chemical approach. However, as described in some of the chapters of this book, from a physiological perspective, it would be more appropriate to start to treat the sample with digestive enzymes in order to remove the compounds that would be released in the digestive tract (whether they are further absorbed or not) so as to obtain a residue containing the compounds that would effectively reach the colon. Such approaches have been used on a few occasions, and we feel that their use should be encouraged in order to obtain data of more physiological relevance.

Once the residue is obtained, either by adopting the chemical or the physiological approach, since, by definition, NEPP cannot be extracted from the food matrix using any typical solvent, a hydrolytic treatment must be performed to release them. Currently, the procedures recommended prior to the analysis of these compounds include harsh conditions of acidity and temperature, as these are the only ones that have been proven to release NEPP from the food matrix. We should remark that, although these conditions are not physiological, they emulate in a chemical way what happens in the large intestine (*i.e.*, NEPP release by the action of microbiota).

However, these methods have different side effects: (i) there may be a loss of the less stable compounds once they are released from the food matrix; (ii) no information on the original associations between PP and the cell wall is obtained; and (iii) new phenolic-derived artefacts, not originally present in the sample, may form during the treatment. In the case of NEPA, these effects are even more relevant, since there is not only depolymerisation of PA – which would release monomeric catechins – but also a chemical transformation, producing a solution with anthocyanins and phlobaphenes as an amorphous solid.

Due to all the limitations associated with strong chemical hydrolysis, enzymatic treatments may be preferable alternatives to chemical procedures, although assays to date have shown that the enzymes tested are much less efficient than chemical hydrolysis. Nevertheless, this should be explored further since, as described, enzymes in the large intestine are indeed able to carry out these transformations. Other approaches would be the application of mild chemical hydrolysis combined with a nitrogen atmosphere (key for the hydrolysis of proteins) or the use of techniques such as steam explosion or pressurised hot water; some such preliminary attempts have been made, but they have not yet been systematically tested.

The hydrolysates obtained by any of the procedures described can be used to determine NEPP content as the sum of individual PP released from the food matrix in the case of hydrolysable PP (HPP) and individual PA after depolymerisation of NEPA. In both cases, since the hydrolysates obtained are under strong acid conditions and at very high concentrations of different salts, they should be treated in order to preserve high-performance liquid chromatography (HPLC) columns or mass spectrometry (MS) ionisation sources. Thus, different solid-phase extraction procedures have been suggested before samples are injected into HPLC or HPLC/MS systems. Nevertheless, they still need to be improved in order to guarantee no PP losses during this processing.

It should never be forgotten that all of these chemical transformations prevent us from knowing what form the NEPP originally adopted in the food matrix and also from using the extracts to evaluate potential biological activities. This is because the hydrolysates obtained do not represent the original compounds present in the sample. Therefore, there is a need to develop new physical methods, similar to those currently used in the field of polymers, which can be used to characterise NEPP as they are originally present in the food matrix or to characterise the PP released from the food matrix, without chemical modifications. For this, structural techniques such as X-ray diffraction, Raman spectroscopy, nuclear magnetic resonance or Fourier-transform infrared spectroscopy should be tested. Also, MS techniques capable of analysing the samples in their solid state, such as matrix-assisted laser desorption/ionisation–time of flight (MALDI-TOF) MS or Fourier-transform–ion cyclotron resonance (MALDI-FT-ICR) should also be assayed to assess their capacity to characterise the original structures in the food matrix. Finally, some preliminary studies using electron microscopy showed the association of NEPA with the cell wall; further studies should be encouraged and developed.

NEPP Databases and Dietary Intakes

The NEPP contents of commonly consumed plant foods and beverages have already been characterised (a necessary step prior to determining dietary intakes). However, there are still food items whose NEPP contents have not been determined. For instance, as stated in the chapter on tropical fruits, data on NEPP in these foods are still scarce, despite their having been shown to be an important source of EPP, and it may therefore be assumed – based on existing preliminary data – that they also contain a high proportion of NEPP. A comprehensive evaluation of NEPP in traditional and non-traditional tropical fruits should be carried out.

This should also be applied to less consumed plant foods in the different groups (fruit, vegetables, nuts, legumes and seeds), since some of them may emerge as particularly rich sources of NEPP. Indeed, since many aspects may affect the PP content of food (agronomical practices, variety, processing, *etc.*), it would be very useful to obtain new NEPP content data for other varieties or origins of the foods already characterised, since they could allow us to establish potential ranges of variation of NEPP for a given food.

Moreover, large amounts of data on NEPP contents in foods could be used as the basis for a specific online database of NEPP contents in foods. Such databases already exist for EPP (United States Department of Agriculture Database, Phenol-Explorer and others), but they should be widened to include NEPP in order to allow for the complete PP picture to be drawn. Such a tool should be developed according to current recommendations for this kind of scientific resource: clear identification of the analytical methods used, enough data to provide representative mean values and traceability of all the original data, among others.

All of these data would allow for the consumption of certain food items to be promoted due to their NEPP content, as well as NEPP dietary intakes in different populations to be assessed better. To date, total dietary intake data for NEPP have only been reported for Spain. It is evident that such studies should be extended to other European and non-European countries. Indeed, preliminary data on NEPP intake from some specific food groups in some European countries and in Mexico indicate that there may be significant differences between countries. Moreover, according to the latest developments in nutrition, the use of dietary questionnaires should be replaced, when possible, by diet biomarkers. These are metabolites that are independently associated with the consumption of certain foods or food products that also contain NEPP, so the circulation levels of these metabolites could be used to determine NEPP intake. The ultimate aim of these assessments would be to enable their use in epidemiological studies in order to identify possible associations between NEPP intake and health outcomes.

It should be borne in mind, however, that not only common food items are rich sources of NEPP. As emphasised in several chapters, many materials discarded during food processing contain NEPP in significant amounts. Their use as functional ingredients would add value to the raw materials and also allow the development of a new generation of functional foods. Moreover, in order to develop these kind of products, we should test not only foods and derived materials, but also other vegetable materials (bark, shrubs, *etc.*), since they could also become important sources of NEPP.

Finally, most studies of the NEPP contents of foods are performed on raw foods, which does not always reflect the way they are consumed. Current data on the effects of processing on NEPP show that some heating techniques may release them from the food matrix. But data on the effects of processing and storage on NEPP are still scarce; future studies should emulate the real conditions of consumption of the different food groups and the effects of the different treatments on NEPP contents and profiles.

Development of New NEPP-rich Products

As stated above, the relevance of NEPP as bioactive ingredients opens up the possibility of developing new products of different natures: foods, dietary supplements, cosmetics, *etc.* However, some technological aspects need to be addressed in order to allow such developments on a large scale.

In particular, there is a need to develop new procedures that allow NEPP concentrates to be obtained that contain over 80% NEPP, which could ensure that the effects observed can be attributed specifically to NEPP. Some preliminary patents, reported in this book, have been developed along these lines. They show that enzymatic treatments of samples in order to remove substance other than NEPP may be a suitable approach for this purpose. However, more research is needed, which should also focus on obtaining different types of NEPP concentrates.

In the case of HPP, such concentrates may also present high percentages of DF, since, as already discussed, the two are intrinsically associated and constitute a single complex. In this case, strategies should be orientated – besides removing other food constituents – towards identifying dietary sources in which this complex presents high percentages of PP. Also, biotechnology approaches could allow us to increase the presence of NEPP associated with DF through additional associations between small PP and the reacting positions of the DF chain.

Finally, the exploration of new NEPP-rich products should consider not only how to obtain NEPP concentrates, but also the best ways to include them as ingredients in foods, dietary supplements or pharmaceutical products. Here again, some specific aspects need to be addressed, such as influence on sensory aspects, as well as the low solubility associated with polymeric structures. Different strategies such as micronisation, the development of nanovesicles containing NEPP, spray drying from NEPP suspensions or the design of effervescent products should be tested.

Scientific Dissemination

To date, research on NEPP has been carried out by a very specific group of researchers, mostly coming from the fields of EPP or DF, which are very restricted areas. Nevertheless, in order to really advance knowledge of NEPP, researchers from other areas and with different backgrounds should be incorporated (biochemists, analytical chemists, epidemiological and clinical researchers, experts on formulation, *etc.*). So, there is still a pressing need to increase the dissemination of the concept and potential applications of NEPP in the scientific community. As previously explained, the substitution of the term NEPP for 'macromolecular antioxidants', among other reasons, aims to make the concept easier to understand for those who do not work in the very restricted field of dietary polyphenols. In order to extend this concept, specific international symposia or workshops on the topic should be planned. We also hope that this book will be a major step forward in this sense, making an overview of our current knowledge on NEPP available to many researchers from other fields.

Furthermore, the relevance of NEPP due to their health-related properties should also be specifically expressed and explained to health professionals so that they can consider these compounds in future clinical practice. In order for that to happen, it is necessary that the promising results concerning these compounds are sustained through more clinical trials.

Finally, the term 'macromolecular antioxidants' should be explained to and adopted by the general public. Currently, there is considerable interest in foods and food constituents with health-related properties among consumers, while at the same time they are usually exposed to much misleading information. Thus, the concept and properties of macromolecular antioxidants should be specifically disseminated to journalists and media specialised in nutrition and health. This would allow them to transfer to the general public validated and sound information. In this way, the field of macromolecular antioxidants would definitively move away from a very restricted number of researchers into a much wider field of research, development and innovation, and finally to whole populations, where these long-ignored bioactive compounds will ultimately exert their health-related properties.

Subject Index